Civil Engineering and Architecture

WORKBOOK

Steven O'Hara
John Phillips
Carisa Ramming

Oklahoma State University
School of Architecture

Australia • Brazil • Japan • Korea • Mexico • Singapore • Spain • United Kingdom • United States

Civil Engineering and Architecture: Workbook

Steven O'Hara, John Phillips, Carisa Ramming

Vice President, Careers & Computing: Dave Garza

Director of Learning Solutions: Sandy Clark

Senior Acquisitions Editor: James DeVoe

Managing Editor: Larry Main

Product Manager: Mary Clyne

Editorial Assistant: Aviva Ariel

Vice President, Marketing: Jennifer Ann Baker

Marketing Director: Deborah Yarnell

Senior Market Development Manager: Erin Brennan

Senior Brand Manager: Kristin McNary

Senior Production Director: Wendy A. Troeger

Production Manager: Mark Bernard

Content Project Manager: David Barnes

Senior Art Director: Bethany Casey

Cover Image: © iStockphoto.com/Henryk Sadura

© 2014 Delmar, Cengage Learning

ALL RIGHTS RESERVED. No part of this work covered by the copyright herein may be reproduced, transmitted, stored, or used in any form or by any means graphic, electronic, or mechanical, including but not limited to photocopying, recording, scanning, digitizing, taping, Web distribution, information networks, or information storage and retrieval systems, except as permitted under Section 107 or 108 of the 1976 United States Copyright Act, without the prior written permission of the publisher.

> For product information and technology assistance, contact us at
> **Cengage Learning Customer & Sales Support, 1-800-354-9706**
> For permission to use material from this text or product,
> submit all requests online at **www.cengage.com/permissions**
> Further permissions questions can be e-mailed to
> **permissionrequest@cengage.com**

ISBN-13: 978-1-4354-4165-1

ISBN-10: 1-4354-4165-6

Delmar
5 Maxwell Drive
Clifton Park, NY 12065-2919
USA

Cengage Learning is a leading provider of customized learning solutions with office locations around the globe, including Singapore, the United Kingdom, Australia, Mexico, Brazil, and Japan. Locate your local office at: **international.cengage.com/region**

Cengage Learning products are represented in Canada by Nelson Education, Ltd.

To learn more about Delmar, visit **www.cengage.com/delmar**

Purchase any of our products at your local college store or at our preferred online store **www.cengagebrain.com**

Notice to the Reader
Publisher does not warrant or guarantee any of the products described herein or perform any independent analysis in connection with any of the product information contained herein. Publisher does not assume, and expressly disclaims, any obligation to obtain and include information other than that provided to it by the manufacturer. The reader is expressly warned to consider and adopt all safety precautions that might be indicated by the activities described herein and to avoid all potential hazards. By following the instructions contained herein, the reader willingly assumes all risks in connection with such instructions. The publisher makes no representations or warranties of any kind, including but not limited to, the warranties of fitness for particular purpose or merchantability, nor are any such representations implied with respect to the material set forth herein, and the publisher takes no responsibility with respect to such material. The publisher shall not be liable for any special, consequential, or exemplary damages resulting, in whole or part, from the readers' use of, or reliance upon, this material.

Printed in the United States of America
1 2 3 4 5 6 7 16 15 14 13 12

Contents

Preface xiii

CHAPTER 1   DEFINITIONS AND HISTORY OF CIVIL ENGINEERING AND ARCHITECTURE   1

Background: Definition and History of Civil Engineering	2
PROBLEM SET **1.1:** Definition and History of Civil Engineering	2
Background: Civil Engineering of Structures	2
PROBLEM SET **1.2:** Civil Engineering of Structures	2
Background: Definition of Architecture	4
PROBLEM SET **1.3:** Definition of Architecture	4
Background: History of Civil Engineering and Architecture	4
A Journey to the Ancient world 3200 BC to AD 337	5
Ancient Egyptian	7
250 BC–AD 1500	8
Greek Architecture 448–432 BC Parthenon in Athens	9
Engineering in Ancient Times	9
Ancient Roman Empire	10
Gothic Architecture 12th to 16th Century/AD1140–1520	11
Renaissance Architecture 1400–1600	12
Famous Buildings	13
PROBLEM SET **1.4:** History of Civil Engineering and Architecture	14
Background: Famous Architects	15
PROBLEM SET **1.5:** Famous Architects	16
Background: Major Developments in Civil Engineering Impact Land Development	17
Properties of Materials	17
Innovation of Material and Process	17
PROBLEM SET **1.6:** Major Developments in Civil Engineering Impact Land Development	17

CHAPTER 2   CAREERS   18

Background: Civil Engineering as a Career	19
Education, Training, and Skills Needed to Become a Civil Engineer	19
Stages of Career Development to Become a Licensed Civil Engineer	21
PROBLEM SET **2.1:** Civil Engineering as a Career	22
Background: Architecture as a Career	23
Education Training and Skills Required to Become an Architect	23
Stages of Career Development to Become a Licensed Architect	25
PROBLEM SET **2.2:** Architecture as a Career	28

iii

Background: Careers Related to Civil Engineering and Architecture	29
PROBLEM SET **2.3:** Careers Related to Civil Engineering and Architecture	29
Background: Civil Engineers, Architects, and Related Agencies Work Together in the Design and Build of a Structure	29
PROBLEM SET **2.4:** Civil Engineers, Architects, and Related Agencies Work Together in the Design and Build of a Structure	29
Background: Agencies and Organizations Involved in the Design and Development of a Structure	30
PROBLEM SET **2.5:** Agencies and Organizations Involved in the Design and Development of a Structure	30

CHAPTER 3 RESEARCH, DOCUMENTATION, AND COMMUNICATION   31

Background: Research is Vital to Land Development	32
PROBLEM SET **3.1:** Research is Vital to Land Development	34
Background: Documentation for Land Development	35
Architectural Brief and Program for Non-Residential Development	36
Development of a Single Family Residence	37
Keeping a Journal	38
Architectural Sketchbooks	38
Bubble Diagrams	39
Difference between Working and Presentation Drawings	41
PROBLEM SET **3.2:** Documentation for Land Development	46
Background: Communication Is Essential to Achieving Shared Project Vision and Construction Productivity	47
PROBLEM SET **3.3:** Communication Is Essential to Achieving Shared Project Vision and Construction Productivity	50

CHAPTER 4 ARCHITECTURAL DESIGN   51

Background: Residential and Commercial Architecture	52
Urban Design at CityCenter, Las Vegas	52
PROBLEM SET **4.1:** Urban Design at CityCenter, Las Vegas	53
Whole-Building Design	53
PROBLEM SET **4.2:** Whole-Building Design	55
Identifying Building Features	55
PROBLEM SET **4.3:** Identifying Building Features	56
Determining Architectural Style	56
PROBLEM SET **4.4:** Identifying Building Features	59

CHAPTER 5 SITE DISCOVERY FOR VIABILITY ANALYSIS   60

Background: How Land Is Described	61
Property Research	61
Legal Descriptions	62
Rectangular Survey System	64
Lots and Blocks	66
PROBLEM SET **5.1:** How Land Is Described	67
Background: Topography and Contour Maps	70
Contour Intervals	70
Soils	72
PROBLEM SET **5.2:** Topography and Contour Maps	73

Background: Rules and Regulations	73
Building Codes	73
Local Ordinances and Zoning	74
Restrictive Covenants	78
PROBLEM SET 5.3: Rules and Regulations	80
Background: Environmental Regulations	80
Environmental Impact	80
Floodplains	81
Wetlands	81
Storm Water Permits	82
Coastal Zones	83
Endangered Species	83
Tree Protection	83
Contamination and Containment	83
Historical and Cultural Resources	83
PROBLEM SET 5.4: Environmental Regulations	83
Background: Other Site Considerations	84
Climate	84
Utilities and Municipal Services	87
Traffic Flow	87
Demographics and Target Market	88
PROBLEM SET 5.5: Other Site Considerations	88
Background: Site Visit	89
PROBLEM SET 5.6: Site Visit	89
Background: Construction Cost Estimate	89
PROBLEM SET 5.7: Construction Cost Estimate	90
Background: Viability Analysis	90
PROBLEM SET 5.8: Viability Analysis	91
Background: Best and Highest Use	91
PROBLEM SET 5.9: Best and Highest Use	91

CHAPTER 6 SITE PLANNING   92

Background: Problem Identification – Programming	93
Needs Analysis	93
PROBLEM SET 6.1: Problem Identification – Programming	96
Background: Site Characteristics	96
Land Surveys	97
Conventional Surveys	98
Topographic Survey	100
PROBLEM SET 6.2: Site Characteristics	100
Background: Soils Investigation	102
Soil Classification	102
PROBLEM SET 6.3: Soils Investigation	108
Background: Concept Development	110
Buildable Area	111
Site Orientation	111
PROBLEM SET 6.4: Concept Development	114

CHAPTER 7 SITE DESIGN   115

Background: Accessibility	116
PROBLEM SET 7.1: Accessibility	117

Background: Parking and Circulation	117
PROBLEM SET **7.2**: Parking and Circulation	119
Background: Grading and Drainage	120
Grading Plan	120
Cut and Fill	120
Storm Water Management	123
Low Impact Development	128
PROBLEM SET **7.3**: Grading and Drainage	128
Background: Utilities	130
Water Supply	130
Wastewater Management	133
PROBLEM SET **7.4**: Utilities	134
Background: Site Plan	135
PROBLEM SET **7.5**: Site Plan	135
Background: Permits	141
PROBLEM SET **7.6**: Permits	143

CHAPTER 8 ENERGY CONSERVATION AND DESIGN 144

Background: Early Building Design	145
PROBLEM SET **8.1**: Early Building Design	146
Building Design and Energy	146
PROBLEM SET **8.2**: Building Design and Energy	148
Energy Conservation	148
Electrical and Lighting	148
Energy Conservation Through Lighting Design	149
Electric Utility Costs	150
PROBLEM SET **8.3**: Energy Conservation	152
ANSI/ASHRAE/IESNA 90.1-07 Lighting Requirements	152
Application Areas	152
Automatic Lighting Shutoff	153
Space Lighting Control and Separate Controls	153
Determining Building Power Limits	153
Additional Lighting Power	157
Exterior Lighting Control and Lighting Efficacy Requirements	158
Exterior Building Lighting Power	158
PROBLEM SET **8.4**: ANSI/ASHRAE/IESNA 90.1-07 Lighting Requirements	160
Passive Solar Design	162
PROBLEM SET **8.5**: Passive Solar Design	163
Active Solar Design	163
Photovoltaic Cells	163
Solar Collector Systems	163
PROBLEM SET **8.6**: Active Solar Design	164

CHAPTER 9 RESIDENTIAL SPACE PLANNING 165

Background: Research and Investigation Defines Project Parameters, Specifications, and Limitations	166
Budget Defines Building Options	166
Site Investigation Determines Building Potential	168
Client Survey and Interview Leads to Project Vision	168
Space Standards Guide the Development of Comfortable and Effective Spaces	168

PROBLEM SET 9.1: Research and Investigation Defines Project Parameters, Specifications, and Limitations	169
Analysis Leads to Optional Residential Space Planning	171
The Ability to Visualize Space Is Key to Space Planning	171
Space Adjacencies Analyzed for Convenience and Comfort	174
Efficiency Is Measured by Projecting Traffic Flow	175
PROBLEM SET 9.2: Analysis Leads to Optional Residential Space Planning	176
Major Development Considerations when Planning Functional and Appealing Spaces	178
Planning within a Budget	178
Planning Living Areas	178
Working and Service Areas	178
Sleeping Area	180
Storage and Closets	180
Stairs and Hallways	180
Planning a Garage	183
Porches, Sunrooms, Decks, Patios, and Balconies	183
PROBLEM SET 9.3: Major Development Considerations When Planning Functional and Appealing Spaces	183

CHAPTER 10 COMMERCIAL SPACE PLANNING   185

Background: Commercial Spaces are Designed for Safety, Accessibility, Flexibility, and Sustainability	186
Adjacency Matrix	193
PROBLEM SET 10.1: Commercial Spaces are Designed for Safety, Accessibility, Flexibility, and Sustainability	193
The Space Plan Allows Safe and Efficient Circulation of Building Occupants	194
PROBLEM SET 10.2: The Space Plan Allows Safe and Efficient Circulation of Building Occupants	195
Public Spaces Are Located for Visibility and Accessibility	196
Exits and Entrances Provide Safety and Accessibility	196
Access to Other Floors	196
Public Restrooms	198
PROBLEM SET 10.3: Public Spaces Are Located for Visibility and Accessibility	198
A Sustainable Space Plan Provides Flexibly, Adaptably, and Affordability	199
Ceiling Heights	199
Space Planning and Energy Consumption	199
PROBLEM SET 10.4: A Sustainable Space Plan Provides Flexibly, Adaptability, and Affordability	199
Workplace Environments Influence Worker Performance	200
An Open Versus a Closed Office Plan	200
Planning Space for Building Equipment and Utilities	200
Building Support Spaces	202
Spaces are Identified by Their Name, Number, and Square Footage	202
Room Numbers	202
PROBLEM SET 10.5: Workplace Environments Influence Worker Performance/Planning Space for Building Equipment and Utilities/ Spaces Are Identified by Their Name, Number, and Square Footage	203

CHAPTER 11 DIMENSIONING AND SPECIFICATIONS — 205

Background: Dimensioning and Annotations Follow Drawing Conventions for Universal Understanding — 206
Symbols — 206
Text and Dimensions — 208
Notes — 210
Abbreviations — 211
PROBLEM SET **11.1**: Dimensioning and Annotations Follow Drawing Conventions for Universal Understanding — 211
Dimensions and Annotations are Used to Communicate Specifications Necessary for Construction — 217
Dimension Placement is Influenced by Wall Construction, and Material — 217
PROBLEM SET **11.2**: Dimensions and Annotations are Used to Communicate Specifications Necessary for Construction — 222
Elevation Views Provide a Preview of the Finished Construction — 228
Section Views Allow Dimensioning and Annotation of Hidden Materials and Components — 228
Details — 228
Schedules Provide Numerous Specifications without Complicating a Drawing — 231
PROBLEM SET **11.3**: Elevation Views Provide a Preview of the Finished Construction/Section Views Allow Dimensioning and Annotation of Hidden Materials and Components/Schedules Provide Numerous Specifications without Complicating a Drawing — 231

CHAPTER 12 BUILDING MATERIALS AND COMPONENTS — 233

Background: The History of Building Materials — 234
Construction Materials Fundamentals — 234
Material Families — 234
Material Properties — 236
Materials and Building Codes — 238
Fire Resistance — 239
Specifying Materials – The Master Format — 241
PROBLEM SET **12.1**: Construction Material Fundamentals — 241
Background: Material Selection Criteria — 242
Strength — 242
Constructability — 242
Durability — 243
Expense — 243
Availability — 243
Environmental Concerns — 243
Summary of Material Selection Criteria — 245
PROBLEM SET **12.2**: Material Selection Criteria — 245
Background: Major and Emerging Construction Materials — 246
Concrete — 246
Masonry — 248
Wood — 252
Steel and Other Metals — 254
Glass — 255
Finish Materials — 256
PROBLEM SET **12.3**: Major and Emerging Construction Materials — 257

CHAPTER 13   FRAMING SYSTEMS RESIDENTIAL AND COMMERCIAL   259

Background: Residential Construction	260
Basic Foundations for Residential Construction	260
Basic Floor Framing for Residential Construction	263
Basic Wall Framing Systems for Residential Construction	267
Basic Roof Framing for Residential Construction	270
PROBLEM SET 13.1: Residential Construction	272
Background: Commercial Construction	273
Basic Foundations for Commercial Construction	273
Basic Floor Framing for Commercial Construction	274
Basic Wall Framing for Commercial Construction	278
Basic Roof Framing for Commercial Construction	279
PROBLEM SET 13.2: Commercial Construction	282

CHAPTER 14   STRUCTURAL SYSTEMS: WHAT MAKES A BUILDING STAND?   283

Background: Loads	284
Dead and Live Loads	284
Snow Load	289
Lateral Loads	290
Load Combinations	290
Tributary Width and Tributary Area	291
Load Path	292
PROBLEM SET 14.1: Loads	292
Background: Statics	293
Forces	294
Shear Force and Bending Moment	298
PROBLEM SET 14.2: Statics	304
Background: Mechanics	306
Stress	306
Strain	308
Elastic and Plastic Behavior	308
PROBLEM SET 14.3: Mechanics	309
Background: Section Properties	310
Structural Design	310
Allowable Strength Design (ASD) vs. Load and Resistance Factor Design (LRFD) in Steel Structures	311
Structural Steel	311
Tension Members	311
Compression Members	312
Bending Members	316
PROBLEM SET 14.4: Structural Design	321
Background: Foundations	325
Shallow Foundations	325
Deep Foundations	327
PROBLEM SET 14.5: Foundations	328

CHAPTER 15   PLANNING ELECTRIC CODES   330

Background: Common Terms and Properties of Electricity	331
Direct and Alternating Current	331

PROBLEM SET **15.1:** Common Terms and Properties of Electricity	332
Background: Power and Energy	332
Determining the Building Needs	333
Calculating the Electrical Loads	335
Determining the Types of Electrical Service	338
Coordination of Other Design Aspects in the Building	339
Preparation of Electrical Plans and Specifications	339
PROBLEM SET **15.2:** Power and Energy	341
Background: Electrical Panel Location	342
PROBLEM SET **15.3:** Electrical Panel Location	342
Background: Lighting	342
The Design Process	343
Programming	343
Design Development	344
Construction Documents	344
Lighting Types	344
Selecting Light Sources and Equipment	345
Lighting Controls	346
Manual Switching	346
Multi-Level Switching	347
Zoning	348
Dimming	349
Occupancy Sensors	349
Photosensors	350
Determining the Proper Quantity of Light	350
Task Lighting	353
PROBLEM SET **15.4:** Lighting	353

CHAPTER 16   PLANNING FOR PLUMBING   354

Background: Quality and Quantity of Water Supply	355
PROBLEM SET **16.1:** Quality and Quantity of Water Supply	356
Background: Location of the Water Supply	356
Codes	358
PROBLEM SET **16.2:** Location of the Water Supply	359
Background: Plumbing Materials and Process	360
PROBLEM SET **16.3:** Plumbing Materials and Process	362
Background: Communication of the Water Supply System	362
Diagramming the Water System	364
Strategies for Water Conservation	366
PROBLEM SET **16.4:** Communication of the Water Supply System	367
Background: Communication of the Waste Removal System	368
Venting of the Waste System	368
Diagramming the Waste Removal System	368
PROBLEM SET **16.5:** Communication of the Waste Removal System	368
Background: Storm Water Drainage	369
PROBLEM SET **16.6:** Storm Water Drainage	369

CHAPTER 17   INDOOR ENVIRONMENTAL QUALITY AND SECURITY   370

Background: Energy	371
PROBLEM SET **17.1:** Energy	371
Background: Factors Influencing Energy Consumption	372

Energy Codes	374
PROBLEM SET 17.2: Energy Codes	377
Background: Calculating Heating and Cooling Loads	378
PROBLEM SET 17.3: Calculating Heating and Cooling Loads	380
Background: Methods of Heat Transfer	382
PROBLEM SET 17.4: Methods of Heat Transfer	382
Background: Heating Systems	382
Heat from solar radiation	385
Heat Pumps	385
Boilers	385
Furnace	386
Air Handling Units and Roof Top Units	387
Air Conditioning	387
Ventilation	387
PROBLEM SET 17.5: Heating Systems	387
Background: Security and Protection	387
Fire and Smoke Protection	388
Command Center, Monitor and Control	388
Intelligent Buildings	389
PROBLEM SET 17.6: Intelligent Buildings	389

CHAPTER 18 LANDSCAPING 390

Background: Problem Identification	391
Needs Analysis	391
Site Analysis	391
Rules and Regulations	392
The Function of the Landscape	402
PROBLEM SET 18.1: Problem Identification	402
Background: Conceptual Planning	406
Principles of Design	406
Bubble Diagram	407
Concept Plan	407
PROBLEM SET 18.2: Conceptual Planning	407
Background: Sustainable Landscaping	408
Climate Zones	409
PROBLEM SET 18.3: Sustainable Landscaping	410
Background: The Details: Choosing Landscape Elements	411
Hardscape	411
Softscape	411
Irrigation	411
PROBLEM SET 18.3: Sustainable Landscaping	411
Background: Final Landscape Plan	413
PROBLEM SET 18.4: Final Landscape Plan	414

CHAPTER 19 VISUAL COMMUNICATION OF DESIGN INTENT 416

Background: Presentation Graphics Support Design Visualization	417
Pictorial Drawings and Design Visualization	417
PROBLEM SET 19.1: Presentation Graphics Support Visualization	421
Background: Audience and Purpose Influence Presentation Graphics	423
Colorful Elevation Views	423
Enhancing Realism of Presentation Graphics	423

Building Design Software Illustrates Climate Conditions   424
PROBLEM SET **19.2:** Audience and Purpose Influence Presentation Graphics   426
Background: Communication of Interior Details   426
PROBLEM SET **19.3:** Communication of Interior Details   426
Background: Models Provide Perspective and Analysis   427
PROBLEM SET **19.4:** Models Provide Perspective and Analysis   428

CHAPTER 20 FORMAL COMMUNICATION AND ANALYSIS   429

Background: Formal Presentation of a Building Proposal   430
PROBLEM SET **20.1:** Formal Presentation of a Building Proposal   430
Background: Presentation Purpose, Audience, and Logistics   430
PROBLEM SET **20.2:** Presentation Purpose, Audience, and Logistics   430
Background: Brainstorming Content and Format   431
PROBLEM SET **20.3:** Brainstorming Content and Format   431
Background: Gather and Prepare Presentation and Support Materials   431
Organization of Documents   432
PROBLEM SET **20.4:** Gather and Prepare Presentation and Support Materials   433
Background: Communication/Implementation of the Presentation   433
The Rehearsal   433
The Presentation   433
Communicating Features and Benefits   433
Presentation Components   433
PROBLEM SET **20.5:** Communication/Implementation of the Presentation   434
Background: Presentation Closure   434
PROBLEM SET **20.6:** Presentation Closure   434
Background: Feedback Supports Presentation Analysis   434
PROBLEM SET **20.7:** Feedback Supports Presentation Analysis   434
Modification or Supplementation to Achieve Desired Outcome   434

Preface

This workbook was developed to support *Civil Engineering and Architecture* with real-world, hands-on activities that build foundational skills for architects and civil engineers and provide opportunities to apply those skills in more challenging projects. The authors have combined their years of experience with Project Lead the Way's® Civil Engineering and Architecture curriculum to produce a resource brimming with:

- Hands-on, directed research and design activities
- Drawing and sketching practice
- Math support
- Team development and collaborative projects
- Open-ended problems and projects to provide greater challenges

Features of this Workbook

This text was developed to complement and support Project Lead the Way's® Civil Engineering and Architecture curriculum and can be used to support any project-based course for these disciplines. The following features are built into each unit to help students apply the design process to achieve productive results.

BACKGROUND

Background sections help students develop and review the knowledge they need to perform the activities that follow.

EXERCISES

The core of this workbook is dozens of hands-on exercises that build essential engineering and design skills, from proper application of the math and physics of architecture and engineering, to brainstorming, drawing, modeling, and presentation.

Problem Sets

Units conclude with problem sets for additional practice at carrying levels of rigor.

Civil Engineering and Architecture with Project Lead the Way, Inc.

This workbook is part of a series of learning solutions that resulted from a partnership forged between Delmar Cengage Learning and Project Lead the Way, Inc. in February 2006. As a non-profit foundation that develops curriculum for engineering, Project Lead the Way, Inc. provides students with the rigorous, relevant, reality-based knowledge they need to pursue engineering or engineering technology programs in college.

Project Lead the Way® curriculum developers strive to make math and science relevant for students by building hands-on, real-world projects in each course. To support Project Lead the Way's® curriculum goals, and to support all teachers who want to develop project/problem-based programs in engineering and engineering technology, Delmar Cengage Learning is developing a complete series of texts to complement all of Project Lead the Way's nine courses:

- Gateway to Technology
- Introduction to Engineering Technology
- Principles of Engineering
- Digital Electronics
- Aerospace Engineering
- Biotechnical Engineering
- Civil Engineering and Architecture
- Computer Integrated Manufacturing
- Engineering Design and Development

To learn more about Project Lead the Way's ongoing initiatives in middle school and high school, please visit www.pltw.org.

Acknowledgments

The authors and publisher wish to thank everyone who assisted in the development of this workbook, especially the following reviewers and Master Teachers who provided valuable input:

Todd Benz
Pittsford Central School District
Pittsford, NY

Eric Dunn
Sinclair Community College
Dayton, OH

Eric Fisher
Hamilton Heights High School
Arcadia, IN

Wendy A. Ku, PhD
Simsbury High School
Simsbury, CT

David Lynch
Memorial High School
Madison, WI

Peter Tucker
Triad High School
Troy, IL

The authors and publisher owe special thanks to Cynthia Mondgock and iD8TripleSSSPress for their patient help developing the manuscript, gathering art, and assisting the authors in the publishing process.

CHAPTER 1
Definitions and History of Civil Engineering and Architecture

Skills List

After completing the problems in this chapter, you should be able to:

- Define Civil Engineering and Architecture

- Recognize and categorize basic structural systems

- Understand accomplishments through engineering and architecture during the early civilization

- Possess a general knowledge of Egyptian, Greek, Roman, Gothic, and Renaissance architecture

- Discuss various famous architects

BACKGROUND

It is critical to understand what defines a civil engineer, along with the responsibilities, history, and dynamics of the profession.

Definition and History of Civil Engineering

Civil engineering is one of the oldest branches of engineering that covers planning, design and construction processes, and maintenance of the man-made and natural built environment. These structures include bridges, dams, canals, tunnels, roads and highways, and water treatment facilities. Civil engineering is a vast field of engineering; there are many divisions within the field of civil engineering. These categories include: construction engineering, environmental engineering, geotechnical engineering, structural engineering, transportation engineering, urban and community planning, water resources, and surveying.

Problem Set 1.1   Definition and History of Civil Engineering

Problem 1-1   When was the term "civil engineering" first used and why?
Problem 1-2   What are some examples of civil engineering projects?
Problem 1-3   Given several definitions of civil engineering, how would you define civil engineering in your own words?
Problem 1-4   Who was the first recognized civil engineer? Provide an example of their credited work.
Problem 1-5   What are the subcategories of civil engineering?
Problem 1-6   Find and photograph three examples of civil engineering works in your hometown. Which subcategories would these fall under?

BACKGROUND

Civil Engineers are responsible for developing structural systems that are the cornerstones of today's built environment.

Civil Engineering of Structures

During any design phase, Civil Engineers must consider all types of loading. This loading may consist of vertical loads (gravity loads) or horizontal loads (lateral loads). There are several basic structural systems that are utilized to resist these loading conditions safely. Based on the load, some systems may be more suitable than others. These systems consist of column and beam, corbel and cantilever, arch and vault, truss and space frame, and tensile structures.

Problem Set 1.2   Civil Engineering of Structures

In problems 1-7 through 1-13, state which structural systems were used in the design. Some of the images might contain more than one structural system.

PROBLEM 1-7

PROBLEM 1-8

PROBLEM 1-9

PROBLEM 1-10

PROBLEM 1-12

PROBLEM 1-11

PROBLEM 1-13

Civil Engineering of Structures   3

BACKGROUND

Architecture is often referred to as an art, but it is also a science and encompasses form as well as function.

Definition of Architecture

Architecture needs to be pleasing and meet the needs of the client and fit into the environment (built and natural). Basic geometric shapes can visually define a piece of architecture. Architectural geometry may consist of intersecting rectangles, prisms, and cylinders, for example. Figures 1-1 and 1-2 show two very famous museums with two very different architectural design solutions.

FIGURE 1-1 *Musee de Louvre, Paris, France.*

FIGURE 1-2 *Solomon R. Guggenheim Museum, New York, Ny.*

Problem Set 1.3   Definition of Architecture

Problem 1-14   Define architecture in your own words.
Problem 1-15   How many different geometric shapes do you recognize in figure 1-1?
Problem 1-16   How many different geometric shapes do you recognize in figure 1-2?

BACKGROUND

The history of civil engineering and architecture is extremely vast and still quite mysterious to historians and archaeologists. Understanding the progression of engineering and architecture is critical to advancing in new technologies.

History of Civil Engineering and Architecture

The history of civil engineering and architecture is constantly changing with new technologies and recent discoveries. It is often difficult to dissect the full story of a structure as it is often left for assumptions and guessing. Technology such as thermoluminescence and carbon dating can provide a general date of construction, and clues such as tools, pottery, and skeletal remains help determine the use and the culture.

Example 1.1   Research and document the ancient Neolithic city of Çatalhöyük, Turkey.

As one of the oldest and best preserved sites, Çatalhöyük, shown in figure 1-3, is believed to date back to 6500 BC. The structure incorporates eleven building levels that express an urban housing environment. Çatalhöyük was first discovered in the 1950s and was excavated during four different periods in the early 1960s by James Mellaart. The housing portion of the complex is made up of dense, one-story rooms with the only access from ladders to the roof. Each house contained wooden platforms and benches, and the dead were buried under the platforms in the floor. In addition to the houses, approximately forty structures, similar in nature to the houses, are believed to be sanctuaries. This theory is based on the elaborate decorations found within. More research and excavation is currently underway at Çatalhöyük.[1]

FIGURE 1-3 *Çatalhöyük.*

A Journey to the Ancient world 3200 BC to AD 337

This time period is home to ancient Neolithic settlements, tombs, temples, ritual structures, and defenses. The use of stone was prevalent in these structures.

Beginning with settlements, many drystone dwellings have been discovered. One of the most famous, Skara Brae, is shown in figure 1-4. This structure, dating back to 2500 BC, was constructed with stacked rock rather than mud or clay-clad structures. The Rinyo settlement on Orkney Island and the Yoxie Settlement in Shetland provide further examples.

FIGURE 1-4 *Skara Brae.*

Thousands of megalithic tombs were built during this era that fall into the categories of passage-graves and gallery-graves. Some tombs, referred to as "longbarrows" were also discovered in England. Two vastly different

History of Civil Engineering and Architecture   5

examples of these tombs are shown in figures 1-5 and 1-6. Figure 1-5 depicts Bryn Celli Ddu located on the Isle of Anglesey, Wales, is an example of a passage-grave, and figure 1-6 is of a gallery-grave in Esse, Brittany, France (La Roche-aux-Fees).

FIGURE 1-5 *Bryn Celli Ddu.*

FIGURE 1-6 *Gallery-grave in Esse, Brittany, France.*

FIGURE 1-7 *Temples with trefoil plans.*

Temples and ritual structures, although fewer in number than tombs, are scattered all over Europe and are representative of early architectural buildings with specific purpose. Temples with trefoil plans (overlapping rings) were found dating back to 2700 BC. An example is shown in figure 1-7.

Ritual structures were also prevalent during this time and consist of upright stones placed in elongated or circular patterns. These structures may also be referred to as "henges." The most famous is Stonehenge, but several other spectacular henges serve as reminders of these prehistoric builders and astronomers. The henge at Callanish still stands on Isle of Lewis, Scotland (figure 1-8).

The beginnings of civil engineering appear during this era with the first defensive structures. The Clickhimin fort shown in figure 1-9 is an example of a Shetland drystone defense used to monitor the sea.[2]

FIGURE 1-8 *The henge at Callanish.*

FIGURE 1-9 *The Clickhimin fort.*

Recent discoveries and investigations

Ancient settlements and temples are still being discovered. First dismissed as broken tombstones in the 1960s, Turkey's Gobekli Tepe (figures 1-10 and 1-11) is now believed to be the world's first temple. The hillside was revisited in 1994, and excavation began shortly after. Carbon dating studies of the site and the surrounding area date the temple to 9000 BC, when the people had yet to craft metal tools or pottery. The site consists of roughly 20 circles (one measuring 65 feet in diameter) made up of intricately carved stone pillars that range in height up to 16 feet. Only a few of these circles have been unearthed, but the rest are known to exist, due to ground-penetrating radar and geomagnetic surveys. Since no evidence of permanent residence has been found at the site, it is assumed that the ruin is a place of worship. Perhaps the most interesting theory that has transpired from the discovery of Gobekli Tepe is that of the abilities of the people. It was believed that in order for a civilization to be organized enough to build temples, they must first learn farming and live in settled societies. However, the people who constructed Gobekli Tepe left no indication of their farming abilities and were thought to be hunters and gatherers.[3]

FIGURE 1-10 *Turkey's Gobekli Tepe.*

FIGURE 1-11 *Turkey's Gobekli Tepe.*

Ancient Egyptian

In great contrast to the neolithic temples and tombs just discussed, Ancient Egyptian architecture was expressed mainly with pyramids and temples.

Tomb architecture was extremely important to ancient Egyptians, who believed very heavily in the after-life. The early tombs were rectangular in plan with flat roofs and are known as "mastabas." As tombs evolved into the form of a pyramid, they began as stacked mastabas. This is observed in the form of the Step Pyramid of Zoser seen in figure 1-12.

The most famous set of pyramids lies in Giza. Six pyramids scatter the landscape with the statue of the Great Sphinx keeping watch. This collection of Egyptian architecture is shown in figure 1-13.

FIGURE 1-12 *Step Pyramid of Zoser.*

FIGURE 1-13 *Pyramids in Giza.*

Figure 1-14 shows the progression of pyramid construction in Giza with the mastaba and Queen's Pyramid in the foreground and the Pyramid of Cheops beyond.

Although pyramids are usually synonymous with Egyptian architecture, large elaborate temples were built for deified pharaohs and ancient gods. An example of one of these massive temples, the Temple for Queen Hatshepsut, is shown in figure 1-15.

FIGURE 1-14 *Timeline of pyramids in Giza.*

FIGURE 1-15 *Temple for Queen Hatshepsut.*

History inspires modern-day designs

Almost all early American architecture can be traced back to some European influences. Major cities, such as Boston, New York, and Philadelphia, flourished with the English stone terrace-style homes. Pierre L'Enfant introduced a plan for Washington, D.C., based on European urban planning (diagonal axes), and the Greek Revival was introduced. Although this "Greek Revival" period lasted between 1818 and 1850, it still occurred over 2,000 years after the Parthenon was built on the Acropolis. Figure 1-16 shows the U.S. Treasury Building (1839–1869) in Washington, D.C., designed by Robert Mills.

Other history revivals occurred, such as the Gothic, Egyptian, Russian, Romanesque, Spanish Colonial, and Mediterranean, to name a few.

FIGURE 1-16 *U.S. Treasury Building.*

Example 1.2 Find and document a history-inspired modern work of architecture.

As with the Luxor hotel in Las Vegas, Egyptian architecture has made its mark in Memphis, Tennessee. In 1991, construction was completed on what is now the third largest pyramid in the world. This pyramid, named "The Great American Pyramid," stands 321 feet tall. The stainless steel clad structure boasts over 500,000 square feet of useable space and is currently utilized as an arena. The pyramid is a smaller-scale replica of the Great Pyramid of Cheops and comes complete with a statue of Ramses guarding the entry.

250 BC–AD 1500

The Great Wall of China was initially constructed in three segments by different states as a border to prevent hostility from the north. Centuries later, the country was unified, and the segments

FIGURE 1-17 *The Great American Pyramid.*

were combined into the longest wall in the world, at over 4,000 miles long. Originally, the walls were built from readily available materials; stones in the mountainous regions, and rammed earth in the plains regions. The wall has been rebuilt over the years with brick, strengthened and stretched to heights ranging from 23 feet to 46 feet at strategic locations. The base of the wall is 20 feet thick and tapers to 16 feet at the top.

Greek Architecture 448–432 BC Parthenon in Athens

The style and number of elements used in Greek architecture will often give the purpose of the structure as well as when it was constructed. Simply the number of columns (even or odd) will indicate its age, and the style of columns match the usage. Geometry was also used to distinguish between temples (which were rectangular in plan) and circular, commemorative buildings.

FIGURE 1-18 *Parthenon.*

When one thinks of the architecture of Greece, the Acropolis and its many structures are likely the first image to come to mind. The most famous of the ruins is the Parthenon, which was built for the Greek goddess Athena. The Parthenon has served many purposes over Greece's rocky history; it has served as various religious sanctuaries as a Christian church, a Latin church, and a mosque. The Parthenon is shown in figure 1-18.

Columns used in today's architecture

Columns are one of the most important elements of architecture and are used for not only structural purposes but also aesthetic reasons as well. The three classical orders of columns are shown in figure 1-19. These increase in detail and complexity from Doric to Ionic to Corinthian. Columns will be discussed in more detail in Chapter 4.

Engineering in Ancient Times

Ancient engineers accomplished many remarkable feats. There are many bridges, roads, canals, pyramids, and aqueducts that are a thousand years old and are still standing.

The early Romans created engineering marvels with their aqueducts. Initially, water was piped down into valleys and then back up the far side. Lead pipes were not strong enough to withstand the water pressure, and the result was the development of the aqueduct. These massive structures are not only impressive engineering solutions but also beautiful architectural elements as well. A breathtaking example remains near Nimes, France, at Pont du Gard (figure 1-20).

FIGURE 1-19 *Orders of Columns.*

FIGURE 1-20 *Pont du Gard.*

In addition to aqueducts, ancient engineers built bridges that are still in use. One of the oldest surviving bridges in Rome, Pons Fabricius, was built in 62 BC and is shown in figure 1-21.

FIGURE 1-21 *Pons Fabricius.*

Ancient Roman Empire

The Roman Empire left behind its own beautiful structures to rival those of ancient Greece and Egypt. Temples, basilicas, amphitheatres, and arches are some of the most famous structures in the world.

As discussed in the text, the Pantheon in Rome is one of the most impressive temples still standing. Its dome was a huge engineering accomplishment, and the span of 142 feet was unchallenged for over a thousand years. Several architectural elements and forms are combined to create this interesting temple. Figure 1-22 shows the portico and rotunda.

FIGURE 1-22 *Pantheon.*

In addition to the temples, several monumental arches tower over the city of Rome. One of these, the Arch of Titus, is a load bearing masonry structure that dates back to 81 AD and is the oldest of the triumphal arches in Rome. This arch is shown in figure 1-23.

FIGURE 1-23 *Arch of Titus.*

One of the most recognizable pieces of architecture left by the ancient Romans is the Roman Coliseum (figure 1-24). Construction began in 70 AD and lasted for twelve years. The arena is a free-standing structure that was used for gladiatorial contests and animal hunts. Although the flooring inside the arena is no longer, the exterior has been well preserved, surviving major hazards as earthquake and fire.

FIGURE 1-24 *Roman Coliseum, Rome.*

Gothic Architecture 12th to 16th Century/AD1140–1520

Gothic architecture built upon elements already in use in structures, mainly the arch, to develop its own personality resulting in delicate cathedrals. This architecture is characterized by pointed arches, rose windows, spires, and flying buttresses. The pointed arch proved essential in Gothic architecture as it provided no fixed height-to-span ratio and made it possible to carry heavier loads more efficiently. An example of the Gothic arch is shown in figure 1-25.

Another very notable symbol of the Gothic era is the rose window, sometimes called a "wheel window." These windows are circular in nature and typically made of very colorful stained glass. The wheel reference comes from the resemblance to the spokes of a wheel. Figure 1-26 is an ornate rose window from Notre Dame Cathedral in Paris, France.

FIGURE 1-25 *Gothic arch.*

FIGURE 1-26 *Rose window.*

With the interior arches comes the flying buttress as a support system. Many Gothic structures utilize these masonry arches against exterior walls to resist thrust. Figure 1-27 shows flying buttresses radiating off of a cathedral in France.

FIGURE 1-27 *Flying buttresses.*

Renaissance Architecture 1400–1600

The Renaissance Period followed closely after the Gothic Architecture Period but varied immensely. The features of this era are found in the columns, arches, vaults, domes, and ceilings. The columns of the Renaissance utilized the classical orders illustrated in figure 1-19. Similar to the Gothic Period, the Renaissance Period is known for its use of arches. However, the Renaissance went back to the semi-circular arch in comparison to the Gothic pointed arch. The vaults were lacking ribs, and domes were utilized on the exterior of the building. Figure 1-28 shows Bramante's Tempietto in Rome. This small temple dates back to 1502.

Another glaring contrast to the Gothic style is the addition of ceilings within the churches. These ceilings were painted with beautiful murals. One of the most famous is Michelangelo's Sistine Chapel. Figure 1-29 shows a painted ceiling inside the Vatican in Rome.

FIGURE 1-28 *Tempietto.*

FIGURE 1-29 *Painted ceiling.*

Famous Buildings

Several famous buildings have been discussed in the previous sections, ranging from churches to palaces, tombs, and temples. A building may be considered famous for many reasons. A building may be the first to accomplish something: it may be the oldest or have the greatest span or the longest cantilever, or it may be the tallest in the world or the first to utilize a specific building material. The fame might not have anything to do with the architectural or engineering achievements at all but might relate to famous events that have occurred: Olympic Games, the World's Fair, tragedies, and joyous events. Figure 1-30 illustrates the Crystal Palace, which was constructed for the first World's Fair (Great Exhibition) of 1851. This is one of the largest glass buildings ever constructed.

FIGURE 1-30 *Crystal Palace, London (relocated and later destroyed by fire, 1936).*

History of Civil Engineering and Architecture

Problem Set 1.4   History of Civil Engineering and Architecture

Problem 1-17	Define thermoluminescence and explain why it is used.
Problem 1-18	Research and document the ancient city of Jarmo, Iraq.
Problem 1-19	Research and document the Neolithic village of Ain Ghazal, Jordan.
Problem 1-20	Research and document the Mehrgarth settlement in Pakistan.
Problem 1-21	Research and document the Harappa Civilization in Pakistan.
Problem 1-22	Research and compare the megalith passage-graves in New Grange, Dowth, and Knowth. All are located in Ireland.
Problem 1-23	Research and document the megalith gallery-grave of the West Kennet Long Barrow in England.
Problem 1-24	Document examples of prehistoric structures that utilize a trefoil plan.
Problem 1-25	Research and document the megalith henge of Avebury, England.
Problem 1-26	List three examples of early prehistoric forts.
Problem 1-27	Find an example of another recent discovery, and document what advancements were made.
Problem 1-28	What term was given to describe the first Egyptian tombs?
Problem 1-29	Amazingly, the Egyptian pyramids were constructed of stone without the knowledge of the pulley. Research and describe what methods and tools were utilized in construction.
Problem 1-30	Who was the builder of the Step Pyramid?
Problem 1-31	Observe the architecture of Boston College in Massachusetts. What influences do you see in the buildings on campus?
Problem 1-32	In addition to the Parthenon, what other temples still exist on the Acropolis in Athens?
Problem 1-33	How does the temple Erechtheion compare to the Parthenon?
Problem 1-34	What structural element is utilized in aqueduct design?
Problem 1-35	Research and document the Segovia aqueduct in Spain.
Problem 1-36	Research and document the ancient bridge of Augustus, Rimini, Italy.
Problem 1-37	Compare and contrast the temple Parthenon to the temple Pantheon.
Problem 1-38	Why is the Pantheon such a notable structure?
Problem 1-39	What triumphal arches still exist in Rome?
Problem 1-40	What structural systems do you see in the Roman Coliseum (figure 1-24)?
Problem 1-41	Which classical column orders from figure 1-19 can you identify in figure 1-24?
Problem 1-42	Describe why Canterbury Cathedral in Canterbury, England (shown in figure 1-31), would be an example of Gothic Architecture.

FIGURE 1-31 *Canterbury Cathedral.*

Problem 1-43 Research the Renaissance Architecture Period, and name two architects from this era.

Problem 1-44 Research and document St. Peter's Basilica in Rome. Why is this structure such a strong example of Renaissance Architecture?

Problem 1-45 What famous structure was "temporarily" built for the 1889 World's Fair in Paris, France?

Problem 1-46 What Seattle, Washington, skyline staple was built for the World's Fair in 1962?

Problem 1-47 Research and document the Olympic Stadium in Montreal, Canada. Why is the roof structure unique? What problems have plagued this building?

Problem 1-48 Research and document the Olympic Park in Munich, Germany. What historical tragedy overshadowed the beauty of the park?

BACKGROUND

An architect's fame may come from success, controversy, or failures. Designing with the public in mind and having the result displayed for all to see make anonymity nearly impossible.

Famous Architects

The text documents several famous architects over time, including the most well known name of Frank Lloyd Wright. Wright is credited with works all over the country, such as the Guggenheim Museum in New York, shown in figure 1-2, and for his Prairie Style residences. One interesting fact not known to many is that Wright's first employee, Marion Mahony Griffin, was the first licensed female architect and a contributor on many of his works.

Designing functional and interesting structures may bring notoriety from the architectural community, but becoming known to the general public may require more. American architect Michael Graves accomplished this feat by designing an entire line of household items for Target retail stores. Graves is known for his sometimes whimsical, yet sophisticated, style of buildings. Figure 1-32 shows the St. Coletta School in Washington, D.C., constructed for children with disabilities.

FIGURE 1-32 *St. Coletta School.*

Another famous architect who parallels Frank Ghery in creative forms of architecture is Iraqi-born Zaha Hadid. Hadid was the first woman to receive the Pritzker Architecture Prize in 2004. Her design for the National Museum of 21st Century Arts in Rome is shown in figure 1-33.

FIGURE 1-33 *National Museum of 21st Century Arts.*

Problem Set 1.5   Famous Architects

Problem 1-49   What famous architect is credited with the Seattle Public Library (shown in figure 1-34)? What other famous buildings are linked to this architect?

FIGURE 1-34 *Seattle Public Library.*

Problem 1-50   Research architect Paul Rudolph. How would you describe his style? What are some of his famous buildings?

Problem 1-51   Who was the first recipient of the Pritzker Architecture Prize? What year was this awarded?

Problem 1-52   Research architect Julia Morgan. What makes her influential as a female architect?

Problem 1-53   Research architect Norma Merrick Sklarek. What makes her influential as a female architect?

Problem 1-54   What architecture student is responsible for the design of the Vietnam War Memorial in Washington, D.C.? (Note: The designer was only 21 at the time!)

Problem 1-55   Name a famous architect from your region. List some of their works.

The success of urban planning depends greatly on the contributions of civil engineers.

Major Developments in Civil Engineering Impact Land Development

Urban infrastructure can greatly influence the way land is used. With the development of new highways, bridges, and roads, the use of the surrounding land is quick to follow. Easy access to work, shopping, and home is always desirable. Civil Engineers also work to control water flow to make land useable.

Properties of Materials

Much has already been discussed about building materials and how they have affected the evolution of design through ancient times. Accessibility as well as tools and technology were the limiting factors to early builders as well as current designers. More on materials will be discussed in Chapter 12.

Innovation of Material and Process

The text discusses the advancements of materials and process in road construction. Another continuously evolving material process is the mix design of concrete. Various admixtures have been developed to affect the concrete in many different ways, such as strength, curing time, and weather versatility. In steel construction, the carbon levels may be altered to achieve higher-strength steels as well. These advancements have led to more possibilities in building design that were not achievable before.

Problem Set 1.6 Major Developments in Civil Engineering Impact Land Development

Problem 1-56	Think of your hometown or a surrounding city. List any new road construction or highway improvements that have resulted in commercial or residential building.
Problem 1-57	Use the Internet to research concrete admixtures. Which ones will affect the strength?
Problem 1-58	What advancements have been made in wood construction?

Endnotes

1. http://www.catalhoyuk.com/history.html Leick, Gwendolyn. *A Dictionary of Ancient Near Eastern Architecture*. New York: Routledge, 1988.
2. Fletcher, Banister; Cruickshank, Dan, *Sir Banister Fletcher's A History of Architecture*, Architectural Press, 20th edition, 1996 (first published 1896).
3. Curry, Andrew. "Gobekli Tepe: The World's First Temple?" *Smithsonian*. Nov. 2008. Smithsonian.com. 11 Aug. 2010. < http://www.smithsonianmag.com/history-archaeology/gobekli-tepe.html>.
4. Fletcher, Banister; Cruickshank, Dan, *Sir Banister Fletcher's A History of Architecture*, Architectural Press, 20th edition, 1996 (first published 1896).
5. Fletcher, Banister; Cruickshank, Dan, *Sir Banister Fletcher's A History of Architecture*, Architectural Press, 20th edition, 1996 (first published 1896).
6. Simpson, Teresa R. "The Pyramid Arena." About.com. <http://memphis.about.com/od/historyandfacts/qt/pyramid.htm>.
7. Fletcher, Banister; Cruickshank, Dan, *Sir Banister Fletcher's A History of Architecture*, Architectural Press, 20th edition, 1996 (first published 1896).
8. Fletcher, Banister; Cruickshank, Dan, *Sir Banister Fletcher's A History of Architecture*, Architectural Press, 20th edition, 1996 (first published 1896).
9. Fletcher, Banister; Cruickshank, Dan, *Sir Banister Fletcher's A History of Architecture*, Architectural Press, 20th edition, 1996 (first published 1896).
10. Fletcher, Banister; Cruickshank, Dan, *Sir Banister Fletcher's A History of Architecture*, Architectural Press, 20th edition, 1996 (first published 1896).
11. Fletcher, Banister; Cruickshank, Dan, *Sir Banister Fletcher's A History of Architecture*, Architectural Press, 20th edition, 1996 (first published 1896).

CHAPTER 2
Careers

Skills List

After completing the problems in this chapter, you should be able to:

- Be familiar with educational requirements to become a civil engineer or architect

- Be familiar with the internships involved in becoming a civil engineer or architect

- Be familiar with the testing procedures involved in becoming licensed as a civil engineer or architect

- Be aware of career options closely related to civil engineering and architecture

- Understand the difference between traditional design and integrative design

- Be aware of other agencies involved in project design

BACKGROUND

Several factors must be considered when pursuing civil engineering as a career, such as education, training, and licensure.

Civil Engineering as a Career

Civil engineering is a popular career choice, since it is a profession needed in nearly every community. Civil Engineers may be utilized in various fields, such as environmental, structural, transportation, and wastewater.

Education, Training, and Skills Needed to Become a Civil Engineer

Becoming a Civil Engineer begins in high school, where requirements must be met before admission into higher education becomes a reality.

The following is an example of general admission requirements into a large university:

- An ACT score of 24/36 or a SAT score of 1090/2400

  OR

- High school GPA of 3.0/4.0 (unweighted) and ranked in the top 33.3% of class

Once admitted to a college or university, many civil engineering programs require specific admissions into a professional school program, in addition to general admission into the university.

The following are requirements from a university's civil engineering professional school program:

For admission into the civil engineering professional school, which is a requirement to take upper-level department courses, the following must be met:

- Completion of at least 60 college-level semester credit hours
- Completion of at least 12 college-level semester credit hours from the program's university
- Completion of at least 53 college-level semester credit hours from the pre-professional school courses
- Completion of Calculus I, II, and III; General Physics; General Chemistry; four Engineering Science courses; English Composition I and II with a grade of "C" or better
- An overall GPA of 2.3/4.0 or greater from the program's university

When selecting a civil engineering degree program, it is important to research not only the program's focus but also the type of degree, the length of the program, and whether the program is accredited or not.

The following is an example of university coursework for a civil engineering student (course descriptions vary with school):

First Year	
Fall Semester	**Spring Semester**
ENGR 1412 Visual Basic (Programming)	ENGR 1332 Engineering Design
CHEM 1414 General Chemistry	PHYS 2014 General Physics
ENGR 1111 Introduction to Engineering	MATH 2153 Calculus II
MATH 2144 Calculus I	GEN ED History
ENGL 1113 English Composition I	HIST 1103 American History
POLS 1113 American Government	

Second Year	
Fall Semester	**Spring Semester**
ENSC 2213 Engineering Thermodynamics	ENSC 2143 Engineering Strength of Materials
ENSC 2113 Engineering Statics	ENSC 2123 Engineering Dynamics
PHYS 2114 General Physics	CIVE 3813 Environmental Engineering Science
MATH 2163 Calculus III	ENSC 2613 Engineering Electrical Science
SPCH 2713 Introduction to Speech	MATH 2233 Differential Equations
GEN ED History	CIVE 3614 Surveying

Third Year	
Fall Semester	**Spring Semester**
CIVE 3713 Geotechnical Engineering	CIVE 3633 Transportation Engineering
CIVE 3413 Structural Analysis	CIVE 4711 Basic Soils Laboratory
ENSC 3233 Engineering Fluid Mechanics	STAT 4073 Engineering Statistics
IEM 3503 Engineering Economics	CIVE 3523 Concrete
GEN ED Social Science	CIVE 3833 Hydraulics
CIVE 3843 Hydrology	CIVE 3623 Engineering Materials

Fourth Year	
Fall Semester	**Spring Semester**
CIVE 4273 Construction Engineering and Management	CIVE 4833 Unit Operations
CIVE 3513 Steel	CIVE 4043 Senior Design
CIVE 4042 Engineering Practice	CIVE Elective
CIVE Elective	CIVE Elective
ENGL 3323 Technical Writing	GEN ED Natural Science

A college or university's program accreditation status plays a large role in meeting requirements for licensing as a civil engineer. The Accreditation Board for Engineering and Technology (ABET) evaluates a program to ensure that quality standards are met. These standards are established by the engineering profession. More may be found at www.abet.org.

FIGURE 2-1 Concrete canoe.

FIGURE 2-2 Concrete canoe.

The following is an example of a university program's ABET Educational Objectives:
The Bachelor of Science in Civil Engineering degree program prepares graduates who:

- Contribute to society through the practice of civil engineering in a variety of contexts
- Possess the technical background (including the six major areas of civil engineering: construction management, environmental, geotechnical, water resources, structural, and transportation) necessary to be adaptable and successful in the civil engineering profession
- Have the ability to advance within their profession, including attaining professional licensure and positions of leadership
- Exhibit life-long learning, including the pursuit of advanced degrees

Stages of Career Development to Become a Licensed Civil Engineer

Just as preparing for a degree in civil engineering began in high school, preparing for a civil engineering career begins in college. Upon completion of the civil engineering program at a college or university, the college graduate must take the Fundamentals of Engineering Exam (FE Exam) and pass in order to obtain licensure within the field. The exam is multiple-choice and is broken into morning and afternoon sessions. The morning exam is a general exam for all engineering disciplines, and the afternoon session corresponds to the examinee's engineering discipline. The National Council of Examiners for Engineering and Surveying (NCEES) administers the FE exam. Additional information regarding the FE exam may be retrieved from the following website: www.ncees.org.

The following is provided by NCEES and is a breakdown of the civil engineering module for the FE exam based on topic area and approximate exam content:

Morning Session	
Mathematics	15%
Engineering Probability and Statistics	7%
Chemistry	9%
Computers	7%
Ethics and Business Practices	7%
Engineering Economics	8%
Engineering Mechanics (Statics and Dynamics)	10%

Morning Session	
Strength of Materials	7%
Material Properties	7%
Fluid Mechanics	7%
Electricity and Magnetism	9%
Thermodynamics	7%

Afternoon Session	
Surveying	11%
Hydraulics and Hydrologic Systems	12%
Soil Mechanics and Foundations	15%
Environmental Engineering	12%
Transportation	12%
Structural Analysis	10%
Structural Design	10%
Construction Management	10%
Materials	8%

After graduating with an accredited degree and successfully completing the FE exam, one must gain work experience under a licensed professional engineer. This internship varies by state but is generally four or more years. After this internship is completed, eligibility for taking the Principals and Practice of Engineering Exam (PE Exam) is achieved. The application process for this exam varies by state and may be found at www.ncees.org. Currently, five different civil engineering exams are available: Construction, Geotechnical, Structural, Transportation, and Water Resources and Environmental Control Systems. Similar to the FE exam, all examinees take the same morning exam and a discipline-specific afternoon exam. According to NCEES, licensed engineers are open to more career options, such as stamping and sealing drawings, bidding on government contracts, becoming principal of a firm, performing consulting services, and offering services to the public.

Problem Set 2.1   Civil Engineering as a Career

For problems 2-1 through 2-4, refer to a civil engineering degree program at a university of your choice.

Problem 2-1   What are the general university admission requirements (for example, minimum grade point average, entrance exam scores)?

Problem 2-2   Does the civil engineering degree program have any additional entrance requirements?

Problem 2-3   How long is the program, and what options exist? Four- or five-year bachelor's, master's, doctorate?

Problem 2-4   Is this degree program ABET accredited?

Problem 2-5   After graduating from an ABET accredited civil engineering degree program and successfully completing the Fundamentals of Engineering Exam, what title is bestowed?

Problem 2-6   Research the Principles and Practice of Engineering Exam at www.ncees.org. What are the requirements for your state?

Problem 2-7   What professional engineering exam types are available for civil engineers?

Problem 2-8   Interview a civil engineer, and ask questions about their education and career paths to becoming a licensed civil engineer.

BACKGROUND

Several factors must be considered when pursuing architecture as a career, such as, education, training, and licensure.

Architecture as a Career

The education, training, skills, and effort required to become an architect are very unique and require much forethought and planning.

The American Institute of Architects (AIA) suggests exploring architecture as a career by considering the following:

- **Discover architecture**

    It is diverse and multifaceted and has many opportunities for specialization—become familiar with the options

- **Be interested**

    In the design of the built environment and public space

- **Ask questions**

    Contact your local AIA chapter; talk with architects; observe buildings and construction sites; visit schools, and speak with architecture students

- **Prepare for professional education**

    Develop a broad interest in the arts and humanities and a solid background in the physical sciences, and math

- **Learn communication skills**

    Writing, speaking, freehand drawing

- **Read**

    Books and magazines on architecture and design

Education Training and Skills Required to Become an Architect

General college admission requirements are the same for any degree path chosen. However, admission into a certain degree program or professional school may vary drastically.

The following are requirements from a university's architecture professional school program:

For admission into the architecture professional school, which is a requirement to take upper-level department courses, the following must be met:

- Completion of at least 55 credit hours of coursework that count toward the degree
- Completion of Intro to Architecture; Architecture Design Studios I, II, and III; Architectural History; Architectural Systems; Calculus I; General Physics; Engineering Statics; and English Composition I with a grade of "C" or better
- A GPA of 2.8/4.0 in required courses

Selecting an architectural degree program requires some research and personal preference, since programs vary vastly. As with civil engineering, it is important to research not only the program's focus but also the type of degree, the length of the program, and whether the program is accredited or not. Design studio atmosphere will also play a major role in selection, since many universities have programs centered around the culture of the design studio, how many studios are required and offered, and the amount of time expected for each studio.

The following is an example of university coursework for an architecture student (course descriptions vary with school):

First Year	
Fall Semester	**Spring Semester**
ARCH 1112 Introduction to Architecture	ARCH 1216 Arch Design I (Studio)
MATH 2144 Calculus	ENGL 1213 Freshmen Composition II
ENGL 1113 Freshmen Composition I	PHYS 1114 General Physics
POLS 1113 American Government	ARCH 2003 Architectural History I
GEN ED Basic Level Social Science	

Second Year	
Fall Semester	**Spring Semester**
ARCH 2116 Arch Design II (Studio)	ARCH 2216 Arch Design III (Studio)
ENSC 2113 Engineering Statics	ARCH 2263 Architectural Systems
HIST 1103 American History	GEN ED Diversity
GEN ED Basic Level Natural Science	GED ED Advanced Level Social Science

Third Year	
Fall Semester	**Spring Semester**
ARCH 3116 Arch Design IV (Studio)	ARCH 3216 Arch Design V (Studio)
ARCH 3263 Architectural Materials	ARCH 3134 Thermal/Life Safety
ARCH 3323 Steel	ARCH 3223 Timbers
ARCH 2203 Architectural History II	ARCH 3262 Computer II
ARCH 3252 Computer I	

Fourth Year	
Fall Semester	Spring Semester
ARCH 4116 Arch Design VI (Studio)	ARCH 4216 Arch Design/Development VI (Studio)
ARCH 3433 Acoustics/Lighting	ARCH 4263 Seminar
ARCH 4123 Concrete	ARCH 5293 Project Management
ARCH History/Theory Elective	Controlled Elective

Fifth Year	
Fall Semester	Spring Semester
ARCH 5117 Arch Design VIII (Studio)	ARCH Elective
ARCH 5193 Architectural Management	ARCH Elective
Controlled Elective	ARCH Elective History/Theory
Controlled Elective	Controlled Elective

FIGURE 2-3 *Architecture design studio.*

Stages of Career Development to Become a Licensed Architect

Architectural programs are held to accreditation standards similar to those of engineering programs. In the case of architecture programs, the authorizing agency is the National Architectural Accrediting Board (NAAB). NAAB works to ensure that a program meets the standards deemed appropriate for an architectural education. Most state registration boards require that applicants for licensure complete a degree from a NAAB accredited program. More information may be found about accreditation at www.naab.org.

After graduation, an internship period is necessary before sitting for the licensing exams. This internship and the Intern Development Program (IDP) are outlined by the National Council of Architectural Registration Boards (NCARB). NCARB consists of the licensing boards from all fifty states, the District of Columbia, Guam,

Puerto Rico, and the U.S. Virgin Islands. NCARB is responsible for enforcing national standards for architectural licensure within the United States. More information about NCARB may be found at www.ncarb.org.

The IDP training requirements are listed below:

Category A: Design and Construction Documents	
Training Area	Hours
1. Programming	80
2. Site and Environmental Analysis	80
3. Schematic Design	120
4. Engineering Systems Coordination	120
5. Building Cost Analysis	80
6. Code Research	120
7. Design Development	320
8. Construction Documents	1,080
9. Specifications and Materials Research	120
10. Document Checking and Coordination	80
Core Minimum Hours Required	2,200
Additional Core Hours Required in Training Areas 1–10	600
Core Minimum Hours Required	**2,800**

Category B: Construction Contract Administration	
Training Area	Hours
11. Bidding and Contract Negotiation	80
12. Construction Phase—Office	120
13. Construction Phase—Observation	120
Core Minimum Hours Required	320
Additional Core Hours Required in Training Areas 11–13	240
Core Minimum Hours Required	**560**

Category C: Management	
Training Area	Hours
14. Project Management	120
15. Office Management	80
Core Minimum Hours Required	200
Additional Core Hours Required in Training Areas 14–15	80
Core Minimum Hours Required	**280**

Category D: Related Activities	
Training Area	Hours
16. Professional and Community Service	80
Core Minimum Hours Required	80
Additional Core Hours Required in Training Area 16	160
Core Minimum Hours Required	**80**

Summary	
Category A: Design and Construction Documents	2,800
Category B: Construction Contract Administration	560
Category C: Management	280
Category D: Related Activities	80
Total Core Minimum Hours Required from Categories	3,720
Elective Hours from Any Category	1,880
Total IDP Training Hours Required	**5,600**

Another great resource for information on the IDP, licensing, and careers is the American Institute of Architects (AIA). The AIA is a professional organization that sponsors continuing education, sets the industry standard for contract documents, and publishes online papers and journals. More information may be found at www.aia.org. The AIA has also established a career website www.ARCHcareers.org with information on becoming an architect, as well as, other careers that focus on architectural skills.

Once the required internship period has been met, the application process for the licensing exams may begin. The Architect Registration Examination (ARE) tests on topics related to the health, safety, and welfare of the public.

NCARB lists the following checklist of items to consider before applying for the exam:
Review your state or jurisdiction's training requirement and conditions such as:

- Does your jurisdiction allow you to take the ARE before completion of the IDP?
- What is your board's required training period? Can this period be reduced if you satisfy the IDP training requirement in less time?
- How many years in "the office of a registered architect" are required?
- Must you satisfy your board's education and training requirements prior to the examination? After the examination?
- Are references required? Who may be used as a reference?

The ARE consists of seven parts that consist of multiple-choice, check-all-that-apply, and fill-in the-blank questions. They also contain a segment of computerized sketches. These exams may be taken in any order. More information on the ARE exam may be found at www.ncarb.org/ARE.

The seven divisions of the ARE exam are described by NCARB and are listed below:

- **Programming, Planning & Practice**   The application of project development knowledge and skills relating to architectural programming; environmental, social, and economic issues; codes and regulations; and project and practice management
- **Site Planning & Design**   The application of knowledge and skills of site planning and design including environmental, social, and economic issues, project and practice management
- **Building Design & Construction Systems**   The application of knowledge and skills of building design and construction, including environmental, social, and economic issues, project and practice management
- **Schematic Design**   The application of knowledge and skills required for the schematic design of buildings and interior space planning
- **Structural Systems**   Identification and incorporation of general structural and lateral force principles in the design and construction of buildings
- **Building Systems**   The evaluation, selection, and integration of mechanical, electrical, and specialty systems in building design and construction
- **Construction Documents & Services**   Application of project management and professional practice knowledge and skills, including the preparation of contract documents and contract administration

Problem Set 2.2   Architecture as a Career

For problems 2-9 through 2-12, refer to an architectural degree program at a university of your choice.

Problem 2-9	What are the general university admission requirements (for example, minimum grade point average, entrance exam scores)?
Problem 2-10	Does the architecture degree program have any additional entrance requirements?
Problem 2-11	How long is the program, and what options exist? Four- or five-year bachelor's, master's, doctorate?
Problem 2-12	Is this degree program NAAB accredited?
Problem 2-13	Why is choosing an accredited program important?
Problem 2-14	What internship program is available for intern architects? What organization sponsors this program?
Problem 2-15	What categories must be fulfilled as a requirement for the Intern Development Program?
Problem 2-16	How many total hours are required for the Intern Development Program?
Problem 2-17	What are the seven divisions of the Architect Registration Examination?
Problem 2-18	Interview an architect, and ask questions about their education and career paths to becoming a licensed architect.

BACKGROUND

The built environment encompasses a large variety of career options beyond civil engineering and architecture.

Careers Related to Civil Engineering and Architecture

Many careers exist that are closely related to the fields of civil engineering and architecture. For example, surveyors, interior designers, computer aided design technicians, landscape designers, building inspectors, code officials, and construction managers are all closely related to the civil engineering and architecture professions. The AIA provides an extensive list of alternate careers at www.ARCHcareers.org.

Problem Set 2.3   Careers Related to Civil Engineering and Architecture

For problems 2-19 through 2-21, choose a career closely related to civil engineering and architecture.

Problem 2-19   Research careers similar to civil engineering and architecture. Which do you find most interesting? Why?

Problem 2-20   What are the education and training requirements for the career in problem 2-19?

Problem 2-21   Referencing problem 2-19, interview someone currently working in this field. Was this career what you anticipated? What were you surprised to learn?

BACKGROUND

For a project to be successful, all associated parties need to work together to achieve the best possible end result.

Civil Engineers, Architects, and Related Agencies Work Together in the Design and Build of a Structure

Traditionally, the design process has initiated with the architect selecting the layout and exterior of the building. After this is complete, the engineer would design the structural requirements, and other designers would complete the site development and mechanical systems. More recently, an integrative design process has come to light. This process involves all parties working together from the very beginning to provide input toward the building design. This method of design is explored more extensively in Chapter 4.

Problem Set 2.4   Civil Engineers, Architects, and Related Agencies Work Together in the Design and Build of a Structure

Problem 2-22   How does integrative design differ from traditional design?
Problem 2-23   What is a design charrette?
Problem 2-24   What does the text define as "mechanicals"?
Problem 2-25   What are the four key elements of the integrated design process?

BACKGROUND

Other considerations in the design of the built environment may encompass the actual environment.

Agencies and Organizations Involved in the Design and Development of a Structure

The location of a project will play a major role in the design of a project. Protecting the environment and accounting for environmental effects on a structure are a large part of smart design. Some agencies to consider during design are listed below:

- Department of Environmental Conservation (DEC)
- Environmental Protection Agency (EPA)
- Army Corps of Engineers
- Local agencies

Problem Set 2.5   Agencies and Organizations Involved in the Design and Development of a Structure

Problem 2-26   Refer to the Environmental Protection Agency's website, www.epa.org for data on your region of the country. Are there any featured stories from this region?

Problem 2-27   Refer to the Environmental Protection Agency's website, www.epa.org. Enter your zip code, and report the current air quality.

Problem 2-28   What local agencies may play a role in the design of a project in your area?

CHAPTER 3
Research, Documentation, and Communication

Skills List

After completing the problems in this chapter, you should be able to:

- Identify the components that help achieve a successful building project

- Utilize resources to aide you in the research of a building project

- Understand the aspects of documentation as it pertains to a building project

- Realize the importance of communication between all the parties involved in a building project

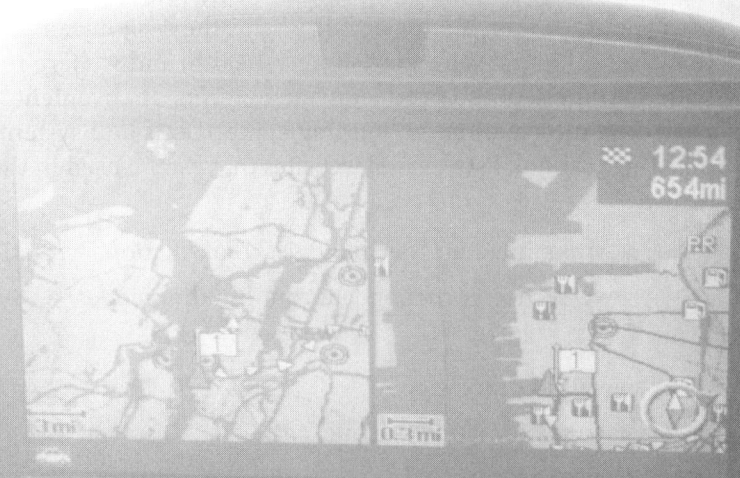

BACKGROUND

Research for a building project involves many resources and informs those involved in the project how, what, why, where, and when a project can be built.

Research is Vital to Land Development

For a building project to be successful, vast amounts of research is required to make sure that the final product will be correctly designed as well as financially sound. Much of this research is a requirement of the financial institution funding the project and of the local governing agencies where the project will be constructed. Without the proper research, the project could be a failure, and all parties involved in the project could be adversely affected financially. Research will indicate which building codes, zoning laws and regulations are in effect for the geographic location of a project. Building codes, building regulations, and zoning have been put into effect to make certain that the constructed building is both safe and well designed for the occupants.

Building Codes: Most cities have adopted a set of codes developed by the International Code Council (ICC), which originated in 1994 through the combination of the Southern Building Code Congress International (SBCCI), the International Conference of Building Officials (ICBO), and the Building Officials and Code Administrators International (BOCA). It was the goal of ICC to develop a single set of building codes for the United States.

Building Regulations: Cities often will amend the building code by adding additional requirements that pertain to specific conditions encountered in the local area. These vary from city to city but can include requirements for occupancy or for construction materials and methods. For example, in California the building code has been amended extensively due to the amount of seismic activity in the state. The amendments give additional design and construction requirements to enhance the safety of buildings.

Zoning: Cities decide on what uses for buildings can occur in parts or zones of the city. There is typically a map of the city with the allowed use and development of an area indicated called a "zoning map." This ensures that conflicting building uses will not occur next to each other. Zoning includes design limitations such as setbacks and the maximum building height for a building site. Refer to figure 3-1 for an example of a zoning map showing part of Stillwater, Oklahoma.

Communities use the building code, local zoning, and regulations in a manner so that the building project will enhance and add value to the geographic area of the project.

With the advent of the Internet, there is an ever-expanding list of resources available on websites that are commonly used in the research, design, and the documentation process for a building project. The use of these websites can give instant access to information for use by the design team that previously might have taken many hours to find. This availability helps the design team as well as the building code officials who must verify that the building design meets safety, health, and welfare requirements. Online resources also allow building material suppliers and product manufacturers a convenient way to communicate with the project design team, providing information on building components and systems that can be used in the construction of a building. The textbook lists several websites that can be used in the research process, along with several others listed below:

- www.architectural-technologist.co.uk/dictionary/dictionary.htm
- www.reedconstructiondata.com/building-codes/
- www.aia.org/
- www.buildipedia.com/
- www.constructionweblinks.com/
- www.arcat.com/
- www.products.construction.com/

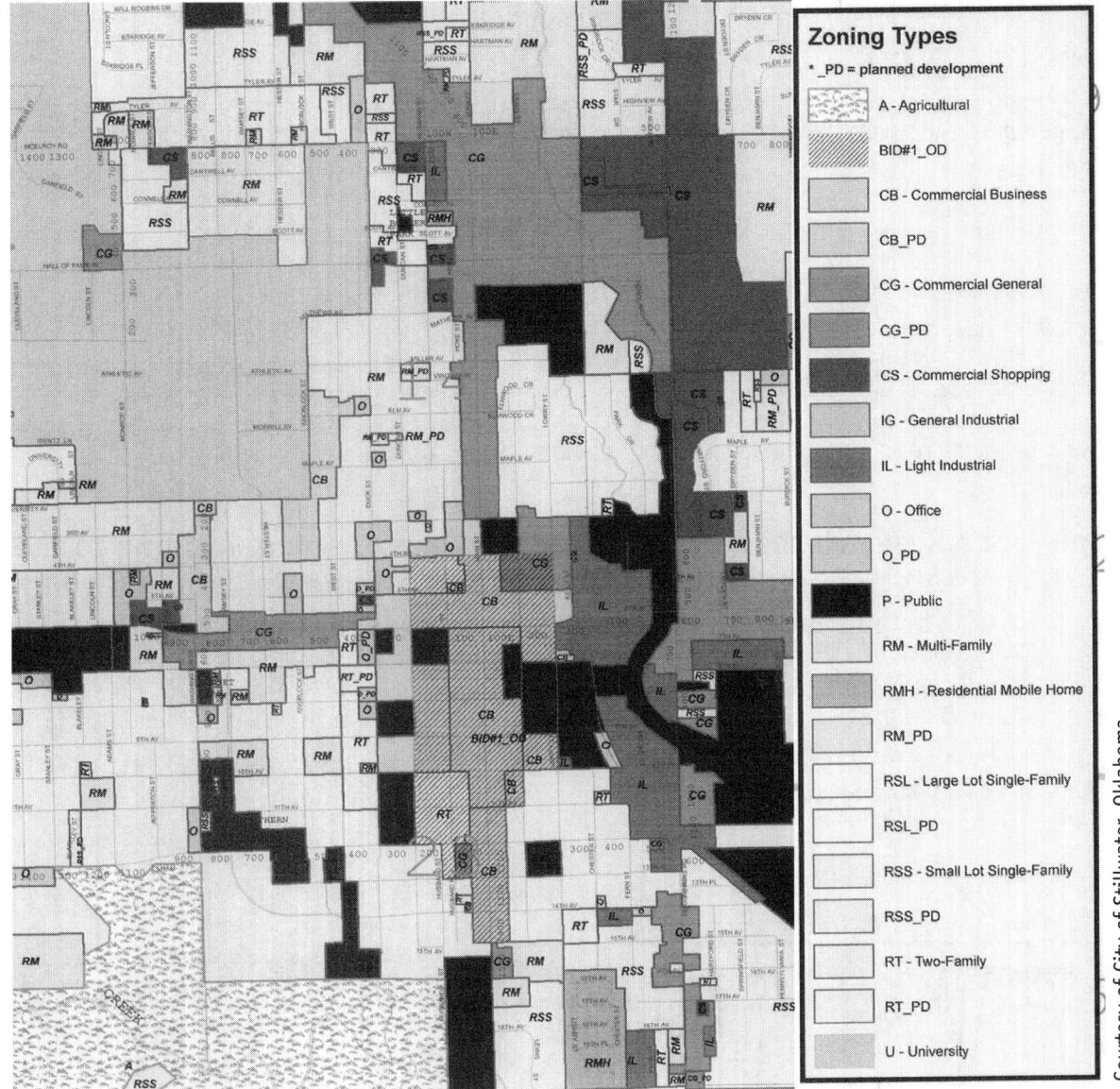

FIGURE 3-1 *Zoning map.*

Example 3.1 Research each of the websites listed above, and indicate what information is available at each site.

www.architectural-technologist.co.uk/dictionary/dictionary.htm

This website includes a list of definitions for commonly occurring terms from the architecture, engineering, and construction fields.

www.reedconstructiondata.com/building-codes/

This website allows one to research the building code in effect for each state and for certain cities within each state.

www.aia.org/

This website promotes architecture in America and is a valuable resource for both the general public and members of The American Institute of Architects.

www.buildipedia.com/

Research is Vital to Land Development

This website provides resources for materials and systems used in design and construction.

www.constructionweblinks.com/

This website lists a series of links to sites that pertain to building codes, regulations, materials, and systems.

www.arcat.com/

This website provides building materials, manufacturers' specifications, CAD Details, and BIM elements for use in the design process.

www.products.construction.com/

This website gives resources, links, and specifications for materials and systems.

Example 3.2 **Research the website listed in the textbook pertaining to the ADA Accessibility Guidelines to determine the required widths for parking spaces and for access aisles.**

ADA Chapter 5, section 502 gives the requirements for parking spaces that require accessibility. Refer to figures 3-2 and 3-3 for images of these requirements. From the ADA document, the required width for a car is 96" and for a van is 132". The width for the access aisle that is required to be adjacent to the parking space is 60".

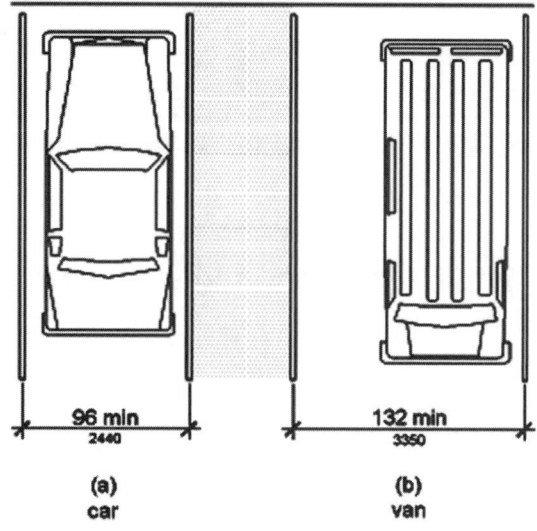

FIGURE 3-2 *ADA parking.*
Source: ADA and ABA Accessibility Guidelines for Buildings and Facilities, Chapter 5: General Site and Building Elements.

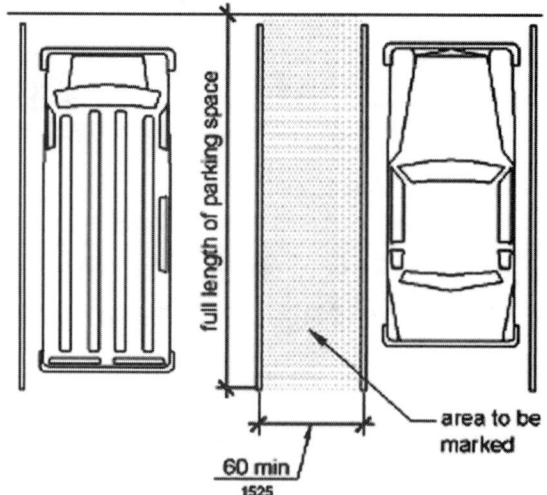

FIGURE 3-3 *ADA parking.*
Source: ADA and ABA Accessibility Guidelines for Buildings and Facilities, Chapter 5: General Site and Building Elements.

Problem Set 3.1 Research is Vital to Land Development

Problem 3-1   From the textbook, list and discuss the three components of a successful building project.

Problem 3-2   From the textbook, list the topics for a particular building site that should be explored before beginning a project.

Problem 3-3   What three organizations worked together to form the International Code Council (ICC)?

Problem 3-4   Using online resources, determine the governing building code, fire/life safety code, and electrical code currently in effect for design and construction in Dallas, Texas.

Problem 3-5   Using online resources, determine the governing building code, mechanical code, and energy code currently in effect for design and construction in Long Beach, California.

Problem 3-6   Using online resources, determine the governing building code, accessibility code, and elevator code currently in effect for design and construction in Syracuse, New York.

Problem 3-7   Using online resources, determine the definition of "concrete."

Problem 3-8   Using online resources, determine the definition of "louver."

BACKGROUND

Documentation includes a record of all information pertaining to a project, from initial concepts to final design and to construction administration. Documentation is used to provide a comprehensive record of the choices available and the decisions made during the design and construction process.

Documentation for Land Development

Documentation for a building project is extensive and includes, but is not limited to, the following topics:

- Architectural Brief
- Project Program
- Maps, Surveys, and Legal Descriptions of the property
- Construction Drawings
- Project Specifications
- Sketchbooks and Journals
- Legal Documents
- Feasibility Studies
- Research Studies (building code, building regulations, zoning)
- Permit Applications
- BIM — Building Information Model (BIM)

Each of these types of documents is often the result of many hours of work by the project team of architects and engineers over a period of months or even years.

Example 3.3   List and describe the four categories included in Project Documentation as defined in the textbook. Project Documentation gives a complete history of the building process through all phases of design and construction and includes:

Drawings: From concepts sketches to final construction documents, the drawings for a project document the evolution of the design and development of the building.

Studies: The viability of a project must be justified before it can receive financial backing from investors. Studies are often used to document the need for a project within a community, and can be based on location, population, or demographics to name just a few categories.

Permits: Legal documents required by the governing jurisdiction where the project is to be built. Obtaining these permits often results from a formal review of the project design.

Legal Proceedings: The legal proceedings are required to make sure that all parties involved in the project are legally bound to perform the job for which they have been hired. This may include the mortgage documents from the financial institute, the contracts with the design architects, engineers, and construction companies, to the certificates of occupancy from the building code officials, to name a few.

Architectural Brief and Program for Non-Residential Development

Integrated Design is often employed for a development project in which the project architect gathers a team of consultants to determine requirements for the project. This process involves research and helps identify requirements for the development to arrive at goals and concepts for a project. The results of this work forms the Architectural Brief for the project, which is a description of a solution to a problem or need that can be solved with the development in question, and typically includes the following categories:

- Summary: An overall statement for the goals of the project
- Historical Background: How the potential project came into existence
- Analysis & results of study: A list of the constraints and requirements for a project, including building codes, building regulations, and zoning codes, plus site analysis which gives vital information for the project, such as traffic patterns, surrounding context and utility locations.
- Scope of Work: A clear, concise explanation of what will be included in the project.
- Utility Requirements: A list of required utility services for the project, including any special systems or components unique to the design.
- Feasibility Studies: Results of a study showing that the development will be financially successful for investors.
- Project Schedule: A time line for the project from start to finish, with intermediate dates and deadlines for phases of the project.
- Cost Estimate: A detailed and comprehensive estimate for the cost of the project to be used in determining financial requirements for the project.

In many cases, an Architectural Brief is the first step in achieving financial interest in a development project. Without documentation showing the viability of the project, investors will not be interested in providing funding for the project. Once interest arises, the team can further develop the initial concepts for the project and produce a Program for the development project. The work required for the Program often takes many hours to complete. The project program can take on many forms but generally conforms to the following breakdown of sections:

- Title Sheet
- Signature Page
- Executive Summary
- Purpose and Scope
- Site Analysis
- Building Requirements
- Space Utilization
- Building Systems Criteria
- Cost Analysis

Once interest in a project occurs, it is truly the Program that sells and is ultimately responsible for the success of a land development project.

Example 3.4 Discuss the information included in each of the sections of a project Program.

Title Sheet: A title describing the name and location for the project.

Signature Page: Information indicating the design team that is responsible for producing the document.

Executive Summary: A general statement consisting of a paragraph or two that gives an overview of the project location, size, and use.

Purpose and Scope: A description of the proposed project indicating the extent of the development and why it is an advantageous opportunity.

Site Analysis: Information pertaining to the location of the project, including traffic (pedestrian and vehicular) patterns, climate (temperatures, rain and snow fall, sun angles), surrounding context, and building setbacks and utility easements.

Building Requirements: A list of requirements for which the building design must satisfy. These can include building regulations set forth by the governing bodies where a project is to be constructed as well as specific design aspects of the project which must be included in order to meet the owner's requirements for the project.

Space Utilization: A detailed description of the spaces in the project along with their use and a breakdown of the floor areas for each.

Building Systems Criteria: A list of the specific systems and equipment that is to be included in the project, including any that is unique to the project.

Cost Analysis: An analysis of the cost of the project, including land purchases, site work, any environmental cleanup or demolition of existing structures that may be required, and the cost of constructing the new project.

Development of a Single Family Residence

A family residence is often less complex in nature than a commercial development, and the documentation necessary is also less complex. There are certain types of systems that are unique to residential construction. For example, wastewater and storm drainage must be removed from the property on which the residence will set. If connections to sewer and storm drain systems are not available, a septic system will be necessary for the building site. It is required that a percolation test be performed by a civil engineer to determine whether the soil present is sufficient to allow the septic system to function properly. Residential construction has its own building code published by the ICC, titled the International Residential Code (IRC). The IRC is set up with guidelines to be used in the construction process, hence there is not always a need for architects and engineers for this type of building. There are many sources available for the design of a residence, including:

- Local builders that can provide plans for residences that they will build for you
- Specific magazines that provide a catalog of house plans that can be purchased. These sets of plans include most of the information needed to construct a residence
- Online websites that offer a catalog of house plans
- An architect who can be hired to design a custom home

It is debatable whether any of these options is more economically feasible than the others. The selection of which option to use is often a function of the type of home desired, the anticipated budget for the home, the timeframe anticipated from start to final construction, and the experience of the builder who will be constructing the residence.

Keeping a Journal

To keep a record of the many events that happen during a building project, architects and engineers often keep journals. This allows the team member to document important information pertaining to the design and development of a particular project. A journal is not a legal document, but it does give the designer a reference of their thoughts and processes used on a project. An example of a journal entry is shown in figure 3-4.

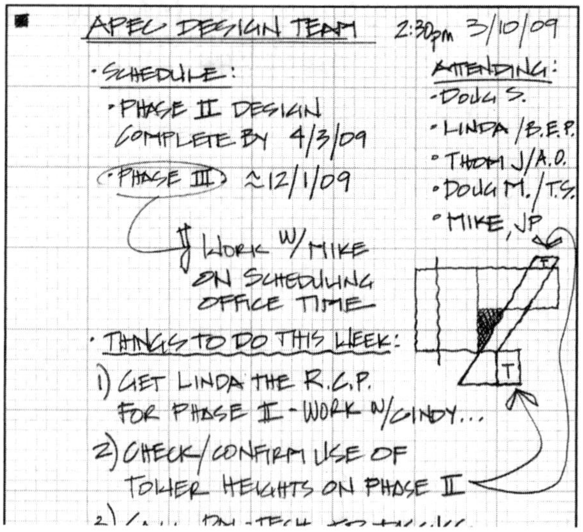

Information kept in a journal should pertain to the project and can include:

- Research findings
- Meeting notes and phone conversations that include the date, time, location, person contacted, and description of discussion. This information should be included for each journal entry.
- Addresses, phone numbers, and email addresses
- Sketches of potential ideas to be incorporated into the project

FIGURE 3-4 Journal entry.

Example 3.5 Give an example of an architect's journal entry for a meeting involving the owner, architect, and contractor in which pricing for a certain finish material was discussed.

May 4, 2010 / 2:00 pm / Oakfield residence jobsite meeting

Present at meeting: Owner Doug Oakfield, contractor Steve Scott and architect Sarah McKenzie

Discussion: Meeting held at site to discuss roofing choices for residence, with samples of potential choices brought to the site. Asphalt shingles are least expensive and readily available, but Doug wants something grander. Slate tiles would be the owner's choice, though Steve pointed out that the structure is not currently designed for the extra weight of this type of roofing. To choose this option, construction will be delayed while the structure now in place is upgraded to support the added weights. In addition, the cost of the slate tiles is double what has been budgeted. Doug made the decision not to go with this option. The third option is a metal panel roofing system. The weight of the system is approximately equal to that of asphalt shingles, and is readily available, however, the cost is 25% more than what is budgeted for the cost of the roof. Doug has agreed that the choice of a metal panel roof system is a good one, and he wants to use it on the project. A meeting will be set up next Monday to go over texture and color options in our office at 10:00 am. Steve is to have a cost estimate for the system to me by Friday of this week.

A journal is a very important aspect of project documentation and can become crucial when an earlier decision in the project needs to be discussed or verified. It often can save time and money when it can be used to validate decisions made during the design and construction of a building project.

Architectural Sketchbooks

A sketchbook is similar to a journal in that it is not a legal portion of the documentation, but it does document the design process and notes important thoughts, observations, and ideas that can be easily referenced throughout the project. It can document elements to be considered for use on a project, such as those shown in figures 3-11A and 3-11B in the textbook or building section studies as shown in figure 3-5.

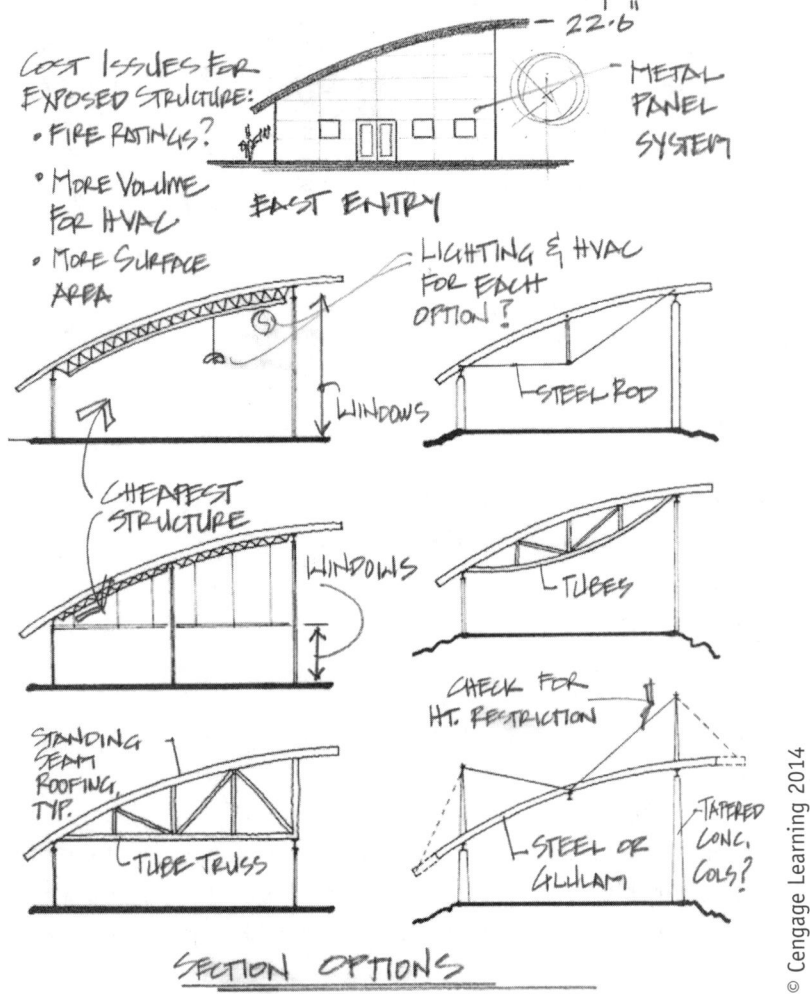

FIGURE 3-5 *Sketchbook entry.*

Bubble Diagrams

From the project program, a list of the required rooms and their floor areas are obtained. Using the list of rooms and the floor area of each, a bubble diagram can be drawn. The bubble diagram is a graphic representation that shows how each of the rooms in a project should relate to each other. This is often the beginning of the planning on a project and is a method that allows the designer to quickly look at several options pertaining to architectural planning.

Example 3.6   Sketch a bubble diagram for a residence with the following rooms and floor areas showing access between spaces:

- Entry hall   80 sq. ft.
- Living room   250 sq. ft.
- Dining room   200 sq. ft.
- Family room   500 sq. ft.
- Kitchen   400 sq. ft.
- Pantry   100 sq. ft.
- Bedrooms   3 × 200 sq. ft. each
- Master bedroom   350 sq. ft.
- Master closet   200 sq. ft.

- Bathrooms   3 × 100 sq. ft. each
- Master bath   200 sq. ft.
- Office   175 sq. ft.
- Laundry room   100 sq. ft.
- Garage   750 sq. ft.

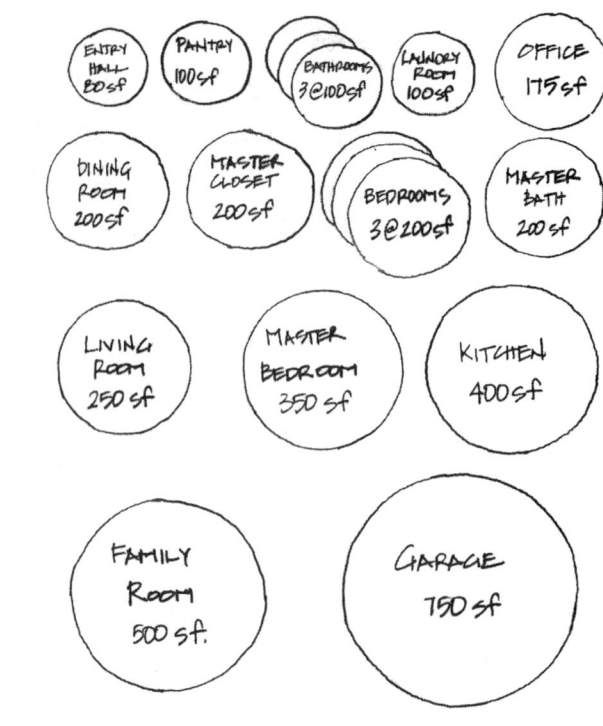

The first step is to draw each of the rooms to a set scale based on the floor areas. The rooms can be drawn as geometric shapes representing the rooms, typically as circles or squares, as shown in figure 3-6. During the process of constructing the bubble diagram, the room shapes can be cut out for easy manipulation and study of the location of the spaces. This allows for quick manipulation of the relationships between the spaces so that multiple solutions may be considered.

FIGURE 3-6 *Rooms and floor areas.*

The next step is to arrange the rooms in a manner that shows their relationship or access to one another. Some rooms in the building need to have direct access to each others, while others need separation. The traffic pattern of any residence is critical and should be thoroughly planned out prior to the final placement of the rooms. Lines are drawn connecting rooms that will have access to each other in the final design, as shown in figure 3-7.

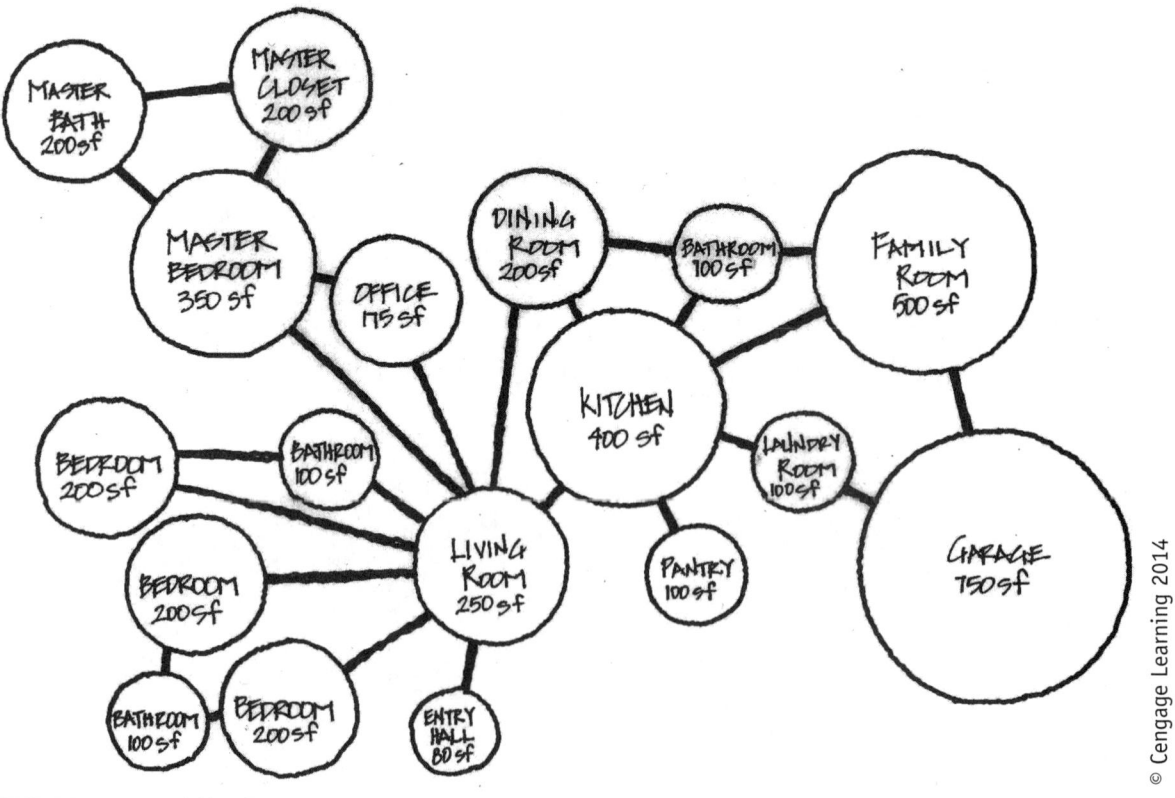

FIGURE 3-7 *Bubble diagram.*

40   Chapter 3: Research, Documentation, and Communication

Difference between Working and Presentation Drawings

Architectural drawings can be categorized as either Presentation Drawings or Working Drawings:

Presentation Drawings: These are drawings that help explain the general concepts of the project. They give realistic images of what the final project will look like and are often used in promoting the project to the general public. These drawings include two-dimensional plans, elevations and sections, and three-dimensional perspectives showing interior and exterior spaces of the project. As shown in figures 3-8 and 3-9, these drawings include people and furniture to give a sense of how a visitor or occupant would interact within the spaces of the project.

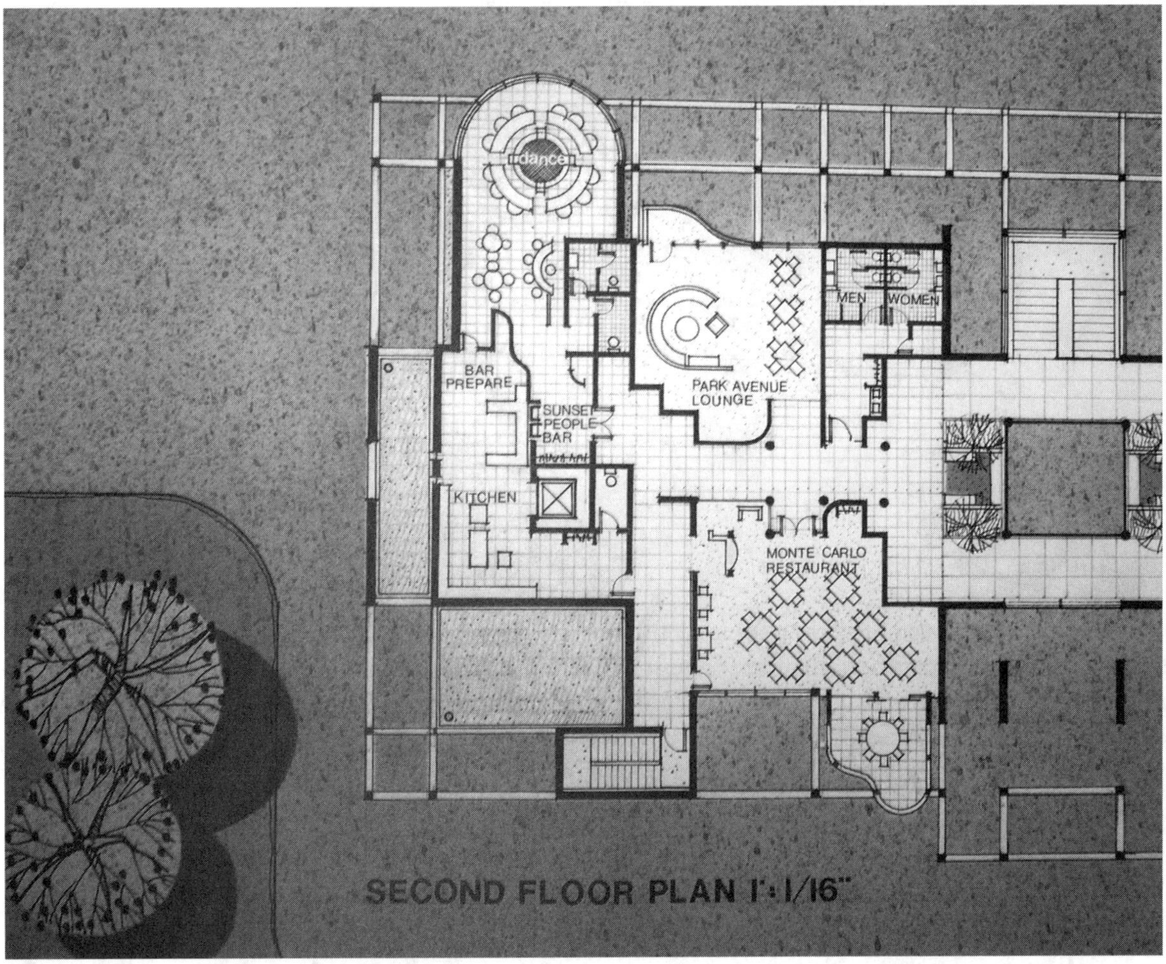

Reproduced with permission from *OSU School of Architecture.*

FIGURE 3-8 *Presentation plan drawing.*

Reproduced with permission from *OSU School of Architecture.*

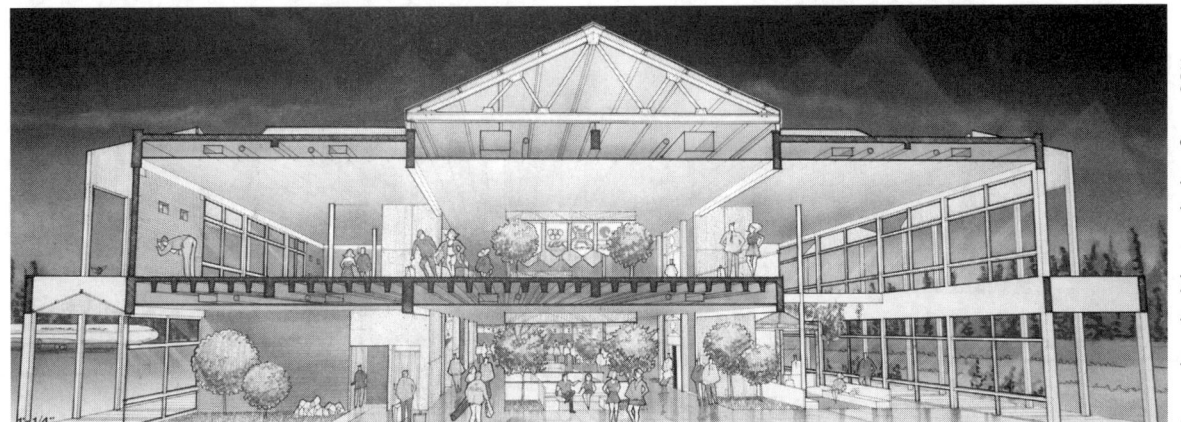

FIGURE 3-9 *Presentation perspective drawing.*

Working Drawings: These are the drawings that explain specific requirements for the project. Working drawings are technical in nature and are used to communicate exactly how the project is to be built. Each of these is drawn to scale, and typically is drawn in two-dimensions, though for more complex projects, such as those by architect Santiago Calatrava, some sections and details are also drawn in three-dimensions. Working drawings include plans such as the one shown in figure 3-10.

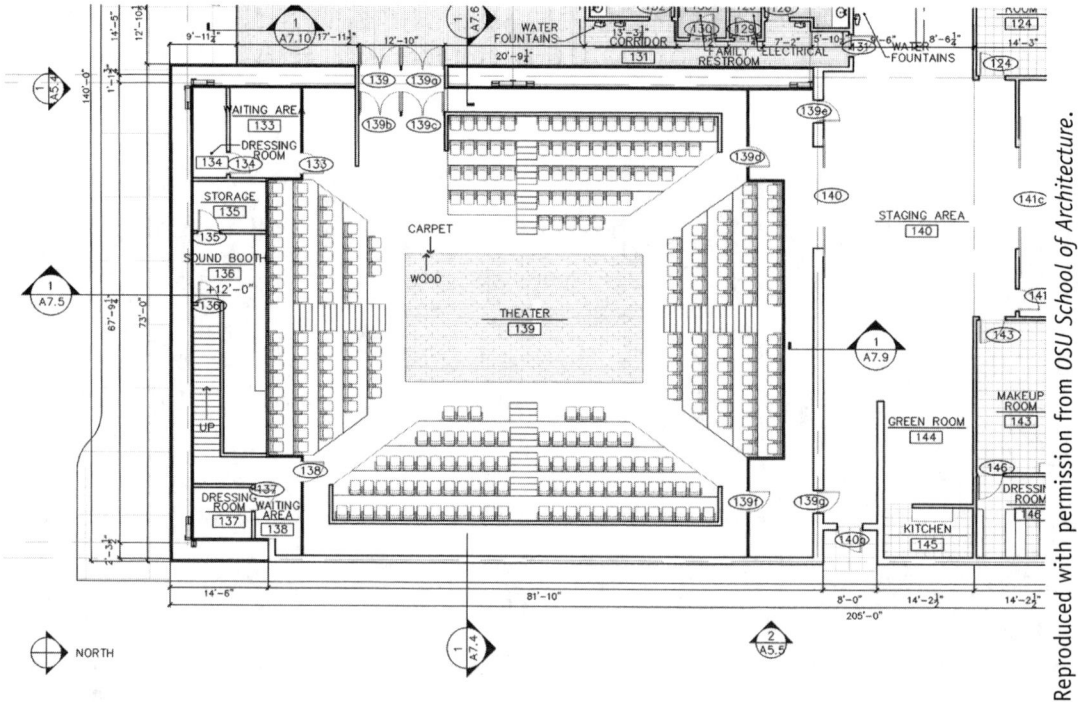

FIGURE 3-10 *Working drawing plan.*

Note that this plan does not show people or furniture, which is typically a rule for working drawings. One exception to this rule is that when furniture is built as part of the project, it is included in the drawings. Working drawings do include dimensions for use by the builder and include permanent fixtures in the building. In addition, working drawings include sections and details, as shown in figure 3-11. Sections and details look at specific parts of the project and zoom in to get a closer look at how it is to be constructed.

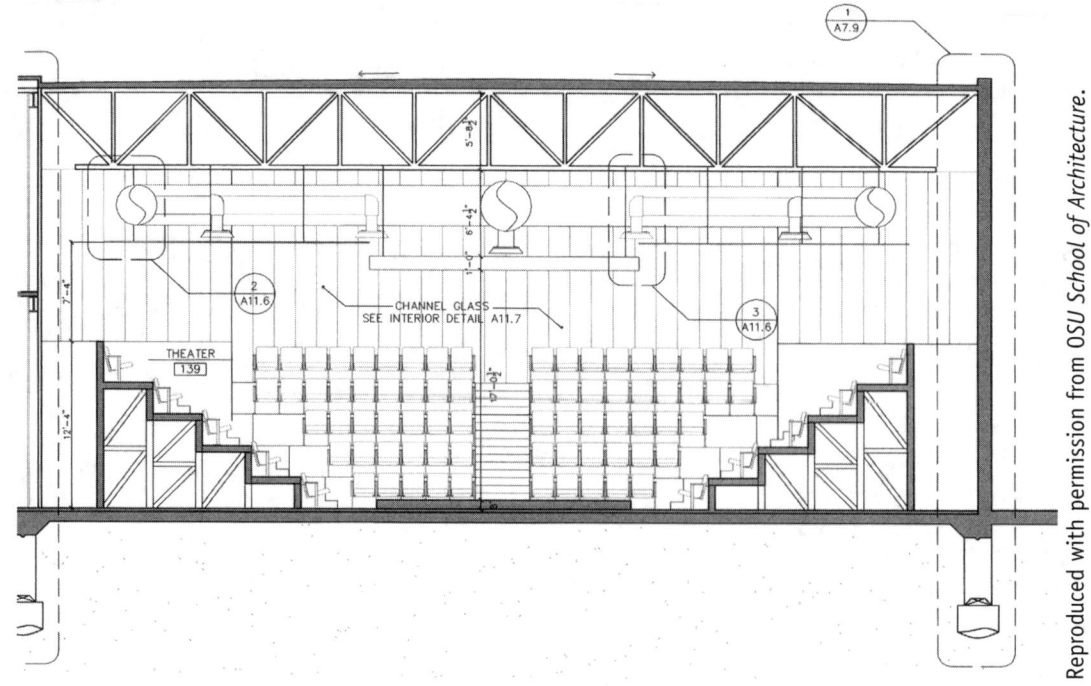

FIGURE 3-11 *Working drawing section detail.*

42   Chapter 3: Research, Documentation, and Communication

Drawing Scale and Sheet Size

Though the format for Presentation Drawings can vary, they often conform to the American National Standards Institute (ANSI) standard sheet size. The size of these sheets varies from 8 1/2" × 11" up to 34" × 44", as shown in the textbook figure 3-15. Working drawings typically conform to the ANSI standard sheet sizes and commonly consist of C size (18" × 24") or larger, depending on the size of the project and the scale used for the drawings. The scale of the drawing is an accurate representation of the actual proportions of the building reduced to a size that can easily be used for the design and construction of a building project. Scales vary based on the size of the project, the type of drawings being produced, and whether the drawings are architectural, engineering, or building component/system drawings. Drawings can be drawn at any scale, however, common scales for architectural working drawings include:

- 1/16" = 1'-0" - Used for site plans
- 1/8" = 1'-0" - Used for plot, architectural, structural, mechanical, electrical, and fire protection plans
- 1/4" = 1'-0" - Used for regular floor plans, building elevations, and building sections
- 3/4" = 1'-0" - Used for enlarged sections and details

In determining the size of the sheets to use for working drawings, the size of the plan must be taken into consideration. If possible, it is advantageous to scale the plans so it will fit on one drawing sheet. In some cases, this is not possible, and the plan has to be split between two or more sheets. When checking to see whether a plan will fit on a sheet, the working drawings sheet layout must be considered. Working drawings also require additional information (besides just the plan) such as:

- Sheet title block
- Sheet border
- Drawing title and scale
- Dimensions

Each of these requirements for a working drawing sheet reduces the available space on which to place the plan. Refer to figure 3-12 for an example of a working drawing sheet.

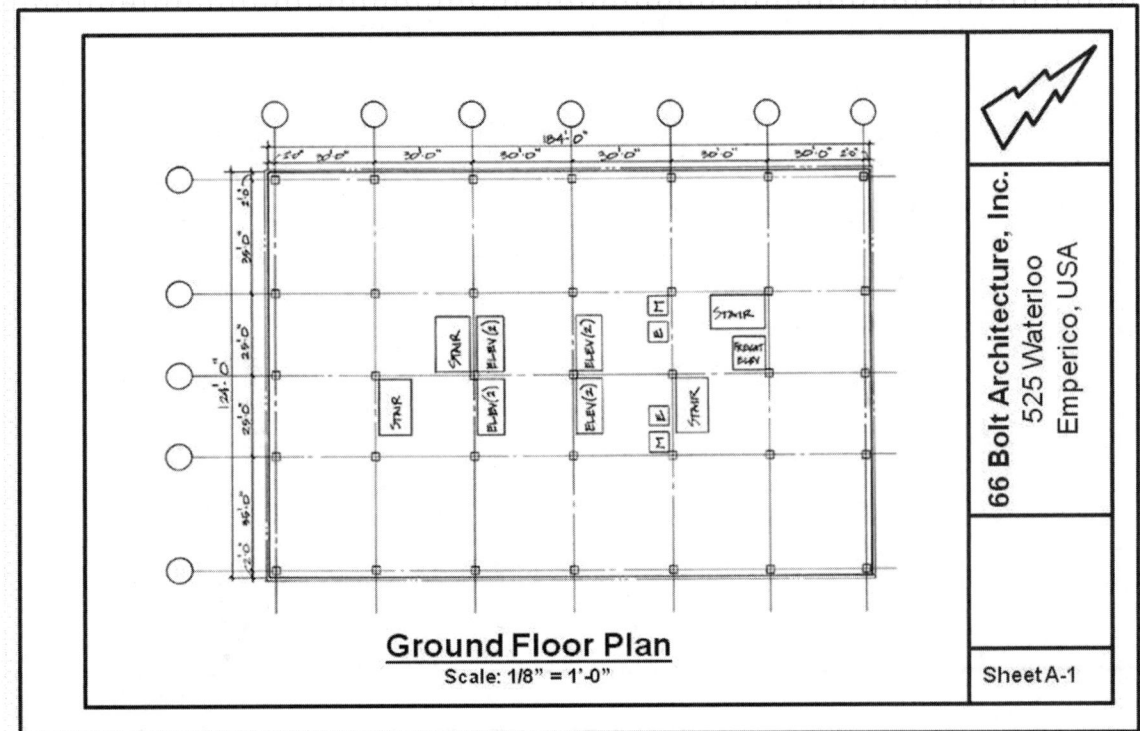

FIGURE 3-12 *Sheet layout with borders for a working drawing.*

Example 3.7 Determine the ANSI standard sheet size required for a plan with the following conditions:

- Overall actual project dimensions in plan are 120'-0" × 80'-0"
- Scale used for plan drawings: 1/4" = 1'-0"

Typically, a plan is oriented on a drawing sheet such that the longer dimension is placed horizontally. The dimensions for the scaled down plan will be:

Required drawing width for plan: 120 ft (1/4 in/ft) = 30"

Required drawing height for plan: 80 ft (1/4 in/ft) = 20"

From text figure 3-15, sheet size D or E would work based on the plan dimensions only. However, in addition to the plan size, the drawing title and scale, plan dimensions, and sheet border must be included when sizing the plan sheet. To make sure we have enough room for all of the information to be included, select sheet size E: 34" × 44"

Example 3.8 Determine the actual dimensions and area from a scaled drawing with the following conditions:

- A measured height of 1-1/2" on a drawing with a scale of 1/8" = 1'-0":
  Height = 1.5 in(1 ft/(1/8 in)) = 12'-0"
- A measured length of 12" on a drawing with a scale of 3/16" = 1'-0":
  Length = 12 in(1ft/(3/16 in)) = 64'-0"
- The area of a room that measures 8" × 6" on a drawing with a scale of 1/4" = 1'-0":
  Area = 8 in(1ft/(1/4 in)) * 6 in(1ft/(1/4 in)) = 32 ft * 24 ft = 768 ft^2

Example 3.9 Show a sketch of a working drawing sheet layout including the border and title block with information about the project and the architecture firm included on the drawing.

The border/title block for a drawing sheet varies among architecture firms and sheet sizes. Although a generic title block may be used for the drawing sheets, many architecture firms have custom title blocks designed for their projects. For the larger size, such as size D or E, it is common to have the working drawing sheet layout as shown in figure 3-13.

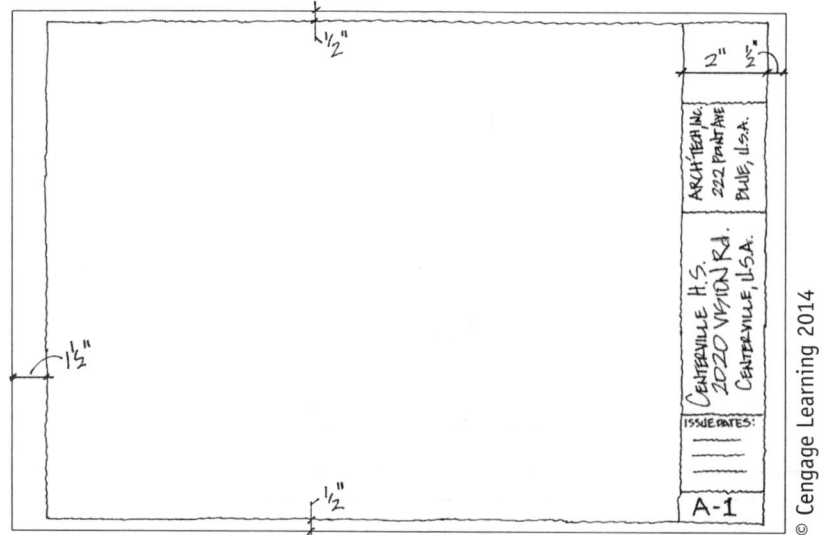

FIGURE 3-13 *Working drawing sheet layout.*

A set of working drawings is typically bound at the left edge of the sheets, along the shorter dimension of the sheets. This can be accomplished with staples, or with a small bolting system for sets with a large number of sheets. It is important that room along the left hand border be provided to allow the bounding of the set to occur. As an example, the border edge might be placed 1 1/2" from the edge of the sheet to allow room for the binding. Without this space, a portion of the drawing could be within the binding and not readable. For the remaining three sides, 1/2" could be used for the space between the edge of the sheet and the title block line. The title block itself could be sized 2" in width and extends the full height of the sheet between the borders. Note that these dimensions can vary based on standards set by the design team, or by requirements set forth by the municipality in which the building is to be constructed. Information on the project would be given at the right end of the sheet and includes:

- Project title and location
- Architecture firm name and location
- Issue date of drawings
- Sheet identification
- Space for the architect to seal and sign the drawings

Plans

Plans are required for all building projects and are used to provide a set of instructions of how the project is to be constructed. The number of drawing sheets required for a project depends on the size of the project. Often, only a few are required for residential projects, while hundreds may be required for commercial projects. Because of the number of drawings required for projects and the different team members that create the drawings, an identification system is necessary to organize the working drawings. This identification system uses a combination of letters and numbers. Although there are no set standards, typical references for working drawings may include:

- A: Architectural
- C: Civil, Site and/or Environmental
- S: Structural
- M: Mechanical
- E: Electrical
- P: Plumbing
- FP: Fire Protection
- T: Telecommunications
- L: Landscaping

Sheet identification includes a letter first, followed by a sheet number. The letter identifies the division of the team that is responsible for the drawings, and the number indicates which sheet it is within the division, as shown in figure 3-14.

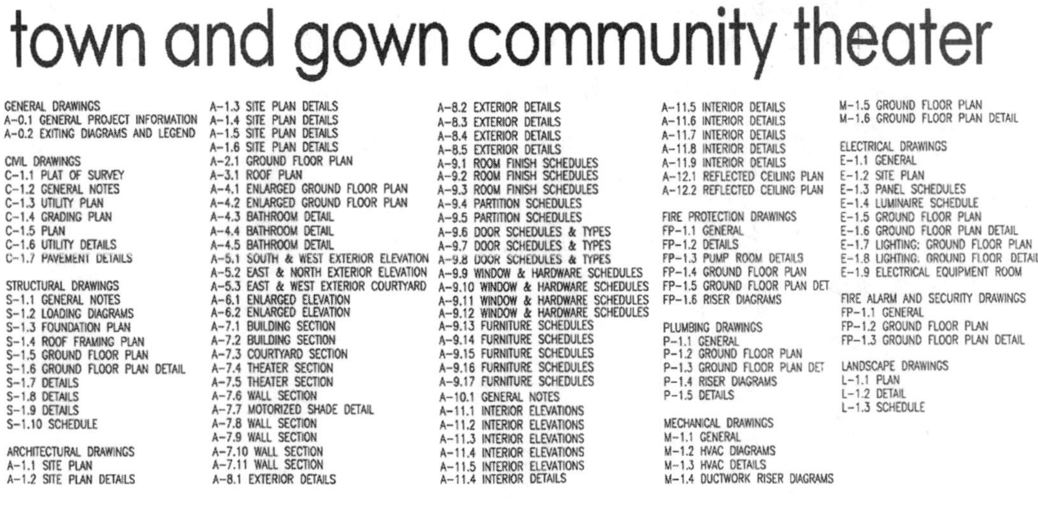
town and gown community theater

GENERAL DRAWINGS
A-0.1 GENERAL PROJECT INFORMATION
A-0.2 EXITING DIAGRAMS AND LEGEND

CIVIL DRAWINGS
C-1.1 PLAT OF SURVEY
C-1.2 GENERAL NOTES
C-1.3 UTILITY PLAN
C-1.4 GRADING PLAN
C-1.5 PLAN
C-1.6 UTILITY DETAILS
C-1.7 PAVEMENT DETAILS

STRUCTURAL DRAWINGS
S-1.1 GENERAL NOTES
S-1.2 LOADING DIAGRAMS
S-1.3 FOUNDATION PLAN
S-1.4 ROOF FRAMING PLAN
S-1.5 GROUND FLOOR PLAN
S-1.6 GROUND FLOOR PLAN DETAIL
S-1.7 DETAILS
S-1.8 DETAILS
S-1.9 DETAILS
S-1.10 SCHEDULE

ARCHITECTURAL DRAWINGS
A-1.1 SITE PLAN
A-1.2 SITE PLAN DETAILS
A-1.3 SITE PLAN DETAILS
A-1.4 SITE PLAN DETAILS
A-1.5 SITE PLAN DETAILS
A-1.6 SITE PLAN DETAILS
A-2.1 GROUND FLOOR PLAN
A-3.1 ROOF PLAN
A-4.1 ENLARGED GROUND FLOOR PLAN
A-4.2 ENLARGED GROUND FLOOR PLAN
A-4.3 BATHROOM DETAIL
A-4.4 BATHROOM DETAIL
A-4.5 BATHROOM DETAIL
A-5.1 SOUTH & WEST EXTERIOR ELEVATION
A-5.2 EAST & NORTH EXTERIOR ELEVATION
A-5.3 EAST & WEST EXTERIOR COURTYARD
A-6.1 ENLARGED ELEVATION
A-6.2 ENLARGED ELEVATION
A-7.1 BUILDING SECTION
A-7.2 BUILDING SECTION
A-7.3 COURTYARD SECTION
A-7.4 THEATER SECTION
A-7.5 THEATER SECTION
A-7.6 WALL SECTION
A-7.7 MOTORIZED SHADE DETAIL
A-7.8 WALL SECTION
A-7.9 WALL SECTION
A-7.10 WALL SECTION
A-7.11 WALL SECTION
A-8.1 EXTERIOR DETAILS
A-8.2 EXTERIOR DETAILS
A-8.3 EXTERIOR DETAILS
A-8.4 EXTERIOR DETAILS
A-8.5 EXTERIOR DETAILS
A-9.1 ROOM FINISH SCHEDULES
A-9.2 ROOM FINISH SCHEDULES
A-9.3 ROOM FINISH SCHEDULES
A-9.4 PARTITION SCHEDULES
A-9.5 PARTITION SCHEDULES
A-9.6 DOOR SCHEDULES & TYPES
A-9.7 DOOR SCHEDULES & TYPES
A-9.8 DOOR SCHEDULES & TYPES
A-9.9 WINDOW & HARDWARE SCHEDULES
A-9.10 WINDOW & HARDWARE SCHEDULES
A-9.11 WINDOW & HARDWARE SCHEDULES
A-9.12 WINDOW & HARDWARE SCHEDULES
A-9.13 FURNITURE SCHEDULES
A-9.14 FURNITURE SCHEDULES
A-9.15 FURNITURE SCHEDULES
A-9.16 FURNITURE SCHEDULES
A-9.17 FURNITURE SCHEDULES
A-10.1 GENERAL NOTES
A-11.1 INTERIOR ELEVATIONS
A-11.2 INTERIOR ELEVATIONS
A-11.3 INTERIOR ELEVATIONS
A-11.4 INTERIOR ELEVATIONS
A-11.5 INTERIOR ELEVATIONS
A-11.4 INTERIOR DETAILS
A-11.5 INTERIOR DETAILS
A-11.6 INTERIOR DETAILS
A-11.7 INTERIOR DETAILS
A-11.8 INTERIOR DETAILS
A-11.9 INTERIOR DETAILS
A-12.1 REFLECTED CEILING PLAN
A-12.2 REFLECTED CEILING PLAN

FIRE PROTECTION DRAWINGS
FP-1.1 GENERAL
FP-1.2 DETAILS
FP-1.3 PUMP ROOM DETAILS
FP-1.4 GROUND FLOOR PLAN
FP-1.5 GROUND FLOOR PLAN DET
FP-1.6 RISER DIAGRAMS

PLUMBING DRAWINGS
P-1.1 GENERAL
P-1.2 GROUND FLOOR PLAN
P-1.3 GROUND FLOOR PLAN DET
P-1.4 RISER DIAGRAMS
P-1.5 DETAILS

MECHANICAL DRAWINGS
M-1.1 GENERAL
M-1.2 HVAC DIAGRAMS
M-1.3 HVAC DETAILS
M-1.4 DUCTWORK RISER DIAGRAMS
M-1.5 GROUND FLOOR PLAN
M-1.6 GROUND FLOOR PLAN DETAIL

ELECTRICAL DRAWINGS
E-1.1 GENERAL
E-1.2 SITE PLAN
E-1.3 PANEL SCHEDULES
E-1.4 LUMINAIRE SCHEDULE
E-1.5 GROUND FLOOR PLAN
E-1.6 GROUND FLOOR PLAN DETAIL
E-1.7 LIGHTING: GROUND FLOOR PLAN
E-1.8 LIGHTING: GROUND FLOOR DETAIL
E-1.9 ELECTRICAL EQUIPMENT ROOM

FIRE ALARM AND SECURITY DRAWINGS
FP-1.1 GENERAL
FP-1.2 GROUND FLOOR PLAN
FP-1.3 GROUND FLOOR PLAN DETAIL

LANDSCAPE DRAWINGS
L-1.1 PLAN
L-1.2 DETAIL
L-1.3 SCHEDULE

FIGURE 3-14 *Project sheet numbering system.*

Example 3.10 Using figure 3-14, identify the sheet(s) within set of working drawings where the following can be found:

- Pump room details for the fire protection system: Sheet FP-1.3
- Ground floor lighting plan: Sheets E-1.7 & E-1.8
- HVAC diagrams for the mechanical system: Sheet M-1.2
- Architectural building sections: Sheets A-7.1 & A-7.2
- Structural details: Sheets S-1.7, S-1.8, & S-1.9
- Grading plan: Sheet C-1.4

Problem Set 3.2 Documentation for Land Development

Problem 3-9   List all sources from which someone can obtain drawings for a residential project, and discuss the conditions under which each of these sources might be utilized.

Problem 3-10 Develop a journal entry for today's date and current time of a meeting among the architect, electrician, and contractor for a commercial project. The meeting topic is the selection of motion-activated light switches for the project. Include discussions on rooms that require these switches, as well as cost and delivery schedule.

Problem 3-11 Provide a sketchbook entry for a unique detail on the entry façade of your school building. Make notations on the sketch indicating scale, materials, and details that are important to that particular part of the building.

Problem 3-12 Determine the actual dimensions or area for the following scaled conditions:
- A measured height of 3 1/4" on a drawing with a scale of 3/16" = 1'-0"
- A measured length of 28 1/2" on a drawing with a scale of 1/4" = 1'-0"
- The area of a room that measures 10 1/2" × 7 1/4" on a drawing with a scale of 1/8" = 1'-0"

Problem 3-13 Determine the actual dimensions for the following scaled conditions:
- A measured height of 9" on a drawing with a scale of 3/8" = 1'-0"
- A measured length of 8 3/4" on a drawing with a scale of 1/16" = 1'-0"
- The dimensions of a room that measures 6 3/8" × 5 3/16" on a drawing with a scale of 1/8" = 1'-0"

Problem 3-14 Draw a bubble diagram for a community center project with the following rooms and floor areas showing required access between the rooms:
- Offices: 1 × 250 sq. ft. and 3 × 150 sq. ft.
- Restrooms: 2 × 300 sq. ft. and 1 × 100 sq. ft.
- Locker rooms: 2 × 1,500 sq. ft.
- Meeting halls: 2 × 1,800 sq. ft.
- Meeting rooms: 3 × 500 sq. ft.
- Kitchen: 600 sq. ft.
- Pantry: 150 sq. ft.
- Storage room: 2,000 sq. ft.
- Basketball court: 6,800 sq. ft.

- Entry vestibule: 150 sq. ft.
- Storage room: 1,200 sq. ft.
- HVAC room: 1,000 sq. ft.

Problem 3-15 Use the textbook to list those who use working drawings during the construction process for a development project.

Problem 3-16 Using figure 3-14, identify the sheet(s) within the set of working drawings where the following can be found:
- Electrical panel schedules
- Bathroom details
- Roof framing plan
- Ductwork riser diagrams
- Site plan
- Grading plan

Problem 3-17 Using figure 3-14, identify the sheet(s) within the set of working drawings where the following can be found:
- Reflected ceiling plans
- Landscaping details
- Riser diagrams for the plumbing system
- Pavement details
- Room finish schedules
- Interior elevations of the building

Problem 3-18 Referring to figure 3-14, discuss why the Architectural drawings have been placed in the order indicated. What is the advantage of placing them in this order?

Problem 3-19 On an 11" × 17" sheet of paper, draw a sheet layout for a working drawing. Provide a border and title block with all of the information typically found there.

BACKGROUND

Communication for a building project involves all members of the project team and can take on many forms. Proper communication is key in allowing everyone involved in the project to better understand the project goals and to be able to address issues as they arise on the project. Without proper communication, the success of any project may be in jeopardy, putting all involved with the project at risk.

Communication Is Essential to Achieving Shared Project Vision and Construction Productivity

Communication starts at the inception of a project with an owner contacting an architect to discuss a potential project. It continues in various forms beyond the project design and construction, and on through the life cycle of the project. Communication is the responsibility of all parties involved in a project, and only through proper communication can it be successful.

Communication can take on many forms, including:

- Building Information Modeling (BIM)
- Meetings
- Phone conference calls

- Letters
- Memos
- Emails
- Progress Sets: Schematic design and Design development
- Construction documents – Including working drawings and specifications
- Addenda
- Change orders
- Shop drawings
- RFIs – Requests for information (RFI)
- Independent Inspection Agency Reports
- Site observation reports

Example 3.10 Give a description for each of the forms of communication used on a project.

Building Information Modeling (BIM): The use of BIM during the design and construction of a project allows everyone involved in the project to have a three-dimensional model that can be utilized for design, coordination, and pricing.

Meetings: Face-to-face interaction where ideas can be expressed and issues can be addressed.

Phone conference calls: Two or more people involved in the project discussing issues. These are very helpful and solve many small issues that constantly arise on any project.

Letters: For important and legal issues, often a letter is necessary to convey issues to all parties involved.

Memos: Less official than letters, memos can be used to convey information to the design team.

Emails: Overwhelmingly helpful in helping to keep the design process moving in the right direction. Emails are often used to deliver files for use on a project.

Progress Sets: Physical sets of in-progress working drawings that are issued periodically through the design phase of a project. These sets of drawings allow those involved in the design process to have a current set of information for use in furthering the project design.

Construction Documents: The final documents, both drawings and specifications, issued for construction. These are used to obtain building permits, for pricing or bidding purposes, and as a set of documents from which the project is constructed. These are also referred to as "working documents."

Addenda: At times, changes to projects need to be made after the construction documents have been issued. Addenda are used to communicate changes to the construction documents. Issuing addendum is a valid method to make changes to the construction drawings for the period of time before the final bids, or pricing of the project has been established.

Change Orders: If changes to a project occur once the final pricing for a project has been established, a change order is required. This communication alters the contract to include any changes for the project.

Shop Drawings: From the set of working documents, manufacturers of components for the construction of a project often submit drawings for review by the design team. These drawings indicate what is to be fabricated for the project and are reviewed for any errors or omissions by members of the design team. Shop drawings are typical forms of communication for pre-fabricated building components, such as steel and concrete structures, doors and windows, and built-in furnishings for a project.

RFI's – Requests for Information: Often, questions arise during the construction of a project. When a question arises, the contractor issues an RFI to the design team. The design team is charged with answering the question or issue outlined in the RFI. At times, this involves a simple answer to clarify confusing information on the drawings. At other times, the design team must issue additional plans, sections, or details to answer the RFI. An answered RFI becomes part of the working drawings and could result in a change order being issued for the project.

Independent Inspection Agency Reports: Inspection of installed components of a project (structure, mechanical and plumbing systems) are often required to be performed as part of the construction process. These reports are sent to the architect and distributed to the team members. Inspection reports determine whether a system or component has been installed per the construction documents.

Site Observation Reports: During the construction process, architects and engineers will visit the site and observe what is taking place. Reports of these visits are used to bring any deficient issues observed to the attention of the project team, owner, and contractor.

Example 3.11 Taking on the role of Architect, give an example of a Site Observation Report for a construction site visit with the architect, electrician, and general contractor present.

Site Observation Report: Thompson's Motor Showroom

Date: June 6, 2009

Time: 2:00 - 3:00 pm

Present:

Doug Barkley / Architects Now, Inc.

Sarah Mills / Mills Electrical

Jean Wilson / Total Construction, Inc.

Observation:

We met at the site to observe the ongoing installation of electrical conduit for the project. Sarah indicated that the installation was one week behind schedule due to delivery schedule issues for the circuit breakers. She indicated that once they arrive (she has been told they will be here in the next two days), her team will work extra hours to get them installed and to get back on schedule. Jean stated that this delay has not impacted the schedule significantly and that if the work can be completed within the next week, everything should be fine. Sarah assured us that they can meet that deadline for finalizing the electrical installation. The work appears to be going well, though two separate locations were indicated as improperly installed, according to Jean. Sarah agreed and called her workers over to discuss the issue. The repairs for these locations began while we were at the site and should be corrected by the end of the day.

One other issue discussed was a wall in the break room where no electrical outlets are indicated to be installed according to the working drawings. Jean is to submit a Request for Information (RFI) for this issue, and Doug will have an answer for it later today.

Problem Set 3.3 Communication Is Essential to Achieving Shared Project Vision and Construction Productivity

Problem 3-20   Which of the forms of communication listed in this section of the workbook occur during the design phase of a project?

Problem 3-21   Which of the forms of communication listed in this section of the workbook occur during the construction phase of a project?

Problem 3-22   Taking on the role of architect for a residential project, give an example of a Site Observation Report that mentions:

- Correct installation of tile in the bathrooms
- Incorrect paint colors in the hallways
- Broken windows found in the kitchen
- Landscaping that does not use the plants listed on the construction documents

CHAPTER 4
Architectural Design

Skills List

After completing the problems in this chapter, you should be able to:

- Identify and discuss the differences between residential and commercial buildings

- Understand the whole-building approach to design and construction

- Be able to identify features that are used in building design

- Identify and discuss elements and principles that make up various architectural styles

BACKGROUND

Distinct differences exist that separate residential from commercial architecture. Residential architecture often requries less documentation and has less regulation imposed on its design and construction than commercial architecture. Commercial architecture typically involves a more complex process of design, documentation, and construction due to the financial cost of these types of projects as well as regulations pertaining to the health, safety, and welfare of its occupants.

Residential and Commercial Architecture

There are distinct differences dividing the categories of residential and commercial architecture. Residential architecture primarily involves buildings that may be used as single-family homes or apartments. Commercial construction is related to buildings used for businesses or for places of assembly, such as churches or convention centers. Additionally, commercial architecture can include manufacturing and industrial facilities. Refer to table 4-1 for a list of differences between residential and commercial architecture. Note that these traits are typical for residential and commercial architecture, and they can vary based on city and state regulations.

Residential Architecture	Commercial Architecture
The construction documents are not required to be signed by a registered architect	The construction documents must be signed by a registered architect
A geotechnical report is typically not required	A geotechnical report is required to give recommendations for the design of the building foundation
Residential foundation design is typically based on building code design values	
A registered professional engineer is not required for the design of the structure	The structure is required to be designed by a registered professional engineer
There are fewer life safety issues involved in the design process	Skilled labor is often required for construction of specific systems or materials
Typically there is less liability for those involved in the project than in commercial architecture	Custom made systems and finishes are often used in the design
Less skilled labor is required for construction	Must conform to the Americans with Disabilities Act (ADA)
Is not required to conform to the Americans with Disabilities Act (ADA)	

TABLE 4-1 *Differences between residential and commercial architecture.*

The number of people involved in the design process will also vary between residential and commercial architecture. Residential architecture often involves fewer people, due to the smaller size, smaller cost, and smaller number of people who will be using the building. Commerical architecture involves larger buildings for the most part, resulting in larger cost, and a larger number of people who will be using the building. This number results in more stringent requirements for health, safety, and welfare of the building occupants, which requires additional review of the design and construction by building code officials. As a general rule, as a building project increases in size and complexity, more people will be involved with design and construction.

Urban Design at CityCenter, Las Vegas

For an example of commercial architecture, let's take a look at a large project. CityCenter is a mixed-use facility that is located on the Las Vegas strip in Paradise, Nevada. It is an approximately 17,000,000 sq. ft. complex that covers 67 acres of land. Due to its immense size and various uses, several architects were needed to design the project. At the time of its opening on December 16th, 2009, it was the largest privately funded architecture project ever constructed in the United States. A vast amount of information about this project is available online. A primary source for information about this project is the website for CityCenter, found at www.citycenter.com. Explore the site to understand the many features of the project.

Example 4.1 Using the CityCenter website, give an overview of the amenities that are available for this project.

CityCenter sits on 67 acres of land, includes approximately 17,000,000 sq. ft. of space with four hotels totaling nearly 6,000 hotel rooms; a residence tower with 2400 apartments; 500,000 sq. ft. of retail, dining, and entertainment; restaurants featuring world-famous chefs; more than 320,000 sq. ft. of convention space, and a casino. The overall project has achieved Leadership in Energy & Environmental Design (LEED) certification with some of the hotels achieving LEED Gold certification. CityCenter was designed by a team of eight world-renowned architecture firms, each responsible for individual buildings within the complex.

Example 4.2 Using the CityCenter website, research and discuss the portion of the complex designed by the architecture firm Pelli Clarke Pelli Architects.

The firm Pelli Clarke Pelli Architects designed CityCenter's ARIA resort and Casino, a 61-story building with 4,004 guestrooms, including 568 suites. The facilities include 16 restaurants, 10 bars/lounges, and four swimming pools. It features an 80,000 sq. ft. spa, 300,000 sq. ft. of meeting space, and an 1,800-seat theater housing a Cirque du Soleil show. In addition, the facility has received LEED Gold Certification from the U.S. Green Building Council.

Problem Set 4.1 Urban Design at CityCenter, Las Vegas

Problem 4-1 Using the CityCenter website, research and discuss the portion of the complex designed by the architecture firm Helmut Jahn.

Problem 4-2 Using the CityCenter website, research and discuss the portion of the complex designed by the architecture firm Studio Daniel Libeskind.

Problem 4-3 Using the CityCenter website, research and discuss the portion of the complex designed by the architecture firm Kohn Pedersen Fox.

Problem 4-4 Use the website for CityCenter to determine the number of hotel rooms available within the entire facility.

Whole-Building Design

Whole-building design is a process in which the entire design team, as well as the owners, investors, contractors, governmental agency, and building users are involved in the design of the project. This method of design has advanced in recent years, mainly due to the technology that is available to enhance the process of building design, construction, and utilization. *"The goal of 'Whole Building' Design is to create a successful high-performance building by applying an integrated design and team approach to the project during the planning and programming phases"* (www.wbdg.org).

Each party on the team has a common goal to design and construct the best building for all involved.

Example 4.3 List the seven resources of technology researched in whole-building design.

- People – Design team, construction team, and occupants.

- Information – Learn from past successes and failures when designing, constructing, and utilizing a building project.

- Materials – What materials are available to be used in constructing the building project?

- Tools and machines – What is available for research, design, construction, and use of a building project?
- Capital – Used to fund and construct a building project.
- Energy – How much energy (work) is needed to complete a building project?
- Time – How long does it take to research, design, construct, and utilize a building project?

Example 4.4 Utilize the Whole Building Design website to list and describe the types of spaces that a well-designed office building would include.

A well-designed office building would include four classifications of spaces:

- Office Spaces – Including private and semi-private rooms for the employees, and conference rooms to be used for meetings.
- Employee and Visitor Support Spaces – Including lobbies and atrium spaces, cafeteria and/or break rooms, gathering spaces such as auditoriums, public and private bathrooms, vending areas, and parking.
- Administrative Support Spaces – Including office space for owners and building administrators.
- Operation and Maintenance Spaces – Including janitor closets, mechanical/electrical/plumbing support spaces, information technology support space, and storage space for tenants.

The Building America website (www.eere.energy.gov/buildings/residential/ba-index.html) is part of a program sponsored by the U.S. Department of Energy and can be used during the whole building design process. The website acts as a clearing house for industry-based research that supports and encourages the development and implementation of new energy technologies for existing and new homes. Included on the website are many topics relating to all aspects of energy efficient construction, with detailed explanations included to allow the user a better understanding of the systems and processes. It is a valuable resource for the whole building design team when researching state-of-the-art construction methods and techniques.

Example 4.5 Using the Building America website, list and describe the types of geothermal heat pump systems.

On the Building America website, information can be found by searching for 'geothermal'. A geothermal heat pump, or ground source heat pump, is a system that is becoming widely accepted in both commercial and residential construction. Geothermal is a renewable energy source that is highly efficient and used for both heating and cooling of spaces, and heating of water. Its efficiency comes from using the relatively constant temperature of the Earth as a heat source or depository to heat and cool a building. The website describes and gives images for two classifications of geothermal heat pump systems:

- Closed-Loop Systems – This system is cost-effective for new residential installations that have adequate land surrounding the building. The surrounding land is required to provide space for a series of pipes that will be placed in the ground. The Building America website lists three types of Closed-Loop Systems, including Horizontal, Vertical, and Pond/Lake.
- Open-Loop Systems – This system uses a body of water for the heat-exchange process. This body can be a well or a pond/lake, as long as it is an adequate supply of clean water. In addition, building codes and regulations must be met regarding groundwater discharge as this system takes water from the source, runs it through the heat exchange unit, and then returns it to the source.

Problem Set 4.2 Whole-Building Design

Problem 4-5   List the five goals of whole-building design.

Problem 4-6   List the five key areas of human and environmental health recognized by the Leadership in Energy and Environmental Design (LEED).

Problem 4-7   Using the Whole Building Design Guide website, define and discuss the Integrated Design Approach.

Problem 4-8   Using the Whole Building Design Guide website, define and discuss the Integrated Team Process.

Problem 4-9   Using the Whole Building Design Guide website, list and define the eight Design Objectives of whole building design.

Problem 4-10   What agencies are listed as participants for the Whole Building Design Guide website?

Use the Building America website for the following problems:

Problem 4-11   List the goals of the Building America program.

Problem 4-12   List the benefits for Homeowners of the Building America program.

Problem 4-13   List the benefits for Builders and Developers of the Building America program.

Problem 4-14   Determine the number of Building America projects that have been constructed in your state.

Problem 4-15   What is the EnergySmart Home Scale (E-Scale), and what do the numbers stand for in the rating scale?

Problem 4-16   List and define the climate regions used in the Building America program.

Identifying Building Features

Buildings are composed of a series of design elements and features. When observing a building, these features might help one to identify its purpose. There are many features that can be used in building design, with textbook figure 4-5 listing many of these by name and providing illustrations and descriptions of each. Additionally, at the website www.wikipedia.org/wiki/Architectural_glossary, a Glossary of Architecture exists that greatly expands upon the information given in the textbook. On this website, the terms are defined, and for many of the terms there are links to web pages that will further explain and illustrate the feature. By using these resources, one can become familiar with the terms and features prevelant in architecture. It is important to be able to identify and understand the many features that make up a building when observing, analyzing, and designing architecture.

Example 4.6 Refer to figure 4-1 and identify six of the architectural features shown.

This building illustrates the Georgian style of architecture, and the portion shown is known as the "Portico." It includes many of the features listed in textbook figure 4-5 and on the website. Some of these include:

- Columns – vertical elements that transfer floor and/or roof loads to the base of the structure
- Capitals – the ornate top of the columns; those shown are of the modified Corinthian style
- Entablature – located directly above the columns
- Plinth – the stone block upon which the columns sit
- Balustrade – the stone post and rail system at the perimeter of the roof
- Keystone – the stone segment that is placed at the peak of the arch above the entry doors

FIGURE 4-1 *School of Architecture, Oklahoma State University; Stillwater, OK.*

There are specific features of a building that give insight into the style of architecture used in the design process. Windows can vary greatly among different styles of architecture, with many different configurations possible. The website www.wikipedia.org/wiki/Window#Types_of_windows gives examples of many window styles used in construction. Similarly, there are many reasons for using different roof styles on a building. Some of these include the style of architectural design used for the building, the climate, and whether attic space is needed in the building. The website www.wikipedia.org/wiki/Roof#Roof_shapes is a good source to use when exploring the different roof styles. It gives examples of many roof types and discusses roof construction in detail.

Problem Set 4.3   Identifying Building Features

Problem 4-17   Identify eight architectural features using the textbook and online resources for the building you are in currently.

Problem 4-18   Define and sketch the soffit of a building.

Problem 4-19   Define and sketch a brise soleil for a building.

Problem 4-20   Identify the window types used for the building you are in currently.

Problem 4-21   Sketch and label the components of a Jalousie window.

Problem 4-22   Sketch and label the components of a Casement window.

Problem 4-23   Define the term "roof pitch," and write an example problem to calculate the pitch of a roof.

Problem 4-24   Identify five different roof outer layer materials commonly used, and list the country where this type of roof is found.

Problem 4-25   Define and draw a sketch for four different roof styles.

Determining Architectural Style

Both the Elements of Design and Principles of Design play crucial roles in the final evaluation of the success or failure of a building design. The textbook provides explanations and examples for elements in figure 4-13 and for principles in figure 4-14. When observing a building, these elements or features help us identify the architectural style of the building. These include building proportion, overall shape and size, roof type, building materials, and placement of doors and windows, to name a few. There are factors beyond the physical building that influence the architectural style. These include cultural and environmental responses, societal and political concerns, and the economy in the community where the building is constructed. The building features given in textbook figure 4-5 can be used to help identify the architectural style of a building, as certain features are common to specific architectural styles. For example, Art Deco, which was a prominent architectural

FIGURE 4-2   *Chrysler Building; New York, NY.*

style in the early 20th century, has certain characteristics that can be used to identify the style. Some of these include smooth wall surface of stucco, stone and/or metal; simplified and streamlined forms; geometric designs and projections enhancing a vertical emphasis; and machined construction materials for features that are decorative in nature. Refer to figure 4-2 for an image of the Chrysler Building in New York City. It is one of the most recognizeable Art Deco buildings in the world. Consider the characteristics for the Art Deco style, and see which of them apply to the Chrysler Building.

An educational website, developed by Northern Arizona University Professor Dr. Tom Paradis, can be used as a resource in identifying architectural styles that are present in America. This website can be accessed at www.architecturestyles.org. It includes photographic examples of each of the styles, in addition to listing information on identifying features and background for each of the sytles. It must be noted that often buildings do not fall into a specific architectural style, but rather have characteristics of several styles. This can make the process of identifying the architectural style of design difficult, and in such cases the predominant style should be identified while noting the influences of other styles included in the design.

Example 4.7 Identify characteristics of the Craftsman style of architecture, and give two examples.

The Craftsman style of architecture was popular from the late 1890s until the 1930s. It can be identified by several characteristics, including:

- Double-hung windows
- Hand-crafted woodwork and stonework
- Low-pitched, hipped, or gabled roof and deep overhanging eaves
- Tapered, square columns used to support the roof
- Decorative brackets under eaves with exposed rafters
- A front porch placed beneath an extension of the main roof

One example of the Craftsman style is the Hyland Hotel in Monticello, Utah, shown in figure 4-3. It was built by Joseph Henry Wood from 1916 to 1918 as a single-family residence. In 1924, it was adapted for use as a hotel. At that time, the home's four upstairs bedrooms were converted into nine guest rooms. In 1994, the Hyland Hotel became a building on the National Register of Historic Places.

Another example of the Craftsman style is the Gamble House in Pasadena, California shown in figure 4-4. It is known as America's masterpiece of the Craftsman style of architecture.

FIGURE 4-3 *The Hyland Hotel, Monticello, Utah.*

FIGURE 4-4 *The Gamble House, Pasadena, California.*

The three-story house was originally a winter residence for David and Mary Gamble. The Craftsman style of architecture was focused on using natural materials, giving much attention to the details and aesthetics of the building, and priding itself on craftsmanship. The house was jointly deeded to the city of Pasadena and the School of Architecture at the University of Southern California in 1966 and was placed on the National Register of Historic Places in 1977. A couple of interesting facts about the house exist. First, each year two USC architecture students attending their fifth year at USC live in the house full-time. Second, the Gamble House is utilized in the Back to the Future movie trilogy (1985–1990) as the mansion of Dr. Emmett Brown.

Example 4.8 **Identify the architectural style used in the design of the Tribune Tower, located in Chicago, and indicate characteristics of the architectural style used in the design of the building.**

By referring to figure 4-5, one can see that the Tribune Tower has many characteristics that place it in the Gothic Revival style of architecture. Also referred to as Neo-Gothic, this style began in England in the 1740s and became popular in the early nineteenth century. Characteristics of the Gothic Revival style include grouped chimneys, pinnacles and battlements, steeply pitched gable roofs, pointed-arch windows or quatrefoil with clover-shaped windows. Other characteristics include the use of asymetrical floor plans, verandahs, hood molds over windows, and gingerbread trim along eaves and gable edges. Many of these characteristics are visible on the Tribune Tower, which was designed by New York architects John Mead Howells and Raymond Hood. It was completed in 1925 and is listed as a Chicago Landmark.

FIGURE 4-5 *Tribune Tower, Chicago, Illinois.*

Problem Set 4.4   Identifying Building Features

Problem 4-26   Identify characteristics of the International style of architecture, and give two examples of the style.

Problem 4-27   Identify characteristics of the Georgian style of architecture, and give two examples of the style.

Problem 4-28   Identify characteristics of the Spanish Revival style of architecture, and give two examples of the style.

Problem 4-29   Identify characteristics of the Prairie style of architecture, and give two examples of the style.

Problem 4-30   Identify the architectural style used in the design of the Empire State Building in New York City, and indicate characteristics of the architectural style used in the design of the building.

Problem 4-31   Identify the architectural style used in the design of the San Francisco City Hall, and indicate characteristics of the architectural style used in the design of the building.

Problem 4-32   Identify the architectural style used in the design of the United States Capitol Building in Washington, D.C., and indicate characteristics of the architectural style used in the design of the building.

Problem 4-33   Identify the architectural style used in the design of the Seagram Building in New York City, and indicate characteristics of the architectural style used in the design of the building.

Problem 4-34   Take a photograph of a building in your community. Identify the architectural style used in the design of the building, and indicate characteristics of the style included in the design.

CHAPTER 5
Site Discovery for Viability Analysis

Skills List

After completing the problems in this chapter, you should be able to:

- Determine the various legal descriptions of land

- Understand physical features that affect land use

- Know how to research rules and regulations that control land use

- Be aware of environmental regulations on land use

- Determine climate, demographics, and other site considerations that control land use

- Know what information to determine on a site visit

- Determine a basic cost estimate and viability analysis

BACKGROUND

Architects and engineers must be able to select building sites that meet the needs of the project and the client. The following problems address the objectives stated in the textbook. The selection of a site must consider many variables and be evaluated for best use.

How Land Is Described

The description of land starts with determining what property is included in the site. The following sections cover the documents that are needed to legally describe a site.

Property Research

The deed and plat for a site is the legal document that gives ownership to the property and identifies the location of the property. Basic size, easements, and rights-of-way are shown on the plat. Figure 5-3 in the textbook is a copy of a deed, and figure 5-4 is the plat for the same property. The deed and plat are used together to provide the property lines of the site. Figures 5-1 and 5-2 are the deed and plat for a property in Stillwater, Oklahoma. The County Clerk's office in Oklahoma records and maintains the deeds. Figure 5-3 is a map showing the location of the subdivision within section 28.

KNOW ALL MEN BY THESE PRESENTS:

That WILLIAM P. DAWKINS and LANNIE P. DAWKINS, husband and wife

_____, parties of the first part,

in consideration of the sum of Ten and no/100-------------------------------------- dollars

and other valuable considerations, in hand paid, the receipt of which is hereby acknowledged, do hereby grant, bargain, sell and convey unto STEVEN E. O'HARA and PHYLLIS A. O'HARA,

husband and wife   2801 Black Oak   Stillwater, OK 74074

as joint tenants and not as tenants in common, with the right of survivorship, the whole estate to vest in the survivor, parties of the second part, the following described real property and premises situate in

Payne County, State of Oklahoma, to-wit:

Lot Three (3) in Block Four (4), QUAIL RIDGE DEVELOPMENT, a subdivision of NE/4 SW/4 Section 28, Township 19 North, Range 2 East of Indian Meridian, Payne County, Oklahoma.

together with all the improvements thereon and the appurtenances thereunto belonging, and warrant the title to the same.

TO HAVE AND TO HOLD said described premises unto the said parties of the second part, as such joint tenants, and to the heirs and assigns of the survivor, forever, free, clear and discharged of and from all former grants, charges, taxes, judgments, mortgages and other liens and incumbrances of whatsoever nature. SUBJECT to easements, restrictions, covenants, zoning ordinances, and mineral reservations of record.

FIGURE 5-1 *2801 Black Oak Dr. Warranty Deed.*

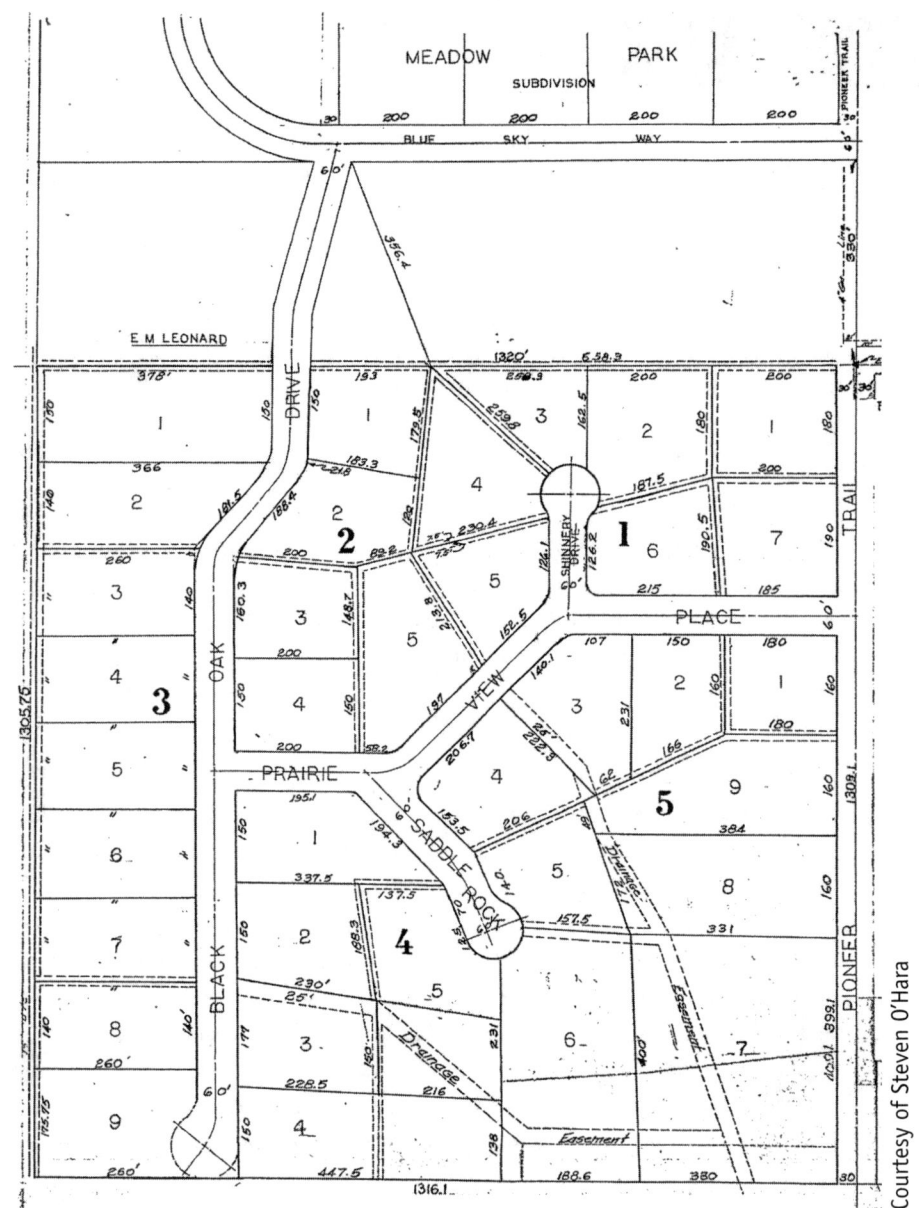

FIGURE 5-2 *Quail Ridge plat.*

The property is described as Lot Three in Block Four of the subdivision also known as 2801 Black Oak. There is a 25-foot drainage easement on the north side of the property and a 7.5-foot easement on the east side. The 7.5-foot easement is typical of the entire addition and is indicated in figure 5-3.

Legal Descriptions

The legal description on the deed in figure 5-1 is as follows:

Lot Three (3) in Block Four (4), QUAIL RIDGE DEVELOPMENT, a subdivision of NE/4 SW/4 Section 28, Township 19 North, Range 2 East of the Indian Meridian, Payne County, Oklahoma.

This is given in the rectangular survey system, which is discussed later in this chapter. The legal description used in the deed shown in figure 5-3 of the textbook uses the metes and bounds system, which is discussed in the following section.

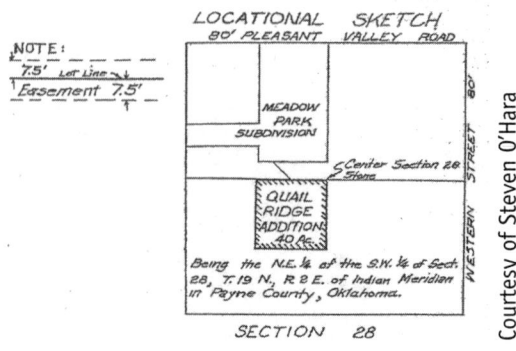

FIGURE 5-3 *Section 28.*

Metes and Bounds

The metes and bound system is used in the original 13 colonies and other areas of the country that were surveyed prior to 1805. This system uses distance and direction between property corners (metes). It may also include less-accurately-described boundaries (bounds). The following is the description of the property in figure 5-4.

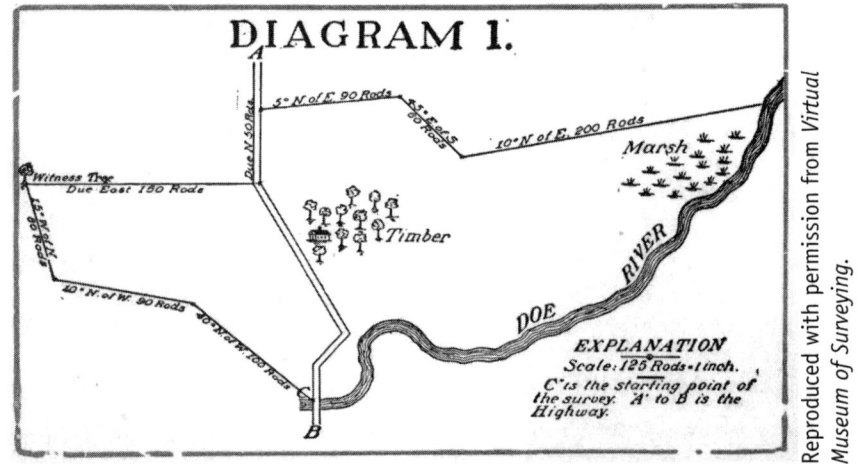

FIGURE 5-4 *Metes and Bounds.*

Beginning at a stone on the Bank of Doe River, at a point where the highway from A to B crosses said river (see point marked C); thence 40 degrees North of West 100 rods to a large stump; then 10 degrees North of West 90 rods; thence 15 degrees West of North 80 rods to an oak tree (see Witness Tree); then due East 150 rods to the highway; thence following the course of the highway 50 rods due North; then 5 degrees North of East 90 rods; thence 45 degrees of South 60 rods; thence 10 degrees North of East 200 rods to the Doe River; thence following the course of the river Southwesterly to the place of beginning.

Example 5.1 Describe the property in figure 5-4 using the metes and bounds system. Use all angles starting with north or south and distances in feet where possible.

The following are the steps to describe the property. Remember from the textbook that a rod is equal to 16.5 feet.

Beginning at a stone on the Bank of Doe River, at a point where the highway from A to B crosses said river (see point marked C)

- Point C on the figure is taken as the point of beginning (POB).

  thence 40 degrees North of West 100 rods to a large stump

- N 50° W 1650 feet

  then 10 degrees North of West 90 rods

- N 80° W 1485 feet

  thence 15 degrees West of North 80 rods to an oak tree (see Witness Tree)

- N 15° W 1320 feet

  then due East 150 rods to the highway

- N 90° E 2475 feet

  thence following the course of the highway 50 rods due North

How Land Is Described

- N 825 feet

    then 5 degrees North of East 90 rods

- N 85° E 1485 feet

    thence 45 degrees of South 60 rods

- S 45° E 990 feet

    thence 10 degrees North of East 200 rods to the Doe River

- N 80° E 3300 feet

    thence following the course of the river Southwesterly to the place of beginning

- SW and W along the river to the POB

Example 5-1 has two landmarks that might not even be in existence. It also has a road and a river that may have changed courses. This could lead to some confusion over property rights.

Rectangular Survey System

The rectangular survey system was adopted in 1805 to provide for more uniformity in property boundaries. The property in the deed shown in figure 5-1 will be used to describe the rectangular survey systems components. The principal meridians and base lines are used as a reference point from which townships are located. Figure 5-12 in the textbook shows the United States map of the principal meridians and base lines. Figure 5-5 is a map showing the Indian Meridian and Base Line used in Oklahoma.

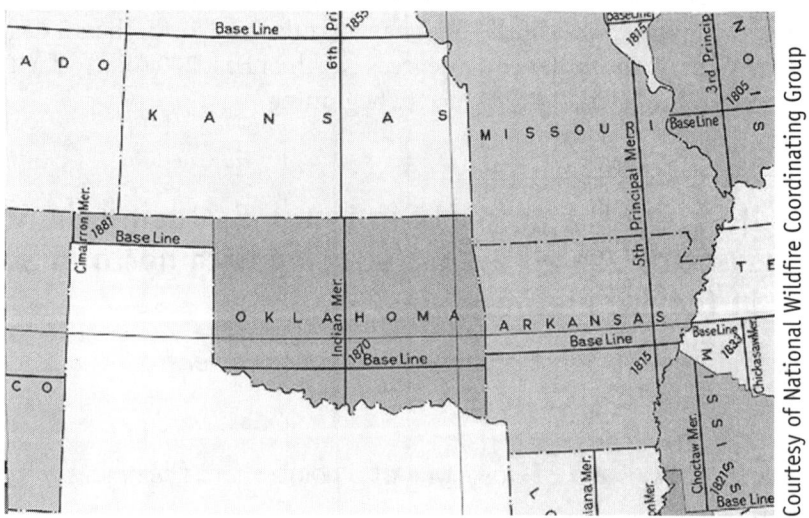

FIGURE 5-5 *Meridians and base lines.*

The Indian Meridian in longitude 97° 14′ 30″ west from Greenwich, extends from the Red River bordering Texas to the south boundary of Kansas, and, with the base line in latitude 34° 30′ north, governs the surveys in Oklahoma east of 100° west longitude from Greenwich (all of Oklahoma except the Oklahoma panhandle). This line was chosen arbitrarily as part of the land survey of 1870 conducted by E.N. Darling and T.H. Barrett, at an arbitrary point about one mile south of Fort Arbuckle (about six miles west of present Davis, Oklahoma). From this initial point, the north-south line was designated as the Indian Meridian and the East-West line was designated as the Indian base line.

Townships are 6 miles by 6 miles located from the principal meridians and base lines. The Stillwater Township is the 19th township north and 2nd township west of the Indian meridian. Figure 5-6 is a map of Payne County, in which the Stillwater Township is located. Note that the Indian meridian runs through western Payne County and that the numbers 18 and 19 are shown on the map.

A map of the Stillwater Township is shown in figure 5-7. The nearly white section on the map is section 28, in which the property is located.

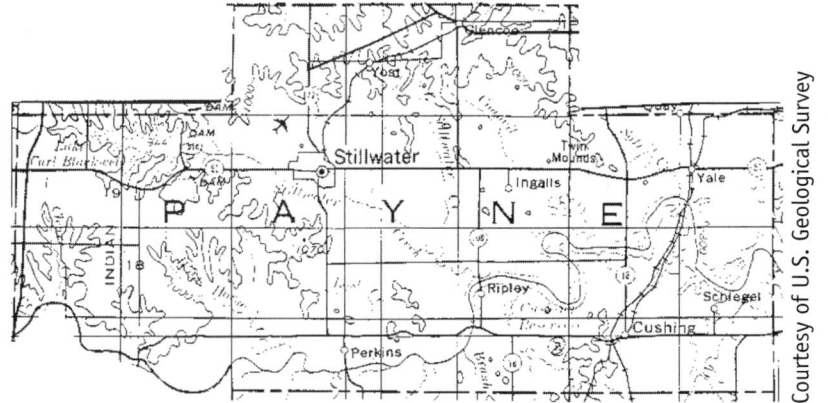

FIGURE 5-6 *Payne County townships.*

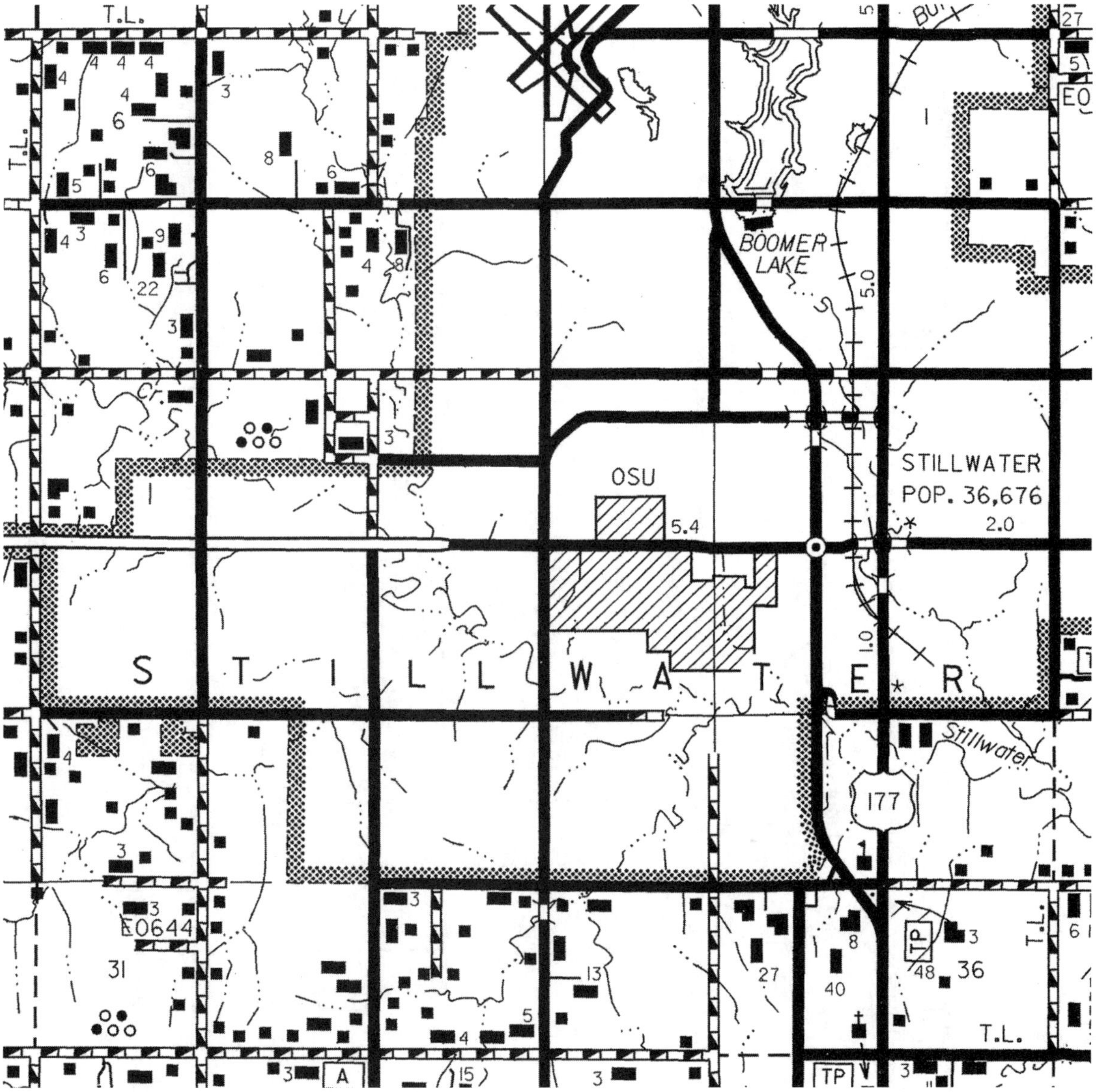

FIGURE 5-7 *Stillwater township.*

How Land Is Described   65

Lots and Blocks

The property on the deed is Lot Three (3) in Block Four (4). A survey of the property is shown in figure 5-8.

Section 28 was shown in figure 5-3. The Quail Ridge Addition is the NE/4 of the SW/4 of Section 28.

The entire legal description from the deed was "Lot Three (3) in Block Four (4), QUAIL RIDGE DEVELOPMENT, a subdivision of NE/4 SW/4 Section 8, Township 19 North, Range 2 East of the Indian Meridian, Payne County, Oklahoma." Sometimes many developments are designed as part of a long-range planning for the municipality. This particular addition was part of many planned additions and is shown in figure 5-9.

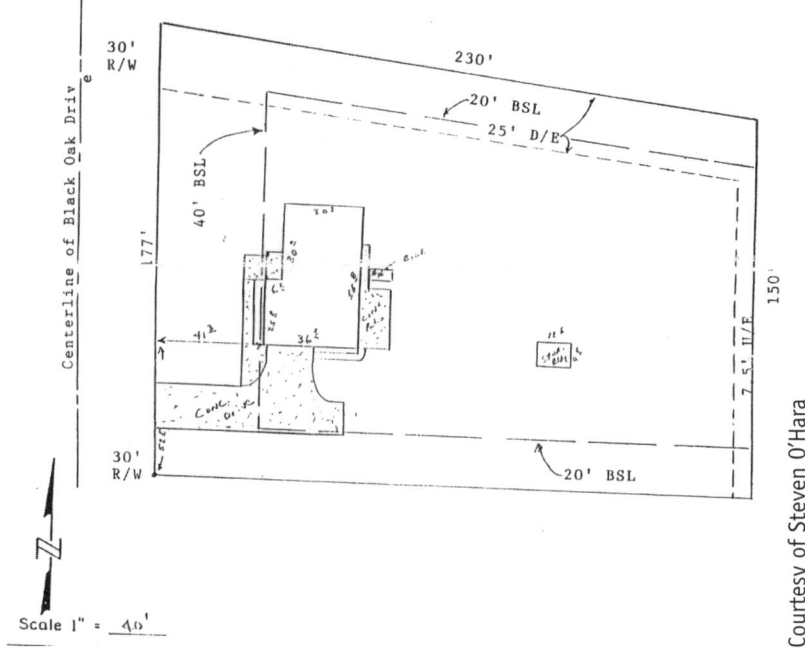

FIGURE 5-8 *2801 Black Oak survey.*

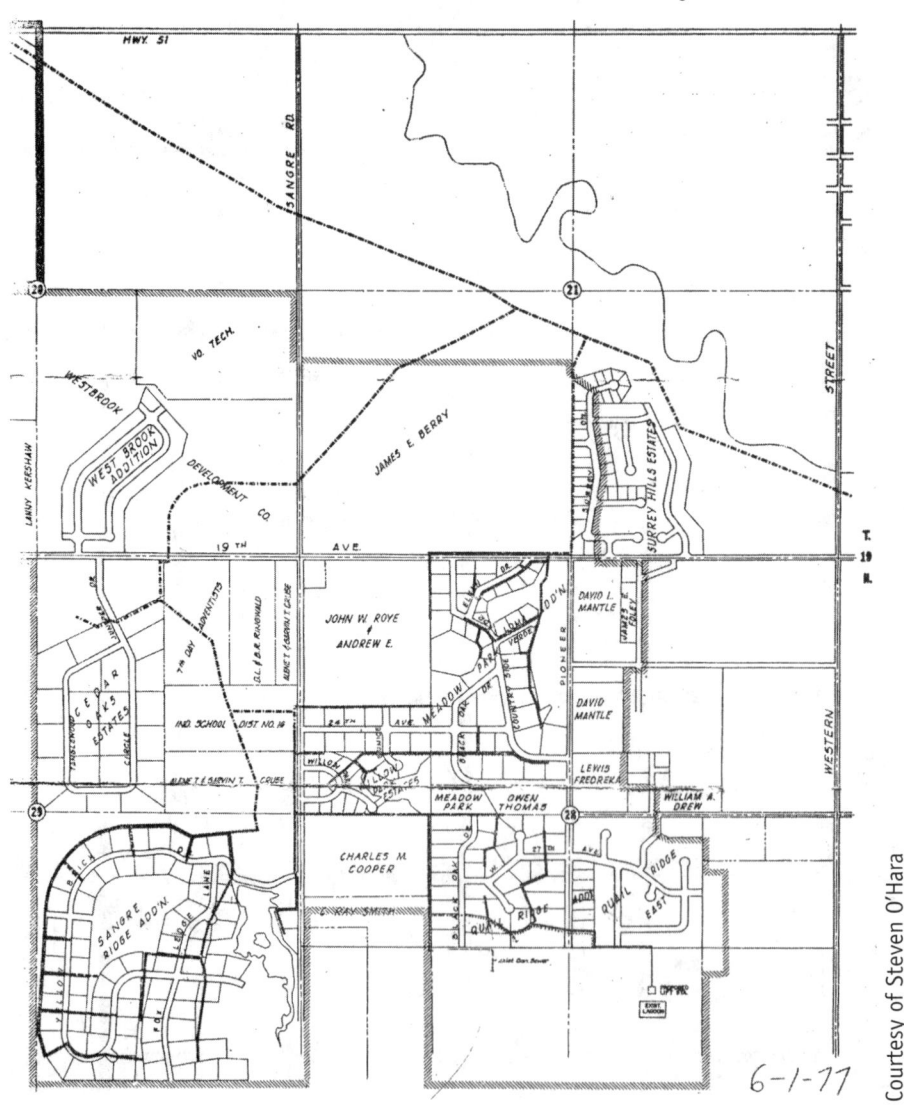

FIGURE 5-9 *Planned additions.*

66   Chapter 5: Site Discovery for Viability Analysis

Problem Set 5.1 How Land Is Described

Problem 5-1 What is the legal description for the property covered in the deed shown in figure 5-10?

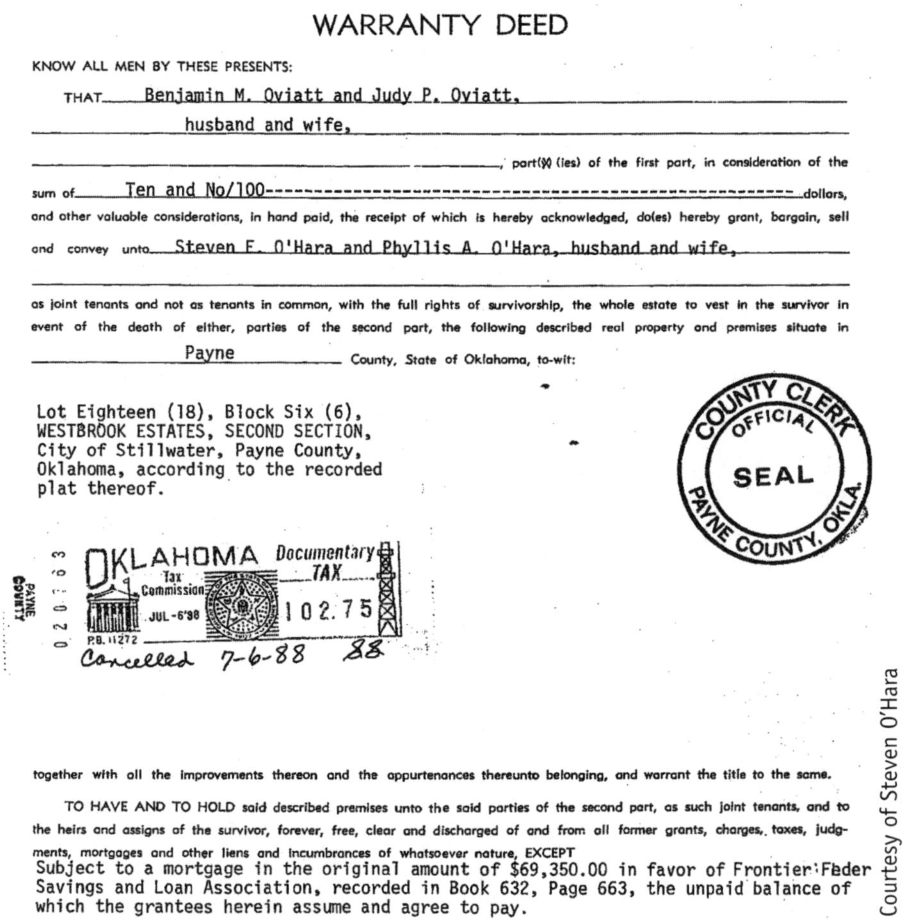

WARRANTY DEED

KNOW ALL MEN BY THESE PRESENTS:

THAT Benjamin M. Oviatt and Judy P. Oviatt, husband and wife,

part(x) (ies) of the first part, in consideration of the sum of Ten and No/100 dollars, and other valuable considerations, in hand paid, the receipt of which is hereby acknowledged, do(es) hereby grant, bargain, sell and convey unto Steven E. O'Hara and Phyllis A. O'Hara, husband and wife,

as joint tenants and not as tenants in common, with the full rights of survivorship, the whole estate to vest in the survivor in event of the death of either, parties of the second part, the following described real property and premises situate in Payne County, State of Oklahoma, to-wit:

Lot Eighteen (18), Block Six (6), WESTBROOK ESTATES, SECOND SECTION, City of Stillwater, Payne County, Oklahoma, according to the recorded plat thereof.

COUNTY CLERK OFFICIAL SEAL PAYNE COUNTY, OKLA.

OKLAHOMA Tax Commission JUL-6'88 Documentary TAX 102.75 P.B. 11272

PAYNE COUNTY 020163

Cancelled 7-6-88 88

together with all the improvements thereon and the appurtenances thereunto belonging, and warrant the title to the same.

TO HAVE AND TO HOLD said described premises unto the said parties of the second part, as such joint tenants, and to the heirs and assigns of the survivor, forever, free, clear and discharged of and from all former grants, charges, taxes, judgments, mortgages and other liens and incumbrances of whatsoever nature, EXCEPT
Subject to a mortgage in the original amount of $69,350.00 in favor of Frontier Feder Savings and Loan Association, recorded in Book 632, Page 663, the unpaid balance of which the grantees herein assume and agree to pay.

Courtesy of Steven O'Hara

FIGURE 5-10 *1514 Dublin Drive Warranty Deed.*

Problem 5-2 Does the legal description in figure 5-10 completely locate the property? Why or Why not?

Problem 5-3 The subdivision plat for the property in problem 5-10 is shown in figure 5-11. What additional information does the plat contain about the legal description?

Problem 5-4 Do the plat and the deed from problems 5-2 and 5-3 provide a complete legal description of the property? Why or Why not?

Problem 5-5 The Westbrook Estates Second Section was plated in 1978. If you look at figure 5-9, you can see that it is not shown. It was added on the west side of the Westbrook Addition. Does this help to complete the legal description of the property? Why or Why not?

Problem 5-6 Describe the property lines in Lot Eighteen (18) in Block Six (6) by metes and bounds. Assume only the western property line is not true north, south, east, or west. Use figure 5-12 for dimensions and angles.

Problem 5-7 Draw the Westbrook Estates Second Section given the following metes and bounds.

- POB
- N 1° 16' W 1618.97'

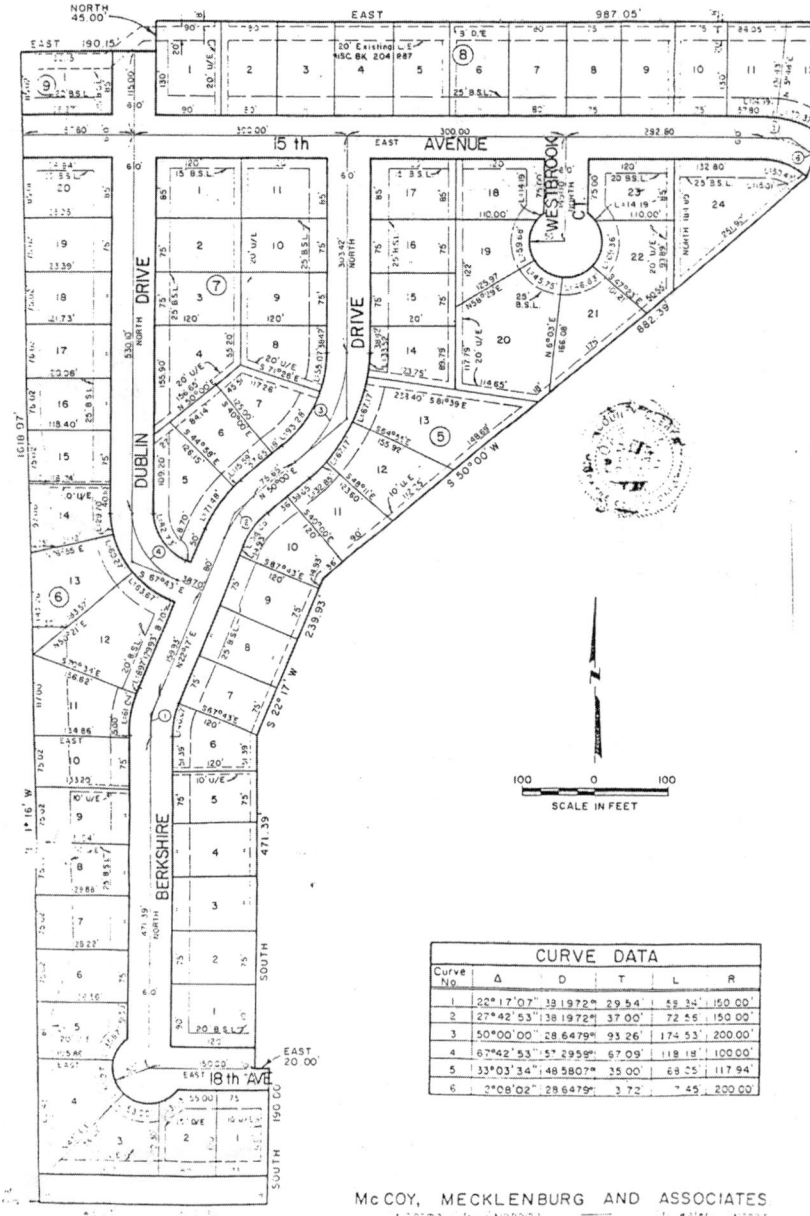

FIGURE 5-11 *Westbrook Estates plat.*

- EAST 190.15'
- NORTH 45.00'
- EAST 987.05'
- S 19° 11' W 225.49'
- S 50° 00' W 882.39'

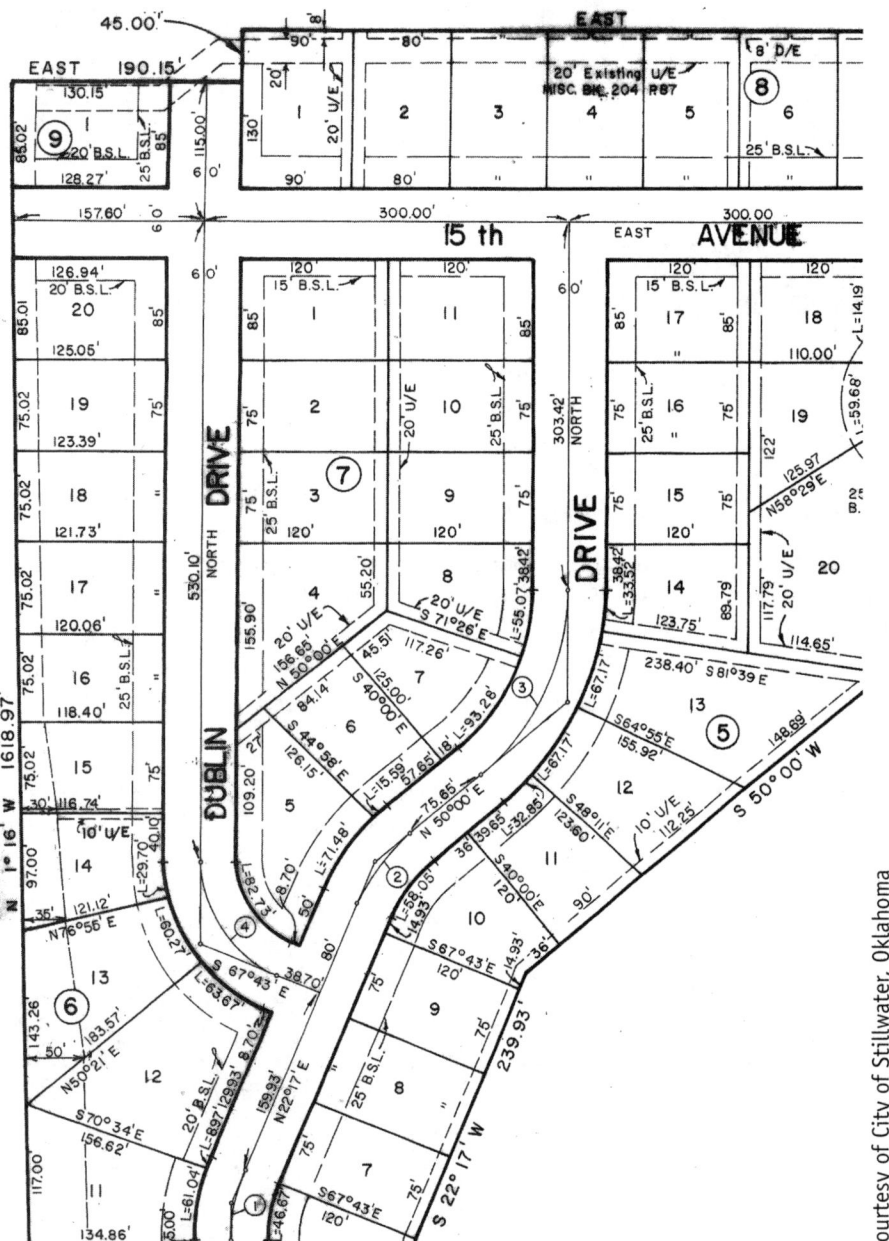

FIGURE 5-12 *Plat for problem 5-6.*

- S 22° 17' W 239.93'
- SOUTH 471.39'
- EAST 20.00'
- SOUTH 190.00'
- WEST 320.32' to the POB

Problem 5-8   What would be the legal description of the Sangre Ridge Addition shown in figure 5-9?

Problem 5-9   What would be the legal description of the property owned by John W. & Andrew E. Roye?

How Land Is Described   69

Problem 5-10 How many acres are there in the NE/4, SW/4, Sec. 28, T 19 N, R 2 E, I.M. (A.K.A. Quail Ridge Addition)?

Problem 5-11 Describe the easements for the property described in Figure 5-1 and 5-2.

Problem 5-12 Describe the easements for the property described in Figure 5-10 and 5-12.

BACKGROUND

The shape of land has an impact on how it can be used. Care must be taken to provide proper drainage away from and around building structures as well as onto surrounding property. The selection of a site should always consider the physical lay of the land.

Topography and Contour Maps

The U.S. Geological Survey under the U.S. Department of the Interior provides topographic maps. They can be found online at http://store.usgs.gov/. The maps are downloadable in 7.5 and 15-minute sizes. The 7.5-minute sizes are slightly larger than a township but do not correspond directly. The maps are interactive, with some being digital. Figure 5-13 show an overall contour map of the Stillwater Township from a 1981 survey. A 2010 survey shows more features and will be used in subsequent sections.

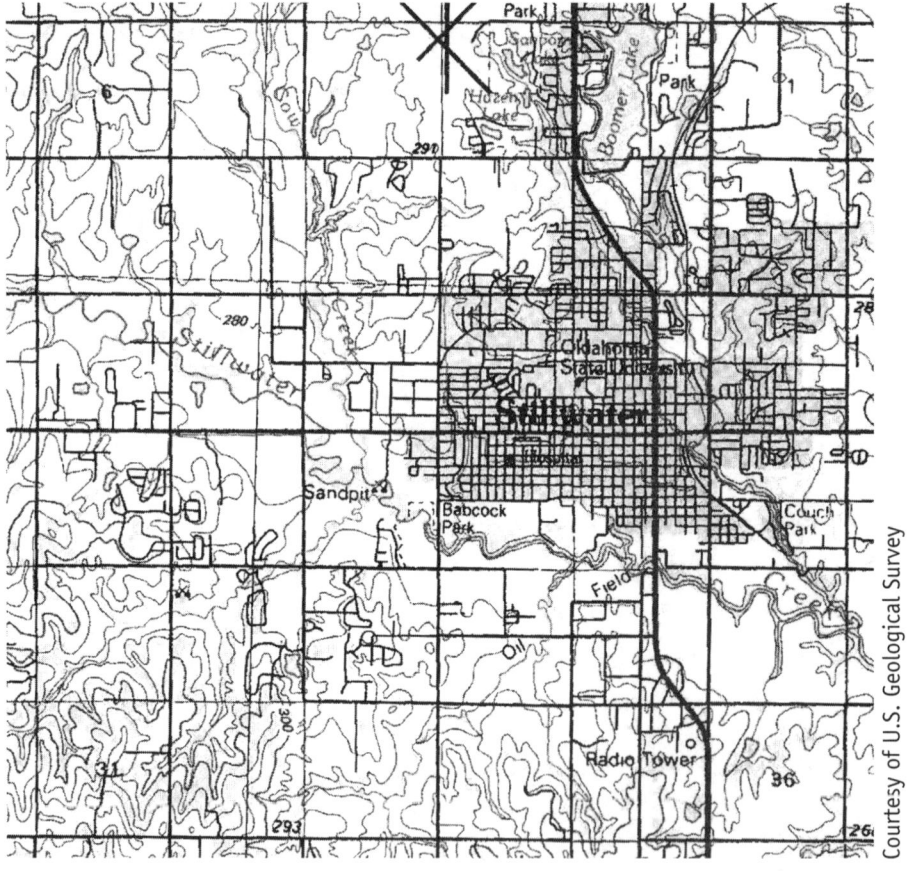

FIGURE 5-13 *Stillwater South topographic map.*

Contour Intervals

The contour interval used in Figure 5-13 is 10 meters. Normally, contour maps in U.S. customary units have intervals of 10 feet. Figure 5-14 is a portion of a contour map with 10 foot contour intervals and some geographic features.

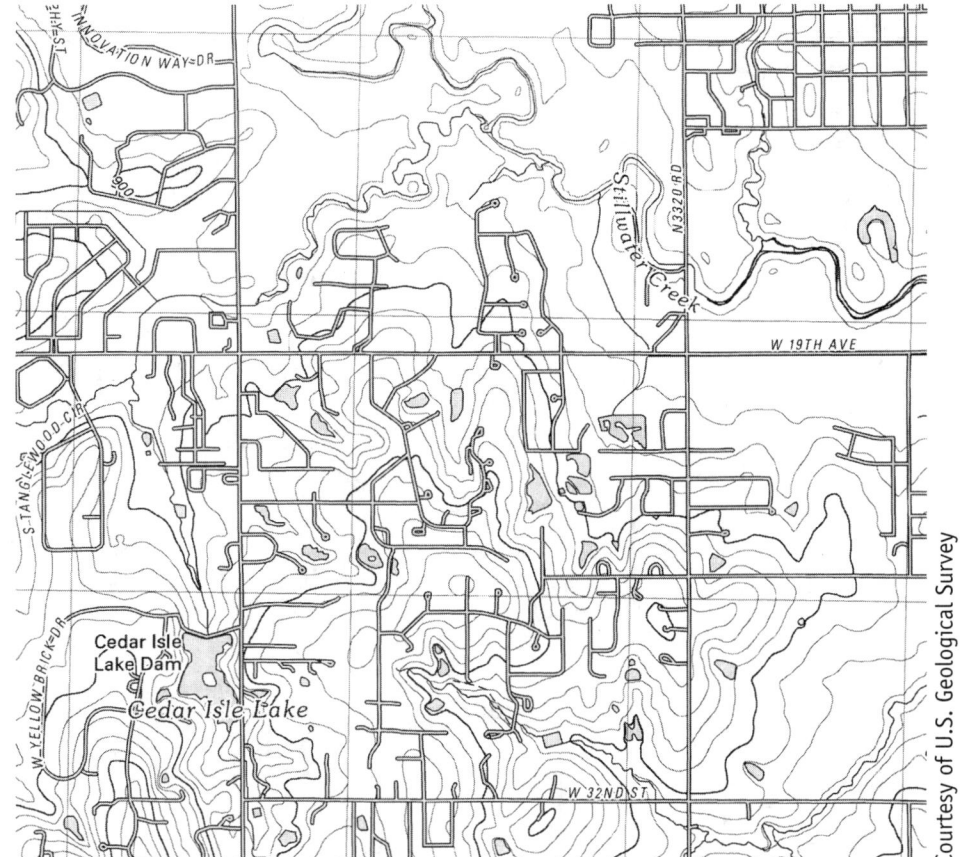

FIGURE 5-14 *Section 28 topographic map.*

Section 28 of the Stillwater Township is detailed in Figure 5-15. The main 50-foot contours are labeled, and the location of the site is indicated. The elevation of the structure on the site is 973.7 feet above sea level. The contour at 970 bisects the site in the north/south direction.

Slope and Gradient

The slope is the change in elevation divided by the length. The gradient is the slope measured as a percentage. The following formula can be used to calculate the slope gradient.

$$\text{Slope Gradient} = \frac{\text{Change in Elevation}}{\text{Length}} \times 100 \text{ percent}$$

Example 5.2 Determine the slope gradient on the south side and in the east/west direction for the site shown in figure 5-15.

An estimate for the elevation would be approximately 974 feet at the street and 958 on the east of the site. The length of the site on the south side is 228.5 feet from figure 5-2. The slope gradient is calculated as follows:

$$\text{Slope Gradient} = \frac{974 - 958}{228.5} \times 100 \text{ percent} = 7.0 \text{ percent}$$

The site appears to slope most on the east side, so if the structure is on the west side, regrading might not be required.

Topography and Contour Maps   71

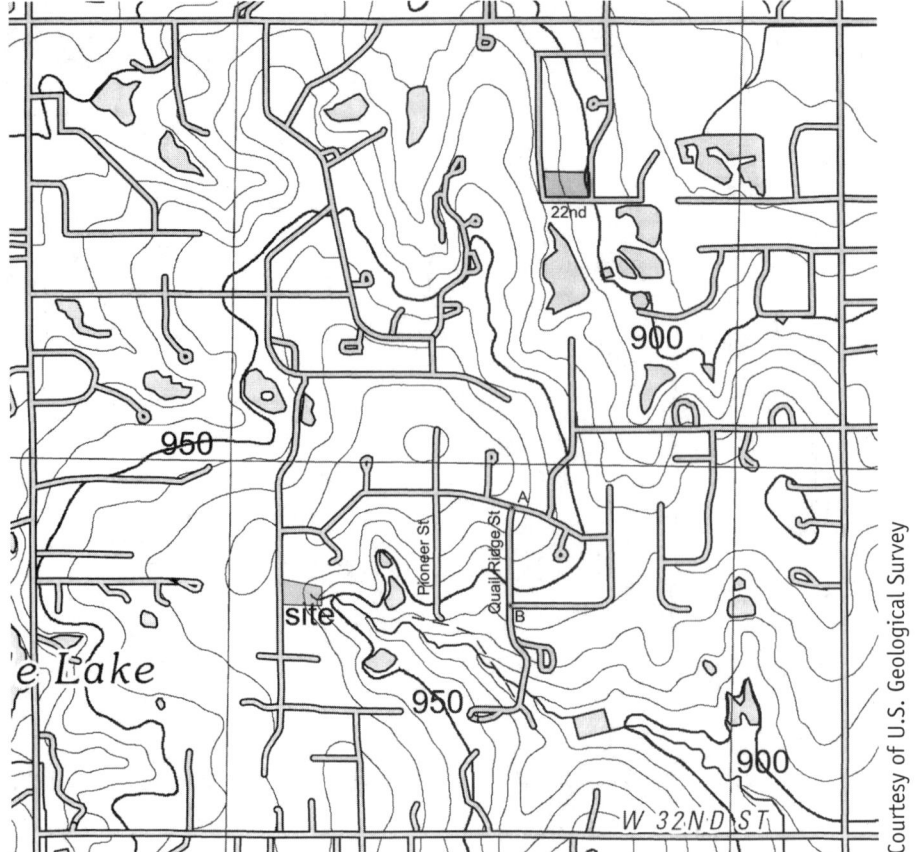

FIGURE 5-15 *Section 28 topographic map.*

Soils

The soil conditions at the site must be suitable to support loads and provide for a stable structure. The soils at the site should be investigated for adequacy. One quick resource is the Web Soil Survey (WSS), provided by the Natural Resources Conservation Service (NRCS). It may be accessed online at http://websoilsurvey.nrcs.usda.gov. A soil survey of the site is shown in figure 5-16. The data areas indicate as 3 and 4 a Coyle loam with 3 to 5 percent slope, and area 66 is a Masham silty clay loam with 5 to 20 percent slopes. This is consistent with the slope gradient calculated previously. The soil report also indicates in areas 3 and 4 that a building without a basement is not limited, but a basement may be limited by the depth to the soft bedrock. In area 66, the shrink-swell potential, slope, and depth to soft bedrock all limit construction, so it would be inadvisable to build in that area.

FIGURE 5-16 *2801 Black Oak soil survey.*

Problem Set 5.2 Topography and Contour Maps

Problem 5-13   What is the approximate slope gradient along the 22nd street side of the area shown in figure 5-17, if the length is 300 feet? Use the contours shown in figure 5-15.

FIGURE 5-17   *Soil survey for problem 5-13.*

Problem 5-14   Would the slope gradient in problem 5-13 require regrading?

Problem 5-15   If the soil for the site in problem 5-13 was Mulhall loam, 3 to 5 percent slopes, eroded, would that be a limitation?

Problem 5-16   What is the slope gradient between points A and B on Quail Ridge Street if the length is 650 feet? Use the contours shown in figure 5-15.

Problem 5-17   Would the slope gradient in problem 5-16 require regrading?

Problem 5-18   If the soil for the area between Pioneer St and Quail Ridge Street in figure 5-15 was Mulhall loam, 3 to 5 percent slopes, would that be a limitation?

BACKGROUND

Many different organizations control the use of land. Architects and engineers must be able to research the necessary organizations that govern land use.

Rules and Regulations

The rules and regulations that control the use of land are normally governed by the local municipality. This is generally the city that the site is within, or it may be the state if the land is in the country. The local municipality will have its own local ordinances and zoning regulations. Also, the national building code is normally adopted by the local municipality. In addition, restrictive covenants may be part of planning development in which a site is situated. A restrictive covenant is a contract between landowners in a development to control land use.

Building Codes

The International Code Council (ICC) publishes many codes and references. The two primary codes for buildings are the International Building Code (IBC) and the International Residential Code (IRC). There are a broad range of issues covered in these codes, and many of them are covered in the textbook and workbook. The primary issues covered in the workbook will be building egress, fire protection, structural, electrical, plumbing, and mechanical systems. The local municipality usually adopts these codes as their standards.

Local Ordinances and Zoning

Most local municipalities have a code controlling the use of land. A portion of the Land Development Code for Stillwater, Oklahoma, is shown in figure 5-18. This describes the general purpose of the code. The entire code contains sections covering zoning, street naming, lot sizes, use categories, parking, landscaping, and many others.

CHAPTER 23. LAND DEVELOPMENT CODE

ARTICLE 1. GENERAL PROVISIONS

Section 23.1. Citation and codification.

This ordinance, including all maps and publications made a part hereof and by reference, is adopted pursuant to the authority granted by the Charter of the City of Stillwater, and shall be known and cited as the "Land Development Code of the City of Stillwater, Oklahoma" and codified as "Chapter 23 of the Stillwater City Code".

Section 23.2. Purpose.

For the purpose of promoting the health, safety, morals, and the general welfare of the community, the City Council of the City of Stillwater, Oklahoma may regulate and restrict the height, number of stories, and size of buildings and other structures, the percentage of lot that may be occupied, the size of yards, courts and other open spaces, the density of population, and the location and use of buildings, structures and land for trade, industry, residence or other purposes.

Municipal regulations as to buildings, structures and land shall be made in accordance with a comprehensive plan and be designed to accomplish any of the following objectives:

A. To lessen congestion in the streets;
B. To secure safety from fire, panic and other dangers;
C. To promote health and the general welfare, including the peace and quality of life;
D. To ensure the provision of adequate light and air;
E. To prevent the overcrowding of land;
F. To promote historical preservation;
G. To avoid undue concentration of population;
H. To facilitate the adequate provision of transportation, water, sewerage, schools, parks and other public amenities.

FIGURE 5-18 *Land development code.*

One of the primary portions of the code governs the land-development process. These are the required steps needed for certain types of land development, and a chart is shown in figure 5-19. This chart determines who must be notified of an action. It also shows the parties that must review and act on the documents presented for the action.

The review parties are the City Administration (Admin), Board of Adjustment (BOA), Planning Commission (PC), and City Commission (CC).

Example 5.3 Determine the notification requirements and review process when you are planning on submitting a preliminary plat of a site.

According to the table in figure 5-19 for a preliminary plat submittal, all property owners within 300 feet of the site must be notified. The preliminary plat must be submitted for approval to the City Administration and the Planning Commission.

Another primary portion of the code covers the Zoning Ordinance that describes what types of buildings can be built within certain zones of the city. A portion of the zoning map containing section 28 of the Stillwater Township is shown in figure 5-20. The majority of the section is zoned Large Lot Single Family (RSL). The areas that are other than RSL are indicated on the map.

Development Services Application/Review Process							
	Notification			Review and Action			
Type of Application/Review	Property Owners within 300 feet	News paper ad	Post Sign on Site	Admin	BOA	PC	CC
Zoning Map Amendment	●	●	●	●		●	●
Text Amendment		●		●		●	●
Closing	●	●		●			●
Commercial Use By Right				●			
Correction of Errors in Plats				●		●	●
Lot Split				●			
Minor Subdivision/Commercial Minor Subdivision				●			
Preliminary Plat	●			●		●	
Final Plat				●		●	●
Site Plan: Minor Amendment				●			
Specific Use Permit	●	●		●		●	●
Planned Unit Development: Preliminary	●	●	●	●		●	●
Planned Unit Development: Final				●			
Improvement Plans				●			
Drainage Plan/Study				●			
Special Exception	●	●			●		
Variance	●	●			●		
Appeals				●	●		

FIGURE 5-19 *Development Services application/review process.*

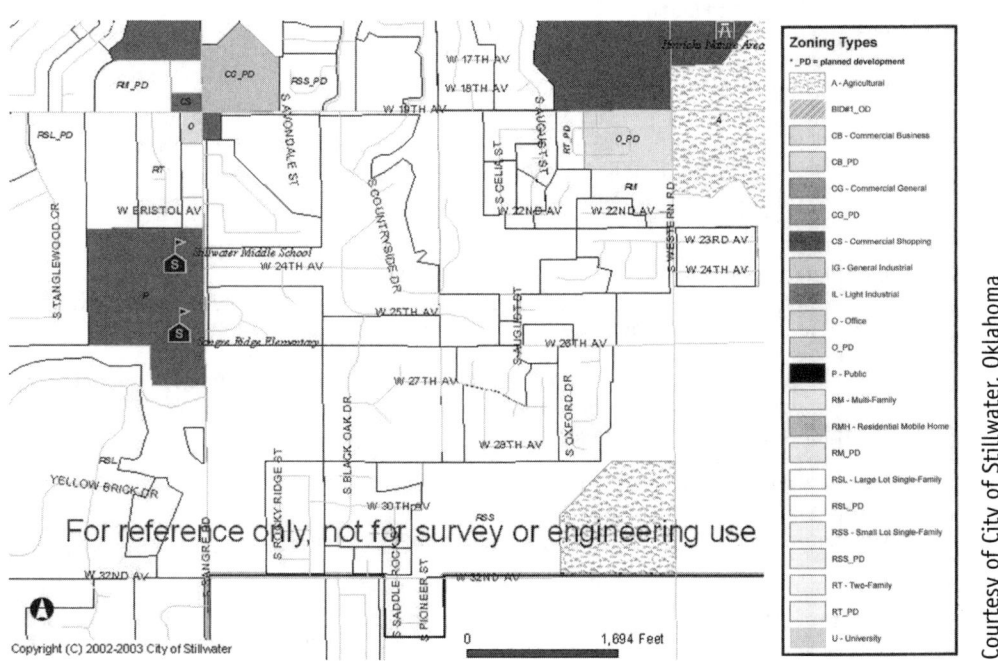

FIGURE 5-20 *Zoning map.*

The zoning ordinance will classify and regulate each of the uses of land it describes. A portion of the ordinance is shown in figure 5-21. This portion simply describes what is covered in the ordinance.

Rules and Regulations 75

Division 2. Regulations And Boundaries.

Section 23.130. Classifications and regulations.

(A) This Section classifies and regulates the use of land, buildings, and other structures as hereinafter set forth. The regulations are necessary to promote the health, safety, convenience, and welfare of the inhabitants of the City by dividing the City into districts and regulating therein the use of the land as to:

(1) Use of buildings;
(2) Size of buildings and other structures;
(3) Location of buildings and other structures;
(4) Coverage of land by buildings;
(5) Size of yards and open spaces;
(6) Density of population; and
(7) Conditions applicable to specified uses of land.

Courtesy of City of Stillwater, Oklahoma

FIGURE 5-21 *Land classification.*

Many of the site requirements are covered in the land development code, including building sizes and setbacks. Each zoning type will have specific uses and restrictions covered in the ordinance. Figures 5-22 and 5-23 show the portions of the ordinance that cover Large Lot Single Family (RSL) and Small Lot Single Family (RSS) Residential Districts, respectively.

Section 23.136. RSL (Large Lot Single Family Residential) District.

(A) *Permitted by Right*:
Conventional Single Family
Residential Design Manufactured Homes

(B) *Specific Use Permit*:
Accommodation: Bed & Breakfast only
Boarding/Rooming Houses
Child and Adult Care Services
Educational Services
Churches and Religious Institutions

(C) *Lot Size Requirements*:
(1) Minimum Lot Area: 10,000 square feet
(2) Minimum Lot Width: 75 feet
(3) Minimum Lot Depth: 125 feet

(D) *Bulk Regulations*:
(1) The maximum structure height as measured from the finished floor elevation of the first floor to the highest point of the roof: Thirty-five (35) feet
(2) Setbacks
(a) Minimum Front Yard: Twenty (20) feet from all property boundaries abutting a right-of-way or road/access easement.
(b) Minimum Side Yard (each):
(i) Residential Structures: Ten (10) feet
(ii) All other uses: Twenty (20) feet
(c) Minimum Rear Yard: Twenty-five (25) feet
(3) Maximum Lot Coverage: Thirty-five (35) percent

(E) *Exceptions*:
(1) For all structures, the minimum side yard for this district shall be increased an additional three (3) feet for each adjacent story above the first story.
(2) For development on property that abuts any local street that currently has 100 feet of right-of-way, the front and exterior side yard setbacks may be reduced by up to fifty percent (50%).
(3) Not more than one principal structure shall be located on a lot, unless specifically exempted.

Courtesy of City of Stillwater, Oklahoma

FIGURE 5-22 *RSL.*

Section 23.137. RSS (Small Lot Single Family Residential) District.

(A) *Permitted Uses*:
Conventional Single Family
Residential Design Manufactured Homes

(B) *Specific Use Permit*:

Accommodation: Bed and Breakfast only
Child and Adult Care Services
Educational Services
Townhomes
Churches and Religious Institutions

(C) *Lot Size Requirements*:
(1) Minimum Lot Area: 5,000 square feet
(2) Minimum Lot Width: 50 feet
(3) Minimum Lot Depth: 100 feet

(D) *Bulk Regulations*:
(1) The maximum structure height as measured from the finished floor elevation of the first floor to the highest point of the roof: Thirty-five (35) feet
(2) Setbacks
(a) Minimum Front Yard: Twenty (20) feet from all property boundaries abutting a right-of-way or road/access easement.
(b) Minimum Side Yard:
(i) Residential Structures: Five (5) feet
(ii) All other uses: Fifteen (15) feet
(c) Minimum Rear Yard: Twenty (20) feet
(d) Corner lots-optional setback requirements:
(i) Minimum Front Yard-Twenty (20) feet from all property boundaries abutting a right-of-way or road access easement;
(ii) Minimum Side Yard abutting a secondary right-of-way or road/access easement not opposite the Front Yard-Fifteen (15) feet;
(iii) Minimum Rear Yard-Twenty (20) feet;
(iv) Minimum Side Yard abutting another lot or parcel-Five (5) feet.
(3) Maximum Lot Coverage: Forty (40) percent

(E) *Exceptions*:
(1) All structures: the minimum side yard for this district shall be increased an additional three (3) feet for each adjacent story above the first story.
(2) Property that abuts any local street that currently has one hundred (100) feet of right-of-way: the front and exterior side yard setbacks may be reduced by up to fifty percent (50%).
(3) No interior side yard is required on one (1) side for zero lot-line dwellings. The opposite side yard for zero lot-line dwellings shall not be less than eight (8) feet.
(4) Townhomes are exempt from the lot size and lot coverage requirements; any exterior building(s) shall have a minimum side setback of eight (8) feet.
(5) Exterior buildings shall have a minimum side setback of eight (8) feet.
(6) Not more than one (1) principal structure shall be located on a lot, unless specifically exempted by this Chapter.

FIGURE 5-23 *RSS.*

In addition to the RSL and RSS, the Two Family Residential (RT) and Multi-Family (RM) Districts are for two or more families, respectively. RT districts have similar restrictions as RSS with a minimum lot size of 7,500 sq. ft., minimum lot width of 60 feet, and minimum lot depth of 100 feet. The RM district has many more uses than the other residential districts, with a minimum lot size of 7,500 sq. ft., minimum lot width of 60 feet, and minimum lot depth of 125 feet,. The maximum density for the RM district is 25 dwelling units per acre unless a special use permit is maintained, and then the maximum density is 40 units per acre.

Example 5.4 Compare the lot size requirements for RSL and RSS zoning types.

According to figure 5-22 for an RSL zone, the minimum lot size is 10,000 sq. ft., with a minimum width of 75 feet and minimum depth of 125 feet. According to figure 5-23, for an RSS zone, the minimum lot size is 5,000 sq. ft., with a minimum width of 50 feet, and minimum depth of 100 feet.

Example 5.5 Does the residence shown in figure 5-8 meet the setback requirements of RSL zoning type?

According to figure 5-22, for an RSL zone the minimum front yard setback is 20 feet, and the minimum side yard setbacks are 10 feet. The actual front and side yard setbacks shown for the residence are 40 and 20 feet, which are more than adequate.

Many other site considerations are covered in the ordinance, such as parking. Figure 5-24 shows a portion of the minimum parking requirements. The requirement for RSL, RSS, and RT zoning types are all two parking spaces per dwelling unit. In multi-family units with one, two, three, and four bedrooms the number of parking spaces is one, two, 2.33, and three per dwelling unit.

Use	Motor Vehicles	Bicycles: % of total number of required motor vehicle parking spaces
Airport	One space per 100 sq. ft. of waiting room	NA
Art Gallery/Museum	One space per 300 sq. ft.	5%
Athletic Field	One space per 5,000 sq. ft. of land area	15%
Automatic Teller Machine	Two spaces per machine	NA
Auction House	One space per four patron seats	5%
Auditorium	One space per each six seats or nine linear feet of fixed benches, or space per 45 sq. ft. of floor area without fixed seats	5%
Auto Parts Store	One space per 400 sq. ft. of leasable area, plus one space per employee on the maximum work shift	5%
Automobile Sales and Rentals	One space per 500 sq. ft. of gross floor area	5%
Automobile Repair/Automobile Body Shop	Four spaces per 1,000 sq. ft. of floor area, with repair space for motor vehicles not counted as parking space	NA
Bank	One space per 300 sq. ft. of gross floor area, plus four per teller station within the bank	10%
Bank with Drive In	Parking standard for banks (see above), plus queue capacity	10%
Bar or Tavern	One space per 100 sq. ft.	10%
Barber Shop or Beauty Parlor	Two spaces, plus one per chair	10%
Bed and Breakfast	One space per guest bedroom	5%
Billiard Hall	One space per 10 persons of the rated capacity	10%
Bingo Parlor	One space for three seats (based on design capacity) or one per 100 sq. ft. of total floor area, whichever is greater	10%
Boarding House, Dormitory, or Camp	Two parking spaces per three persons for which sleeping accommodations are provided	10%

FIGURE 5-24 *Parking code.*

Restrictive Covenants

Covenants differ from codes in that they are adopted by the planning development. This could be part of a development or a residential community. A portion of the covenant for the Quail Ridge Development is shown in figures 5-25 and 5-26.

P AND L DEVELOPMENT CO. INC.   RESTRICTIONS AND COVENANTS FOR QUAIL RIDGE DEVELOPMENT

TO

DATED: November 1st, 1965.
FILED: NOV. 1, 1965 at 1:30 p.m.

The Public.

RECORDED: Book 164 Misc. R. Page 363-366.

STATE OF OKLAHOMA   )
) SS:
COUNTY OF PAYNE   )

KNOW ALL MEN BY THESE PRESENTS:

     THAT on this 1st day of November, 1965, for the purpose of carrying out a uniform plan for the improvement and sale of QUAIL RIDGE DEVELOPMENT in Payne County, Oklahoma, being a subdivision of the Northeast Quarter (NE/4) of the Southwest Quarter (SW/4) of Section Twenty-eight (28), Township Nineteen (19) North, Range Two (2) East of the Indian Meridian, in Payne County, State of Oklahoma, according to the plat thereof, recorded on the 1st day of November, 1965 in Volume 164 of the Miscellaneous Records at Page 361, in the official records of the County Clerk of Payne County, Oklahoma. The P and L Development Co., Inc., the owner of all property located in said Quail Ridge Development, desires to restrict the use and development of the property located in Quail Ridge Development in order to insure that it will be a high class restricted residential property development.

FIGURE 5-25 *Quail Ridge covenants.*

     3. No building shall be located on any lot or tract nearer than forty (40) feet to the front lot line. No building shall be located on any lot nearer than thirty (30) feet to the side street line as said street line is located on said plat. No building shall be located nearer than twenty (20) feet to interior lot line. The sum of side yards and the distance between buildings shall be a minimum of forty (40) feet. No detached garage or other outbuildings shall be located nearer than twenty (20) feet to any side lot line.

     4. For the purpose of this covenant, eaves, steps and open porches shall not be considered as a part of a building, provided, however that this shall not be construed to permit any portion of a building on a lot to encroach upon another lot.

     5. No lot or tract shall be subdivided into smaller building lots or sites.

     6. No business, trade or activity shall be carrier on upon any residential lot or tract, no noxious or offensive activity shall be carried on upon any residential lot or tract, nor shall anything be done thereon which may be or may become an annoyance or nuisance to the neighborhood.

     7. No structure of a temporary character, trailer, basement tent, shack, garage, barn or any other outbuilding shall be used on any lot at any time at a residence either temporarily or permanently.

     8. The ground floor square foot area of any dwelling, exclusive of garage and other outbuildings, shall not be less than 1600 square feet in the case of a one story structure, and the whole house square foot area not less than 2200 square feet in the case of a one and one-half or two story structure. Each structure in the main shall be constructed of brick or stone.

FIGURE 5-26 *Quail Ridge covenants (continued).*

It should be noted that the building setback line prescribed in section 3 of the covenants in figure 5-26 are 40 feet in the front yard and 20 feet in the side yards. This is consistent with the residential site shown in figure 5-8. The covenant is more restrictive than the land development code in this and most cases.

Problem Set 5.3   Rules and Regulations

Problem 5-19   What are the objectives of the Stillwater Land Development Code?

Problem 5-20   Determine the notification requirements and review process, if you are planning on submitting a planned unit development preliminary plat.

Problem 5-21   What zoning type do public schools and parks fall under?

Problem 5-22   What are three of the regulations that are placed on buildings under a zoning ordinance?

Problem 5-23   Can a church be built within a Large Single Family Residential District?

Problem 5-24   Does the plat in figure 5-12 show all of the required RSS district building setbacks for Lot Eighteen (18) in Block Six (6)?

Problem 5-25   Could all the lot sizes in the plat in figure 5-12 be used for an RSL district?

Problem 5-26   What is the maximum roof height allowed in the RSL district?

Problem 5-27   Could an apartment complex with 100 units be built on a site zoned RM with 3.5 acres of land?

Problem 5-28   How many parking spaces would be required for the apartment complex in problem 5-26 if 50 of the units were one bedroom, and the other 50 were two bedroom units?

Problem 5-29   Draw the building setback lines on Lot Six (6) Block Three (3) in figure 5-2 according to the covenant in figure 5-26.

Problem 5-30   Could a two-story residence that is 2,000 sq. ft. with cedar siding be constructed in the Quail Ridge Development according to the covenants?

BACKGROUND

Many additional requirements of land use are controlled by the natural surroundings and the local context. Environmental regulations are developed to protect the natural surroundings. Special zoning and municipal regulations may protect sites and buildings for historical or cultural reasons.

Environmental Regulations

Many environmental regulations may apply to a site. Architects and engineers must understand how to find the environmental information that controls the development of land. Environmental issues cannot be ignored.

Environmental Impact

When land is developed, the impact on the environment must be considered. Using the National Environmental Policy Act (NEPA), a project may require an Environmental Assessment (EA) or an Environmental Impact Statement (EIS). The EA is an assessment of the possible impact of a project on the environment and is used to determine whether an EIS is required. An EIS is required to describe the following:

- The environmental impacts of the proposed action
- Any adverse environmental impacts that cannot be avoided should the proposal be implemented
- The reasonable alternatives to the proposed action
- The relationship between local short-term uses of man's environment and the maintenance and enhancement of long-term productivity

- Any irreversible and irretrievable commitments of resources that would be involved in the proposed action should it be implemented

Figure 5-25 in the textbook can be used as a guide through the NEPA process.

Floodplains

The National Flood Insurance Program (NFIP) under the Federal Emergency Management Agency (FEMA) creates Flood Insurance Rate Maps (FIRMs). The maps can be found online at http://www.fema.gov/hazard/map/firm.shtm or the link given in the textbook. A Special Flood Hazard Area (SFHA) is an area that has a 1 percent chance of being flooded in any given year (100-year floodplain). Over a 30-year period, the life of most mortgages, there is at least a 26 percent chance that this area will be flooded. Zones that begin with the letters V and A are in the SFHA. Zones B, C, D, or X are within the floodplain but not in the SFHA and, therefore, are not considered to be areas requiring flood insurance for structures located in those areas. The following are the SFHA zones:

- A - No Base Flood Elevations determined
- AE - Base Flood Elevations determined
- VE - Coastal flood zone with velocity hazard (wave action); Base Flood Elevations determined
- X - Areas determined to be outside the 0.2 percent annual chance floodplain
- D - Areas in which flood hazards are undetermined, but possible

Figure 5-27 shows the FIRM map containing Section 28 of the Stillwater Township. The hatched areas are in zone AE, and the crosshatched area is zone A. The building on the property in figure 5-8 is approximately 3,950 feet from the floodplain to the northeast.

Of section 28

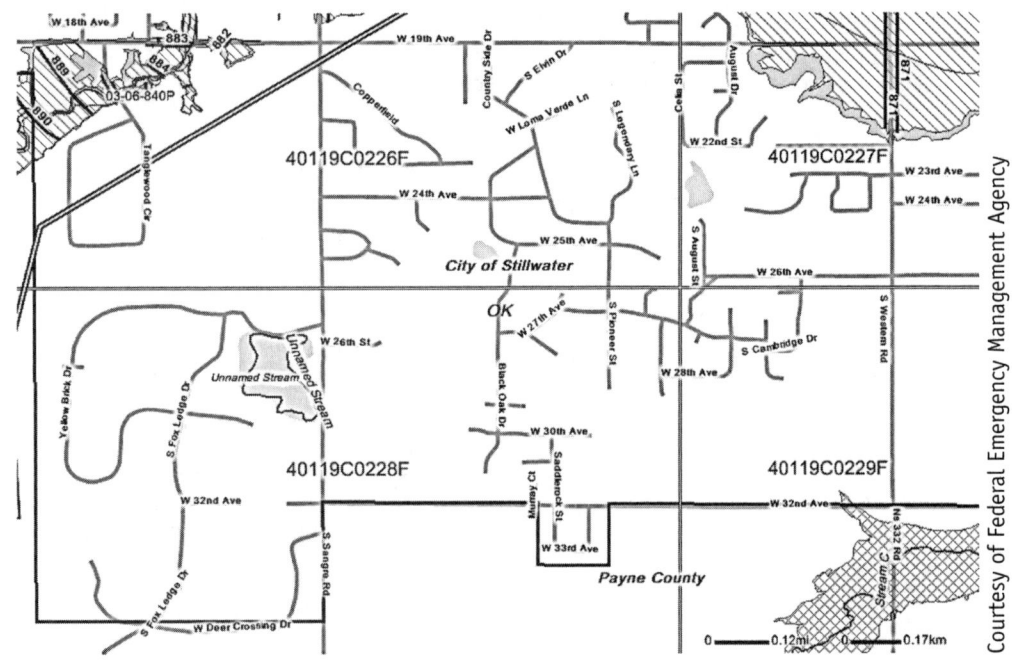

FIGURE 5-27 *FIRM map.*

The Land Development Code for Stillwater, Oklahoma, also contains a section on Flood Hazard Regulation.

Wetlands

The wetlands are protected by the Clean Water Act, and compensation must be provided if it is impacted. The U.S. Fish and Wildlife Services' Wetlands Inventory Map containing Section 28 of the Stillwater Township is shown in figure 5-28. There are numerous freshwater ponds and a few areas containing freshwater forested/shrub, including the Stillwater Creek area to the northeast.

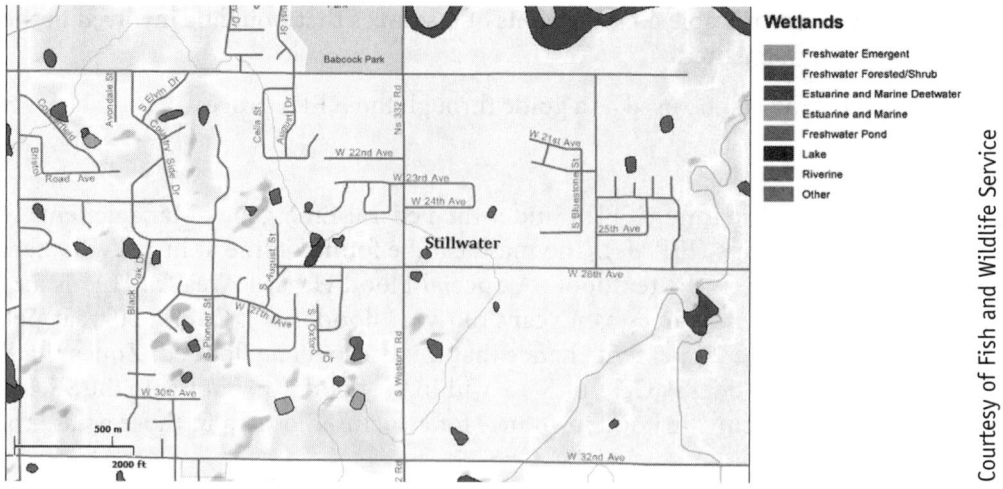

FIGURE 5-28 *Wetlands inventory map of section 28.*

Storm Water Permits

Land development must not adversely affect the flow of storm water onto adjacent properties and into waterways. Most municipalities have Storm Water Management Regulations. Figure 5-29 shows the required submittals for the various land development actions.

Section 23.386. Drainage studies and drainage plans.

(A) Required Submittals: Drainage studies and drainage plans are required to demonstrate compliance with Stillwater City Code, City Drainage Standards, City of Stillwater Standards. The type of land development application dictates the required submittal(s) as shown in the table below. All submittals shall be signed and sealed by a professional engineer licensed by the State of Oklahoma.

Required Submittal	Application Type
Preliminary Drainage Study	Preliminary Plat, Preliminary Planned Unit Development
Final Drainage Study	Final Plat, Final Planned Unit Development, Commercial Use By Right, Earth Change Permit
Drainage Plan	Final Plat, Final Planned Unit Development, Commercial Use By Right, Earth Change Permit
Record Drawings	Upon completion of the drainage improvements and prior to issuance of certificate of occupancy and use or occupancy of a site, development, or other improvement

FIGURE 5-29 *Drainage plans.*

In addition, the Clean Water Act requires a Storm Water Pollution Prevention Plan (SWPPP) as described in the textbook.

Coastal Zones

Coastal areas require additional permits similar to wetlands. The U.S. Fish and Wildlife Services' Wetlands Inventory Map can be used for information on coastal areas.

Endangered Species

The U.S. Fish and Wildlife Service (FWS) maintain a list of the endangered and threatened species and maps of their critical habitats as indicated in the textbook. The Endangered Species Act (ESA) requires that you apply for an Incidental Take Permit if a development poses any risk to an endangered or threatened species.

Tree Protection

Tree ordinances are a normal part of the land development codes. Most codes require replacement of trees that are removed by land development. In addition, many municipalities belong to the Tree City USA program developed by the Arbor Day Foundation. To be a member, the following four standards must be in place:

- A Tree Board or Department
- A Tree Care Ordinance
- A Community Forestry Program with an Annual Budget of at Least $2 Per Capita
- An Arbor Day Observance and Proclamation

Contamination and Containment

Land development and building must not contaminate the environment. The Environmental Protection Agency (EPA) has the power to require a contaminated property to be cleaned up through the Superfund Program.

Brownfields

Environmentally contaminated sites are called "brownfields." In 1996, the EPA developed the Brownfields Program to award grants for cleanup and development of these sites.

Historical and Cultural Resources

Many sites and buildings are protected under the National Historic Preservation Act and may be in the National Register of Historic Places. In addition, a site may be regulated by the local municipality to preserve the context of the surrounding buildings.

Problem Set 5.4   Environmental Regulations

Problem 5-31	What are the two possible actions required for a project by the National Environmental Policy Act?
Problem 5-32	Which of the two actions in problem 5-31 should be performed first?
Problem 5-33	What is the approximate floodplain in the northeast of figure 5-27?
Problem 5-34	What is the probability that a Special Flood Hazard Area will experience a 100-year flood over a 30-year period?
Problem 5-35	What is the primary difference between an "A" zone and a "V" zone under the SFHA?
Problem 5-36	What type of submittal is required for a Preliminary Plat under the Stillwater Land Development code?
Problem 5-37	What are the two levels of risk under the Endangered Species Act?
Problem 5-38	Describe the three sizes of trees protected by the Mt. Pleasant Code of Ordinance in the textbook.
Problem 5-39	What are the standards that must be in place to become a member of Tree City USA?
Problem 5-40	What federal program awards grants for the cleanup and development of contaminated sites?

BACKGROUND

The following are some things to consider in land development and building design.

Other Site Considerations

Architects and engineers should be aware of any physical considerations that affect how land is used. The following are a few more that should be considered.

Climate

The climate can affect the actual materials and construction methods used for a building and how it is situated on the site. Although direct measures, such as temperature and precipitation, are important to consider, more useful quantities are heating and cooling degree days. A degree day is the difference between the outdoor mean temperature over a 24-hour period and a given temperature. Heating degree days (HDD) are a measure of how much (in degrees), and for how long (in days), the outside air temperature was lower than the specific base temperature. They are used for calculations relating to the energy consumption required to heat buildings. Cooling degree days (CDD) are a measure of how much (in degrees), and for how long (in days), the outside air temperature was higher than the specific base temperature. They are used for calculations relating to the energy consumption required to cool buildings. Heating and cooling of buildings will be covered in chapter 17. Generally heating degree days use a mean temperature of 65°F (HDD65), and cooling degree days 50°F (CDD50). The annual degree days are taken as the sum of each day for the entire calendar year. Simple equations to find monthly HDD65 and HDD50 are as follows:

$$HDD65 = (65 - \text{average monthly temperature}) \times \text{number of days in the month}$$

$$CDD50 = (\text{average monthly temperature} - 50) \times \text{number of days in the month}$$

Example 5.6 Table 5-1 contains the average temperatures for Stillwater, Oklahoma, during 2009. Calculate the values of HDD65 for the month of November and CDD50 for the month of June. Remember that both of these months contain 30 days.

JAN	FEB	MAR	APR	MAY	JUN	JUL	AUG	SEP	OCT	NOV	DEC
34.79	46.95	52.86	59.30	66.59	80.96	80.93	77.64	69.29	54.34	52.76	33.65

TABLE 5-1 *Daily average temperature for Stillwater, Oklahoma, in 2009.*

$$HDD65 = (65 - 52.76) \times 30 = 367.2$$

$$CDD50 = (80.96 - 50) \times 30 = 928.8$$

The value of HDD65 for the month of November is 367.2°F and CDD50 for the month of June is 928.8°F.

The American Society of Heating, Refrigerating, and Air-Conditioning Engineers (ASHRAE) Standard lists the 30-year averages for numerous cities in the United States and Canada and throughout the world. For example, Stillwater, Oklahoma, has HDD65 of 4,028 and CDD50 of 4,718. This would indicate a relative balance of heating and cooling required. More detailed degree days may be calculated, such as on a monthly basis to determine loads for a given month. Table 5-2 is a five-year (2005 to 2009) average taken at the Stillwater, Oklahoma, airport using the Bizee Degree Days program at www.degreedays.net.

Month	HDD65	% Est.	CDD50	% Est.
Jan	813	2	65	2
Feb	644	0	80	0
Mar	405	1	239	1
Apr	227	10	380	10
May	79	0	609	0
Jun	8	0	861	0
Jul	4	0	996	0
Aug	4	1	1002	1
Sep	47	0	686	0
Oct	242	0	378	0
Nov	430	0	196	0
Dec	839	1	50	1
Total	3742	1	5542	1

TABLE 5-2 *Degree days for Stillwater, Oklahoma.*

Note that the values for April have a 10 percent est. This may be due to missing or erroneous data. It may also be due to large deviations in the data over the average period used. The values for degree days can be used in selecting materials and systems for heating and cooling buildings. The National Oceanic and Atmospheric Administration (NOAA) summarizes weather data and figures 5-30 and 5-31 are HDD65 and CDD50 maps of the United States.

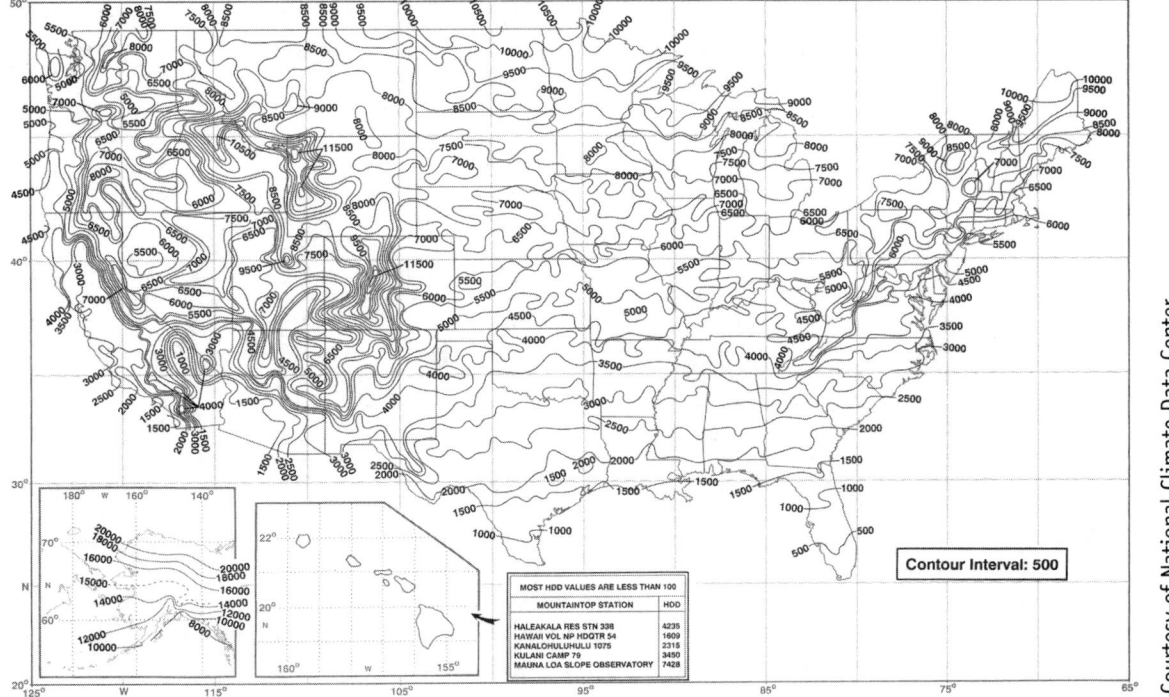

FIGURE 5-30 *Annual heating degree days.*

Other Site Considerations

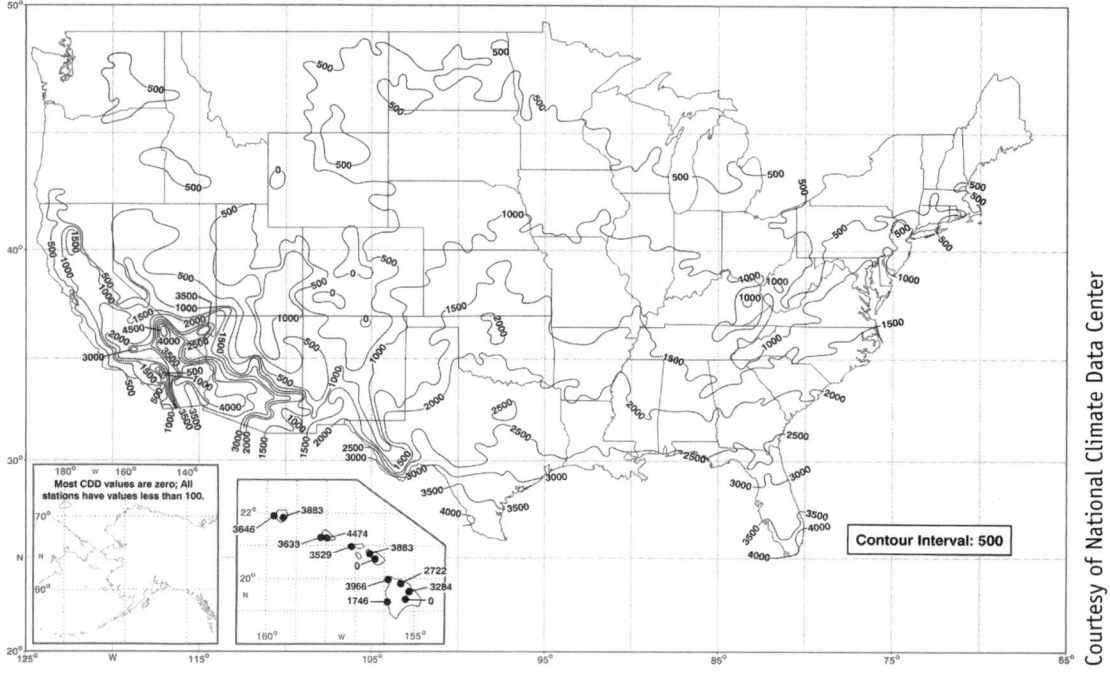

FIGURE 5-31 *Annual cooling degree days.*

Another very important climate consideration is the sun angle. The altitude and the azimuth location of the sun can be found on the U.S. Naval Observatory website www.usno.navy.mil/USNO/astronomical-applications/data-services/alt-az-us. Refer to figure 5-45 in the textbook for the definition of these sun angles. The sun angles for Stillwater, Oklahoma, on June 21, 2010 are given in table 5-3.

Time hr:min	Altitude degrees	Azimuth (E of N) degrees	Time hr:min	Altitude degrees	Azimuth (E of N) degrees
05:00	−2.8	58.1	13:00	75.8	209.0
06:00	8.0	66.7	14:00	66.8	243.0
07:00	19.4	74.6	15:00	55.4	259.1
08:00	31.2	82.3	16:00	43.3	269.4
09:00	43.3	90.6	17:00	31.3	277.7
10:00	55.3	100.9	18:00	19.4	285.4
11:00	66.8	116.9	19:00	8.1	293.3
12:00	75.8	150.8	20:00	−2.8	301.9

TABLE 5-3 *Sun angles for Stillwater, Oklahoma, on June 21, 2010.*

The sun angles can be used to determine overhangs and other shading devices on buildings. They can also be used to provide for optimum positioning of a building on a site.

Microclimate

The primary features that produce microclimates on sites are topography, water, and vegetation. These may be used to provide for positive effect on a building.

Utilities and Municipal Services

Utilities must be located on a site to best determine how to service a building. Figure 5-32 shows the utilities in a portion of the Quail Ridge Development in Stillwater, Oklahoma. All of the utilities in this development are underground and include, water, sewer, and power.

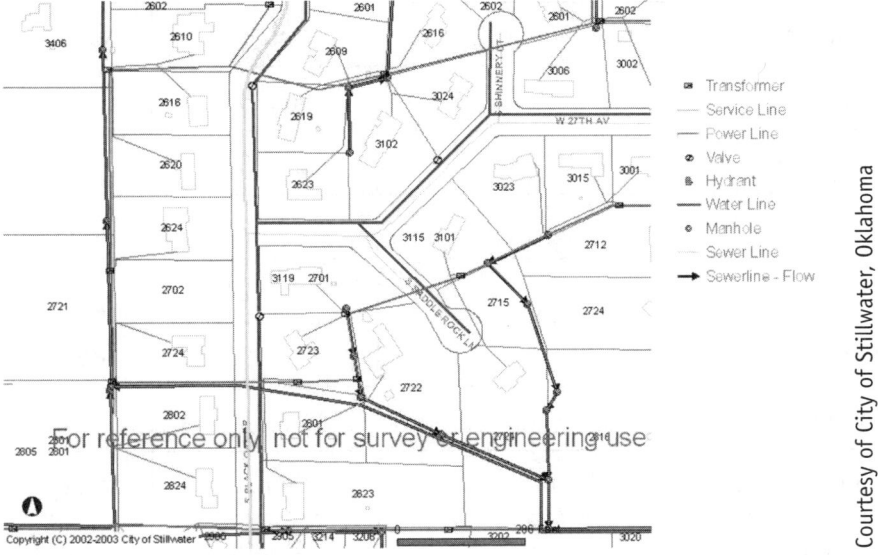

FIGURE 5-32 *Stillwater utilities with GIS.*

Traffic Flow

Traffic count and how traffic flow can be used to provide for proper sizing and configuration of roads. Figure 5-33 shows the traffic counts in the northern portion of Section 28 of the Stillwater Township. Black Oak Dr. has a daily traffic count of 699.

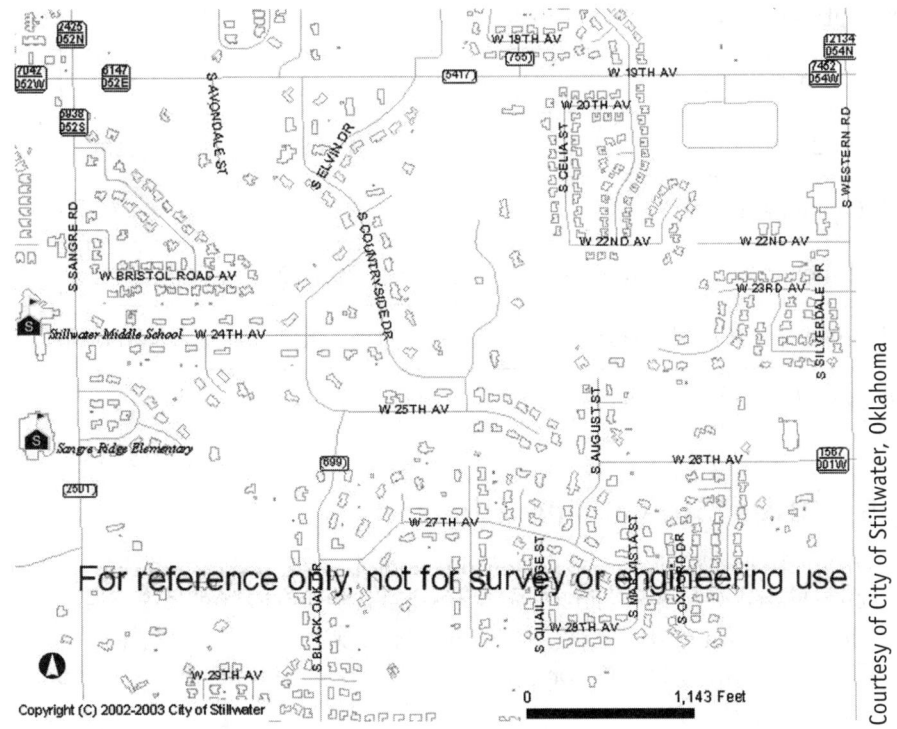

FIGURE 5-33 *Stillwater traffic count with GIS.*

Other Site Considerations

Demographics and Target Market

The demographic of an area like a municipality can influence how successful a given project might be on a site. Many businesses have specific demographics that they look for before they build a project. Table 5-4 shows some of the 2006–2008 census estimate for Stillwater, Oklahoma.

General Characteristics	Number	Percent	U.S.
Total population	42,253		
Male	21,282	51.6	49.3%
Female	20,471	48.4	50.7%
Median age (years)	24.3		36.7%
18 years and over	36,195	85.7	75.5%
65 years and over	3,642	8.6	12.6%
Housing Characteristics	Number	Percent	U.S.
Total housing units	18,205		
Occupied housing units	15,029	82.6	88.0%
Owner-occupied housing units	6.593	43.9	67.1%
Renter-occupied housing units	8,436	56.1	32.9%
Social Characteristics	Number	Percent	U.S.
Population 25 years and over	20,022		
High school graduate or higher		92.3	84.5%
Bachelor's degree or higher		47.0	27.4%
Economic Characteristics	Number	Percent	U.S.
In labor force (16 years and over)	23,646	64.3	65.2%
Median household income in 2008 (dollars)	29,210		52,175
Median family income in 2008 (dollars)	57,128		53,211

TABLE 5-4 *Stillwater, Oklahoma, 2000 census data.*

Problem Set 5.5 Other Site Considerations

Problem 5-41 In which direction do the prevailing winds blow in Oklahoma City, Oklahoma, during the month of May? (Hint: Use the website given in the textbook.)

Problem 5-42 Using the data from table 5-1, calculate the values of HDD65 for the month of December and CDD50 for the month of July. Remember that both of these months contain 31 days.

Problem 5-43 Compare and contrast the HDD65 for the month of November and CDD50 for the month of June from example 5-6 with the values given in table 5-2.

Chapter 5: Site Discovery for Viability Analysis

Problem 5-44 Compare and contrast the average yearly values for HDD65 and CDD50 from ASHRAE with the values given in table 5-2.

Problem 5-45 Compare and contrast the average yearly values for HDD65 and CDD50 from ASHRAE with the values given in figures 5-30 and 5-31.

Problem 5-46 Draw the sun angles for 12:00 p.m. on June 21, 2010 for Stillwater, Oklahoma, using the data in table 5-3.

Problem 5-47 Describe the traffic count at the intersection of W. 19th Avenue and S. Sangre Road given in figure 5-33.

Problem 5-48 Would you consider Stillwater, Oklahoma, a good town to build a trendy restaurant with medium prices that caters to younger patrons? Why or why not?

BACKGROUND

Research is always important in any design project, but hands-on experience can make the research much more useful and informative.

Site Visit

Architects and engineers almost always visit a site to gather information and supplement research. In addition, other technical professionals should perform site inspections, such as soils investigation and surveying. A photographic survey is a valuable resource to understand the surrounding context of a site. The textbook gives a site inspection checklist to be used on a site visit.

Problem Set 5.6   Site Visit

Problem 5-49 What types of maps would be useful when performing a site inspection?

Problem 5-50 What are some of the important features that should be photographed during a site visit?

Problem 5-51 What are some potential site problems to look for during a site visit?

Problem 5-52 What are some potential environmental problems to look for during a site visit?

Problem 5-53 What are some potential traffic problems to look for during a site visit?

Problem 5-54 What type of site test should be performed if a sanitary sewer system is not available on a site?

BACKGROUND

All the research done on a site would be worthless if a project were not economically done. With the changing markets and construction costs, one of the most important steps in project feasibility is a cost estimate.

Construction Cost Estimate

Architects and engineers must be able to perform construction cost estimates at various stages of the design process. Estimates are normally done during the project-planning stage to determine economic feasibility. This is normally performed based on gross assumptions and national averages. The R.S. Means Quick Cost Estimator is a valuable tool for this type of estimate. Once the design of a project is underway, a more detailed estimate should be performed to validate estimated cost. Computer programs and published R.S. Means Cost Data are normally used for these types of estimates. Finally, when a project has been designed, the contractors will make very detailed estimates based on local costs to provide for bidding to be performed on the project. Figure 5-34 is a quick cost estimate for an apartment project in Tulsa, Oklahoma. The medium price of this project is $98.40 per sq. ft.

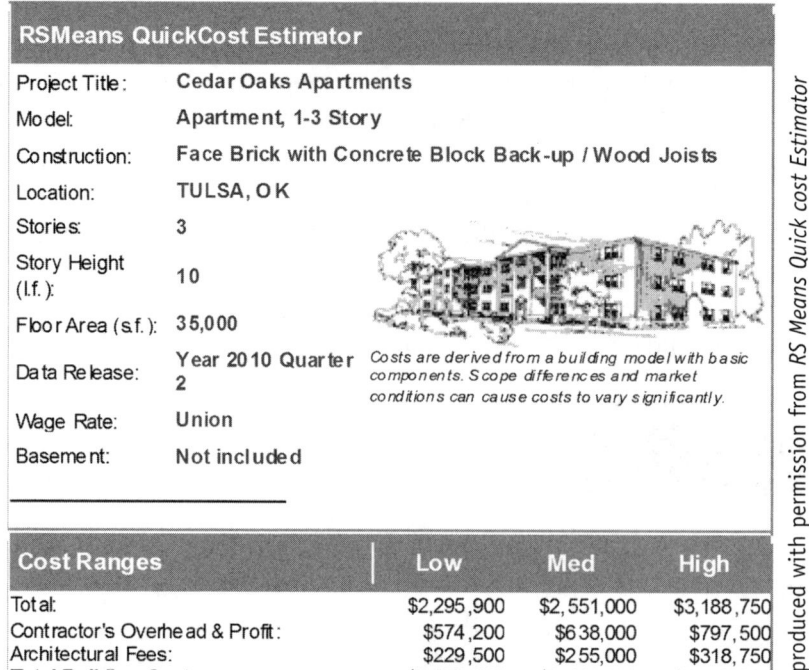

FIGURE 5-34 *RS Means QUICK COST Estimator.*

Problem Set 5.7   Construction Cost Estimate

For each of the following problems, use the R.S. Means Quick Cost Estimator located online at http://www.rsmeans.com/calculator/index.asp.

Problem 5-55   What is the medium cost estimate for a 1–3 story apartment complex with approximately 40,000 sq. ft. in Tulsa, Oklahoma?

Problem 5-56   What is the medium cost estimate for a 1–3 story apartment complex with approximately 50,000 sq. ft. in Tulsa, Oklahoma?

Problem 5-57   Compare and contrast the estimates in problems 5-55 and 5-56 in terms of cost per sq. ft.

Problem 5-58   What is the medium cost estimate for a 1–3 story apartment complex with approximately 40,000 sq. ft. in Tulsa, Oklahoma?

Problem 5-59   What is the medium cost estimate for a 1–3 story apartment complex with approximately 50,000 sq. ft. in Chicago, Illinois?

Problem 5-60   What is the national average medium cost estimate for a 1–3 story apartment complex with approximately 50,000 sq. ft.?

Problem 5-61   Compare and contrast the estimates in problems 5-55, 5-59, and 5-60 in terms of cost per sq. ft.

BACKGROUND

Whether a proposed project can be developed depends on all of the factors that have been discussed in this chapter, as well as others. All of these must be used to determine whether the project is viable.

Viability Analysis

Architects and engineers should consider all reasonable limitations for a project. A viability analysis should be included if a project is legally permissible, physically possible, and financially feasible.

Problem Set 5.8   Viability Analysis

Problem 5-62   What are some of the legal considerations for a proposed project?

Problem 5-63   What are some of the physical considerations for a proposed project?

BACKGROUND

When time and budget permit, a site can be investigated for a number of possible uses. This can lead to an optimum use based on the client's needs.

Best and Highest Use

Architects and engineers should be able to compare and contrast various options on a given project or site. This could include changing various parameters for a given project type or exploring different project types on a site.

Problem Set 5.9   Best and Highest Use

Problem 5-64   Based on the demographics of Stillwater, Oklahoma, what type of dwellings would probably lead to the best investment for a developer?

Problem 5-65   Explore various options for an office development in Oklahoma City, Oklahoma, of approximately 10,000 sq. ft.

CHAPTER 6
Site Planning

Skills List

After completing the problems in this chapter, you should be able to:

- Identify and understand the requirements of project programming
- Identify and understand the different types of site surveys
- Produce conventional survey calculations
- Read a topographic survey and corresponding symbols
- Know how to classify various soil types
- Determine the grain size for a soil sample
- Identify of elements of a soils report
- Be aware of site restrictions and orientations

BACKGROUND

Architects and engineers must be able to assimilate project research and produce useful information to aide in the design. The various aspects of the site need to be analyzed and summarized to make best use of a site. Concepts for site design should be evaluated against the site analysis to determine which concepts will be most successful.

Problem Identification – Programming

The success of a project is normally based on the amount of planning. Many have said that "prior, proper, planning prevents poor performance." Good site design requires careful planning and many considerations. The design must meet the needs of the client and all the rules and regulations required, as well as the constraints and limitation of the design. The viability analysis discussed in chapter 5 will be expanded for site-planning purposes. The research for a project is organized and summarized in a project program. The program for a site normally contains the following:

- Goals and objectives for the project
- Desirable or undesirable site orientations
- Parking required by ordinances and proper layout
- Site access for vehicles and pedestrians
- Existing utilities and services
- Site restrictions and building setbacks
- Site regulations and covenants
- Site topography
- Soil characteristics and foundation recommendations

Building programs will contain more information and are covered in chapter 3. The site analysis is a part of the building program. The table of contents for the Range Elementary School program is shown in figure 6-1.

FIGURE 6-1 *Range Elementary School building program.*

Needs Analysis

The primary measure of success on a project is whether the user's needs have been met. The client should be interviewed or surveyed to provide as much information as possible to the designers. Input should be solicited from all the stakeholders and end-users, including surrounding property owners and municipal administration (see figure 5-19). The project needs should be summarized in a concise statement similar to the one shown in figure 6-2.

The summary should be supported by simple, easy-to-read graphics that properly communicate the information. Figure 6-3 is a map of the elementary school districts before and after the proposed new school.

Critical data should also be summarized in tables that are easy to understand. Figure 6-4 is a summary of the existing schools showing the disparity in population and demographic of the students.

Adjacency Matrix

In addition to the adjacency matrix, as shown in figure 6-4 of the textbook, a bubble diagram should be drawn as the next step in the programming and design process. The bubble diagram should show the general areas to

Analysis

The Stillwater elementary school system is expanding and the current school facilities are being stressed. The proposed new Range Elementary School will be needed to relieve the crowding in the current schools and help buffer the county against future growth.

FIGURE 6-2 *Range Elementary School building program.*

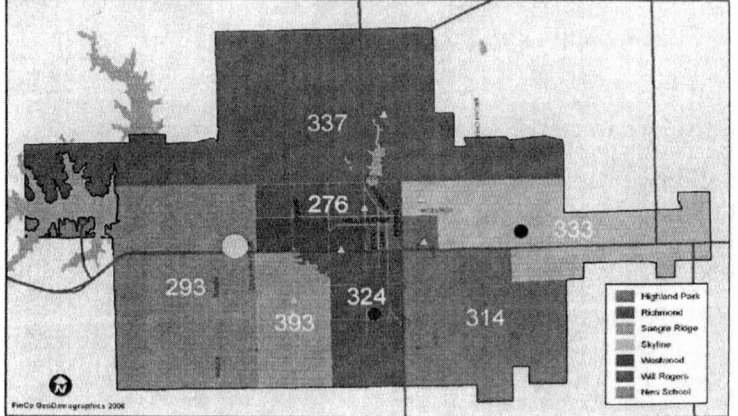

FIGURE 6-3 *School district needs.*

Schools	# of Students	Students per Teacher	% Free & Reduced-price Lunch	% OCCT Test Scores (Math ; Reading) *
Sangre Ridge	599	18	12%	90% ; 94%
Westwood	417	14	46%	79% ; 90%
Will Rogers	430	14	57%	84% ; 86%
Highland Park	382	15	71%	74% ; 86%
Skyline	439	12	49%	67% ; 81%
Richmond	426	15	30%	93% ; 94%

*Oklahoma Core Curriculum Test

FIGURE 6-4 *School district data.*

approximate scale and how they relate to the other areas. Figure 6-5 is a summary of the needs for the overall areas of the elementary school, and figure 6-6 is a bubble diagram derived from those needs. The long dashes indicate semi-private areas, and the short dashes indicate private areas.

Zone	Total SF	Description
Parking	30,000 +	Located in front of the main entry of the school, parking will provide spaces for staff and visitors, and serve as the transition space to the administrative area.
Administration	3,590	All administrative offices and faculty areas will be in this zone, such as the reception area, work room, teacher's lounge, and conference rooms.
Education	36,270	This zone will cover spaces for every general and specialty classroom, as well as the computer labs, music room, and art room. Direct relationships to the gymnasium, library, cafeteria, and recreational zones are desired.
Library	2,925	The library includes spaces for stacks, a circulation desk, storage areas, computer stations, a group/teaching area, and a librarian's office. It is vital that this zone should be highly accessible to all education spaces.
Gymnasium	6,600	A multi-purpose space which will need to directly relate to the cafeteria, education, and recreational zones.
Cafeteria	3,680	Service and classroom access will be the most important relationships to the cafeteria.
Recreation	20,000 +	This zone includes the playgrounds, playing fields, and outdoor classroom. Therefore, it should relate to the gym, classrooms, and cafeteria.
Service	7,460	Direct access to the cafeteria and gymnasium is highly recommended.

FIGURE 6-5 *Program areas.*

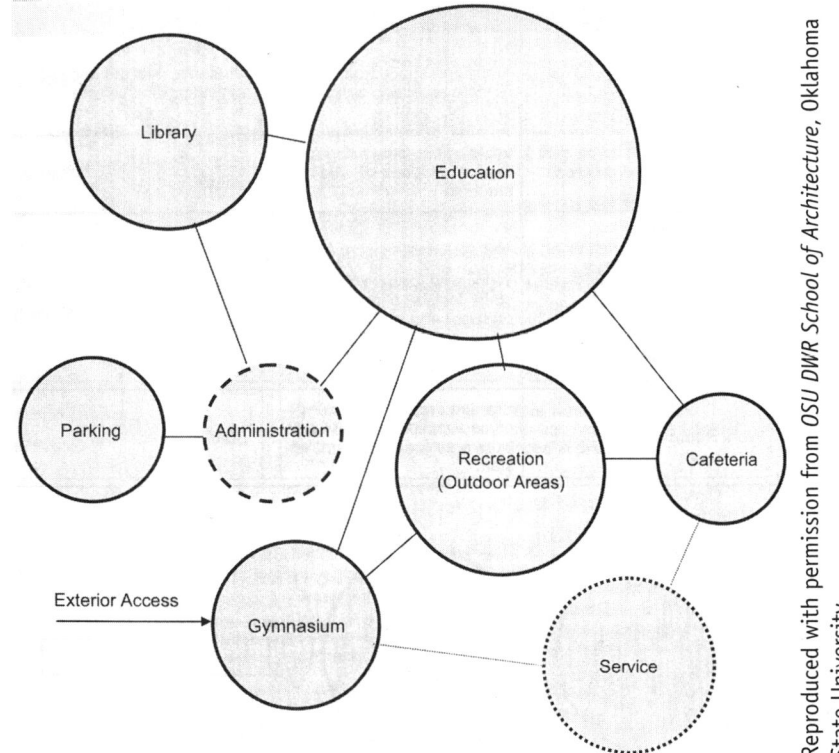

FIGURE 6-6 *Bubble diagram.*

Problem Set 6.1 Problem Identification – Programming

Problem 6-1   What is the name given to all of the people collectively involved in a project?
Problem 6-2   What document is used to summarize and present all the project research?
Problem 6-3   What are some of the site requirements included in the project programming?
Problem 6-4   Who is the most important person in developing a project needs analysis?
Problem 6-5   Create an adjacency matrix using the areas in figure 6-5. Use a key similar to the one shown in figure 6-4 of the textbook.
Problem 6-6   Create a bubble diagram from the adjacency matrix in figure 6-4 of the textbook.

BACKGROUND

It is critical to have a thorough understanding of the site and the surrounding. This will lead to the most effective and proper use of the sites and to the design of the project. Research of a site should include quantitative, qualitative, and contextual information.

Site Characteristics

Architects and engineers should be able to collect various forms of site information used to design and develop a project. This includes the regulations and requirements of the municipality, such as, zoning, as shown in figure 6-7.

The information should also include the general context of the site and its surroundings. Figure 6-8 shows an aerial photograph for an elementary school site. This view can give some insight for the site characteristics and any further investigation needed.

FIGURE 6-7 *Zoning map.*

Chapter 6: Site Planning

FIGURE 6-8 *Aerial photograph.*

Land Surveys

There are many types of land surveys. Some of these were used in chapter 5. Below is a list of some types and examples of surveys.

- Control survey to establish reference locations like figures 5-5 (meridians), 5-6 (townships) and 5-7 (sections)
- Construction survey used to stake out buildings and bridges
- Topographic survey used to show elevations of the ground and other features like figures 5-13, 5-14, and 5-15
- Property survey similar to figure 5-8 and plats like figures 5-2 and 5-11
- Hydrographic survey for mapping water features like figure 5-28
- Route surveys used to map roads

Horizontal and Vertical Datum

The National Spatial Reference System (NSRS) establishes horizontal and vertical locations to be used for surveys. The horizontal location can be the latitudes or longitudes, but more commonly used are the meridians and baseline as discussed in chapter 5. The Indian Meridian used in Oklahoma lies 12 miles east of the Stillwater township and is marked with the sign shown in figure 6-9 on the south side of State Highway 51.

FIGURE 6-9 *Indian Meridian sign.*

Site Characteristics

FIGURE 6-10 *Benchmark photographs.*

Vertical locations are marked with benchmarks that come in many forms, as discussed in the textbook. A couple of examples are shown in figure 6-10 that are located in the Stillwater Township.

You can find information on many of the 1.5 million benchmarks online at the Official Global GPS Cache Hunt Site www.geocaching.com/.

Conventional Surveys

A conventional survey is typically done with an auto level or theodolite to measure vertical heights and horizontal angles. A tape measure or auto level is used to measure horizontal distances.

Vertical Control – Differential Leveling

The vertical position of a location can be found by differential leveling using an auto level and leveling rod. A portion of a leveling rod is shown in figure 6-11. The larger numbers are normally red and indicate feet. The top of each tick mark on the rod indicates even hundredths of a foot, and the bottom of the mark indicates odd hundredth of a foot.

The heights with even hundredth indicated in figure 6-11 are 6.00 and 5.60 feet. Note that the larger 6 on the rod is the foot mark and that the smaller 6 is tenth-of-a-foot mark. The odd readings on the rod indicated are 5.85 and 4.99 feet. The differential leveling method is shown in figure 6-12. The steps of finding vertical location follow and table 6-1 shows a summary of the calculations.

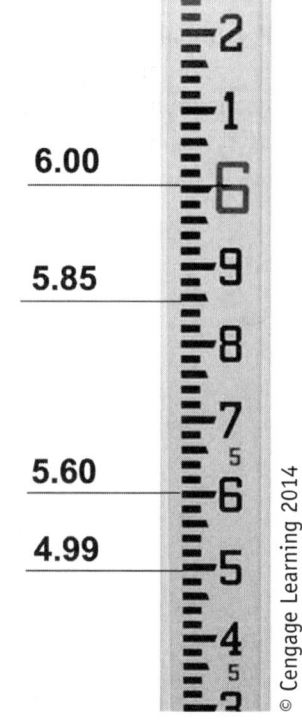

FIGURE 6-11 *Leveling rod example.*

98 Chapter 6: Site Planning

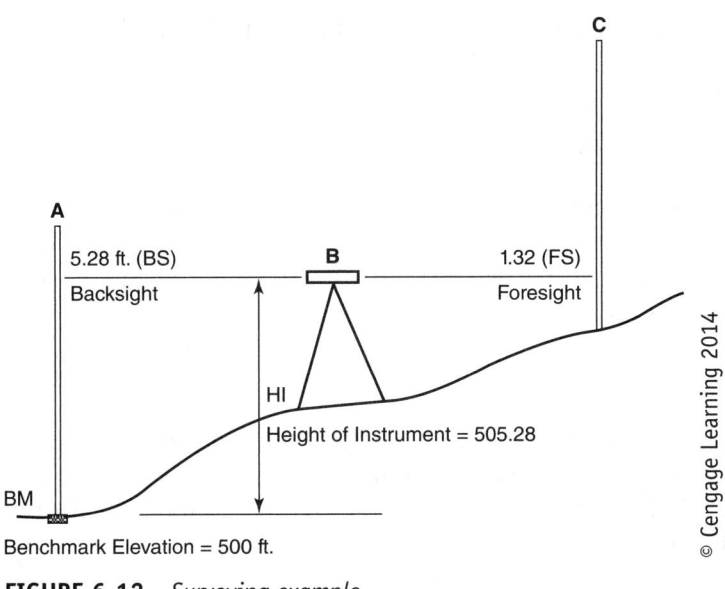

FIGURE 6-12 *Surveying example.*

A benchmark must be used to determine the height of the instrument. The benchmark in this case is at point A and indicated as BM with an elevation of 500 ft. This is recorded on the first line of the table as a known elevation. The auto level is placed at point B, and a reading of the leveling rod placed at point A is taken as 5.28. This is known as the *backsight reading* and is recorded in the first line of the table. The height of the instrument can now be determined and recorded by taking the sum of the benchmark and the backsight reading.

$$HI = BM + BS$$

$$HI = 500.00 + 5.28 = 505.28 \, ft$$

Any other point may now be found by taking another reading. In this case, a reading at point C is taken as 1.32. This is known as *foresight reading*. The reading is recorded in a new line of the table. The elevation of the new point is the height of the instrument minus the foresight reading.

$$Elev = HI - FS$$

$$HI = 505.28 - 1.32 = 503.96 \, ft$$

Point	Backsight	Height of Instrument	Foresight	Elevation	Angle or Distance
PT	BS (+)	HI	FS(-)	Elev.	US/LS
B	5.28	505.28		500	
C			1.32	503.96	1.55/1.09

TABLE 6-1 *Sample surveying data sheet.*

If the instrument needs to be moved, any known point found may be used as a new benchmark.

Horizontal Control

The horizontal distance between two points may be found by using a tape measure or by taking a reading with the instrument. The view through the instrument is shown in figure 6-13. There are horizontal and vertical crosshairs, which are used to take the primary reading.

In addition, there are marks above and below the horizontal crosshair, known as the *upper stadia* (US) and *lower stadia* (LS). These can be used to determine the distance to the reading. The auto level will have a stadia multiplier somewhere on the instrument. Normally, this is 100, which means the difference between the upper and lower stadia readings are multiplied by 100 to get the distance in feet. In table 6-1, the upper

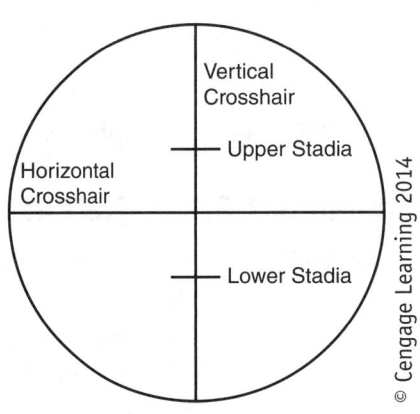

FIGURE 6-13 *Surveying crosshairs and stadia.*

Site Characteristics

and lower stadia readings are shown as 1.55 and 1.09, respectively. The distance to the point C from the instrument is calculated as follows:

$$\text{Distance} = (US - LS) \times \text{multiplier}$$
$$\text{Distance} = (1.55 - 1.09) \times 100 = 46 \text{ ft}$$

The horizontal angle between points may also be read on the auto level. There is normally a gradation of 1 degree marked completely around the auto level. The angle from the backsight to the foresight is recorded in the same column as the stadia readings. Horizontal readings can also be taken with an electronic measuring device (EDM) or a total station, which collects and stores the data in a computer.

Global Positioning Systems

The Global Positioning System (GPS) developed by the U.S. Department of Defense (DOD) uses three satellites to triangulate any position and a fourth satellite to correct timing errors due to the differential distance to the original three satellites. Typical GPS readings are accurate to within several inches, but more accurate readings can be received by the Continuously Operating Reference System (CORS) at www.ngs.noaa.gov/CORS/.

Topographic Survey

Topographic surveys are performed to determine elevations of the ground and mark features on the site. A few examples are shown in figures 5-13, 5-14, and 5-15. The maps shown in those figures are from the U.S. Geological Survey (USGS) and have contour intervals of 10 feet. This would not be accurate enough for construction, so a more detailed survey should be performed by the grid method discussed in chapter 6 of the textbook. A detailed topographic map with a 2-foot contour interval for the Range Elementary project is shown in figure 6-14.

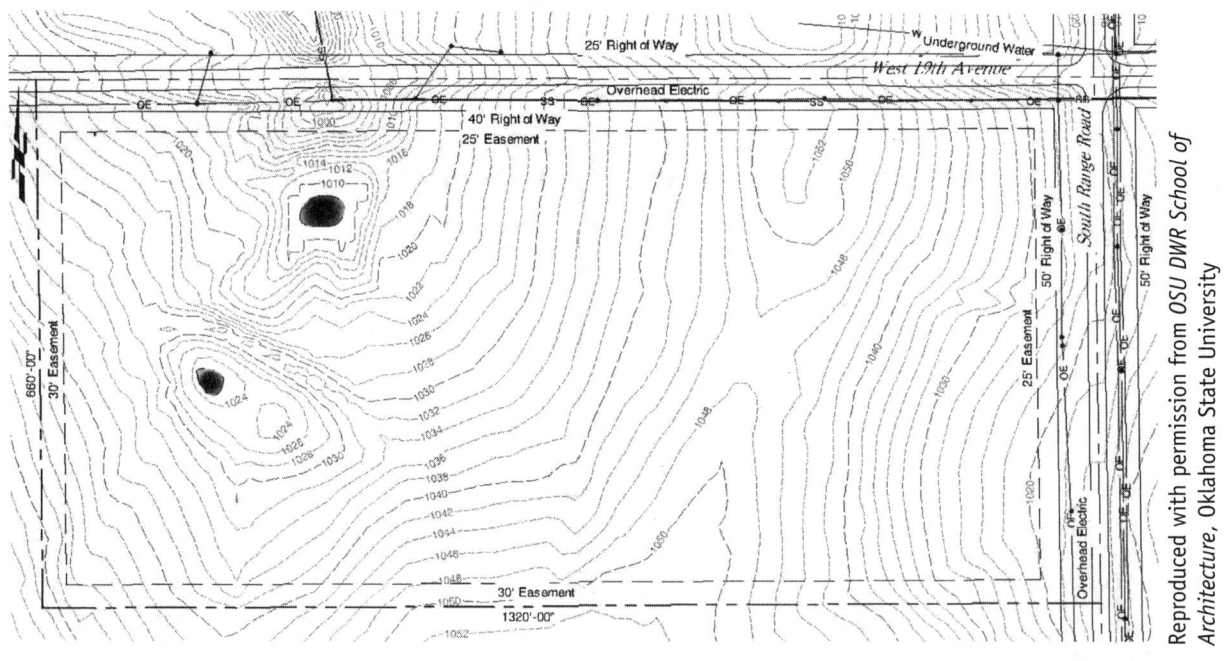

FIGURE 6-14   *Topographic survey plan.*

Problem Set 6.2   Site Characteristics

Problem 6-7   What is the zoning type for the Range Elementary School site? (Hint: See figure 6-7.)

Problem 6-8   Is an elementary school an acceptable use for the zoning in problem 6-7? If yes, how is the use achieved? Hint: See figures 5-23 and 5-19.

Problem 6-9   Which type of survey would you use to locate land elevations and features on a site?

Problem 6-10   Which type of survey locates the boundaries of a site and typically is required for a construction permit?

Problem 6-11   What meridian is used in Missouri, and at what longitude is it located?

Problem 6-12   What are the longitude, latitude, and altitude of benchmark 11B9 shown in figure 6-10?

Problem 6-13   What are the longitude, latitude, and altitude of benchmark P51 shown in figure 6-10?

Problem 6-14   What is the rod level reading in figure 6-15?

Problem 6-15   What are the upper and lower stadia readings in figure 6-15?

Problem 6-16   Determine the height of the instrument at point 1, and foresight elevations for points 2 and 3, missing in table 6-2.

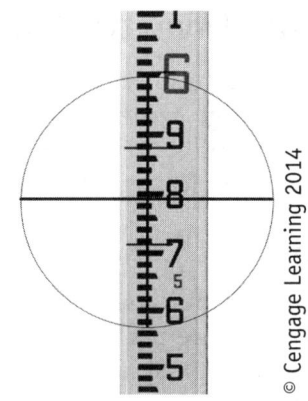

FIGURE 6-15 *Leveling rod and site for problem 6-15.*

Point	Backsight	Height of Instrument	Foresight	Elevation	Angle or Distance
PT	BS (+)	HI	FS(+)	Elev.	US/LS
1	3.40			619	3.54/3.25
2			4.16		4.41/3.91
3			5.65		5.85/5.45

TABLE 6-2 *Sample surveying data sheet.*

Problem 6-17   Determine the distance from the benchmark to the instrument at point 1 using the stadia readings.

Problem 6-18   Determine the distance from the instrument at point 1 to point 3 using the stadia readings.

Problem 6-19   What is the approximate elevation of the ridge between the two ponds on the western side of the site in figure 6-14? Note: the shaded areas on the western side of the site are the ponds.

Problem 6-20   What are the length and width of the buildable area within the easements for the site in figure 6-14?

BACKGROUND

When selecting the proper foundation system to support a building, it is critical to be familiar and have knowledge of the existing conditions. Architects and engineers will need to have an understanding of the different types of soils, including their limitations, prior to planning any site design.

Soils Investigation

A geotechnical engineer utilizes the principles of soil and rock mechanics to investigate subsurface environments. A geotechnical engineer should be used to provide investigation and evaluation of soil conditions on any building project. The extent of the investigation is normally proportional to the size and complexity of the site and the project, but there are a few common properties that are always included.

Soil Classification

Soils can be classified a number of ways depending on their physical and chemical properties. Once a few tests are performed on the soil, it can be classified based on the properties using the Unified Soils Classification System.

Grain Size

A sieve analysis is performed to find the grain size distribution by separating the particles of the soil size groups. The soil is dried and broken down into individual grains and passed through a series of sieves. The size of the openings in the sieve is given in table 6-3, along with the basic soil classification based on grain size. The larger sieve sizes are used to grade gravel, and the smaller sizes are used to grade sand. The finest particles are silt and clay that pass through the No. 200 sieve.

Sieve No.	Sieve Size		Particle Size (in)		Soil Classification Passes Through Sieve	
	in	mm	Upper	Lower		
				12	Boulders	Rock
	12	350	12	3	Cobbles	Rock
	3	75	3	3/4	Coarse Gravel	Soil
	3/4	19	3/4	0.19	Fine Gravel	Soil
#4	0.19	4.75	0.19	0.079	Coarse Sand	Soil
#10	0.079	2.00	0.079	0.016	Medium Sand	Soil
#40	0.016	0.425	0.016	0.0029	Fine Sand	Soil
#200	0.0029	0.075	0.0029		Fines (silt and clay)	Soil

TABLE 6-3 *Sieve sizes and soil particle size classification.*

A simple sieve analysis can be done with just a No. 4 and No. 40 sieve, which are shown individually in figure 6-16. In this analysis, the fine sand and fines are separated out using water. The fines remain suspended in water, and the fine sand settles to the bottom. A little more detailed analysis can be done with No. 4, No. 10, No. 40, and No. 200 sieves, which are shown, stacked together in figure 6-16. The example in the textbook details this process, which includes the 3/4 inch sieve.

The soil grains shown in figure 6-23 of the textbook would pass through the next-larger sieve size than the one it is retained on. The medium sand would pass through the No. 10 sieve and be retained on the No. 40 sieve. The coarse sand would pass through the No. 4 sieve and be retained on the No. 10 sieve. The fine gravel would pass through the No. 3/4 inch sieve and be retained on the No. 4 sieve.

FIGURE 6-16 *Soil sieves.*

Example 6.1 Perform a sieve analysis, and plot the grain size distribution for the data in table 6-4.

Sieve Size	Mass Retained, W_n (g)	Percentage of Sample Retained, R_n	Cumulative Percentage Retained, ΣR_n	Percentage Passing $100 - \Sigma R_n$
3/4 in	158			
No. 4	308			
No. 10	608			
No. 40	652			
No. 200	266			
Pan	8			
Total	2000			

TABLE 6-4 *Grain size distribution.*

First, the percentage retained on each sieve is the mass retained divided by the total mass, expressed as a percentage using the following formula

$$R_n = \frac{W_n}{W} \times 100$$

W_n = Mass of sample retained

W = Total mass of sample

R_n = Percentage of sample retained

Soils Investigation 103

Second, the cumulative percentage retained is obtained by summing all previous percentages retained and that sieve's percentages. Third, the percentage passing is taken as 100 percent minus the cumulative percentage retained. This is summarized in table 6-5.

Sieve Size	Mass Retained, W_n (g)	Percentage of Sample Retained, R_n	Cumulative Percentage Retained, ΣR_n	Percentage Passing $100 - \Sigma R_n$
3/4 in	158	7.9	7.9	92.1
No. 4	308	15.4	23.3	76.7
No. 10	608	30.4	53.7	46.3
No. 40	652	32.6	86.3	13.7
No. 200	266	13.3	99.6	0.4
Pan	8	0.4	100	
Total	2000	100		

TABLE 6-5 *Grain size analysis.*

The values for the percentage passing are then plotted on a log scale as shown in figure 6-17.

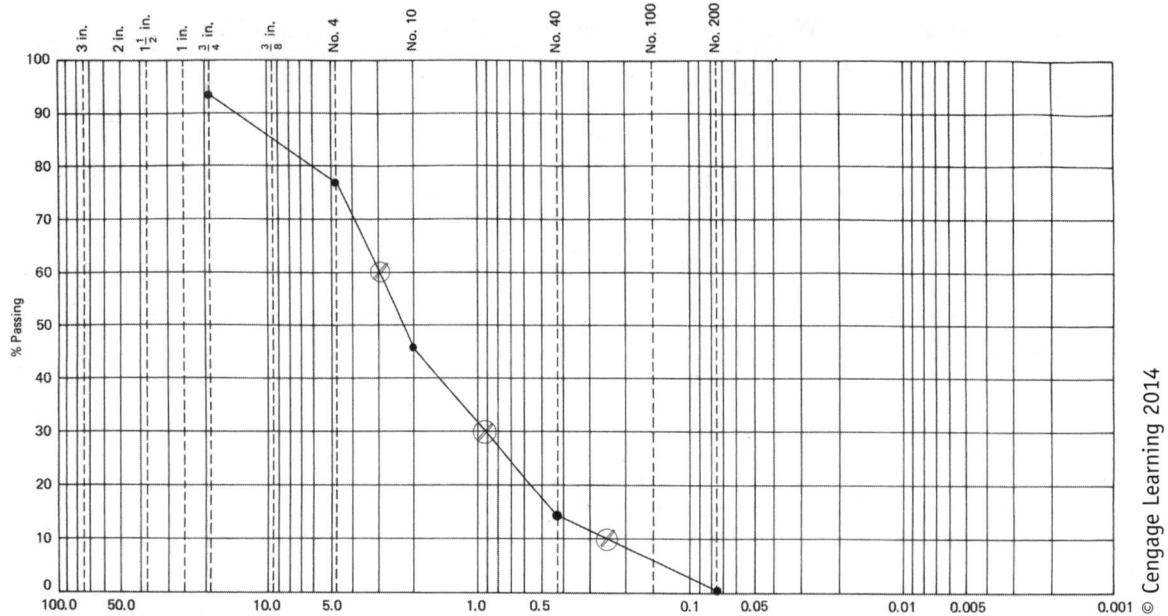

FIGURE 6-17 *Grain size plot for example 6-1.*

There are typically two parameters necessary to classify coarse-grained soils: the coefficient of uniformity (C_u), and the coefficient of curvature (C_c). These coefficients are calculated based on three particle diameters taken from the grain size distribution curve. The following are the two formulas for the coefficients:

$$C_u = \frac{D_{60}}{D_{10}}$$

$$C_c = \frac{(D_{30})^2}{D_{60} D_{10}}$$

D_{10} = particle size finer than 10 percent
D_{30} = particle size finer than 30 percent
D_{60} = particle size finer than 60 percent

Example 6.2 Determine the coefficient of uniformity and the coefficient of curvature for the soil in example 6-1 and figure 6-17.

The diameters corresponding to 60, 30, and 10 percent have been marked with an open circle with values of approximately 3.00, 0.92, and 0.25 mm. The coefficients are calculated as follows:

$$C_u = \frac{D_{60}}{D_{10}} = \frac{3.00}{0.25} = 12$$

$$C_c = \frac{(D_{30})^2}{D_{60} D_{10}} = \frac{0.92^2}{3.00 \times 0.25} = 1.13$$

Plasticity

Fine-grained soil can exist in four different states, as shown in figure 6-18. There is increasing moisture in the content as the soil transforms from the solid to the liquid state. The difference between silt and clay is primarily based on the plasticity index. Clays are cohesive and will tend to stick together by chemical properties. Silts are somewhat cohesionless and will separate easily.

The Atterberg tests are used to determine the plasticity index (PI) of the soil. The plasticity index is the difference between the liquid limit (LL) and the plastic limit (PL) from the following formula:

$$PI = LL - PL$$

The Atterberg limits along with the grain size distribution are used to classify the soil.

Unified Soil Classification System

The Unified Soils Classification System (USCS) uses a two-letter designation to classify 15 types of soil. The first letter tells the general type of soil as follows:

G = gravel
S = sand
M = silt
C = clay
O = organic

The second letter gives supplementary information, depending on the general soil type as follows:

W = well-graded
P = poorly graded
M = silty
C = clayey
L = low plasticity
H = high plasticity

Table 6-6 is used along with the grain size coefficients to classify coarse grained soils. Table 6-6 is used along with the Atterberg limits plotted on figure 6-19 and is used to classify fine-grained soils. Figure 6-19 is known as the Plasticity Chart.

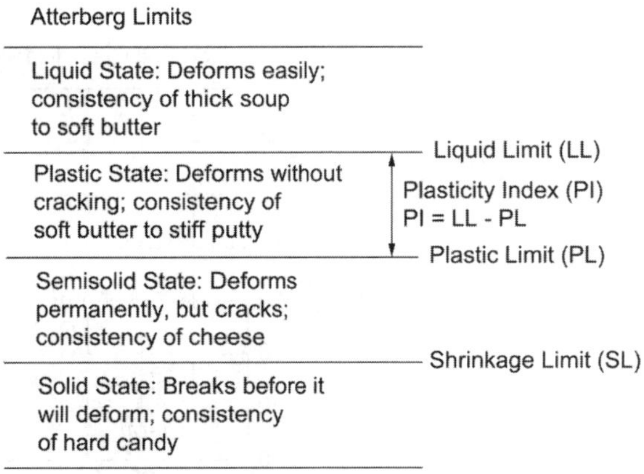

Atterberg Limits

Liquid State: Deforms easily; consistency of thick soup to soft butter
— Liquid Limit (LL)

Plastic State: Deforms without cracking; consistency of soft butter to stiff putty
Plasticity Index (PI)
PI = LL - PL
— Plastic Limit (PL)

Semisolid State: Deforms permanently, but cracks; consistency of cheese
— Shrinkage Limit (SL)

Solid State: Breaks before it will deform; consistency of hard candy

FIGURE 6-18 *Atterberg limits.*

Criteria for Assigned Group Symbol and Group Names Using Laboratory Test				Soil Classification Symbol and Name
Coarse-grained soils: More than 50% retained on No. 200 sieve	Gravels: More than 50% of coarse fraction retained on No. 4 sieve	Clean gravels: Less than 5% fines	$C_u \geq 4$ and $1 \leq C_c \leq 3$	GW - Well-graded gravel
			$C_u < 4$ and/or $1 > C_c > 3$	GP - Poorly graded gravel
		Gravels with fines: More than 12% fines	Fines classify as ML or MH	GM - Silty gravel
			Fines classify as CL or CH	GC - Clayey gravel
	Sand: 50% or more of coarse fraction passes No. 4 sieve	Clean sands: Less than 5% fines	$C_u \geq 6$ and $1 \leq C_c \leq 3$	SW - Well-graded sand
			$C_u < 6$ and/or $1 > C_c > 3$	SP - Poorly graded sand
		Sands with fines: More than 12% fines	Fines classify as ML or MH	SM - Silty sand
			Fines classify as CL or CH	SC - Clayey sand
Fine-grained soils: 50% or more passes the No. 200 sieve	Silts and Clays: Liquid limit less than 50 Low plasticity	Inorganic	PI > 7 & plots on or above A	CL - Lean clay
			PI < 4 & plots below A	ML - Silt
		Organic	$LL_{oven\ dried}/LL < 0.75$ and see inorganic above for plot	OL - Organic clay
				OL - Organic silt
	Silts and Clays: Liquid limit 50 or more High plasticity	Inorganic	PI plots on or above A	CH - Fat clay
			PI plots below A	MH - Elastic silt
		Organic	$LL_{oven\ dried}/LL < 0.75$ and see inorganic above for plot	OH - Organic clay
				OH - Organic silt

TABLE 6-6 *Soil classification chart by Unified Soil Classification System.*

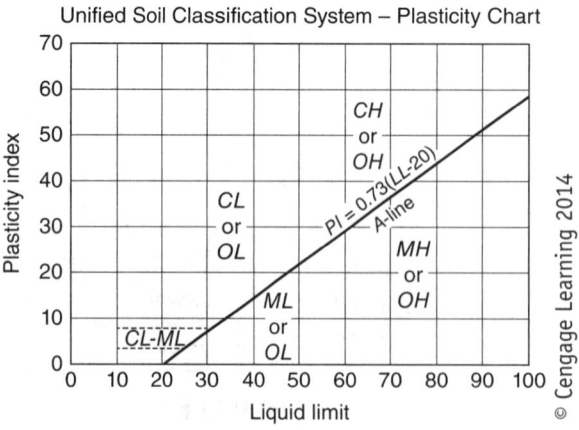

FIGURE 6-19 *USCS plot.*

In addition a highly organic soil that is primarily organic material is classified as Peat (Pt).

Example 6.3 Determine the soil group symbol and name according to the Unified Soils Classification System for the soil in examples 6-1 and 6-2.

To determine the classification of the soil, work from the left column in table 6-6 toward the right in four steps. Use table 6-5 for the cumulative percentages. First, for the No. 200 sieve, 99.6 percent was retained, which is more than 50 percent and indicates a coarse-grained soil. Second, for the No. 4 sieve, 23.3 percent was retained, which is less than 50 percent and indicates sand. Third, the percentage of fines that passed the No. 200 sieve is 0.4 percent, which is less than 5 percent and indicates clean sands. Finally, the coefficient of uniformity and the coefficient of curvature were calculated in example 6-2. The coefficients are $C_u = 12$, which is greater than 6, and $C_c = 1.13$ which lies between 1 and 3. The fifth column indicates a group symbol SW and a group name of well-graded sand.

The previous example was a coarse-grained soil and did not require the use of figure 6-19 or the Atterberg limits. The results in example 6-4 are from a fine-grained soil.

Example 6.4 Determine the soil group symbol and name according to the Unified Soils Classification System for a soil with the given test data. The percentage passing No. 200 sieve was 82 percent, LL = 32 and PL = 18.

Since the percentage passing the No. 200 sieve is 82 percent, which is greater than 50 percent, this indicates that it is a fine-grained soil. The liquid limit (LL) of 32 is less than 50 percent, which indicates low-plasticity silts and clays. The plasticity index (PI) must be calculated and then plotted on the plasticity chart in figure 6-20.

$$PI = LL - PL = 32 - 18 = 14$$

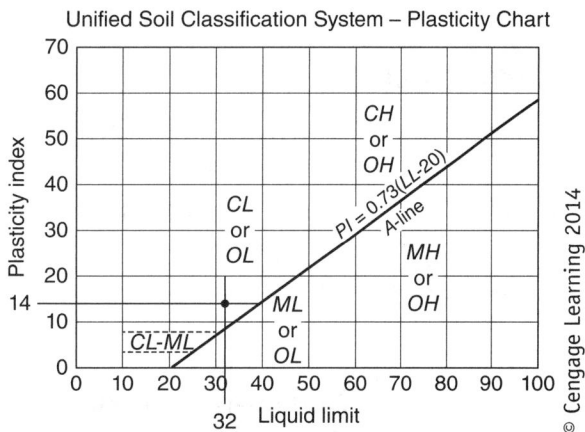

FIGURE 6-20 USCS Plot for example 6-4.

The plasticity index of 14 is greater than 7, and the point plots above the A-line. The fifth column indicates a group symbol CL and a group name of Lean clay. Since there was no oven-dried liquid limit given, information on the organic material is not available.

The Geotechnical Report

A geotechnical report is prepared by a licensed geotechnical engineer and should be done for any building project. A typical geotechnical report may contain the following: the conditions at the site, soil data collected, testing performed, soils properties, and recommendations on foundation systems. The textbook contains a list of information that could be included in the report. One key element of any soils report is the boring log, which is a record of depths where the soil data were collected, and the holes drilled, at the site for exploration. The logs should be plotted on a preliminary site plan in the report. A sample boring log is shown in figure 6-21. The log reports a description of the soil material, including the USCS symbol. Location and type of soil samples are indicated, including the standard penetration test (SPT). Also shown is the water content of the soil and Atterberg limits, where appropriate.

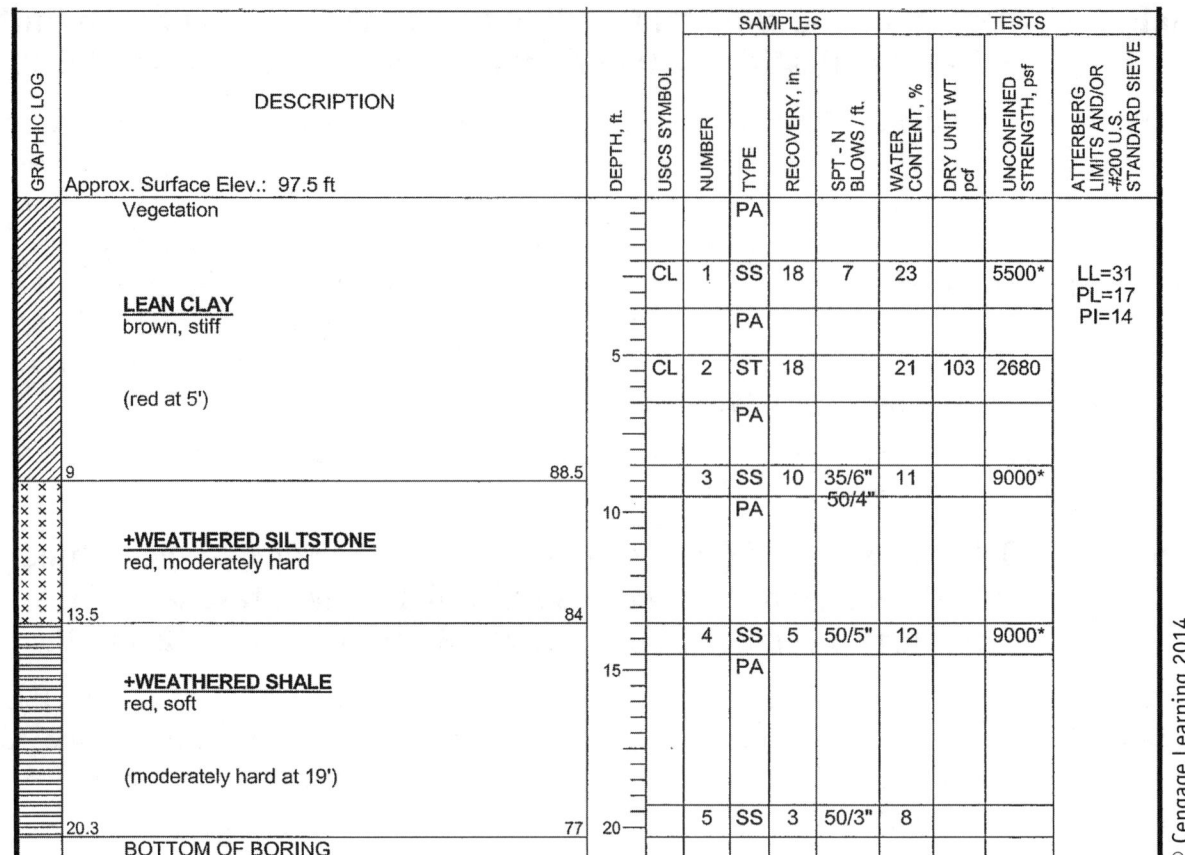

FIGURE 6-21 Boring log.

Problem Set 6.3 Soils Investigation

Problem 6-21 Who is normally hired to perform soil site investigation and testing?
Problem 6-22 What is the size range for coarse gravel?
Problem 6-23 What is the size of No. 4 sieve in inches and millimeters?
Problem 6-24 What is the maximum size for fines?
Problem 6-25 Perform a sieve analysis for the soil in table 6-7.

Sieve Size	Mass Retained, W_n (g)	Percentage of Sample Retained, R_n	Cumulative Percentage Retained, ΣR_n	Percentage Passing $100 - \Sigma R_n$
No. 4	0			
No. 10	40.2			
No. 40	174.8			
No. 200	274.6			
Pan	10.4			
Total	500			

TABLE 6-7 Grain size distribution.

Problem 6-26 Plot the grain size distribution for soil in table 6-8.

Sieve Size	Percentage Passing
No. 4	100
No. 10	92
No. 40	57.1
No. 200	2.1

TABLE 6-8 *Grain size distribution.*

Problem 6-27 Determine the uniformity coefficient and coefficient of curvature for the soil grain size distribution plot in figure 6-22.

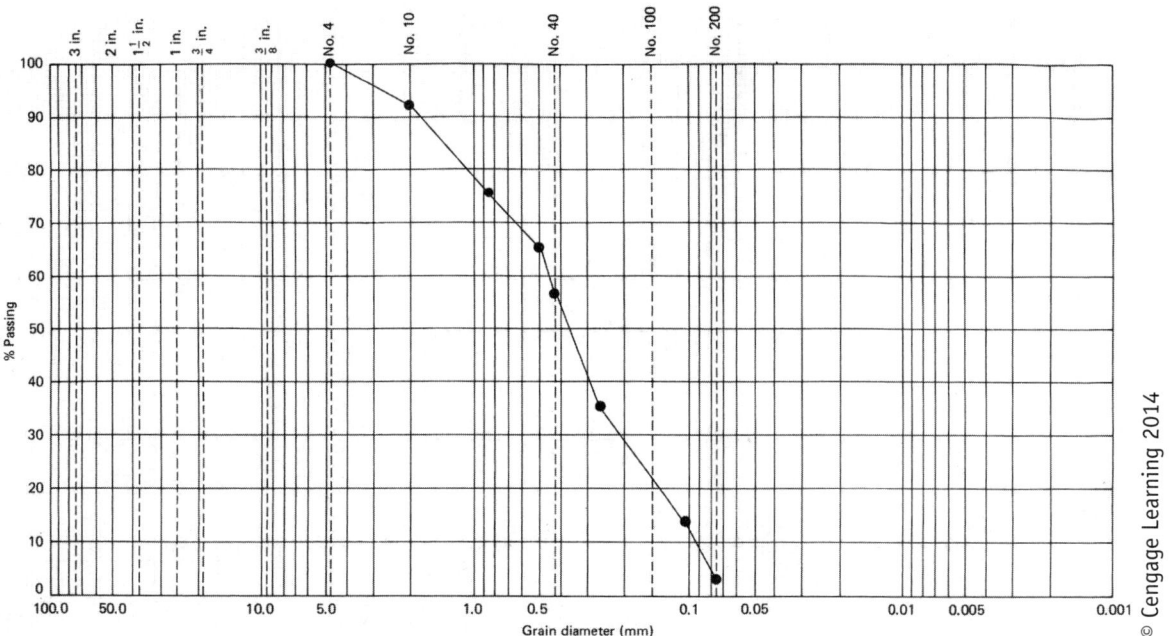

FIGURE 6-22 *Grain size plot for problem 6-27.*

Problem 6-28 What is the Unified Soil Classification System group symbol for well-graded gravel?

Problem 6-29 What is the Unified Soil Classification System group symbol for silty clay?

Problem 6-30 Determine the Unified Soil Classification System group symbol and group name for the soil with a liquid limit of 42 percent, plastic limit of 16 percent, and grain size distribution in table 6-9.

Sieve Size	Percentage Passing
No. 4	100
No. 10	93.2
No. 40	81.0
No. 200	60.2

TABLE 6-9 *Grain size distribution.*

Soils Investigation

Problem 6-31 Determine the Unified Soil Classification System group symbol and group name for the soil with a liquid limit of 30 percent, plastic limit of 12 percent, and grain size distribution in table 6-10.

Sieve Size	Percentage Passing
No. 4	76.5
No. 10	60.0
No. 40	39.7
No. 200	15.2

TABLE 6-10 *Grain size distribution.*

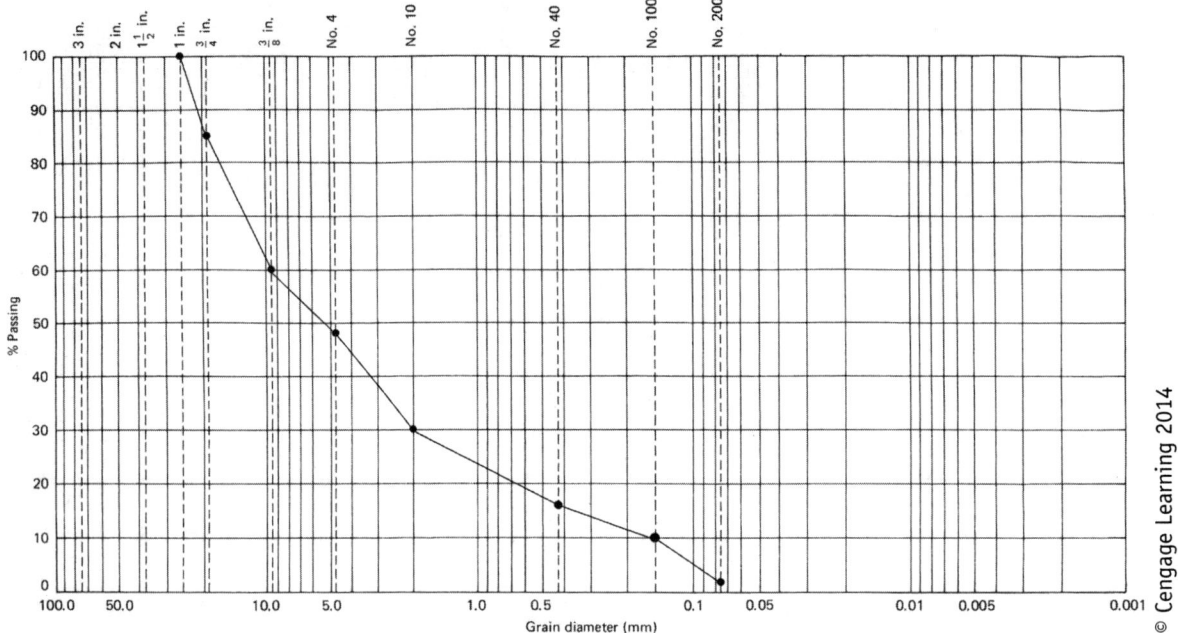

FIGURE 6-23 *Grain size plot for problem 6-32.*

Problem 6-32 Determine the Unified Soil Classification System group symbol and group name for the soil with a grain size distribution in figure 6-23.

BACKGROUND

Once the architect and engineer have collected and analyzed the necessary information for the site, some planning decisions can be developed to aid in the design process. Many factors will drive the design decision and site development.

Concept Development

When developing a concept to meet the goals of a project, the design team begins with a generation of preliminary designs, the program and needs analysis, and a brainstorming session to help generate possible solutions. A design charette, with all the parties involved, is an effective way to generate preliminary ideas and concepts. The photograph shown in figure 6-24 was taken during a three-day charette for the design of the Donald W. Reynolds

School of Architecture at Oklahoma State University. The charette generated dozens of solutions and involved students, faculty, architects, engineers, and university administrators.

FIGURE 6-24 *Charette photo.*

Buildable Area

The buildable area on the site is determined once the property lines, easements, and setback have been located. Other influences on the buildable area may include buffer zones, unbuildable area, and restricted areas. The site plan shown in figure 6-14 indicates all of the easements, rights of way, and utilities. In addition, the western part of the site contains two ponds that would render that portion of the site unbuildable. This would indicate that the best buildable area would be the highest portion of the eastern portion of the site.

Site Orientation

The location of the best buildable area and the other physical features, such as the ponds and major roads, will help develop the site layout and orientation. Figure 6-25 shows a traffic study of the site and indicates the northeast corner of the site to be the major entry point toward the site.

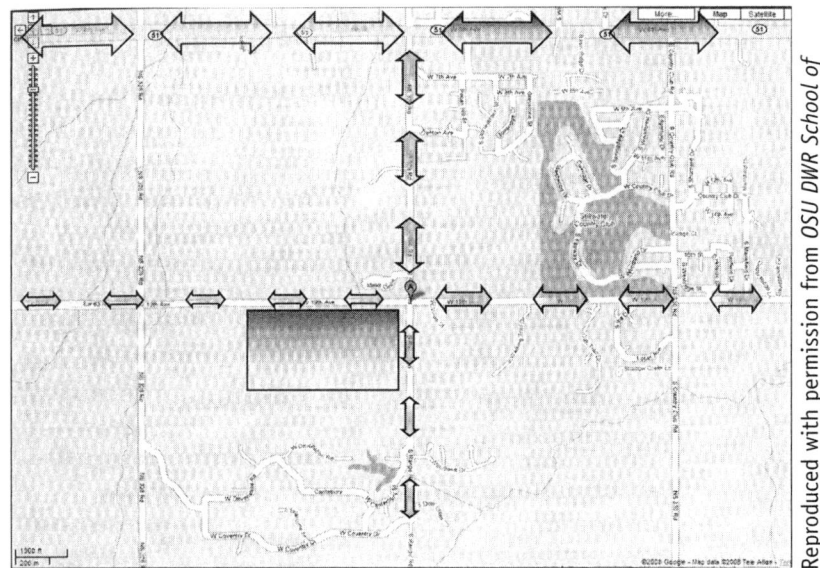

FIGURE 6-25 *Traffic study.*

Loading and parking are best located to the far east of the site closest to the major entry point to the site. The recreation zones could be to the south or west of the building to provide for security. A simple bubble diagram of this is shown on an aerial of the site in figure 6-26. The exact location at this point is not important, rather the general location. An alternate layout could be with the parking on the north of the building, but that would mean that the entry to the building would be on the north. This solution may not be the best for weather considerations.

FIGURE 6-26 *Site design diagram.*

Solar Orientation

The solar study of the site is shown in figure 6-27 with the prevailing winds also indicated. For maximum solar gain in the winter, a long east-west orientation is normally best with overhangs on the south side to provide shading in the summer. The sun angles are also indicated in figure 6-27 to determine the overhangs.

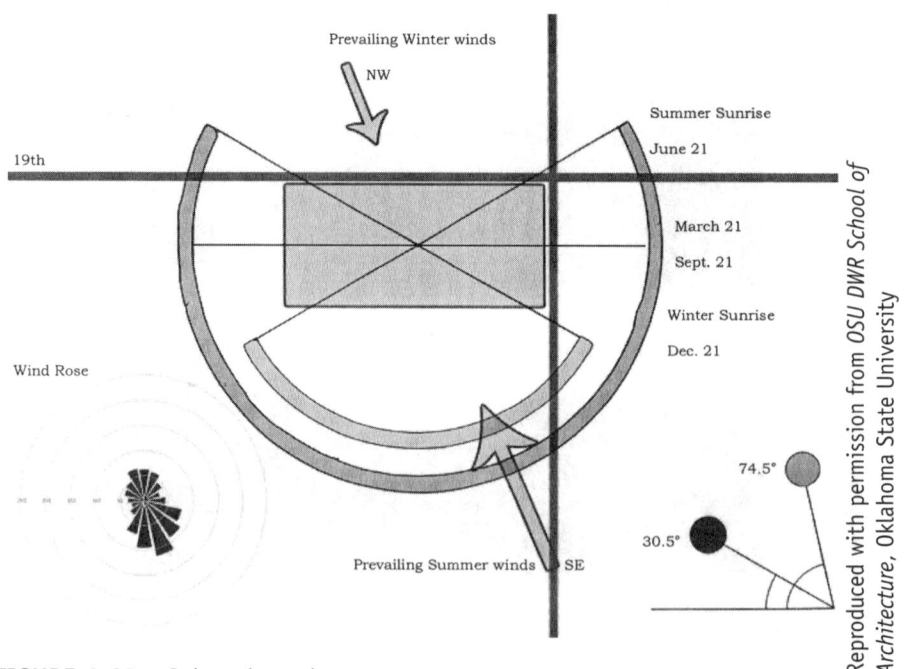

FIGURE 6-27 *Solar orientation.*

112   Chapter 6: Site Planning

Wind Orientation

The wind rose is shown in figure 6-27 and indicated the prevailing winds. In the winter, the cold wind comes from the northwest direction. This cold wind should be avoided when developing a site plan. The summer winds come from the southeast and should be used to provide for passive cooling. In addition, the building could be located down to the southeast to allow the highest point on the site to the northwest to provide for some shielding from the wind out of the northeast.

Sound Orientation

The traffic patterns shown in figure 6-25 indicate that most of the noise will come from the north and that some noise will come from the east. The major point of noise would be the northeast corner of the site. All of this is shown diagrammatically in figure 6-28. This would further indicate that the building should be held back from the north and east sides of the site as much as possible.

FIGURE 6-28 *Sound orientation.*

View Orientation

Building and site orientations should take advantage of desirable views. Undesirable views should be avoided when possible. A photo survey of the site can help inform the orientation process. Figure 6-29 shows a photo survey with views throughout the site and their locations on an aerial photograph. In this case, views are fairly desirable.

Terrain Orientation

As discussed in chapter 5, the terrain should be considered when positioning a building. The site should slope enough to provide for proper drainage, but not too much as to cause excess water runoff. A good rule is to place a structure at a high point on the site and not near areas of steeper slopes. In general, a slope gradient between 1 and 5 is best, but slopes up to 10 percent are acceptable and may require some regrading. The eastern portion of the site is acceptable as indicated by the contours in figure 6-14. The general slope gradient calculated according to chapter 5 is between 2 and 6.

Site Opportunities Map

All of the orientation decisions could be provided on one map indicating the best building location and orientation. This should always be done if there are conflicting indicators. In the case of the Range Elementary project, most of the indicators would support the general location shown in figure 6-26 with the building running longitudinally in the east-west direction. The actual building design should continue using the best orientation of the site development.

FIGURE 6-29 *Views.*

Problem Set 6.4 Concept Development

Problem 6-33   What is a good way to develop numerous design concepts and involve most of the design team and stakeholders?

Problem 6-34   What normally determines the best buildable area on a site that is free of physical limitations?

Problem 6-35   Where should the major entry point for vehicles be located onto a site?

Problem 6-36   What is normally the best solar orientation for a site in the Midwestern United States?

Problem 6-37   What symbolically indicates the prevailing winds for a location?

Problem 6-38   What is commonly the major source for noise entering a site?

Problem 6-39   What type of survey can help exploit the view in and around a site?

Problem 6-40   What is generally the best range of slope gradient to place a building upon?

CHAPTER 7
Site Design

Skills List

After completing the problems in this chapter, you should be able to:

- Be aware of site requirements for persons with disabilities

- Understand the factors to consider when designing parking and traffic circulation

- Determine site grading and the effects on construction cost

- Be able to calculate stormwater runoff

- Determine water supply pressure and how to manage wastewater

- Understand the contents of a site plan

BACKGROUND

Architects and engineers must be able to design a site for the least amount of disruption to the environment, while creating a positive flow of traffic, pedestrians, stormwater, water supply, and other utilities. The site design should be evaluated against the site analysis to determine whether a concept will be successful.

Accessibility

The Americans with Disabilities Act of 1997 (ADA) allows access to public accommodations for people with disabilities. Any public access to a public service must be accessible to those with disabilities. In addition, the Architectural Barriers Act of 1968 (ABA) requires uniform federal compliance for accessibility. The ADA-ABA guidelines have similar requirements to the International Building Code (IBC). This workbook and the textbook will use the ADA Accessibility Guidelines of 2004 (ADAAG), which apply to most commercial facilities. It should be noted that standards and guidelines are not requirements unless the state or local government adopt them by code. The two primary requirements are that accessible routes are designed for easy navigation for persons in wheelchairs and that the accessible parking spaces are designed properly. In chapter 7 of the textbook, some of the basic ADAAG requirements are listed and are summarized here.

- Minimum width of accessible route is 36 inches, but a passing space every 200 feet must be provided for widths less than 60 inches. Passing zones are 60 inches by 60 inches.
- Maximum level change of 1/4 inch or maximum bevel of 1/2 inch (see textbook figure 7-3)
- Maximum slope of route is 1:20 or 1 inch rise for every 20 inches of horizontal run
- Maximum ramp slopes of 1:12 and maximum ramp rise of 30 inches (see textbook figure 7-4)
- Minimum 60 inch landing at ends of ramps (see textbook figure 7-5)
- Maximum ramp rise of 6 inches without handrails
- Ramp curb details must be provided along the accessible route (see textbook figure 7-6)
- Parking spaces per Table 7-1 and located at the shortest accessible route
- Minimum of 1 van accessible space of each 6 accessible spaces
- Car- and van-accessible spaces sized as shown in the textbook figure 7-8

Total Number of Parking Spaces Provided in the Facility	Minimum Number of Required Accessible Parking Spaces
1 to 25	1
25 to 50	2
51 to 75	3
76 to 100	4
101 to 150	5
151 to 200	6
201 to 300	7
301 to 400	8
401 to 500	9
501 to 1000	2 percent of total
Over 1000	20 plus 1 for each 100 over 1000

TABLE 7-1 *Accessible parking requirements.*

Problem Set 7.1 Accessibility

Problem 7-1   What is the minimum width of an accessible route?
Problem 7-2   If the accessible route has a slope of 1:16 is it considered a ramp?
Problem 7-3   How many intermediate landings are required if a ramp rises 8 feet?
Problem 7-4   When are handrails required on a ramp?
Problem 7-5   What is the minimum length of a landing for a ramp?
Problem 7-6   What are the maximum slope and minimum width of a curb ramp?
Problem 7-7   What is the minimum number of accessible parking spaces for a parking lot with 450 parking spaces?
Problem 7-8   How many van-accessible parking spaces would be required for the parking in problem 7-7?
Problem 7-9   What is the minimum number of accessible parking spaces for a parking lot with 950 parking spaces?
Problem 7-10 What is the minimum number of accessible parking spaces for a parking lot with 1450 parking spaces?

BACKGROUND

Architects and engineers must be able to design the parking and vehicular circulation on a site that allows for safety and efficiency. The traffic should flow easily onto, throughout, and off of the site.

Parking and Circulation

Parking requirements may be specified by the local municipality. The textbook covers a general list of considerations that are summarized here. The overall design for parking and circulation is illustrated in the textbook figure 7-9.

- Required number, size, and spacing of entrances/exits for the site (see textbook figure 7-10)
- Number of required parking spaces per local municipality, Stillwater shown in figures 5-24 and 7-1 (also see textbook figure 7-11)

Use	Motor Vehicles	Bicycles: % of total number of required motor vehicle parking spaces
Residential, Single Family and Two Family	Two spaces per dwelling unit	NA
Residential Multi-Family, One Bedroom or Studio	One space per dwelling unit	1 per dwelling unit
Residential Multi-Family, Two Bedroom	Two spaces per dwelling unit	1 per dwelling unit
Residential Multi-Family, Three Bedroom	2.33 spaces per dwelling unit	1 per dwelling unit
Residential Multi-Family, Four Bedrooms	3 spaces per unit	1.25 per dwelling unit
Restaurant	One space per 2.5 seats or 40 sq. ft. of dining and/or drinking area	10%
Restaurant, Fast-Food	One space per three seats or 40 sq. ft. of usable floor area	10%
Retail, Outdoor	One space per 600 sq. ft. of lot area	5%
School, Dance	One space per 300 sq. ft.	10%
School, Elementary	One space per classroom and each 1.5 employees, plus off-street loading space for at least two school buses	25%
School, Junior High	Two spaces per classroom	40%
School, High School	Four spaces per classroom	60%
School, Trade	One space per four students	20%
Shopping Center	One space per 200 sq.ft. gross leasable area	10%

FIGURE 7-1 *Parking code.*

- Size of parking space based on configuration in figure 7-2 and table 7-2 (also see textbook figure 7-12)
- Accessibility for parking as previously discussed (see textbook figures 7-7 and 7-8)
- Aisle width for one- and two-way traffic as shown in figure 7-2
- Pedestrian walkways generally 3 feet or 5 feet wide for one- and two-way flow (see textbook figure 7-13)
- Special access for fire trucks, buses, garbage trucks, and other vehicles

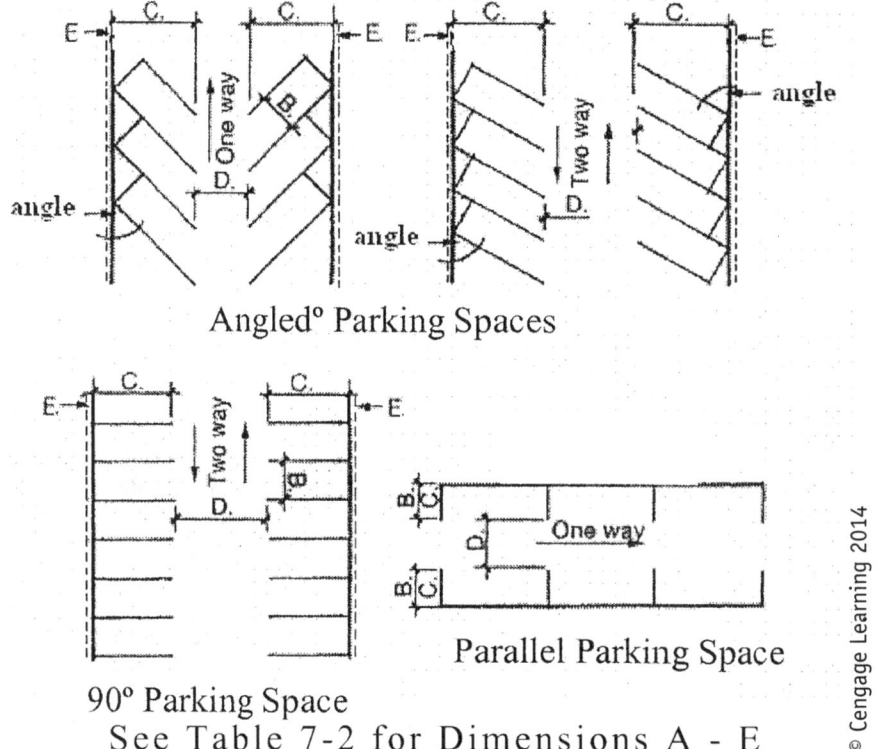

FIGURE 7-2 *Parking dimensions.*

Type of Space	Parking Angle, A	Minimum Width (ft) B	Minimum Length (ft) C	Minimum Drive Width (ft) D	Maximum Overhang (ft & in) E
Standard Car Space	45°	9	19	13	1' 5"
	60°	9	20	13	1' 8"
	90°	9	18	24	2' 0"
Parallel Space	0°	8	24	12	0
Small Car Space	45°	8' 6"	17	12	1' 5"
	60°	8' 6"	18	12	1' 8"
	90°	8' 6"	16	24	2' 0"

TABLE 7-2 *Parking space size requirements.*

118 Chapter 7: Site Design

- Requirements for loading areas
- Allowance for waste disposal and collection (see textbook figure 7-14)
- Drainage and stormwater management
- Landscaping requirement, Stillwater point system shown in figure 7-3, with minimum landscape area equal to 9 percent of the developed area for parking greater than 20 spaces.
- Adequate lighting design for the site (see chapter 15)

> (4) To provide flexibility in designing the best landscape plan for each site, no specific materials or locations are required for landscaping by these regulations.
>
> (a) Developments shall provide landscape material from the list below, in any combination, such that the total points received for the development equal at least fifty (50) points per one hundred (100) square feet in all areas.
>
> (i) Berms, minimum thirty (30) inches tall-5 points per 10 linear feet.
>
> (ii) Turf--10 points per one hundred (100) square feet.
>
> (iii) Vegetative ground cover (other than natural turf or sod)--25 points per one hundred (100) square feet of cover.
>
> (iv) Small shrubs (mature height of four (4) feet)-25 points each.
>
> (v) Large shrubs (mature height over four (4) feet)-75 points each.
>
> (vi) Trees of one-inch caliper or greater at the time of planting which are classified as: Small--100 points; medium--150 points; large--200 points.
>
> (vii) Preservation of existing trees will earn the same base points as new trees based on the size classification of tree. For trees over four-inch caliper, an additional 20 points will be awarded for each inch of caliper over four (4), up to double the point value of the tree. (For example: Saving a tree which is classified in the large tree category will earn 200 points; if the tree measures to be eight-inch caliper, an additional 80 points will be awarded.) The diameter of the tree shall be measured at four and one-half (4 1/2) feet above the ground.
>
> (viii) Any plant materials installed to meet the requirements of this ordinance which die, or are otherwise removed, must be replaced with plant materials which will earn at least the same number of; points. If the replacement materials are to be located more than twenty (20) feet from where the previous materials had been, a revised landscape plan shall be submitted to the city planner for administrative review and approval.

FIGURE 7-3 *Landscape code.*

Problem Set 7.2 Parking and Circulation

Problem 7-11   What driveway spacing should be used along a street with a speed limit of 45 mph?

Problem 7-12   How many parking spaces are recommended for a 100-bed hospital?

Problem 7-13   Compare the parking requirement for a restaurant seating 120 people from textbook figure 7-11 and figure 7-1 for Stillwater.

Problem 7-14   Compare the parking requirement for a 10,000 sq. ft. shopping center from textbook figure 7-11 and figure 7-1 for Stillwater.

Problem 7-15   What would be the approximate dimensions of one row of two-sided 90° standard parking with a two-way aisle and 20 parking spaces on each side?

Problem 7-16   What would be the approximate dimensions of one row of two-sided 45° standard parking with a two-way aisle and 20 parking spaces on each side?

Problem 7-17   What would be the approximate dimensions of one row of two-sided 45° standard parking with a one-way aisle and 20 parking spaces on each side?

Problem 7-18   What would be the approximate dimensions of one row of two-sided 90° small car parking with a two-way aisle and 20 parking spaces on each side?

Problem 7-19   What is the required landscape area and number of landscape points required for a parking lot 64 feet by 180 feet?

Problem 7-20   How many landscape points are awarded for saving 12-, 18-, and 24-inch caliper trees?

BACKGROUND

Architects and engineers should develop a site in such a way as to not adversely impact the flow of stormwater around and off the site. The topography should be carefully designed as to provide minimum impact on development and the surrounding properties.

Grading and Drainage

The site will almost always need to be recontoured to provide for proper drainage. The process of regrading is done to allow for water to flow away from structures and not cause adverse effects.

Grading Plan

The grading plan shows the existing and proposed contours on a site as shown in figure 7-4 for the Range Elementary School. Existing contours are indicated with dashed lines, and proposed contours are indicated with solid lines.

One way to imagine the flow of water on a site is to move perpendicular to the contours. The slope of a site must not be too flat as to collect stormwater nor too steep such that stormwater causes erosion. The textbook in figure 7-16 gives minimum and maximum slope recommendations for various portions of the site. In chapter 5, the calculation of the slope gradient was introduced, and in chapter 6 a 1 percent to 5 percent gradient was recommended, although steeper slopes are possible. As illustrated in figure 7-17 of the textbook, the site should slope away from a structure at about a 5 percent gradient for 10 feet, then at an allowable slope beyond that point.

Cut and Fill

The removal and addition of soil on a site is known as *cut and fill*. The amount of cutting and filling should try to be balanced. This will help minimize the grading cost. In the case of figure 7-4, it appears that mostly cutting will be requried for the building. This can help in the removal of unwanted soil that could be under the building and reduce fill under it, which is nomally the best solution. Mainly fill will be required in the parking area. If a retention pond is required, fill could be obtained from there.

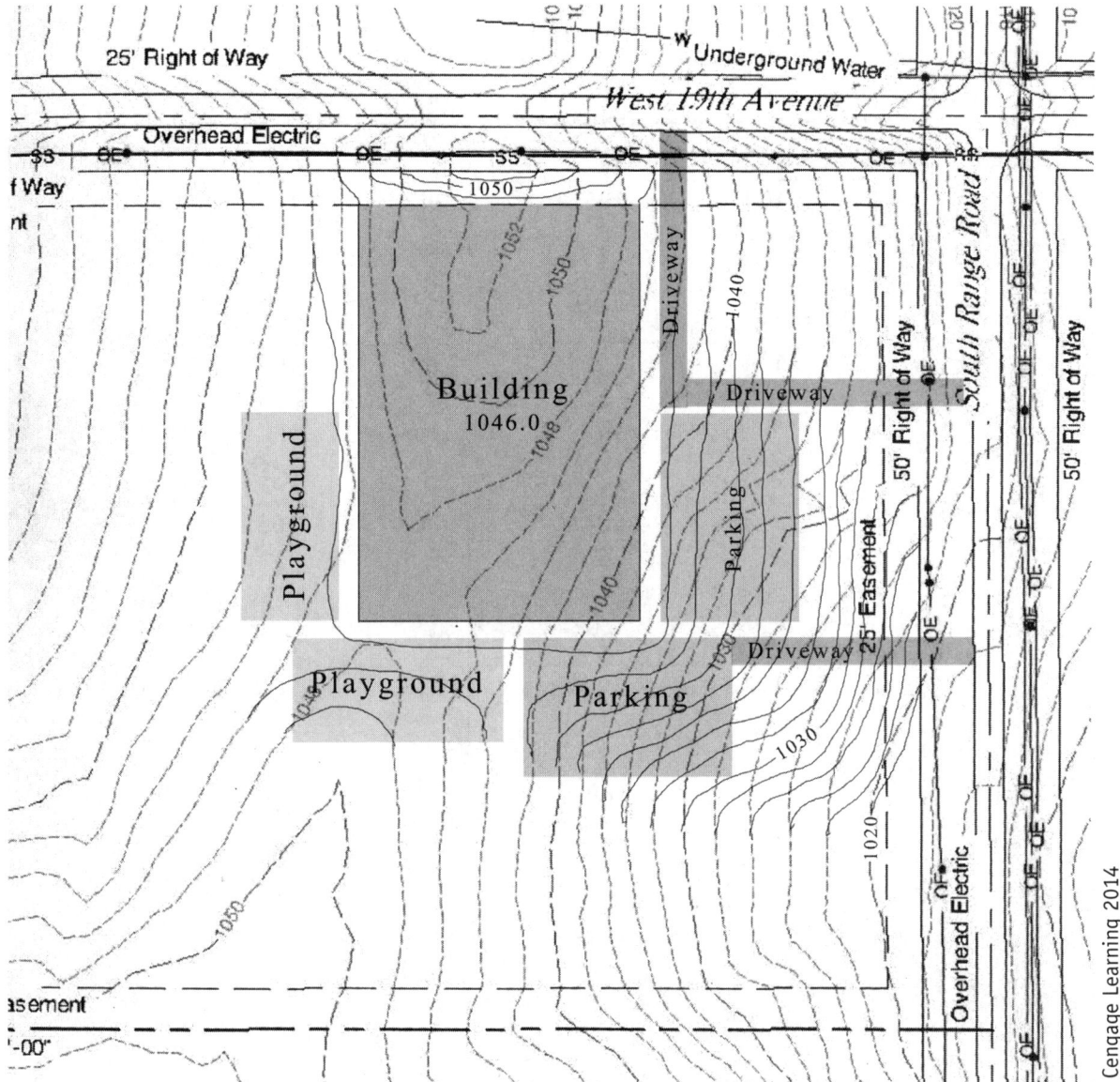

FIGURE 7-4 *Grading plan.*

Example: Cut and Fill Calculations

A portion of the site plan for Range Elementary School is shown in figure 7-5 with a 100-foot grid system on the developed areas. The building pad is at elevation 1046, and the site adjacent to the building can be assumed to be the same for simplicity.

Example 7.1 Determine the approximate amount of fill within the area bounded by grid lines C, D, 4, and 5.

The estimated existing and proposed grade elevations are indicated in figure 7-5 and are summarized in table 7-3.

Grading and Drainage   121

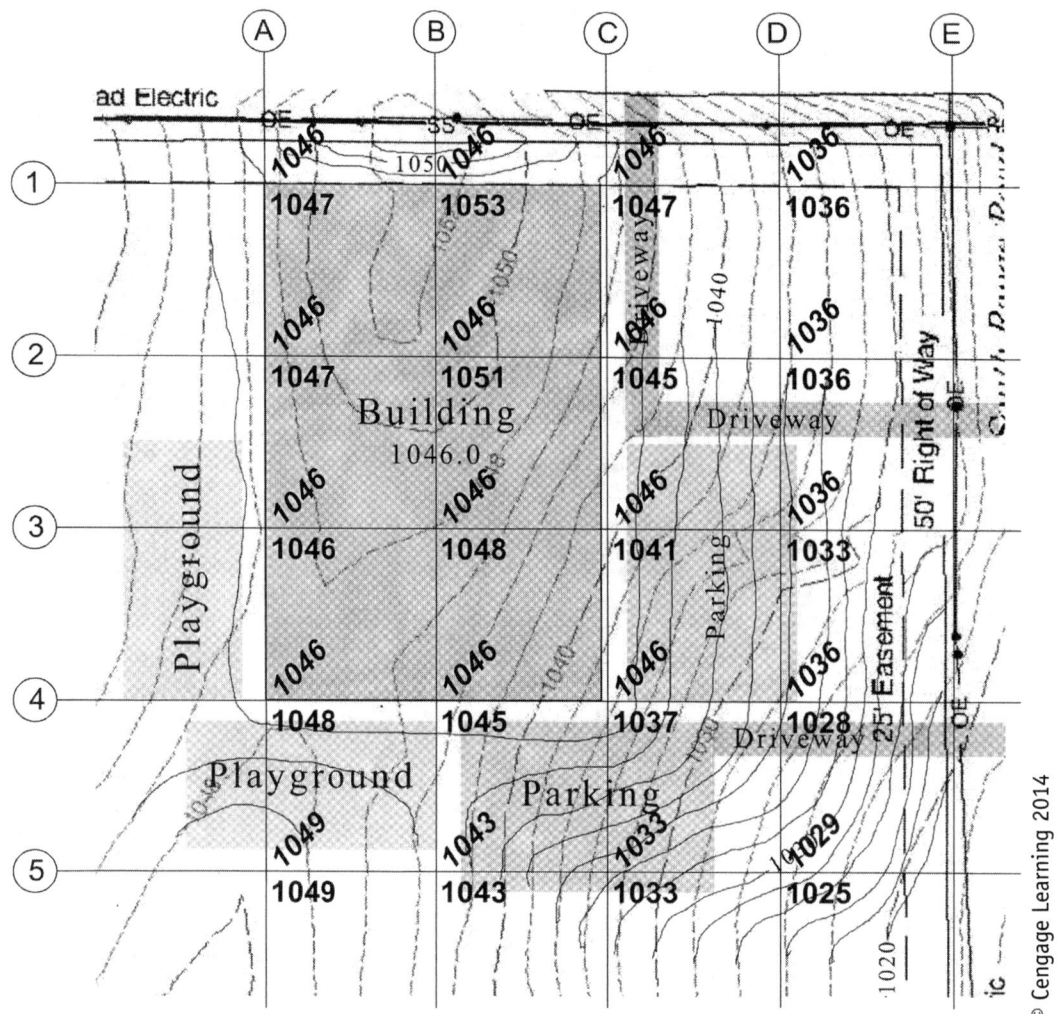

FIGURE 7-5 *Cut and fill plan.*

Grid Area	Existing Elevation		Proposed Elevation	
	C4	C5	C4	C5
	D4	D5	D4	D5
C4	1037	1028	1046	1036
	1033	1025	1033	1029

TABLE 7-3 *Existing and proposed grade elevation.*

The change in elevation at each corner of the area is first calculated using the following formula and recorded in table 7-4. The calculation for the top left grid C4 is shown.

Change in Elevation = existing elevation − proposed elevation = 1037 − 1046 = −9

122   Chapter 7: Site Design

Grid Area	Top Left	Top Right	Average Elevation Change (ft)	Area (sq. ft)	Volume (cu. ft)	Volume (cu. yd)
	Bot. Left	Bot. Right				
C4	−9	−8	−5.25	10,000	−52,000	−1944
	0	−4				

TABLE 7-4 *Cut and fill calculation.*

The average of the 4 values is then calculated and recorded.

$$\text{Average Depth} = \frac{\text{Sum of the Depths}}{4} = \frac{(-9-8-0-4)\,\text{ft}}{4} = -5.25\,\text{ft}$$

The area for this example is 100 ft by 100 ft or 10,000 sq. ft. The volume, V, is taken as the product of the average depth, Δh, and the area, A.

$$V = \Delta h \cdot A = -5.25\,\text{ft}(10{,}000\,\text{ft}^2) = -52{,}500\,\text{ft}^3$$

The volume is converted from cubic feet to cubic yards by dividing by 27.

$$V = -52{,}500\,\text{ft}^3 \left(\frac{1\,\text{yd}^3}{27\,\text{ft}^3}\right) = -1944\,\text{yd}^3$$

A negative number indicates a shortage of fill, so soil must be added to the site.

This calculation should be done for all areas of the site that have cut or fill to determine the total volume required.

Storm Water Management

When a site is developed, the flow of water across and exiting the property is altered. Most municipalities require a drainage study and appropriate drainage plans. Figure 7-6 shows the required drainage submittals, depending upon the type of land development for Stillwater, Oklahoma. The calculation of stormwater is dependent on many variables and can be performed by many methods.

Peak Stormwater Runoff

The Rational Method for calculating stormwater peak flow is commonly used for smaller drainage areas. The general procedure is outlined in the textbook and summarized here.

- Estimate the drainage area in acres.
- Determine the runoff coefficient, C, and correction factor, C_f.
- Estimate the hydraulic length used to determine the time of concentration.
- Determine the time of concentration, T_c, for the drainage area.
- Determine the rainfall intensity using the time of concentration.
- Substitute the drainage area, $C_f C$ value, and intensity into the Rational Formula.

Required Submittals: Drainage studies and drainage plans are required to demonstrate compliance with Stillwater City Code, City Drainage Standards, City of Stillwater Standards. The type of land development application dictates the required submittal(s) as shown in the table below. All submittals shall be signed and sealed by a professional engineer licensed by the State of Oklahoma.

Required Submittal	Application Type
Preliminary Drainage Study	Preliminary Plat, Preliminary Planned Unit Development
Final Drainage Study	Final Plat, Final Planned Unit Development, Commercial Use By Right, Earth Change Permit
Drainage Plan	Final Plat, Final Planned Unit Development, Commercial Use By Right, Earth Change Permit
Record Drawings	Upon completion of the drainage improvements and prior to issuance of certificate of occupancy and use or occupancy of a site, development, or other improvement

FIGURE 7-6 *Drainage code.*

The following sections will cover the calculations using the Range Elementary School project. For the purpose of calculating the drainage, only the eastern one-half of the property will be considered, since the land being developed is within that area.

Drainage Area

The drainage area could be taken as the buildable area excluding the rights of way and/or easements. For simplicity, the total site will be used, which is 1,320 ft east-west by 660 ft north-south, as shown in figure 6-14. The eastern one-half of the site is then 660 ft in both directions. The area must be converted from square feet to acres, and there are 43,560 sq. ft. in an acre.

$$\text{Area} = 660 \text{ ft} (660 \text{ ft}) = 435{,}600 \text{ ft}^2 = 435{,}600 \text{ ft}^2 \left(\frac{1 \text{ acre}}{43{,}560 \text{ ft}^2}\right) = 10 \text{ acres}$$

Runoff Coefficient

The runoff coefficients are a measure of the permeability of the surface that water flows over. Many of these coefficients are given in the textbook table 7-23. The use of pervious concrete pavement can help reduce stormwater runoff. Table 7-5 lists runoff coefficients to be used with pervious concrete. The values in the table may be used for preliminary studies of pervious concrete pavement systems at least 6 in. (150 mm) thick in 2-year and 10-year storms. A minimum value of 0.05 is recommended. Values at the lower end of the range may be used if the pervious concrete pavement system includes a clean stone base. For preliminary calculation, we will use 0.1 for the runoff coefficient.

Infiltration Rate in. / (cm/h)	Runoff Coefficient (Rational method C)
1 in./h (2.5 cm/h) or greater	0.05 to 0.10
0.5 to 1 in./h (1.3 to 2.5 cm/h)	0.10 to 0.20
0.1 to 0.5 in./h (0.3 to 1.3 cm/h)	0.20 to 0.35

TABLE 7-5 *Pervious concrete runoff coefficients.*

The coefficients for each of the developed areas of the Range Elementary project are listed in table 7-6 along with the approximate surface area.

Surface	Area, A (ft²)	(acres)	Coefficient, C
Site (unimproved)		10 acres (gross)	0.1–0.3
Building (roof)	60,000	1.38	0.75–0.95
Parking (concrete)	30,000	0.69	0.8–0.95
Driveways (pervious)	15,000	0.34	0.1–0.2
Playground	20,000	0.46	0.2–0.35

TABLE 7-6 *Runoff coefficients and surface areas.*

The net undeveloped site area is the gross area minus all the developed areas.

$$\text{Area} = 10 \text{ acres} - (1.38 + 0.69 + 0.34 + 0.46) \text{ acres} = 7.13 \text{ acres}$$

The composite runoff coefficient, C, is a weighted average of all the coefficients. The average values for each of the coefficients are used in the calculation.

$$C = \frac{\Sigma C_i A_i}{A} = \frac{7.13(0.2) + 1.38(0.85) + 0.69(0.875) + 0.34(0.15) + 0.46(0.275)}{10} = 0.337$$

C_i = Runoff coefficient of each surface
A = Total area of site
A_i = Area of each surface

Hydraulic Length

The hydraulic length on the site is drawn in figure 7-7 using the method described in the textbook. The approximate hydraulic length for the eastern one-half of the site is shown as 625 feet. This will be used for the rational calculation.

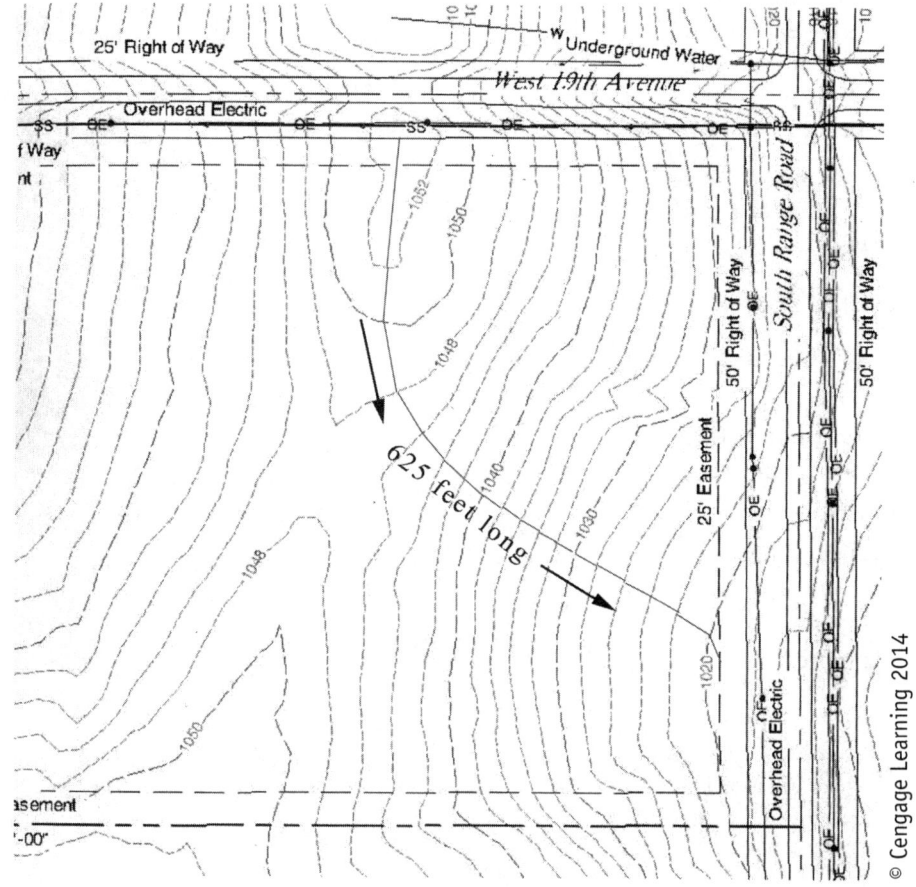

FIGURE 7-7 *Hydraulic length.*

Time of Concentration

The time of concentration is based on hydraulic length, the average ground slope, and the runoff coefficient. The average slope is calculated for the hydraulic length path.

$$\text{Average slope} = \left(\frac{\text{Change in elevation}}{\text{Hydraulic length}}\right) \times 100 \text{ percent} = \left(\frac{1053 \text{ ft} - 1019 \text{ ft}}{625 \text{ ft}}\right) \times 100 \text{ percent} = 5.44 \text{ percent}$$

The values of the hydraulic length, the average slope, and the runoff coefficient determined are plotted on the Seelye chart as shown in figure 7-8. On the left side, a line is drawn from the length in feet through the coefficient to the pivot line. Next, a line is drawn from the pivot through the slope to the time of concentration. In this case, we get 25.5 minutes.

Rainfall Intensity

The rainfall intensity is based on the time of concentration and the frequency of the storm considered. These values are plotted on an Intensity-Duration-Frequency (IDF) chart, which are available online from the National Oceanic and Atmospheric Administration (NOAA) at hdsc.nws.noaa.gov/hdsc/pfds/index.html. An IDF chart for the Range Elementary project is shown in figure 7-9 using a 30-minute time of concentration and a 10-year storm. The rainfall intensity is shown as in/hr.

The Rational Formula

The rational formula is a common method to calculate stormwater runoff. The formula is as follows:

$$Q = C_f C i A$$

Q = Peak rate of runoff in cubic feet per second (cfs)

C_f = Runoff coefficient correction factor (see textbook figure 7-27)

C = Composite runoff coefficient

i = Average precipitation intensity for the time of concentration (in/hr)

A = Watershed area (acres)

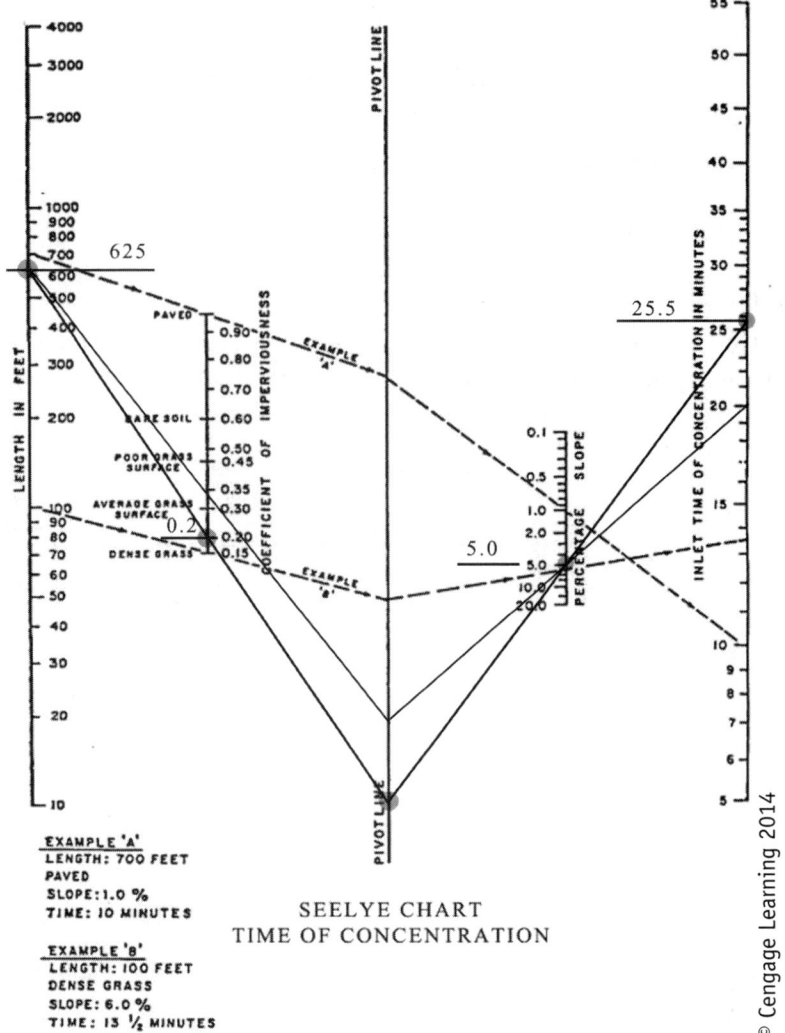

FIGURE 7-8 *Seelye chart.*

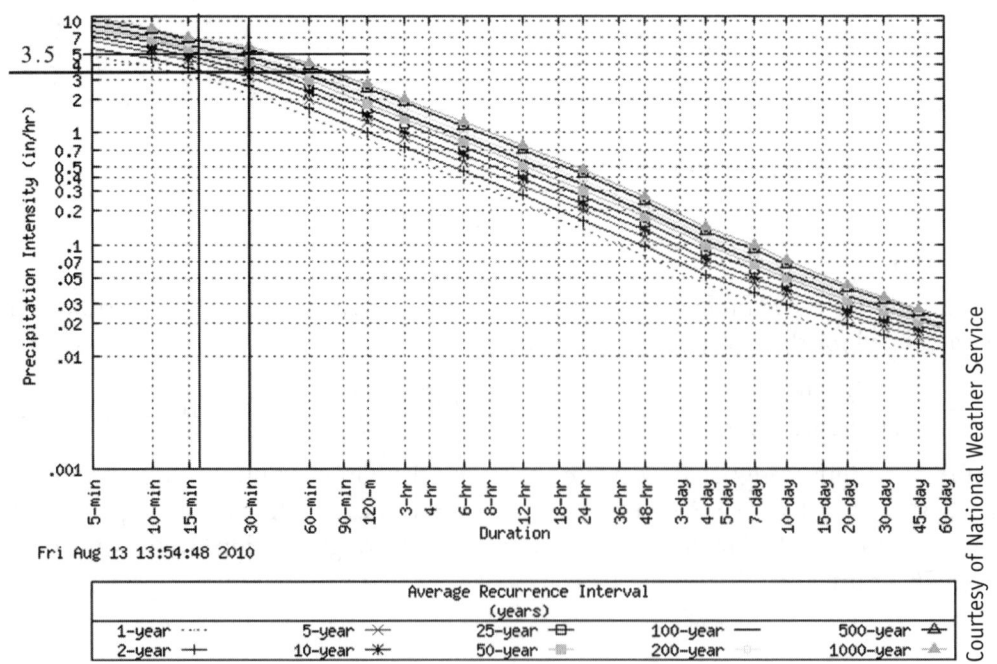

FIGURE 7-9 *Rainfall intensity chart.*

Since we used a 10-year storm, the value of C_f is 1.0 from textbook figure 7-27. The peak runoff rate is then calculated for the site.

$$Q = C_f CiA = 1.0(0.2)3.5 \text{ in/hr } (10 \text{ acres}) = 7.0 \text{ cfs}$$

This is the pre-development runoff on the site for a 10-year storm. Also shown on figures 7-8 and 7-9 are the 25-year post-development values. For figure 7-8, the composite runoff coefficient is 0.35, which yields a time of concentration of 20 minutes. For figure 7-9, the 20-minute duration is plotted on the 25-year storm, and the intensity is 5.0 in/hr. The post-development runoff is the calculated as 19.25 cfs with cf of 1.1. The net increase in peak flow rate is the difference between the post-development and pre-development values. In this case, the increase is 12.25 cfs.

Stormwater Storage and Treatment

The excess stormwater due to land development must be properly stored to reduce flooding. There are many ways to achieve this as outlined in the textbook. A few of the ways are summarized here.

- Retention ponds for sites larger than 25 acres that maintain a permanent pool of water (see textbook figure 7-29)
- Detention ponds for sites at least 10 acres that hold water for approximately 24 hours and drain dry (see textbook figure 7-29)
- Stormwater wetlands for sites larger than 25 acres that are similar to retention ponds but incorporate wetland plants (see textbook figure 7-30)
- Underground storage tanks where area is limited for ponds and wetlands

Estimating the Size of Stormwater Ponds

The general rule of thumb for stormwater management structure is approximately 10 percent of the developed area of the project. With careful planning and the use of environmentally friendly development, this may be reduced to 5 percent. For ponds and wetland, you could calculate the 24-hour difference between the pre- and post-development runoff. If this is not available, another approximation is 0.2 acre-ft/acre of development.

$$V = \frac{0.2 \text{ acre-ft}}{\text{acre}} A$$

V = Volume of stormwater (acre-ft)
A = Drainage area (acres)

Once the volume of the stormwater to be managed is found, the depth of the retention or detention structure is assumed, and the area of detention can be calculated.

$$A_D = \frac{V}{d}$$

A_D = Area of detention facility (acre)
V = Volume of stormwater (acre-ft)
d = Average depth of detention facility (ft)

The size of the detention or retention structure can be found using the textbook figure 7-31 for rectangular or circular areas.

Example 7.2
Determine approximate area of a retention pond required for the 10 acres developed for the Range Elementary Project. The average depth of the pond can be assumed to be 4 feet. The previously calculated 24-hour difference between the pre- and post-development runoff was 1.5 cfs.

The runoff must be calculated for a 24-hour period and then converted to acre-ft.

$$V = \frac{1.5 \text{ ft}^3}{\text{sec}} \left(60 \frac{\text{sec}}{\text{min}}\right)\left(60 \frac{\text{min}}{\text{hr}}\right)\left(24 \frac{\text{hr}}{\text{day}}\right)\left(\frac{1 \text{ acres}}{43,560 \text{ ft}^2}\right) = 2.975 \frac{\text{acre} \cdot \text{ft}}{\text{day}}$$

If the approximate calculation were used, the volume would only be 2 acre-ft/day. The area of the detention facility is calculated for the volume of stormwater.

$$A_D = \frac{V}{d} = \frac{2.975 \text{ acre} - \text{ft}}{4 \text{ ft}} = 0.744 \text{ acres} = 0.744 \text{ acres} \left(\frac{43{,}560 \text{ ft}^2}{\text{acre}}\right) = 32{,}800 \text{ ft}^2$$

If the approximate value were used, the area would be 1/2 acre or 21,280 ft^2.

Low Impact Development

Low Impact Development (LID) can be used to reduce the amount of stormwater runoff. In the Range Elementary Project, pervious concrete was used for the driveways as an LID technique. The textbook explains some of the LID techniques that are summarized here.

- Rain gardens and bioretention cells like wetlands
- Impervious surface reduction and disconnection
- Green roofs that have many advantages (see chapter 18)
- Rain barrels and cisterns that store roof runoff
- Permeable pavers like pervious concrete and other paving systems

Many municipalities have developed LID guidelines like the online reference for Sequim, Washington, in the textbook.

Problem Set 7.3   Grading and Drainage

Problem 7-21   What is the preferred minimum slope for a lawn?

Problem 7-22   What is the recommended slope immediately adjacent to a building?

Problem 7-23   What is the required amount of cut or fill between grids C, D, 4, and 5 for this textbook examples on page 245?

Problem 7-24   Redo the cut and fill calculation between grids C, D, 4, and 5 for figure 7-5 with 50-foot grid spacing instead of the 100-foot grid spacing.

Problem 7-25   What is the required amount of cut or fill under the building in figure 7-5, using the 100-foot grid spacing?

Problem 7-26   What is the difference in the runoff coefficients for sandy and clay soil lawns with a slope of 5 percent?

Problem 7-27   Determine the composite runoff coefficient for the entire 20-acre site of the Range Elementary School project using the materials in table 7-7. Use the average runoff coefficient in the calculation.

Surface	Area, A (ft^2) (acres)
Site (farmland)	20 acres (gross)
Building (roof)	60,000
Parking (asphalt)	30,000
Driveways (asphalt)	15,000
Playground	20,000

TABLE 7-7   Surface materials and surface areas.

Problem 7-28   Determine the approximate hydraulic length and average slope from the high point on the south side to the low point on the north side of the site shown in figure 7-10. The scale of the drawing is 1 inch equals 240 feet.

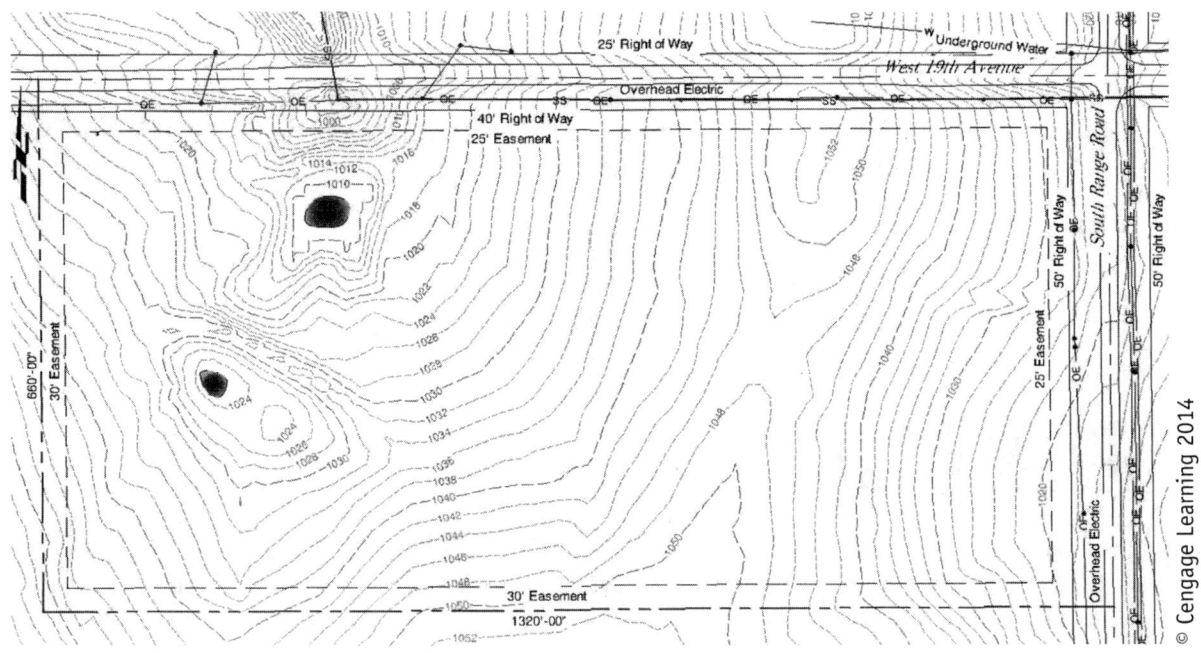

FIGURE 7-10 Topographic plan for problem 7-28.

Problem 7-29   What would be the time of concentration for a site with a hydraulic length of 850 feet, an average slope of 6 percent, and a runoff coefficient of 0.30?

Problem 7-30   What would be the time of concentration for a site with a hydraulic length of 625 feet, an average slope of 5 percent, and a runoff coefficient of 0.25?

Problem 7-31   What is the precipitation intensity for a time of concentration of 30 minutes and a 10-year storm using figure 7-11?

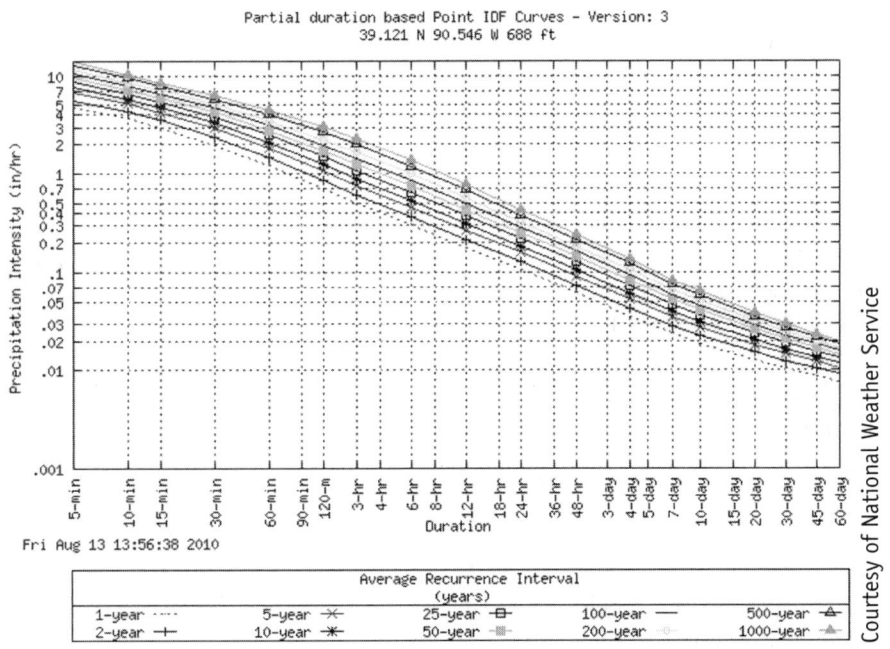

FIGURE 7-11 Rainfall intensity chart for problem 7-31.

Problem 7-32   What is the precipitation intensity for a time of concentration of 15 minutes and a 25-year storm using figure 7-11?

Grading and Drainage   129

Problem 7-33 Determine the stormwater runoff for a pre-development 10-year storm with a composite runoff coefficient of 0.175, precipitation intensity of 3.5 in/hr, and a 20-acre site.

Problem 7-34 Determine the stormwater runoff for a post-development 25-year storm with a composite runoff coefficient of 0.25, precipitation intensity of 5.5 in/hr. and a 20-acre site.

Problem 7-35 What would be the required diameter of a circular retention pond with an average depth of 5 feet on a 20-acre site with a stormwater runoff of 18cfs?

Problem 7-36 What would be the required size of a square retention pond in problem 7-35 using the approximate value of 0.2 acre-ft/acre?

BACKGROUND

Architects and engineers must be able to design a site for the available utilities including water, sewer, power, and communication services. These utilities must be properly located, and the design should consider their location as part of the concept development.

Utilities

The site utilities must be properly located, and the appropriate service providers contacted for their use. In most cases, the utilities will be placed in a right of way or an easement adjacent and parallel to the property line of the site. The local municipality is normally the contact agency for locating utilities. In figure 6-14, the utilities are shown for the Range Elementary School site. The sanitary sewer (SS) is located along the north side, overhead electrical (OE) service is located along the north and east sides, and the underground water (W) service is at the northeast corner of the site. The City of Stillwater uses a graphic information system (GIS) to view utilities and other property information. In figure 7-12, the utilities and streets are shown adjacent to the Range Elementary School site.

Water Supply

The water supply system includes the water source, treatment facilities, storage facilities, and distribution system. In the case of Stillwater, Oklahoma, the water source is Kaw Lake, which is approximately 40 miles north of town. The treatment facility supplies approximately 7 million gallons per day and is stored in water towers throughout the town. One of the newer water towers is just to the north of the Range Elementary School site as seen in figure 6-8. This particular tank supplies a large section of southwest Stillwater including the properties discussed in chapters 5 and 6.

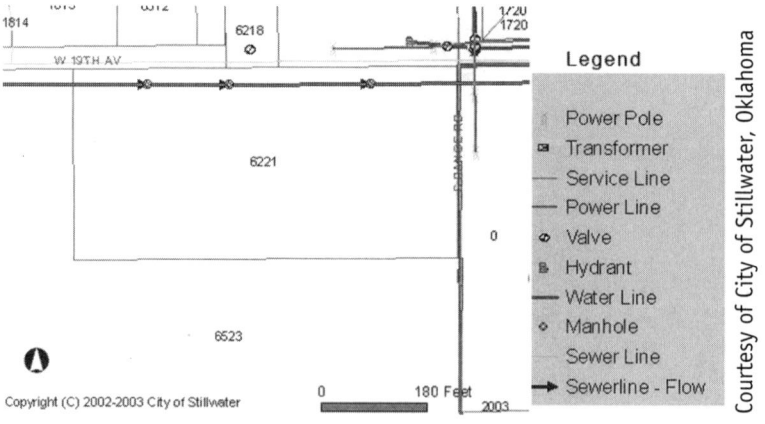

FIGURE 7-12 *Stillwater utilities with GIS.*

Water Pressure

Water pressure is a function of the depth below the water surface. The basic pressure is the specific weight of the fluid times the depth below the fluid surface.

$$p = \gamma h$$

p = fluid pressure (psi or psf)
γ = specific gravity (lb/in^3 or lb/ft^3)
h = depth below the water surface (in or ft)

For water pressure, the specific weight of water is 62.4 lb/ft³. The simplest form of the water pressure formula is the one foot water column.

$$1 \text{ foot of water} = 62.4 \frac{\text{lb}}{\text{ft}^3}\left(\frac{1 \text{ ft}^2}{144 \text{ in}^2}\right) = 0.433 \frac{\text{lb}}{\text{in}^2}\frac{1}{\text{ft}} = 0.433 \text{ psi}$$

This can be equated to 1 psi equal to 2.31 feet of water height. This pressure will act in all directions equally. This is illustrated in the textbook figure 7-37.

Static Head

Many times, the water pressure is described as the depth of the water or the static head. The pressure can then be found as the static head times the pressure of one foot of water. The configuration of the water column is not important in the pressure, just the static head as illustrated in the textbook figure 7-38.

$$\text{pressure} = \text{static head} \times 0.433 \frac{\text{psi}}{\text{ft}}$$

or

$$\text{pressure} = \frac{\text{static head}}{2.31 \frac{\text{ft}}{\text{psi}}}$$

Example 7.3 Determine the maximum water pressure due to the height differential between the 96-foot water tower in figure 6-8, sitting at an elevation of 1,054 and the house in figure 5-8 at an elevation of 974.

The static head is the maximum static height of the water drop. That is the height of the water tower plus the difference in the elevations of the tower and the building.

$$h = 96 \text{ ft} + (1054 - 974) \text{ ft} = 176 \text{ ft}$$

The height can then be multiplied times the pressure for one foot of water to determine the pressure.

$$p = (176 \text{ ft})0.433 \frac{\text{psi}}{\text{ft}} = 76.2 \text{ psi}$$

If the minimum pressure were desired, the diameter of the water tower would need to be known to find the minimum static head. That diameter would be subtracted from the water tower height to determine the minimum static head.

Head Loss

As water flows through the distribution system, it loses pressure due to friction on the pipe walls. In addition to frictional losses, losses will occur in pipe fittings. The Hazen-Williams formula is used to calculate the head loss, h_f.

$$h_f = \frac{10.44 \cdot L \cdot Q^{1.85}}{C^{1.85} \cdot d^{4.8655}}$$

h_{fp} = head loss in feet of water
L = length of pipe in feet
Q = flow of water in gallons per minute (gpm)
C = Hazen-Williams coefficient (see textbook figure 7-41)
d = diameter of the pipe in inches

The design values in the textbook figure 7-41 indicated design values lower than that of clean new pipe to be used in long term calculations. For the fittings, an equivalent length of pipe is used as shown in textbook figure 7-42. The losses due to fittings are minor and will only affect the loss by a few percent in most cases.

Example 7.4 What is the head loss for the property in figure 5-15 labeled "site"? The property is approximately 1 mile from the intersection of 19th and Sangre Rd., which is 2 miles east of 19th and Range Rd., where the water tower is located. The water flows through a 12-inch steel pipe at 150 gallons per minute.

The total length of pipe is 3 miles, and the Hazen-Williams coefficient from textbook table 7-41 for steel pipe is 100.

$$h_f = \frac{10.44 \cdot L \cdot Q^{1.85}}{C^{1.85} \cdot d^{4.8655}} = \frac{10.44 \cdot 3 \text{ miles} (5,280 \text{ ft/mile}) \cdot 150^{1.85}}{100^{1.85} \cdot 12^{4.8655}} = 1.97 \text{ ft}$$

Example 7.5 What is the equivalent length of 12-inch flanged pipe for the fittings shown in table 7-8?

Type of Fitting	Number of Fittings
Regular 90° elbow	4
Line flow tee	16
Branch flow tee	4
Gate valve	8
Globe valve	1

TABLE 7-8 *Pipe fittings.*

The equivalent lengths for each fitting are found in textbook figure 7-41 and recorded in column 3 of table 7-9. The product of the number of fittings and the equivalent length is calculated and recorded in column 4. The sum of the values in column 4 is taken as the total equivalent length.

Type of Fitting	Number of Fittings	Equivalent Length of Fitting (ft)	Total Equivalent Length (ft)
Regular 90° elbow	4	17	68.0
Line flow tee	16	6	96.0
Branch flow tee	4	34	136.0
Gate valve	8	3.2	25.6
Globe valve	1	390	390
Total			715.6

TABLE 7-9 *Pipe fittings.*

If this were included in example 7-3, the equivalent length would only increase by about 4.5 percent.

Total Dynamic Head

The total dynamic head (TDH) is the static head minus all head losses. The dynamic pressure is the pressure due to the dynamic head and is the approximate design pressure.

$$TDH = \text{static head} - \text{head loss}$$

$$\text{dynamic pressure} = TDH \, \frac{0.433 \text{ psi}}{\text{ft}} = TDH \, \frac{\text{psi}}{2.31 \text{ ft}}$$

Wastewater Management

Wastewater is not safe to drink or discharge into the environment, so it must be treated to improve quality. The wastewater is discharged into either a publicly owned treatment system or an onsite treatment facility like a septic system.

Publically Owned Treatment Works

In many larger municipalities, the sanitary wastewater is collected in pipes and fed to a wastewater treatment plant. This is known as *Publicly Owned Treatment Works* (POTW). The system is owned and operated by a state or local municipality. The water first goes through primary treatment that physically removes matter. Secondary treatment uses biological organisms that the eat pollutants. The piping system is gravity driven and must slope to provide adequate flow. In some cases, lift stations can be used to pump wastewater up to a higher level to be gravity fed. The textbook details some common code requirements that are summarized here.

- Pipe size of 4 inches for single family residential; 6 inches multi-family residential, retail, or commercial; and 8 inches for industrial
- Depth of pipe 2 feet below the lower floor level and completely below the frost depth (see figure 13-1)
- Slope of 2 percent minimum or 1/4 in/ft and 10 percent maximum
- Separated from water lines by 10 feet horizontally and at least 18 inches below

The sewer lateral slope from the building to the main line can be calculated as follows:

$$\text{Sewer lateral slope} = \left(\frac{\text{Invert of lateral at building} - \text{Crown elevation of main} - 1/2 \text{ OD}}{\text{Distance from building to sewer main}} \right) \times 100 \text{ percent}$$

1/2 OD = 1/2 the outside diameter of the sewer branch or main line

Example 7.6 The 12-inch sewer main along the center of Black Oak Drive is located at 6 feet below the ground. The elevation of the road is 976 above the sewer main. The lowest level of the residence, in figure 5-8, is at elevation 974. Assume the building is setback 41'-3" from the 30 foot right of way as shown. What is the lowest possible invert elevation at the residence?

The lowest possible invert elevation may be found using a minimum slope of 2 percent (1/4 in/ft). The formula for slope can be rearranged to solve for the invert elevation

Invert of lateral at building = Sewer lateral slope (Distance from building to sewer main) + (Crown elevation of main + 1/2OD)

Invert of lateral at building = 0.02(30 + 41.25) ft + (976 ft − 6 ft + 1/2(1 ft)) = 971.9 ft

This would just barely be below the 2-foot minimum depth below the floor. Also, from figure 13-1, the frost depth is approximately 0.75 m or 20 in.

Onsite and Decentralized Wastewater Treatment Systems

If a POTW is not available at a site, an onsite wastewater treatment system may be used. The most common onsite system used for residential construction is shown in textbook figure 7-46. The system has a septic tank for storage and settlement of solid waste, a drainage field to distribute the liquid waste, and the soil to disperse the liquid. The size of the septic tank must hold at least two days worth of waste and retain the solid. The length and size of the drainage pipes depend on the soil and building code. The soil must be permeable, and a percolation test may be used to determine the rate of water infiltration. In a percolation test, the time it takes the water level to drop a distance in a pre-saturated hole is percolation rate. The Web Soil Survey discussed in chapter 5 can be used to determine whether the soil is suitable for a septic system (websoilsurvey.nrcs.usda.gov). Search for the property of interest, and select an area of interest (AOI), then click "Soil Data Explorer," "Suitability and Limitations for Use," "Sanitary Facilities," "Septic Tank Absorption Fields," and "View Rating." The soils shown in figure 5-16 for the Black Oak property were investigated. The data areas labeled as 3 and 4 are a Coyle loam, and the area labeled 66 is a Masham silty clay loam all have very limted ratings.

Problem Set 7.4   Utilities

Problem 7-37   What is the water pressure at the bottom of an 8-foot-deep pool?

Problem 7-38   Determine the maximum water pressure due to the height differential between the 96 foot water tower in figure 6-8, sitting at an elevation of 1,054 and the southwest corner of the property on 22nd in figure 5-15 at an elevation of 920.

Problem 7-39   What is the major reason for pressure losses in a water distribution system?

Problem 7-40   What is a minor reason for pressure losses in a water distribution system?

Problem 7-41   What is the head loss for the property in figure 5-15 labeled 22nd? The property is approximately 1 mile from the intersection of 19th and Sangre Rd., which is 2 miles east of 19th and Range Rd., where the water tower is located. The water flows through a 16-inch steel pipe at 100 gpm.

Problem 7-42   What is the equivalent length of 8-inch pipe for the fittings shown in table 7-10?

Type of Fitting	Number of Fittings
Regular 90° elbow	6
Regular 45° elbow	2
Line flow tee	10
Branch flow tee	2
Gate valve	10
Angle valve	1

TABLE 7-10   *Pipe fittings.*

Problem 7-43   Determine the dynamic pressure for the property in problems 7-38 and 7-41.

Problem 7-44   Determine the dynamic pressure for the southwest corner of Dublin Drive and 15th Ave. in figure 5-12 at an elevation of 900. The property is approximately half a mile north and one and one-half miles east of 19th and Range Rd where the water tower in problem 7-38 is located. The water flows through an 8-inch steel pipe at 120 gpm.

Problem 7-45   What size pipe should be used for wastewater in an apartment complex?

Problem 7-46   What is the recommended minimum depth below the ground of a wastewater pipe?

Problem 7-47   What is the maximum depth of frost in Lincoln, Nebraska?

Problem 7-48   The 12-inch sewer main along the center of Dublin Drive is located at 6 feet below the ground. The elevation of the road is 900 above the sewer main. The lowest level of the residence at the southwest corner of Dublin Drive and 15th Ave. in figure 5-12 is at elevation 902. Assume the building is at the setback of 20 feet from the 30-foot right of way as shown. What is the lowest possible invert elevation at the residence?

Problem 7-49   What are the three basic components of an onsite septic system?

Problem 7-50   Would the soil at the corner of Celia St. and 22nd St. in Stillwater, Oklahoma, shown in figure 5-17 be adequate for a septic system?

BACKGROUND

Architects and engineers must be able to describe the components of the site needed to construct the project. This includes the buildings, parking, driveway, utilities, and all improvements to the site.

Site Plan

The site plan is the key drawing in locating all the site improvements. It might not contain a lot of detailed information about any particular item, but it must locate and describe all the items relevant to the site. The textbook outlines some of the requirements of a site plan. There are many different types of site plans that can be drawn, including architectural, demolition, civil survey, paving, grading, utilities, and landscape. On smaller projects, they may all be included in one drawing. A legend for site plans is shown in figure 7-13. A portion of a civil survey plan is shown in figure 7-14. A portion of an architectural site plan is shown in figure 7-15. A portion of a grading plan is shown in figure 7-16. A portion of a utilities plan is shown in figure 7-17. These are all from the Donald W. Reynolds School of Architecture Building at Oklahoma State University.

Problem Set 7.5   Site Plan

Problem 7-51   How are existing and new contours normally differentiated on a site plan?

Problem 7-52   Describe some of the features in a civil survey plan.

Problem 7-53   Describe some of the features in an architectural site plan.

Problem 7-54   Describe some of the features in a grading plan.

Problem 7-55   Describe some of the features in a utilities plan.

Problem 7-56   What are other types of plans that are drawn for an entire site?

LEGEND

---- 660 ----	EXISTING MAJOR CONTOUR
---- 662 ----	EXISTING MINOR CONTOUR
—— 660 ——	NEW MAJOR CONTOUR
—— 662 ——	NEW MINOR CONTOUR
——×——	FENCE
—— T ——	TELEPHONE OVERHEAD
—⌇—	POWER LINE OVERHEAD
—— G ——	GAS LINE
—— O ——	OIL LINE
—— PUG ——	POWER UNDERGROUND
—— TUG ——	TELEPHONE UNDERGROUND
—— TVUG ——	TV UNDERGROUND
—— W ——	WATER LINE
—— SS ——	SANITARY SEWER LINE
---- ··· ——▷	FLOW LINE DITCH
▬▬▬▬▬	HAY BALE
—— SF ——	SILT FENCE
IRR	IRRIGATION
RCP	REINF CONC PIPE
CI	CAST IRON
HB	HOSE BIB
EJ	EXPANSION JOINT
CJ	CONSTRUCTION JOINT
R	RADIUS
R/W	RIGHT OF WAY
PAVT	PAVEMENT
UNO	UNLESS NOTED OTHERWISE
SF	SQUARE FEET
DO	DOOR OPENING
OHD	OVERHEAD DOOR
CLR	CLEAR
CICI	CAST IRON CURB INLET
LF	LINEAR FEET
HJ	HAND-TROWELED JOINT
SJ	SAWED JOINT

FH	FIRE HYDRANT
GM	GAS METER
☒	GAS / OIL WELL
SS	SANITARY MANHOLE
S	STORM MANHOLE
FO	FIBER OPTIC MANHOLE
ST	STEAM MANHOLE
P	POWER MANHOLE
T	TELEPHONE MANHOLE
TP	TELEPHONE PEDESTAL
WM	WATER METER
⊚	WATER WELL
⋈	VALVE
CO	CLEANOUT
←	DOWN GUY
⌽	POWER POLE
LP	LIGHT POLE
XFR	TRANSFORMER PAD
BM	BENCH MARK
EP	EMERGENCY PHONE
FDC	FIRE DEPARTMENT CONNECTION
TP	TOP OF PAVEMENT
TC	TOP OF CURB
BC	BOT OF CURB
TS	TOP OF SIDEWALK
TR	TOP OF RIM
TG	TOP OF GRATE
TW	TOP OF WALL
TOF	TOP OF FOOTING
EL	ELEVATION
FL	FLOWLINE
FG	FINISH GRADE
FF	FINISH FLOOR
PVC	POLY VINAL CHLORIDE
PE	POLYETHYLENE
DIP	DUCTILE IRON PIPE
RJ	RESTRAINED JOINT
DB	DRAINAGE BASIN
ARCH	ARCHITECT

FIGURE 7-13 *Site legend.*

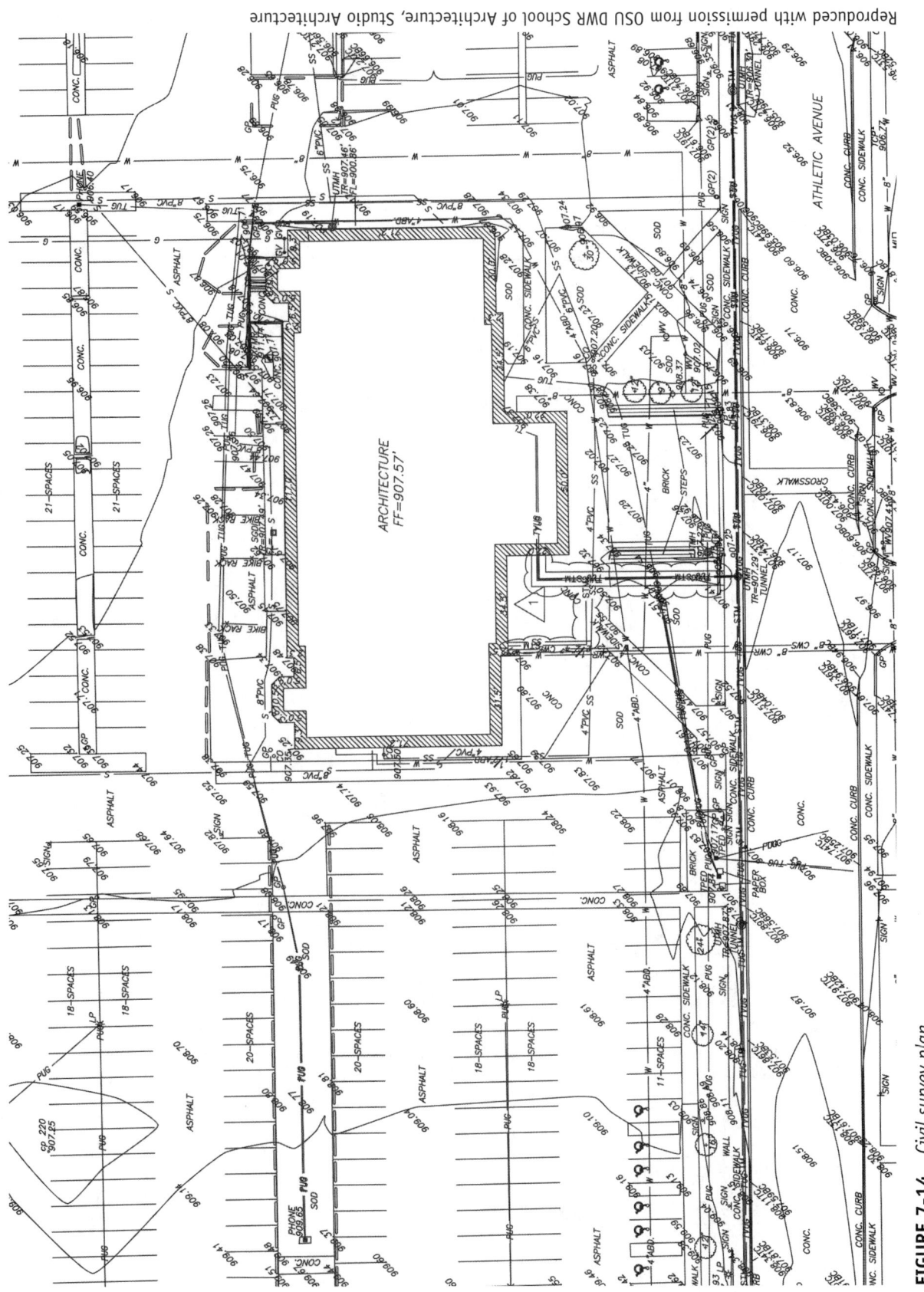

FIGURE 7-14 Civil survey plan.

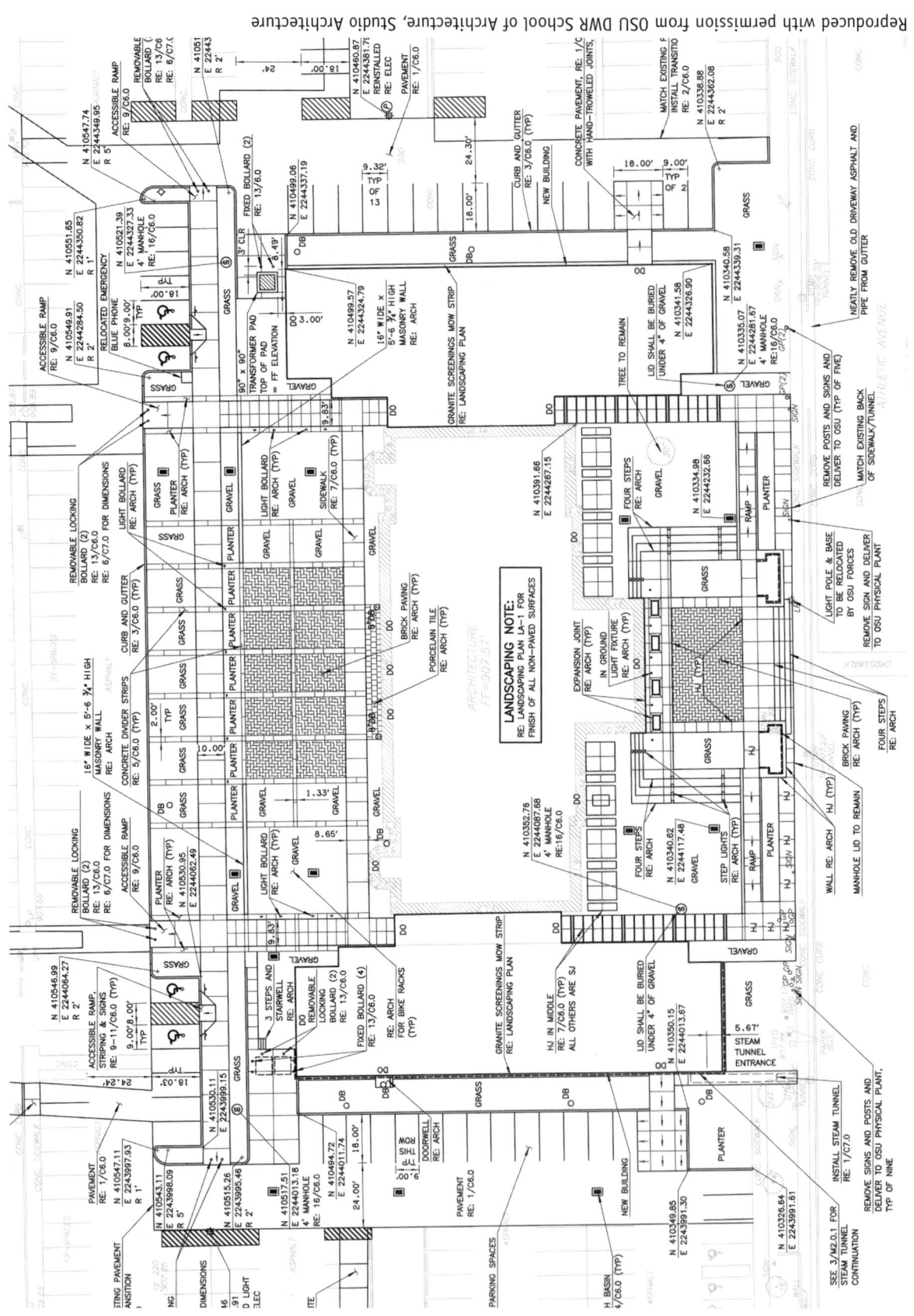

FIGURE 7-15 Architectural site plan.

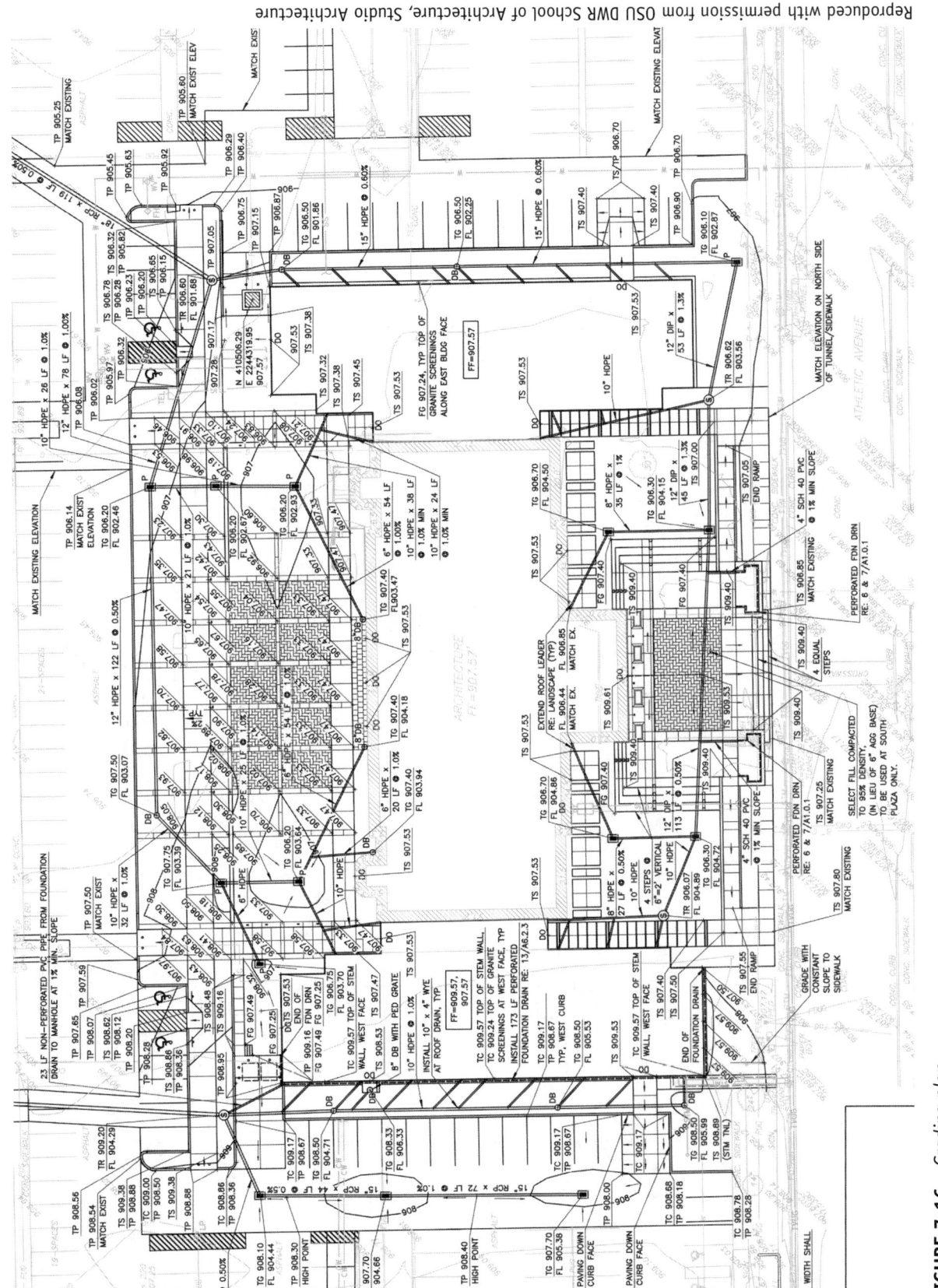

FIGURE 7-16 Grading plan.

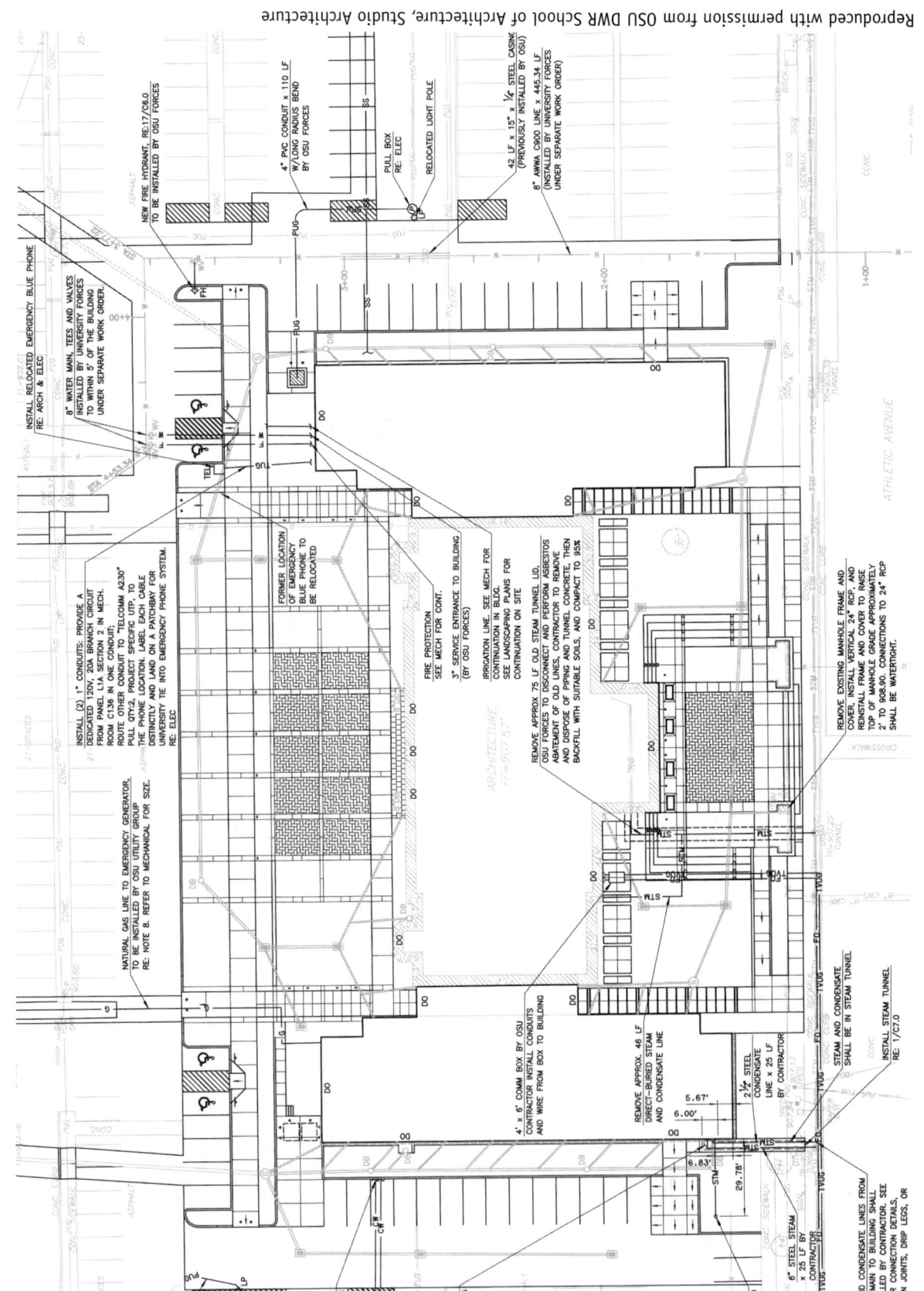

FIGURE 7-17 Utilities plan.

BACKGROUND

Architects and engineers must know what legal documents are required to be obtained to construct projects on a site. They must know which types of operations require documentations and which do not.

Permits

Building and earth-moving permits are typically obtained from the local municipalities. Normally, there are various parts to the permit, including foundation, structural, electrical, mechanical, and fire protection. A portion of the building permit requirement for Stillwater, Oklahoma, is shown in figure 7-18.

Section 23.44. Building permits.

Except as may be otherwise exempted by the Stillwater City Code, it shall be unlawful for any person to erect, construct, alter, move, remodel, or reconstruct any building or other structure, including accessory structures, until the building official has issued a building permit certifying that the plans and intended use of the land, buildings, and other structures conform with this Chapter and all other applicable codes. All building permit applications and submittals shall comply with rules and regulations governing building construction within the City of Stillwater in accordance with:

Chapter 7	Buildings, Articles II and III
Chapter 7	Buildings, Article VII, Stillwater Existing Building Code
Chapter 10	Electricity
Chapter 12	Fire Prevention & Protection, Articles III, V and VI
Chapter 24	Plumbing & Gas

A building permit applicant shall first file written application on a form furnished by the building official for that purpose. Such application shall:

1. Identify and describe the work to be covered by the permit for which application is made.
2. Describe the land on which the proposed work is to be done by legal description, street address or similar description that will readily identify and definitely locate the proposed building or work.
3. Indicate the use and occupancy for which the proposed work is intended.
4. Be accompanied by construction documents and other information as necessary to indicate the work to be completed.
5. Be signed by the applicant, or the applicant's authorized agent.

All building permit applications shall also comply with all state, local and federal rules and regulations, as applicable. No building permit shall be issued until all pending land use applications for the subject property are approved.

Once the building permit application is approved, the appropriate fee shall be paid and the building permit shall be issued.

FIGURE 7-18 *Building permit requirements.*

Residential and commercial permit applications for Stillwater are shown in figures 7-19 and 7-20.

RESIDENTIAL (1&2 Family) Permit Application

☐ New Construction   ☐ Alteration   ☐ Addition   ☐ Remodel   ☐ Accessory   ☐ Storm Shelter   ☐ Modular   ☐ Other

Project Address:_____

FOR OFFICE USE ONLY: FEMA floodplain? ☐ Yes   ☐ No

Lot #:_____ Block #_____ Subdivision:_____ Zoning:_____

IF UNPLATTED PROVIDE COPY OF DEED

Property Owner:_____ Address:_____ Phone:_____

Contractor:_____ Address:_____

Phone Number:_____ Fax Number:_____

OWNER/CONTRACTOR SIGNATURE OF UNDERSTANDING AND AGREEMENT

I hereby certify that the statements in this application and the attachments hereto are accurate and that the property owner has given permission for this work to proceed. I further certify that all construction work under this permit will conform to all applicable ordinances, rules or regulations of the City of Stillwater and that all electrical, plumbing, mechanical, fence, sign and driveway construction shall be performed by contractors licensed by the State of Oklahoma (if applicable) and registered and bonded with the City of Stillwater.

(OWNER)(CONTRACTOR): **SIGNED**_____ DATE_____

(OWNER)(CONTRACTOR): **PRINT**_____ DATE_____

DESCRIPTION:
_____
_____

Number of Stories:_____ # of Bedrooms:_____ # of Bathrooms:_____ # of Water Closets_____

Electric Service (# of amps):_____ Contact Electric Utilities Regarding Cost Evaluation & Installation of Services

Water Meter: ☐ ¾"   ☐ 1"   Work Order#_____

Sewer Service: ☐ Public   ☐ Septic (if Septic, provide copy of Perk Test)

☐ Other_____

Exterior Wall Finish:_____ Roof Covering:_____

Square Footage

Finished: 1st _____ 2nd _____ Unfinished: 1st _____ 2nd _____ Garage: _____

Total Sq. Ft.: _____ Lot Square Footage: _____ % Lot Coverage: _____

Valuation: $_____ (Valuation includes structural, electrical, plumbing, mechanical, interior finish, overhead and profit R108.3.)

(All contractors MUST be licensed and registered with the City of Stillwater and/or the State of Oklahoma)

Plumbing Contractor:_____ Contact:_____ Phone:_____

Mechanical Contractor:_____ Contact:_____ Phone:_____

Electrical Contractor:_____ Contact:_____ Phone:_____

Driveway/Sidewalk Contractor:_____ Contact:_____ Phone:_____

FIGURE 7-19 *Residential permit application.*

I. **ADDRESS:** (LOCATION)____________________

SUBDIVISION____________ LOT________ BLOCK________ ZONING DISTRICT____________

II. TYPE AND COST OF BUILDING

A. ✓TYPE OF IMPROVEMENT	**B. ✓ USE** *(circle **Proposed Occupancy**)*
❑ NEW BUILDING	❑ ASSEMBLY A-1 A-2 A-3 A-4
❑ ADDITION	❑ BUSINESS
❑ ALTERATION ***(complete item K)***	❑ EDUCATIONAL
❑ REPAIR, REPLACEMENT ***(complete item K)***	❑ FACTORY OR INDUSTRY F-1 F-2 H-1 H-2 H-3 H-4 H-5
❑ DEMOLITION	❑ INSTITUTIONAL I-1 I-2 I-3 I-4
❑ MOVING (RELOCATION)	❑ MERCANTILE
C. VALUATION Value of improvement $________ (***Valuation*** *= TOTAL COST includes structural, electrical, plumbing, mechanical, permanent systems, interior finish, materials and labor excluding land value. IBC108.3.)*	❑ RESIDENTIAL R-1 R-2 R-3 R-4 ________# of Buildings ________# of Units
	❑ STORAGE S-1 S-2
	❑ UTILITY- specify ____________

III. SELECTED CHARACTERISTICS OF BUILDING

D. ✓TYPE OF CONSTRUCTION	**F. DIMENSIONS**	
	Occupant Load	
	Number of stories	
❑ TYPE I - A or B (noncombustible)	Total square feet of floor area, all floors	
	Total square feet based on exterior dimensions	
❑ TYPE II - A or B (noncombustible)	Total Land Area *(Trans fee=$0.075/sq ft or $750 minimum-04/15/08)*	
	G. NUMBER OF OFF-STREET PARKING SPACES	
❑ TYPE III - A or B (masonry bearing / combustible framing)	Enclosed	
	Outdoors	
❑ TYPE IV - A or B (heavy timber)	**H. RESIDENTIAL BUILDINGS** Number of bedrooms	
	Number of bathrooms	
❑ TYPE V - A or B (combustible)	Full	
	Partial	
E. COMMERCIAL BUILDINGS - Number of Air Handler, Furnace or Roof Top Units	**I. COMMERCIAL BUILDINGS -** Number of toilets Urinals *(Water Closet Fee=$100 / toilet or urinal)*	

Water Meter Size : ❑ **3/4"** ❑ **1"** ❑ **1-1/2"** ❑ **2"** ❑ **3"** ❑ **4"** ❑ **6"** **Effective February 9, 2009**

Water and/or Sewer Capacity fees are paid at time of permit issuance. **NOTE:** *A separate commercial water meter request and applicable fees wil be paid when a water meter order is requested.*

J. ✓DESCRIPTION *(✓ all that apply)*

❑ EXTERIOR WALL FINISH - ____________	❑ ROOF COVERING - ____________
❑ BUILDING HEIGHT - ____________	❑ FIRE SUPPRESSION SYSTEM **Yes** **No**
❑ SOILS REPORT *(contact Bldg Division for ADDITION requirements)*	❑ LOT COVERAGE ________

K. EXPLANATION OF PROJECT:

Courtesy of City of Stillwater, Oklahoma

FIGURE 7-20 *Commercial permit application.*

Problem Set 7.6 Permits

Problem 7-57 What are some of the types of actions that a require building permit?

Problem 7-58 What are some general items that must be included in a building permit?

Problem 7-59 What are some items that are specific to a commercial building permit application?

Problem 7-60 What are some items that are specific to a residential building permit application?

CHAPTER 8
Energy Conservation and Design

Skills List

After completing this chapter, you should be able to:

- Understand the history of energy use in the United States

- Understand the EPA's Energy Star program and learn how to use the program website

- Understand the costs associated with electric power consumption in a building

- Understand and be able to utilize the American Society of Heating, Refrigerating, and Air-Conditioning Engineers (ASHRAE) lighting design guidelines

- Identify active and passive design elements that can be used to reduce energy consumption

BACKGROUND

Energy conservation is becoming a major concern as it applies to building design. According to the U.S. Energy Information Administration (EIA), in 2009 the United States was the largest consumer of energy in the world, using approximately 20 percent of the world's energy. As developing countries continue to emerge economically, the demand for energy worldwide will greatly increase. Therefore, it is imperative for architects and engineers must find ways to design and construct buildings that utilize energy more efficiently to help address the inevitable worldwide energy shortfall.

Early Building Design

In order to understand the current state of energy use in the United States, it is important to understand the history of energy in this country. Information on this topic can be found on several websites, including:

- U.S. Energy Information Administration: www.eia.doe.gov/kids/energy.cfm?page=4
- The Franklin Institute: www.fi.edu/learn/case-files/energy.html
- Union of Concerned Scientists: www.ucsusa.org/clean_energy/clean_energy_101/a-short-history-of-energy.html

Figure 8-1 gives the breakdown of energy types and the amounts used in the United States in the past 150 years, showing a steep increase in energy consumption in the last 60 years. During that time, petroleum has been used to supply more energy each year than any other source.

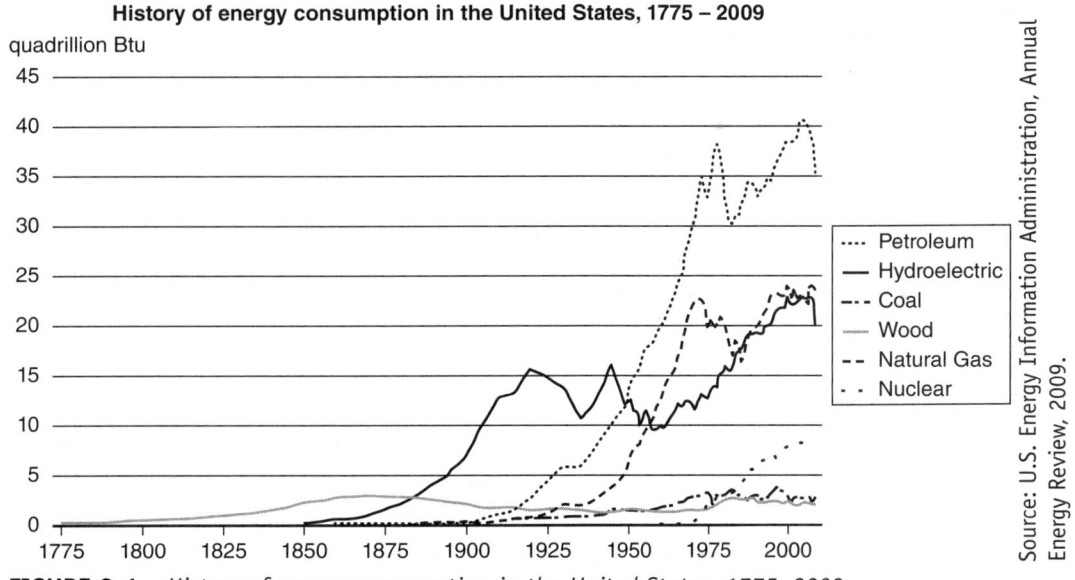

FIGURE 8-1 *History of energy consumption in the United States, 1775–2009.*

In addition to the history of energy use in the United States, it is also important to understand the global use of energy. The U.S. Energy Information Administration website www.eia.gov/ is a clearinghouse of information pertaining to energy supply, demand, and use in the United States and around the world.

A joint venture between the Environmental Protection Agency (EPA) and the U.S. Department of Energy (DOE) has established the Energy Star program for use in energy-efficient design and living. The program website at www.energystar.gov/ provides information on a multitude of topics pertaining to energy efficiency, from tips to save energy at home to a list of Energy Star–qualified home builders. Established by the EPA in 1992, the Energy Star program has become a valueable resource for use in designing for energy efficiency. Explore the website to become familiar with the types of information available, and think about how this information could be utilized to achieve an energy-efficiently designed building.

Example 8.1 Using the Energy Star website, describe the Energy Star Performance Rating system.

Energy Star's rating system rates energy performance of buildings and is based on a 100-point scale, with a rating of 50 being a building with average energy performance. The rating system was originally developed for office buildings in 1999 and since that time has been expanded to evaluate many different types of buildings. The system evaluates several criteria to arrive at a building's rating value and uses:

- A national database of information obtained through surveys performed every four years.
- Energy performance compared to source energy consumption.
- A statistical analysis of reference data to indicate causes of energy consumption.
- Comparisons of the energy performance with all buildings in the database.

Because energy efficiency can be impacted due to unique characteristics in each of the building type, the EPA Energy Star Performance Rating system evaluates and modifies the rating system as necessary for each of the building types.

Problem Set 8.1 Early Building Design

Problem 8-1 Research the U.S. Energy Information Administration website to determine the mission statement of this agency. List products and services provided by this agency.

Problem 8-2 Using the Energy Star website, research and discuss the history of the program.

Problem 8-3 List the types of buildings for which the Energy Star Performance Rating system has been established.

Problem 8-4 List and discuss the four criteria for the Energy Star Performance Rating system.

Problem 8-5 Use the Energy Star website to find the home builder that has built the greatest number of Energy Star–qualified homes in Dallas, Texas. How many Energy Star–qualified homes have been built in the Dallas-Fort Worth-Arlington area to date?

Problem 8-6 Within the Energy Star program, what does it mean for a contractor to incorporate the "100% Builder Commitment Option" into their construction company?

Problem 8-7 List the six features that every Energy Star–qualified home must possess.

Building Design and Energy

Design of building structures at the turn of the 20th century was driven by available materials and their limitations. For example, before the development of air conditioning, cross ventilation was used to cool a building. To make cross ventilation work effectively, limits were placed on building width. This was often accomplished by incorporating light wells into the building design. Light wells worked as both a skylight to bring light into the center of a building, and a means of supplying cross ventilation. Today, light wells are often used in lobby spaces to provide natural light to a building's interior, as shown in figure 8-2. Except in milder climates, light wells have become skylight systems and are sealed to prevent conditioned air from escaping the building.

FIGURE 8-2 *Light well brings natural light into interior spaces.*

Advances in passive design can often be directly related to other events. In the 1970s, two separate energy crises greatly affected every aspect of life in the United States. These crises caused shortages of fuel for vehicles, resulting in increases in fuel prices that, in turn, negatively affected the economy of the U.S for a number of years. As a result, a concentrated effort by the United States occurred to include passive and active design into building construction, and to utilize alternate energy sources to help reduce the country's dependency on foreign energy sources. These alternate energy sources include:

- Nuclear power
- Solar power using photovoltaic systems (refer to figure 8-3)
- Solar water heating (refer to figure 8-3)
- Geothermal heating and cooling systems
- Wind turbine systems

FIGURE 8-3 *Home with both solar water heater and photovoltaic cells.*

One result of the oil crisis in the early 1970s was the approval for construction of the Trans-Alaska Pipeline in 1973. This pipeline allowed access to the vast oil reserves that are present in the state of Alaska. The Department of Energy was created in 1977 by President Jimmy Carter as part of the President's Cabinet. Its responsibility includes the U.S. policies on energy as well as the safety in the handling of nuclear material. In addition, the government passed several acts affecting energy in the United States, including:

- The Geothermal Energy Research, Development and Demonstration Act in 1974
- The Solar Heating and Cooling Demonstration Act in 1974

Building Design and Energy

- The Solar Energy Research, Development, and Demonstration Act in 1974
- The National Energy Act in 1978
- The Natural Gas Policy Act in 1978

Problem Set 8.2   Building Design and Energy

Problem 8-8   Draw a sketch of a home in your community that utilizes passive design, and identify the systems used in the home.

Problem 8-9   Research the use of photovoltaic systems in building design, and explain their advantages and disadvantages.

Problem 8-10   Define the Trans-Alaskan pipeline, and discuss how it has impacted the United States.

Problem 8-11   Discuss the causes of the 1973 oil embargo.

Problem 8-12   List and describe additional alternate energy sources beyond those listed in this section.

Problem 8-13   Research Frank Lloyd Wright's Larkin Building, and describe the passive design elements included in the design.

Problem 8-14   Research the geothermal system, and discuss how it can be used to heat and cool a building.

Energy Conservation

To promote the development of energy conservation, the United States has passed three Energy Policy Acts (EPACTs), in 1992, 2005, and 2007. These acts include provisions for energy conservation through programs such as Energy Star and also provide tax incentives and grants for renewable and non-renewable energy development. As an example, EPACTs guarantee loans to developers of building/energy technologies that do not produce greenhouse gases as a by-product. The Energy Policy Act of 2005 can be accessed on the Environmental Protection Agency website at www.epa.gov/lawsregs/laws/epa.html.

Electrical and Lighting

According to the Energy Star website, in 2009 a typical single family home in the United States had an annual energy bill of approximately $2,200. This annual energy bill can be broken down into categories as shown in figure 8-4. Note that heating, cooling, water heating, and lighting make up over 70 percent of the annual cost. By incorporating passive and active energy systems into the design of a home, these costs can be reduced.

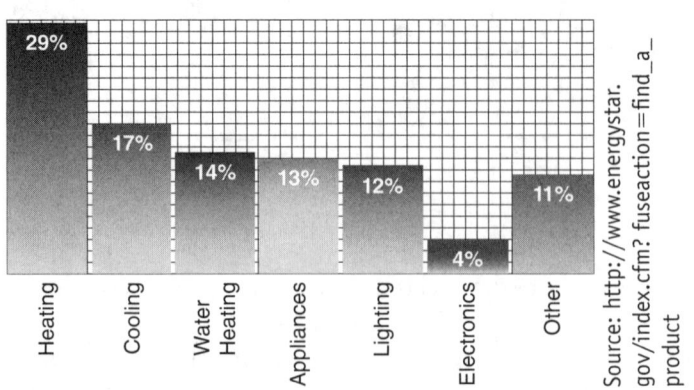

FIGURE 8-4 *Residential energy usage.*

Energy Conservation Through Lighting Design

According to the Energy Information Administration, lighting in a commercial building consumes 21 percent of the total energy used, as seen in figure 8-5.

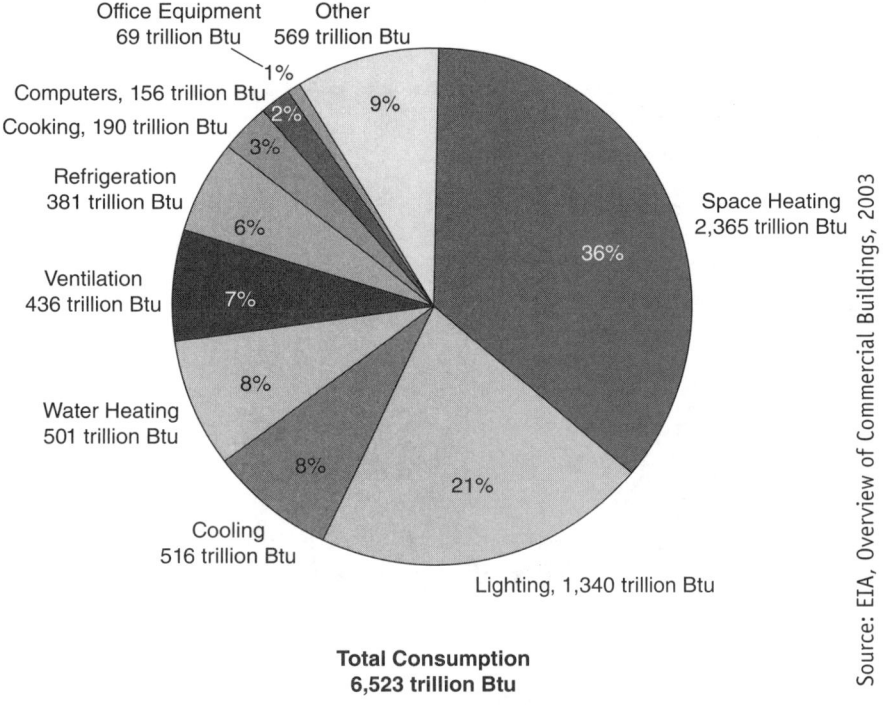

FIGURE 8-5 *Energy consumed by commercial buildings.*

There can be significant energy savings if the lighting within a building can be designed to work more efficiently. Advances in light bulb design include compact fluorescent bulbs and light emitting diode (LED) bulbs, as shown in figures 8-6 and 8-7.

FIGURE 8-6 *Compact flolorescent bulb.*

FIGURE 8-7 *Light emitting diode(LED) bulb.*

Each of these systems saves energy over incandescent light bulbs, with the LED bulbs being the most efficient and also currently the most expensive. In 2007, the Energy Independence and Security Act became law, requiring that incandescent bulbs be phased out of use in the United States. According to this law, starting on

January 1, 2012, the manufacture and import of 100 W bulbs stopped; after January 1, 2013, the manufacture and import of 75 W bulbs will stop; and after January 1, 2014, the manufacture and import of 40 and 60 W bulbs will stop. In addition to more efficient bulbs, dimmers and motion detectors are now being installed in buildings for better efficiency. Motion detectors work in public and private spaces and sense when no one is present in a room, at which time the lights shut off. These sensors, an example of which is shown in figure 8-8, can work for large spaces, such as lecture halls, and for small spaces, such as offices.

FIGURE 8-8 *Multi-function switch.*

To help reduce the amount of energy used for lighting in buildings, the American National Standards Institute (ANSI), the American Society of Heating, Refrigeration, and Air-Conditioning Engineers (ASHRAE), and the Illuminating Engineering Society of North America (IESNA) have jointly developed the ANSI/ASHRAE/IESNA Standard 90.1. This standard, for the construction of new commercial buildings, has limitations on the amount of lighting power that can be included in a building design. The code is written in a format that is easily utilized by both building designers and regulating agencies.

Electric Utility Costs

When designing a building, the project budget is often a major concern. In the current state of building design and construction, in addition to construction costs, lifetime operating costs must also be considered. For most buildings, by installing energy-saving devices in a building initially, the lifetime cost of energy for the building can be greatly reduced. Though the initial cost of installing energy saving devices can be substantial, the payback can be significant when one considers the current cost of energy and that this cost will continue to rise.

The cost of electricity is a function of kilowatt-hours, which is defined as 1,000 watts of electricity used constantly for one hour. An example of one kilowatt-hour is a lamp turned on for 10 hours that uses a 100 watt bulb (10 hours × 100 watts = 1000 watt-hours = 1 kilowatt-hour). The cost of electricity supplied by the power company can be broken into two categories. The first is known as the energy charge and is based on the amount of energy used during a billing period. This charge reflects the actual cost of energy used. In some locations, this charge varies throughout the day, with peak energy-use times costing more.

The second is known as the demand charge, which is a surcharge associated with the maximum energy demand for a building during a set period of time, usually a billing period. The demand charge is imposed by the electric companies to pay for electrical facilities that are capable of producing energy to meet customer's anticipated demand for energy. In hotter climates, the use of large amounts of energy to cool a building can affect the cost of energy throughout the entire year. It is possible for the demand charge to exceed the energy charge in some parts of the country.

For some electric companies, a third charge is known as a ratchet clause, which is included in the cost of electricity. This charge is similar to the demand charge, but instead of basing the peak energy use on a particular billing period, it is often based on a period of one year. It typically states that the minimum demand charge in any month will not be less than a certain percentage of the same charge for the month when the maximum energy was used. This charge can greatly increase the energy costs, since it is applied to each month's billing, even when that amount of energy was not used.

For each of the costs associated with an electric bill mentioned above, it is to the advantage of the building owner to incorporate energy-saving measures in the initial design to reduce the energy costs. Simple changes, such as using compact fluorescent bulbs instead of incandescent bulbs, can result in substantial cost savings. Figure 8-9 provides the cost of electricity for each state in the U.S. for 2008. The costs shown are based on a kilowatt-hour of energy for residential buildings. The average cost of electricity for the entire U.S. in 2008 was 11.26 cents per kilowatt-hour, with costs in individual states ranging from a low of 6.99 cents to a high of 32.5 cents per kilowatt-hour.

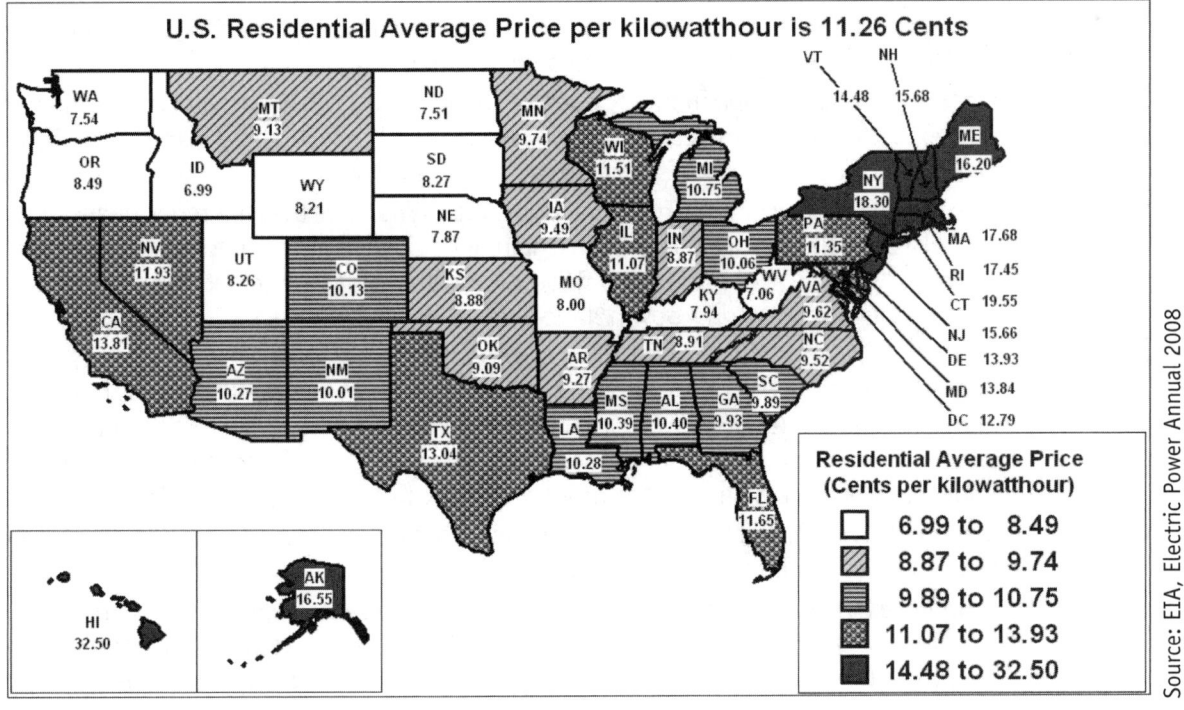

FIGURE 8-9 U.S. residential average price per kilowatthour.

Example 8.2 For a residence located in Denver, Colorado, determine the yearly cost savings of using 13 watt compact fluorescent bulbs instead of 60 watt incandescent bulbs for the following conditions:

- Base calculations on 2008 electricity costs and 52 weeks in a year
- 25 light bulbs
- Each bulb is lit an average of 10 hours a day, 7 days a week

To determine the cost savings, utilize the following steps:

Step one: Determine the cost of electricity in Denver, Colorado, using figure 8-13, as 10.13 cents per kilowatt hour.

Step two: Determine the difference in watts used in a single bulb: 60 watt − 13 watt = 47 watts.

Energy Conservation **151**

Step three: Calculate the cost savings per year based on 25 bulbs lit 10 hours a day, 7 days a week, for 52 weeks based on the difference in watts used:

Savings = $0.1013/kilowatt hour × 47 watts × 10 hours/day × 7 days/week × 52 weeks
× 25 bulbs × (1 kilowatt/1000 watt) = $433.26

Problem Set 8.3 Energy Conservation

Problem 8-15 Identify the twelve topics pertaining to energy production in the United States that are included in the Energy Policy Act of 2005.

Problem 8-16 Identify the seventeen laboratories owned by the U.S. Department of Energy that are included within the jurisdiction of the Energy Policy Act of 2005.

Problem 8-17 What is the yearly electricity cost for a home located in New York City that uses an average of 1,100 kilowatt-hours per month?

Problem 8-18 Determine the average monthly cost of using fifty 100 watt light bulbs that are turned on for 12 hours a day, 5 days a week, in a home located in Houston, Texas. Base the calculation on 52 weeks in a year.

Problem 8-19 Count the number of light bulbs in your home. What cost savings per year could be saved if all bulbs in your home were originally 75 watt incandescent bulbs and were replaced with 20 watt compact fluorescent bulbs? Assume the lights are on an average of 6 hours a day, seven days a week, for 52 weeks.

Problem 8-20 A family is moving from Miami, Florida, to San Diego, California. In Miami, they lived in a 4,800 sq. ft. home, and in San Diego they will live in a 3,600 sq. ft. home. If the Miami home used 19,200 kilo-watt hours of electricity per year, and the electricity use per sq. ft. of home will be the same for both locations, what will be the increase or decrease in the monthly cost of electricity for the family?

Problem 8-21 Determine the five states with the highest cost of electricity for residences. Give reasons why the costs may be higher in these states.

Problem 8-22 Using the U.S. Energy Information Administration website, determine the average annual consumption of electricity for a residence. What states have the highest and lowest average consumptions of electricity for residences, and what are the consumption rates?

ANSI/ASHRAE/IESNA 90.1-07 Lighting Requirements

An energy standard has been produced jointly by the American National Standards Institute (ANSI); the American Society of Heating, Refrigerating, and Air-Conditioning Engineers (ASHRAE); and the Illuminating Engineers Society of North America (IESNA). This standard, titled Energy Standard for Buildings (except low-rise residential buildings), gives guidelines to establish minimum efficiency requirements for buildings, excluding low-rise residential buildings. This standard should be used so that a building design will provide minimum requirements, resulting in an energy-efficient design. The topics pertaining to lighting within this standard will now be discussed.

Application Areas

The spaces for which this standard applies include the following:

- Interior building spaces
- Exterior features, including façades, architectural features, lighted roofs and canopies, entrances and exits, and loading docks
- Exterior grounds lighting that is provided by the building's electrical system

The spaces for which this standard does not apply include the following:
- Emergency lighting that is turned off during a building's normal operations
- Lighting within residential units
- Lighting specifically required by health and/or life safety statutes, regulations, or ordinances
- Decorative lighting systems powered by gas

Automatic Lighting Shutoff

For buildings with areas greater than 5,000 sq. ft., automatic shutoff sensors shall be required for all spaces. The function of these automatic sensors shall meet one of the following conditions:
- A control device that turns off lights at specific times
- A sensor that turns off lighting within 30 minutes of vacancy of space
- A signal is obtained from an alternate control that indicates an area to be unoccupied, resulting in lights being turned off

Exceptions to the requirement for the automatic sensors include:
- Spaces where lighting occurs 24 hours a day
- Spaces where patients are given care
- Spaces where automatic shutoff might endanger security or safety of a space or its occupants.

Space Lighting Control and Separate Controls

Spaces with ceiling-height partition enclosures must have at least one lighting control that independently controls lighting within each space. These controls must be easily accessible and must be located such that building occupants have visual access to the controlled lighting. In classrooms, conference rooms, and break rooms, a control device to automatically turn off the lights within 30 minutes of room being vacated is required. For all other spaces, the control devices must be activated by either a sensor or manually.

Individual separate lighting controls are required for the following cases:
- Accent or display lighting
- Case lighting used for display purposes
- Lighting for hotel and motel guest rooms
- Task lighting, such as under-cabinet or under-shelf lighting
- Non-visual lighting, such as used for food warming or plant growth
- Demonstration lighting

Determining Building Power Limits

The Energy Standard for Buildings (except low-rise residential buildings) limits the total amount of power allowed for lighting within a building. This limit can be determined using one of two methods. The first method is the Building Area Method that considers the total square footage of a building and an average power density value based on the building use in terms of watts per square foot. A partial list of the power density values for use in lighting design is given in table 8-1.

Type of building	Lighting power density(LPD), watts/ft²
Convention center	1.2
Dining: fast food/cafeteria	1.4
Dining: family	1.6
Dormitory	1.0
Hospital	1.2
Library	1.3
Movie theater	1.2
Multi-family residence	0.7
Museum	1.1
Office	1.0
Performing arts center	1.6
Religious building	1.3
Retail	1.5
Warehouse	0.8

TABLE 8-1 *Building area method lighting power densities (from ASHRAE-90.1-07).*

For this method, the maximum power allowance that can be used for lighting can be found from equation:

$$P = A(LPD)$$

where, P = Total power for building lighting
A = Total building floor area, ft²
LPD = Power density based on building type from table 8-1, watts/ft²

Example 8.3 For a 5,000 sq. ft. retail space, determine the maximum number of 100 watt lighting fixtures that can be placed inside the building, using the Building Area Method.

From table 8-1, the maximum Lighting Power Density value for a retail space is 1.5 watts/ft². The total maximum power for the building is found from equation:

$$P = A(LPD) = 5000 \text{ ft}^2 (1.5 \text{ watts/ft}^2) = 7500 \text{ watts}$$

The number of 100 watt lighting fixtures is found from equation:

$$\text{Number of fixtures} = \frac{7500 \text{ watts}}{100 \text{ watts/fixture}} = 75 \text{ lighting fixtures}$$

The second method is the Space-by-Space Method, and also uses the total square footage of a building along with power densities for individual spaces within the building. This method is more time-consuming and also gives greater flexibility. A partial list of the power density values for use with this method is given in table 8-2. The use of this table differs from using table 8-1 in that it lists common spaces as well as building-specific spaces. For the common spaces, the **bold** description is for general use, with specific building types listed directly below these designations. For the Building-Specific spaces, the **bold** description indicates the building type with the building-specific spaces listed directly below.

Common space type	Power density (watts/ft^2)	Building-specific space	Power density (watts/ft^2)
Office-enclosed plan	1.1	**Gymnasium**	
Office-open plan	1.1	Playing area	1.4
Conference rooms	1.3	Exercise area	0.9
Classrooms	1.4	**Library**	
Lobby	**1.3**	Card file/catalog	1.1
For hotel	1.1	Stacks	1.7
For performing theater	3.3	Reading area	1.2
For movie theater	1.1	**Hospital**	
Audience/seating area	**0.9**	Emergency	2.7
For gymnasium	0.4	Recovery	0.8
For convention center	0.7	Nurse stations	1.0
For religious building	1.7	Exam rooms	1.5
For sports arena	0.4	Pharmacy	1.2
For performing theater	2.6	Patient room	0.7
For movie theater	1.2	Operating room	2.2
Dining area	**0.9**	Nursery	0.6
For hotel	1.3	Medical supply	1.4
For motel	1.2	Physical therapy	0.9
For lounge / leisure dining	1.4	Radiology	0.4
For family dining	2.1	**Museum**	
Laboratory	1.4	General exhibition	1.0
Restrooms	0.9	Restoration	1.7
Dressing/locker rooms	0.6	**Religious buildings**	
Corridors	**0.5**	Worship pulpit/choir	2.4
For hospital	1.0	Fellowship hall	0.9
For manufacturing facility	0.5	**Retail**	
Stairs	0.6	Sales area	1.7
Electrical/mechanical	1.5	Mall concourse	1.7

TABLE 8-2 *Space-by-space method lighting power densities (from ASHRAE-90.1-07).*

Example 8.4 Using the Space-by-Space Method, calculate the interior lighting power allowance in an office building with the following specific spaces and corresponding floor areas:

Space	Area (ft²)
Lobby	1,800
Enclosed offices	12,000
Open offices	40,000
Conference rooms	2,800
Break rooms	600
Exit stairs	2,000
Corridors	5,000
Restrooms	1,600
Mechanical/electrical	2,800

TABLE 8-3 *Floor areas.*

From table 8-2, the maximum Lighting Power Density values for the spaces are as follows:

Space	Power density (watt/ft²)
Lobby	1.3
Enclosed offices	1.1
Open offices	1.1
Conference rooms	1.3
Break rooms	1.4
Exit stairs	0.6
Corridors	0.5
Restrooms	0.9
Mechanical/electrical	1.5

TABLE 8-4 *Space lighting power densities.*

To determine the lighting power allowance for the office building interior, multiply the floor area of each space by its corresponding power density value. In equation form:

$$P = \sum[A\,(LPD)]$$

where, P = Total power for building lighting

A = Building floor area for each space, ft^2

LPD = Power density based for each space from table 8-2, watts/ft^2

Space	Area (ft^2)	Power density (LPD) (watts/ft^2)	A (LPD) (watts)
Lobby	1,800	1.3	2,340
Enclosed offices	12,000	1.1	13,200
Open offices	40,000	1.1	44,000
Conference rooms	2,800	1.3	3,640
Break rooms	600	1.4	840
Exit stairs	2,000	0.6	1,200
Corridors	5,000	0.5	2,500
Restrooms	1,600	0.9	1,440
Mechanical/electrical	2,800	1.5	4,200

TABLE 8-5 Floor areas and lighting power densities.

Summing up the right column of table 8-5 gives the maximum power allowance for lighting of 73,360 watts or 73.36 kilowatts.

For both of the methods, the area is the gross area, measured to the exterior of the building when using the Building Area Method, and to the center of walls separating spaces when using the Space-by-Space Method. Once the total power allowance is calculated for a building, it may be distributed through the interior of the building in any manner as long as the allowance is not exceeded.

Additional Lighting Power

In some instances, the maximum allowable lighting power can be exceeded, but only if the additional power is used for designated purposes. These instances include:

- For decorative lighting (i.e., sconces, chandeliers, and art lighting), 1.0 watts/ft^2 can be added.
- For lighting spaces that have visual display terminals (VDT), 0.35 watts/ft^2 can be added as long as the lighting system meets VDT lighting requirements.
- For retail lighting that highlights merchandise, 1.6 watts/ft^2 can be added for general merchandise, or 3.9 watts/ft^2 can be added for merchandise that includes jewelry and accessories, fine clothing, china, and silver. Additionally, the 3.9 watts/ft^2 can be added for art galleries.

Exterior Lighting Control and Lighting Efficacy Requirements

The requirement for exterior lighting includes the control of lighting by photo sensors or time switches except for the following conditions:

- Building exits
- Covered vehicle entrances
- Where lighting for security and safety are required
- Where lighting is required that allows occupants to adapt to changes in lighting level between an interior or covered space and the exterior daylight.

For exterior spaces, efficacy of the lighting fixtures must meet minimum requirements. This minimum requirement is set at 60 lumens per watt for fixtures intended to operate at greater than 100 watts. This requirement restricts the use of lighting fixtures that utilize incandescent or mercury bulbs. As an exception to this requirement, there is no minimum value if lighting fixtures are:

- Controlled with a motion sensor
- Used for security and safety
- Used for building signage
- Part of a historic structure

Exterior Building Lighting Power

For exterior building lighting, the total lighting power allowance is determined using values for exterior systems, a partial list of which is given in table 8-6.

Space	Lighting power density
Drives and parking lots	0.15 watts/ft^2
Walkways < 10 feet wide	1.0 watt/lineal foot of walk
Walkways ≥ 10 feet wide	0.2 watts/ft^2
Plaza areas	0.2 watts/ft^2
Stairways	1.0 watts/ft^2
Main building entrance/exit	30 watts/lineal foot of door width
Other building doors	20 watts/lineal foot of door width
Canopies	1.25 watts/ft^2
Building façades	0.2 watts/ft^2 for each lighted wall, or
Drive-through windows for fast food	5.0 watts/lineal foot for each lighted wall 400 watts per drive-through

TABLE 8-6 *Exterior lighting power densities (from ASHRAE-90.1-07).*

If an independent control device is provided, certain types of lighting are exempt from the power allowances listed in table 8-6. These exemptions include:

- Transportation lighting
- Advertising signage

- Playing field lighting
- Temporary lighting
- Lighting of public monuments or historic landmarks

Example 8.5 For the exterior of a fast food building, determine the lighting power allowance for the following elements:

Exterior space	Area (ft²) or length (ft)
Parking lot	20,000 ft²
Walkways < 10 foot wide	120 lineal ft
12 foot wide walkways	200 lineal ft
Two drive-through windows	————
3 main entrances/exits	8 lineal ft each
2 service doors	5 lineal ft each
Illuminated façade	6,000 ft²
Exterior plaza	1,600 ft²
2 canopies	400 ft² each

TABLE 8-7 *Exterior space areas and lengths.*

In table 8-6 the power densities are given based on units of floor area, surface area, lineal distances, and watts per unit. To find the total exterior power allowance, calculate the lighting power allowance for each element, and then sum the values.

Exterior space	Power allowance using values from table 8-6
Parking lot	20,000 ft² × 0.15 watts/ft² = 3000 w
Walkways < 10 foot wide	120 lineal ft × 1.0 watt/lineal ft of walk = 120 w
12 foot wide walkways	200 lineal ft × 12 ft × 0.2 watts/ft² = 480 w
Two drive-through windows	2 × 400 watts each = 800 watts
3 main entrances/exits	3 × 8 lineal ft × 30 watts/lineal ft = 720 w
2 service doors	2 × 5 lineal ft × 20 watts/lineal ft = 200 w
Illuminated façade	6,000 ft² × 0.2 watts/ft² = 1,200 w
Exterior plaza	1,600 ft² × 0.2 watts/ft² = 320 w
2 canopies	2 × 400 ft² × 1.25 watts/ft² = 1,000 w

TABLE 8-8 *Exterior space power densities.*

Summing the values in the right column gives the maximum power allowance for lighting of 7,840 watts.

Problem Set 8.4 ANSI/ASHRAE/IESNA 90.1-07 Lighting Requirements

Problem 8-23 List the advantages of utilizing the Space-by-Space Method over the Building Area Method when determining lighting power allowances for a building interior using the ANSI/ASHRAE/IESNA 90.1-07 Lighting Requirements.

Problem 8-24 For a 24,000 sq. ft. church, determine the number of 75-watt lighting fixtures that can be placed inside the building based on the lighting power allowance using the Building Area Method.

Problem 8-25 For a 200,000 sq. ft. warehouse, plus an 8,000 sq. ft. office space, determine the total lighting power allowance using the Building Area Method.

Problem 8-26 For an 18,000 sq. ft. library, based on the lighting power allowance using the Building Area Method, determine the yearly cost of electricity of the interior lighting, using energy costs from figure 8-9, if the building is located in California. Assume that the power is turned on for 10 hours a day, 7 days a week over a 52-week period.

Problem 8-27 Using the Space-by-Space Method, calculate the lighting power allowance for a movie theater with the following specific spaces and corresponding floor areas:

Space	Area (ft²)
Lobby	1,400
Enclosed offices	200
Break room	100
Exit stairs	1,000
Corridors	2,000
Restrooms	1,600
Mechanical/electrical	2,500
Theater seating	48,000

TABLE 8-9 *Floor areas.*

Problem 8-28 Using the Space-by-Space Method, calculate the lighting power allowance in a hospital with the following specific spaces and corresponding floor areas:

Space	Area (ft²)	Space	Area (ft²)
Lobby	600	Nurse stations	2,400
Enclosed offices	2,000	Exam rooms	8,000
Conference rooms	300	Pharmacy	2,000
Exit stairs	5,000	Patient room	36,000
Corridors	10,000	Operating room	12,000
Restrooms	2,400	Nursery	4,200
Mechanical /electrical	7,500	Medical supply	6,800
Emergency	9,600	Physical therapy	6,500
Recovery	4,000	Radiology	2,600

TABLE 8-10 *Floor areas.*

Problem 8-29 For the exterior of a small office building, determine the lighting power allowance for the following elements:

Exterior space	Area (ft²) or length (ft)
Entrance canopy	240 ft²
8 building exits	3'-0" each
Illuminated façade	900 ft²
Exterior plaza	1,600 ft²

TABLE 8-11 *Exterior space areas and lengths.*

Problem 8-30 For the exterior of a warehouse building, determine the lighting power allowance for the following elements:

Exterior space	Area (ft²) or length (ft)
Parking lot	100,000 ft²
Walkways < 10 foot wide	40 lineal ft
3 main entrances/exits	6 lineal ft each
20 service doors	10 lineal ft each
Illuminated façade	10,000 ft²
Exterior plaza	400 ft²
Entry canopy	80 ft²

TABLE 8-12 *Exterior space areas and lengths.*

Passive Solar Design

Passive solar design utilizes walls, windows, and floors to collect, store, and distribute the energy of the sun as heat in the winter. Passive solar design also allows for solar heat gain to be minimized in the summer. It does not involve using mechanical and electrical systems to heat and cool. For passive solar design to have the greatest impact on efficiency and cost, it must be considered from the very beginning of the design process. One key to a successful passive solar design is to take advantage of local climates. By considering local climates, decisions can be made on which systems to include in the building.

For any passive design to be successful, it must incorporate all of the following elements:

- Aperture
- Absorber
- Thermal Mass
- Distribution
- Control

Figure 8-10 illustrates these five elements in a building. Further information about each of these passive solar design elements can be found at the Department of Energy website www.energysavers.gov/your_home/designing_remodeling/index.cfm

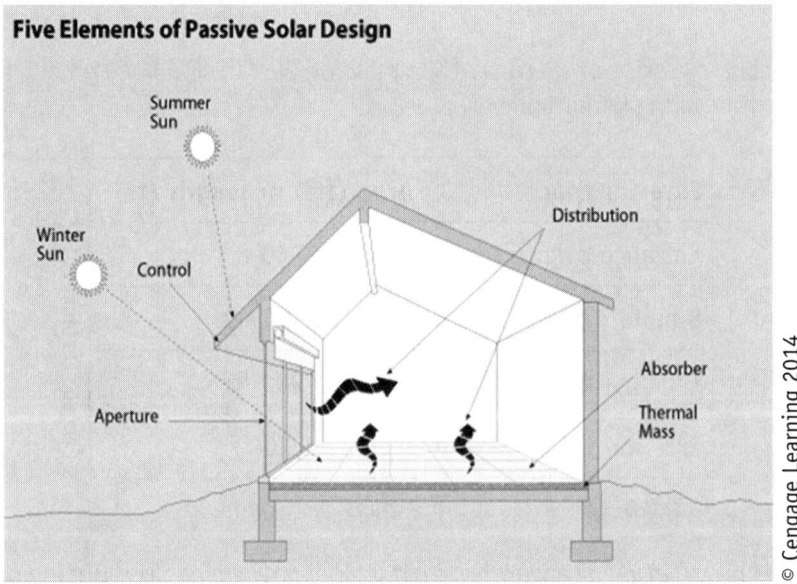

FIGURE 8-10 *Five elements of passive solar design.*

To accomplish a successful passive solar design, basic decisions that must be made include:

- Type of windows
- Window placement in the building
- Type of thermal insulation
- Whether to use thermal mass
- Type of shading

Each of these can potentially have a significant effect on reducing the amount of energy used in a building.

Passive solar energy design involves heat transfer through radiation, conduction, and/or convection. In colder months, the low angle of the sun allows absorption of solar radiation into building floor or wall systems that can be utilized to heat a building. In hotter months, overhangs can shade the building, and the use of convective air currents can ventilate and help cool a building.

Problem Set 8.5   Passive Solar Design

Problem 8-31   What are the three natural heat transfer mechanisms? How can these be incorporated into passive design?

Problem 8-32   List and discuss the five elements of a complete passive solar design.

Problem 8-33   Research and discuss passive solar water heaters, providing and labeling a sketch of the system.

Problem 8-34   Research and discuss a Trombe Wall system, providing and labeling a sketch of the system.

Problem 8-35   Survey your home, and list any passive design elements that have been included. In addition, make a list of passive design elements that could be included to reduce energy use.

Problem 8-36   Research and discuss how landscaping around a building can be utilized as part of the passive solar design system?

Active Solar Design

Active Solar Design utilizes technologies to convert sunlight into usable energy. To increase power efficiencies, active solar systems use electrical and mechanical systems, including fans and pumps. The two prominent categories of active solar power systems are photovoltaic cells and solar collector systems.

Photovoltaic Cells

Photovoltaics, or solar cells, are solid state devices used to convert sunlight into electricity. The study of Photovoltaics involves research related to using photovoltaic cells to produce electricity from sunlight. Solar cells, which are used to form solar panels, are made of negative and positive semiconductors. As sunlight hits the cells, the photons are separated from the sunlight and turned into electricity. A single photovoltaic cell will typically produce from 1 to 2 watts of electricity. These individual cells are combined with other cells to form solar modules. The modules, in turn, are connected in series with other modules to form a solar array. The modules can be built into a solar array to a desired output energy level. In addition to the solar array, structural supports for the arrays must be provided as well as a means of positioning the array such that it receives the optimum solar exposure. The solar arrays produce direct current (DC), and equipment is required to change this energy into a form that can be utilized for specific purposes. This total active solar system is typically referred to as a building integrated photovoltaic (BIPV) system. Information on photovoltaic cells and other active solar systems can be found through the Solarbuzz company website at www.solarbuzz.com. Solarbuzz is a company with a specialty in solar energy research and is a consulting company for the industry with analysts in Asia, Europe, and the United States. The website can be used for researching photovoltaic cells as well as other passive and active solar systems.

Solar Collector Systems

Solar Collector Systems use the sun's energy from light and convert it to energy in the form of heat for use in buildings. The most common types of solar collector systems include:

- Flat plate collectors
- Evacuated tube collectors
- Parabolic trough
- Parabolic dish
- Power tower

The first two systems listed are typically used for generating heat in residential and commercial buildings. The other systems listed are used for generating electricity. Each of these systems can be researched online for more information.

For active solar systems, it is important that solar geometry be taken into consideration. The sun emits a fairly constant amount of solar radiation, known as the solar constant, and the amount that reaches the earth's surface varies with the angle of the sun throughout the year. It is important to know the angle of the sun so that a building, or an active solar system, can be oriented to best take advantage of the sun angles. Online websites, such as Sunposition (www.sunposition.info/sunposition/spc/locations.php) can be used to determine the sun angles throughout the year for specific worldwide locations. Knowing sun angles throughout the year allows the design team to incorporate environmentally conscious elements that utilize passive and active elements, which can result in a reduction in energy needs for a project. For additional information on sun angles, refer to chapter six of this workbook.

Often, passive and active solar systems are included in a building's design. The Solar Umbrella House, shown in figure 8-11, is a residence in Venice, California, that combines both types of systems to arrive at a home that is essentially a net-zero energy home, one that uses no energy when considered over the course of a year.

FIGURE 8-11 *The solar umbrella house.*

Most residences could not reach this result, but by incorporating passive and active solar elements into building design, any home can become more energy efficient, resulting in a reduction of the energy used in this country.

Problem Set 8.6 Active Solar Design

Problem 8-37 Research photovoltaic cells online, and write a short report on the history and development of this active solar energy system.

Problem 8-38 List six materials currently used to manufacture different types of photovoltaic solar cells. Which of these materials gives the most energy-efficient cell?

Problem 8-39 Research and discuss the lifespan of photovoltaic cells. Which type of photovoltaic cell has the longest lifespan?

Problem 8-40 Research online to find three types of photovoltaic cells and the costs associated with each.

Problem 8-41 What are the advantages and disadvantages of solar thermal collectors?

Problem 8-42 What percent of the world's energy is supplied by solar thermal collectors? What percent is supplied in the United States?

Problem 8-43 What is a solar air heat collector? Describe the two categories of this system.

Problem 8-44 Research the Solar Umbrella House, giving an overview of its history as well as a description of the passive and active solar systems included in the design.

CHAPTER 9
Residential Space Planning

Skills List

After completing this chapter, you should be able to:

- Understand average size and costs of residential construction in different parts of the country

- Understand the basic concepts of a home loan, called a mortgage

- Become familiar with the Americans with Disabilities Act (ADA) website

- Understand the concept and use of an adjacency matrix

- Understand minimum dimensions and method of determining area for stairs

BACKGROUND

Historically, space planning for residential construction has been directly related to human proportions. Think about the home in which you live, and how much of its design is based on the human form. The doors and windows are sized and placed for convenient use, as are the shelves, cabinets, and countertops. It makes sense that buildings are designed around those who will be using them. Residential space planning that is properly and successfully laid out allows for flexibility as well as the potential for expansion, each of which can increase the value of the home. Additionally, space planning extends past the exterior walls of residences. The building site, including the topography, orientation for views and solar angles, and surrounding context must be considered. Space planning utilizes the human form to provide comfortable, welcoming rooms in which we live our lives.

Research and Investigation Defines Project Parameters, Specifications, and Limitations

At the start of any building design, basic information must be gathered that will help identify the requirements, limitations, and options that can be considered in the design process. Chapter three of the workbook discusses the types of information that must be gathered at the beginning of a design project. It is critical for an architect to work with the client in understanding their wants and needs when designing their home. One major factor that particularly affects residential design is the project budget.

Budget Defines Building Options

A budget constraint is generally the basis for how a residential space will be designed by the architect and client. Budgets for residential buildings can vary greatly, and the resulting design must accurately reflect the amount of money the client can afford when designing a new home. Average cost of residences per square foot can vary widely, depending on location, size, complexity, selection of finish materials, and many other factors. There are many resources online that will help determine general construction costs. Many of these are subscription sites and are very specific in the detail to which cost estimating for a project is taken. For general information on the average square footage and cost per square foot of residences in each state, a website such as Home-Cost.com can be utilized, available at www.home-cost.com/construction-cost-per-sf.html. It is important to note that the architect uses average costs only as a starting point, and will refine the project budget with input from the clients and local builders.

Example 9.1 Using online resources, determine the average square footage, cost per square foot, and total average construction cost for residences in Arkansas, California, New York, and Ohio.

Using the Home-Cost.com website, the average square footage, and cost per square foot can be determined directly. The total average construction cost can be found from the equation:

Total average cost = square footage × average cost per square foot

The answer to this example is given in table 9-1

State	Average square footage	Average cost/ft^2	Total average cost
Arkansas	2,500	$73.81	$184,525.00
California	2,400	$128.93	$309,432.00
New York	2,750	$122.31	$336,352.50
Ohio	2,250	$87.44	$196,740.00

TABLE 9-1 Average square footage and cost of homes.

Note the differences in the total cost for the average home in these four states. What do you think are factors that influence the difference in total cost? Remember that labor and material costs vary across the country, in addition to the variation in the average square footage for a home in differing states. One important item to realize is that the prices listed here do not include the cost of land, which can vary significantly and in some locations can cost more than the residence.

At the very beginning of a residential project, the architect and client will discuss the amount of money available for use on the project. Often, this amount is broken down into different cost categories, such as financial and professional fees, land acquisition, building construction, and site improvements. By determining the amount of money available for a project, the general budget can be defined, and the architect can begin the design process. The cost of designing and building a house usually requires a homeowner to use a financial instruction to help fund the construction of the residential home. Often, the budget is a function of the homeowner's income, as financial institutions set limits on the amount of a mortgage they will lend, based on a percentage of gross income. A mortgage is a long-term loan, typically from 10 to 30 years in length, for use in purchasing a home. A mortgage payment will consist of four components, including principal, interest, taxes, and insurance, referred to as PITI. As a rule of thumb, the PITI should not exceed 28 percent of the yearly gross income of the homeowners. A mortgage payment is a function of the amount of the loan, the interest rate for the loan, and the duration of the loan. The insurance portion for the mortgage is broken into home replacement insurance and private mortgage insurance (PMI). PMI is required by the financial institution when a homeowner borrows more than 80 percent of the cost of the home, and it is intended as insurance against default of payment of the mortgage. If a homeowner pays a minimum of 20 percent of the home cost upfront, the PMI will typically not be required, which will reduce the monthly payments. To help determine monthly and overall payments, mortgage calculators exist online, such as the one provided by Mortgage Calculator at www.mortgagecalculator.org/. By inputting basic information, the approximate monthly payment for a mortgage may be determined.

Example 9.2 What is the monthly payment for a 15-year versus a 30-year mortgage for a new home, and what is the total cost difference between the two with the following conditions?

- Home value = $275,000.00
- Mortgage amount = $250,000.00
- Mortgage rate = 6.75 percent
- Taxes = 1.3 percent
- PMI = 0.6 percent

Note that because the mortgage is more than 80 percent of the home value, private mortgage insurance (PMI) is required to qualify for the loan. Using the mortgage calculator at www.mortgagecalculator.org/, the monthly payment for a 15-year mortgage is $2,537, and for a 30 year mortgage is $1,947. The total payment for a 15-year mortgage is $456,709, and the total payment for a 30-year mortgage is $700,738 for a difference of $700,735 − $456,709 = $244,026.

Once a budget is determined, the designer can begin to make decisions on the size, complexity, and quality of the design. The style of the architecture can also affect the cost of the building. In chapter 4 of the textbook, architectural styles were defined and discussed. If there is a dominant style of architecture where a project is being built, then it is most cost-effective to design a project within the architectural styles in the area. This cost-effectiveness is due to the familiarity of construction techniques by local builders as well as the availability of construction materials and finishes used in the design.

Construction costs for residences are often the cause of conflict among the client, architect, and builder. It is the job of the architect to make sure the client is realistic about the project budget. Though the client and architect may agree on a general budget, often it can be exceeded based on factors encountered during the design and construction of the project. Some factors that can affect the budget include time delays, changes in residence size, shape, or finish materials, and increases in construction material costs due to a rise in supply prices or due to natural disasters that cause a shortage of building materials.

For the process of space planning for a residential design, the project budget becomes crucial in determining the complexity of the project as well as the style of architecture and the types of finish materials utilized. For a residential project to be successful, the client, architect, and builder must be realistic when considering the budget and the final product of a residence.

Site Investigation Determines Building Potential

An architect must perform research to arrive at the influential factors that will affect the building design. Some of these factors include:

- Adjacent properties (site context)
- Home owner's association regulations (covenants)
- Climatology
- Local building code requirements
- Funding institution requirements

Each of these factors will affect the final design of a residence and must be considered by all involved in the design and construction process.

Client Survey and Interview Leads to Project Vision

An architect needs to gather and consider a multitude of information to arrive at the design for any type of building. For a home design, the process is extremely personal, especially for the client who plans to live in the designed home. Therefore, every aspect of the design is crucial to arrive at a successful design. Once the client's budget is determined, the architect will evaluate the amenities a client wants in their new home. Often, this is accomplished through a survey and interview process to prioritize the client's needs and wants for use in designing the residence. This information will have a direct effect on the final design, helping the architect to consider potential long-term lifestyle changes. For example, does the client have children (and, if so, what are their ages), or are there medical concerns or physical constraints that would affect design decisions? Refer to textbook figure 9-6 for an example of a survey that could be used in the design of a home. The client survey and interview process will help define the design parameters for a home will be used by the architect in the process of design and construction.

Space Standards Guide the Development of Comfortable and Effective Spaces

Architects utilize their experience and knowledge about standard space sizes within a home to arrive at an initial design. The architect must consider the lifestyle of the occupants who will use each space, taking into account their number, size, and any disabilities. Standard space sizes have been developed through anthropometrics (the study of the measurements of the human body). Additionally, ergonomics (the study of equipment and space design to fit the human body, its abilities, and movements) is used in space design. National and local building codes also affect the design of spaces. For example, codes have minimum requirements for window opening size and location in rooms. For the state of New York, windows in habitable spaces must have an area of at least 8 percent of the room square footage for natural light, and at least 4 percent of the room square footage for ventilation. The ratio of these two values allows for double-hung windows to be used that will meet both requirements. Double-hung windows typically are constructed with two equal window pane sizes and are operable so that one half of the window may be opened for ventilation. For stationary windows, another method used to meet the ventilation requirement is an operable door opening to an exterior space.

Example 9.3 You are given the task of designing double-hung windows for a living room with a floor area of 1,200 sq. ft. in the state of New York. There will be eight windows in the room, and each will have a height of 48 inches. Determine the minimum width of each window required for lighting and ventilation.

Total area of windows for room based on natural lighting, A_w = 1,200 sq. ft. (0.08) = 96 sq. ft.

For eight windows, the width of each window must be at least:

Window width = 96 sq. ft. / 8 windows / 4' - 0" height = 3' - 0" = 36"

Planning for Special Needs

Guidelines for designing residences for accessibility can be found at the American Disabilities Act (ADA) website at www.access-board.gov/adaag. Designing for occupants with disabilities often includes considerations for wheelchairs, among other requirements. For any disability, it might take additional research by the architect to determine the requirements and to locate building materials. For example, if a wheelchair will be used within a home, the ADA code requires 36" clear width in hallways, and 32" of clear width at doorways when the door is open at a 90° angle. Additionally, if a door occurs at the end of a hallway, an additional 12" of hallway width must be provided. Bathrooms are another area where special attention must be given, to make sure the room functions properly for the occupants of the residence. Review the ADA website to understand how it can be used as a resource when space planning.

Planning Building Component Specifications and Options

Many decisions must be made by the design team when building a residence. Many involve which products to use in the construction of the home. Much of the decision on products involves budget. Software and the Internet can be used to research standard sizes for building components by downloading manufacturers' product catalogs. Products used in residential construction, including doors, windows, and cabinetry, can be researched on the Internet to determine style and size availability.

Problem Set 9.1 Research and Investigation Defines Project Parameters, Specifications, and Limitations

Problem 9-1   Using online resources, determine the average square footage, cost per square foot, and total average construction cost for residences in Alaska, Colorado, South Carolina, and Washington.

Problem 9-2   Using online resources, determine the average square footage, cost per square foot, and total average construction cost for residences in Delaware, Michigan, Texas, and Wisconsin.

Problem 9-3   Using online resources, determine the average square footage, cost per square foot, and total average construction cost for residences in Connecticut, Illinois, Oklahoma, and Oregon.

Problem 9-4   Using online resources, determine the average square footage, cost per square foot, and total average construction cost for residences in Georgia, Missouri, Nevada, and Wyoming.

Problem 9-5   Using online resources, determine the average square footage, cost per square foot, and total average construction cost for residences in Idaho, Mississippi, New Hampshire, and Utah.

Problem 9-6   What would be the average cost of a 3,200 sq. ft. home located in Honolulu, HI?

Problem 9-7   What would be the average cost of a 1,950 sq. ft. home located in Omaha, NE?

Problem 9-8   What would be the average cost of a 4,250 sq. ft. home located in Nashville, TN?

Problem 9-9   What would be the average cost of a 2,600 sq. ft. home located in Fort Worth, TX?

Problem 9-10 What would be the average cost of a 1,400 sq. ft. home located in Santa Fe, New Mexico?

Problem 9-11 Using an online mortgage calculator, determine the monthly payment for a 30-year mortgage with the following conditions:
- Home value = $200,000.00
- Mortgage amount = $170,000.00
- Mortgage rate = 7.35 percent
- Taxes = 1.5 percent
- PMI = 0.5 percent

Problem 9-12 Using an online mortgage calculator, determine the monthly payment for a 15 year mortgage with the following conditions:
- Home value = $375,000.00
- Mortgage amount = $325,000.00
- Mortgage rate = 5.65 percent
- Taxes = 1.25 percent
- PMI = 0.4 percent

Problem 9-13 Using an online mortgage calculator, determine the monthly payment for a 20-year mortgage with the following conditions:
- Home value = $425,000.00
- Mortgage amount = $210,000.00
- Mortgage rate = 6.10 percent
- Taxes = 1.3 percent
- PMI = 0.0 percent

Problem 9-14 Using an online mortgage calculator, determine the monthly payment for a 10-year mortgage with the following conditions:
- Home value = $650,000.00
- Mortgage amount = $580,000.00
- Mortgage rate = 4.5 percent
- Taxes = 1.6 percent
- PMI = 0.5 percent

Problem 9-15 Using an online mortgage calculator, determine the monthly payment for a 15-year mortgage with the following conditions:
- Home value = $1,800,000.00
- Mortgage amount = $1,250,000.00
- Mortgage rate = 5.125 percent
- Taxes = 1.5 percent
- PMI = 0.0 percent

Problem 9-16 Using online resources, determine the approximate square footage of a house to be built in St. Louis, MO, that an owner could afford, based on average construction costs, using the following criteria:
- Gross income = $10,000.00/month

- 30-year mortgage with mortgage rate = 6.5 percent
- Taxes = 1.2 percent
- PMI = 0.5 percent
- Maximum PITI = 28 percent of gross income
- Down payment of 15 percent of home cost provided by owner

Problem 9-17   Using the client survey in textbook figure 9-6, take on the role of the client, and answer each question for your home. List additional information or conditions not included in the survey that you would provide to an architect for a better understanding of your family's wants and needs.

Analysis Leads to Optional Residential Space Planning

Analysis of a residential building design is a process that should be implemented through all phases of design and construction. If analysis is not continuous, the design can suffer from missed opportunities caused by not exploring all of the available options. Analysis during the research and concept phases can be used to make decisions on the big questions, such as building size, shape, and orientation. Continued analysis through all phases of analysis and design can be used to arrive at a finished product that is efficient, economical, and representative of the final design. Analysis through all phases of design and construction is a necessary process to ensure a quality residential building.

The Ability to Visualize Space Is Key to Space Planning

Architects must be able to visualize the size and shape of each room in a home design, along with the appliances, fixtures, and furniture intended to be placed in the room. The layout for each room must be designed such that the intended use of the room is maximized while allowing for usable traffic flow. When room adjacency is added into the design process, multiple solutions must be considered to arrive at the final design. The initial size of a room in a home design is adjusted throughout the design process. Often, an approximate overall square footage at the start of design is adjusted based on decisions made during the design process, resulting in a revised overall final square footage for a home.

When designing a home, the square footage is based on the gross, or overall, square footage that includes walls but does not include the garage, basement, attic, or porches. The net, or livable, square footage is the gross area minus the floor area taken up by walls. Livable square footage includes spaces with finished floors, walls, and ceilings that have permanent heating and cooling systems.

Wood construction uses framing members called *studs* that are identified according to their nominal dimensions, which are greater than their actual dimensions. For example, a 2 × 4 wood stud has actual dimensions of 1 1/2" × 3 1/2". Refer to table 9-2 for nominal and actual dimensions of wood studs.

Nominal Dimension (inches)	Actual Dimensions (inches)
2 × 4	1 1/2 × 3 1/2
2 × 6	1 1/2 × 5 1/2
2 × 8	1 1/2 × 7 1/4

TABLE 9-2   *Wood stud nominal and actual dimensions.*

In addition to the wood studs, the sheathing applied to each side of the wood framing will take up floor space. Commonly, this sheathing is plywood or gypsum board and has a thickness of 1/2" on each face of the wall. Other items regularly found in residential design that reduce the net floor area include chimneys and mechanical spaces.

Example 9.4 For the plan of a one-story residence shown in figure 9-1, determine the gross and net floor areas if all walls are 2 × 6's with 1/2" sheathing on each face.

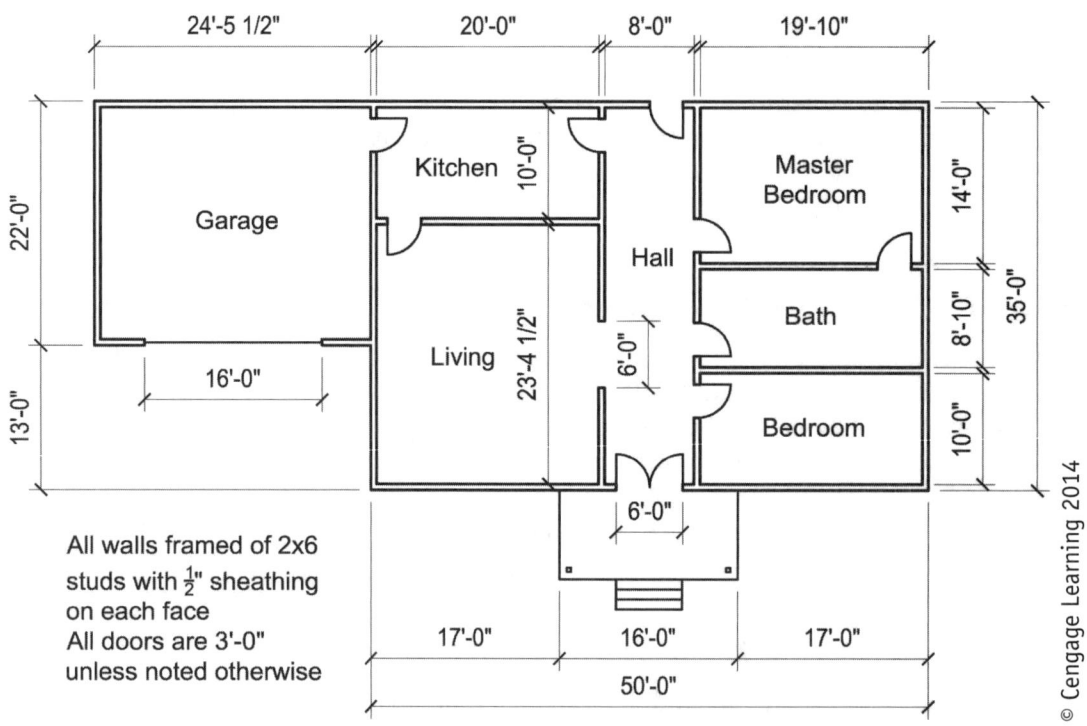

FIGURE 9-1 Home plan for example 9-4.

When calculating the gross area of a residence, use the overall dimensions of the home measured from the outside face of building, but do not include porches or garages. For this example, the gross area is:

$$A_{gross} = 50.0' \times 35.0' = 1,750 \text{ sq. ft.}$$

When calculating the net area of a residence, the area of the walls in plan is subtracted from the gross area. For measuring purposes, the length of the wall should be measured from the center of the wall to achieve an accurate net area. Note that the thickness of a 2 × 6 wall stud is 5 1/2", thus the thickness of all walls is 5 1/2" + 2(1/2") = 6 1/2". Table 9-3 lists the net areas for each room and the doorway in the home.

Room	Area (sq. ft.)
Living room	23' =4 1/2" × 20'−0" = 467.5 sq. ft.
Kitchen	10'−0" × 20'−0" = 200.0 sq. ft.
Hallway	[35'−0" − 2(6.5"/(12in/ft))] × 8'−0" = 271.33 sq. ft.
Bedroom	19'−10" × 10'−0" = 198.33 sq. ft.
Bathroom	19'−10" × 8'−10" = 175.19 sq. ft.
Master bedroom	19'−10" × 14'−0" = 277.67 sq. ft.
Doorways	[6.5"/(12in/ft)] × [2(6'−0") + 7(3'−0")] = 17.88 sq. ft.

TABLE 9-3 Room and doorway net areas.

The net area of the home is found by summing the areas in table 9-3:

A_{net} = 467.5 + 200 + 271.33 + 198.33 + 175.19 + 277.67 + 17.88 = 1,607.9 sq. ft.

Note that the walls in this home account for 1,750 sq. ft. − 1,607.9 sq. ft. = 142.1 sq. ft. of space, or approximately 8.1 percent of the gross area of the home.

When space planning a residence, there are generally accepted minimum sizes for rooms. Table 9-4 gives general sizes for rooms within a residence that can be utilized when initially determining the size of a home. Noted within the table are dimensions for small, medium, and large rooms. Also note that these sizes are typically used as a starting point when space planning, and that each home must be designed based on the client's requirements.

Example 9.5   Determine the overall area of a home with the following rooms:

- Living room – small
- Family room – large
- Kitchen – medium
- Dining room – large
- Bedrooms – three medium and one large
- Full bathrooms – two medium and one large
- Garage – double car

Using the room dimensions given in table 9-4 as gross areas (including allowances for walls), note that the size of the individual rooms in a home can vary. For example, a medium-sized kitchen has typical plan dimensions listed as 12 × 14 and 13 × 15. For this example, we will use the larger dimensions for each category, thus the overall floor area for this home is:

- Living room: 12 ft × 14 ft = 168 sq. ft.
- Family room: 18 ft × 22 ft = 396 sq. ft.
- Kitchen: 13 ft × 15 ft = 195 sq. ft.
- Dining room: 14 ft × 16 ft = 224 sq. ft.
- Bedrooms: 3(12 ft × 14 ft) + 14 ft × 20 ft = 784 sq. ft.
- Full bathrooms: 2(8 ft × 10 ft) + 10 ft × 12 ft = 280 sq. ft.
- Garage: 24 ft × 24 ft = 576 sq. ft.
- Overall floor area: = 2,623 sq. ft.

Room	Small	Medium	Large
Kitchen	10' × 14', 12' × 12'	12' × 14', 13' × 15'	14' × 16', 12' × 22'
Dining	10' × 12'	12' × 14', 12' × 16'	14' × 16'
Bedroom	7' × 10', 8' × 10', 9' × 11'	10' × 12', 12' × 12', 12' × 14'	12' × 16', 14' × 20'
Living room	12' × 12', 12' × 14'	13' × 15', 14' × 16'	14' × 18', 16' × 20', 16' × 22'
Family room	13' × 16'	16' × 18'	18' × 22'
Full bathroom	5' × 7'–6"	8 × 10', 7' × 11'	9' × 12', 10' × 12'
Garage	12' × 24' Single car	22' × 22', 22' × 24', 24' × 24' Double car	28' × 26', 30' × 30'

TABLE 9-4 *Typical room sizes in residences.*

Architects Visualize Spaces in Three Dimensions

In addition to visualizing room plan dimensions, an architect must be able to visualize rooms in three dimensions. The ceiling height and shape are critical in helping to establish the aesthetic, or feel, of a room. In addition, any appliances, fixtures, and furniture intended to be placed in the room must be visualized in the three-dimensional space to see how it will appear and function. Along with ceiling heights, window size and placement can greatly influence a room's aesthetics. Ceilings can vary in height starting at 7' and extending higher. For a particular room size, shape, and use, there are ceiling heights that are appropriate. For example, a 12' ceiling height would not be appropriate for a bathroom that was 5' × 8' in plan, but it would be acceptable for a 16' × 20' living room. Being able to visualize spaces in three dimensions will help the designer make appropriate decisions during the design process.

Space Adjacencies Analyzed for Convenience and Comfort

Interior spaces will fall into one of three categories:

- Living
- Sleeping
- Service area

When space planning for a residence begins, often a bubble diagram shows the individual room sizes and adjacency. As previously discussed in Chapter 3, a bubble diagram is a graphic representation that shows how each of the rooms in a project should relate to each other. Often, this is the beginning of project planning, and it allows the designer to quickly look at different options for room adjacency. Refer to figure 3-7 for an example of a bubble diagram for a home. An adjacency matrix is similar to a bubble diagram in that it gives information on rooms in a building and indicates which rooms should have primary or secondary adjacency to other rooms in the building.

Example 9.6   Develop an adjacency matrix for a residence with the following rooms and floor areas showing access between spaces:

- Entry hall   80 sq. ft.
- Living room   250 sq. ft.
- Dining room   200 sq. ft.

- Family room   500 sq. ft.
- Kitchen   400 sq. ft.
- Pantry   100 sq. ft.
- Bedrooms   3 @ 200 sq. ft. each
- Master bedroom   350 sq. ft.
- Master closet   200 sq. ft.
- Bathrooms   3 @ 100 sq. ft. each
- Master bath   200 sq. ft.
- Office   175 sq. ft.
- Laundry room   100 sq. ft.
- Garage   750 sq. ft.

Each room is shown on the adjacency matrix and is indicated as a primary or secondary adjacency to all other spaces in the home. Refer to figure 9-2 for the adjacency matrix for this example.

Residential Adjacency Matrix

■ : Primary adjacency □ : Secondary adjacency	Entry hall	Living room	Dining room	Family room	Kitchen	Pantry	Bedroom 1	Bedroom 2	Bedroom 3	Master bedroom	Master closet	Bathroom 1	Bathroom 2	Bathroom 3	Master bathroom	Office	Laundry room	Garage
Entry Hall		■																
Living room			■		■		□	□	□	□		■				■		
Dining room					■										■			
Family room					■										■			■
Kitchen						■								□			■	
Pantry																		
Bedroom 1												■						
Bedroom 2													■					
Bedroom 3													■					
Master bedroom											■				■	□		
Master closet																		
Bathroom 1																		
Bathroom 2																		
Bathroom 3																		
Master bathroom																		
Office																		
Laundry room																			■
Garage																		

FIGURE 9-2 *Adjacency matrix.*

The adjacency matrix in example 9-2 is for the same project as the bubble diagram in figure 3-7 from chapter 3, example 3.6. Compare the adjacency matrix and the bubble diagram to determine the similarities and differences of the two sets of information.

Efficiency Is Measured by Projecting Traffic Flow

A well-designed residence will make efficient use of how spaces are organized. For a residence to function efficiently, the traffic patterns of both owners and guests as they use the home must be considered. The design must take the placement of stairs, hallways, and entrances into consideration when studying traffic flow through the home. A home with a well-designed traffic pattern will often have closely related spaces

Analysis Leads to Optional Residential Space Planning

adjacent to each other. For example, placing the kitchen and dining room next to each other is often a good decision. The rooms within the sleeping area are often grouped together, along with a bathroom, and are separated from the living and service areas. Efficient designs reduce the amount of hallway and stair space in the building.

Problem Set 9.2 Analysis Leads to Optional Residential Space Planning

Problem 9-18   Using figure 9-3, calculate the gross floor area for the house.

Problem 9-19   Using figure 9-3, calculate the net floor area for the house if all exterior walls are 12" actual thickness, and interior walls are framed with 2 × 6 members plus 1/2" sheathing on each face.

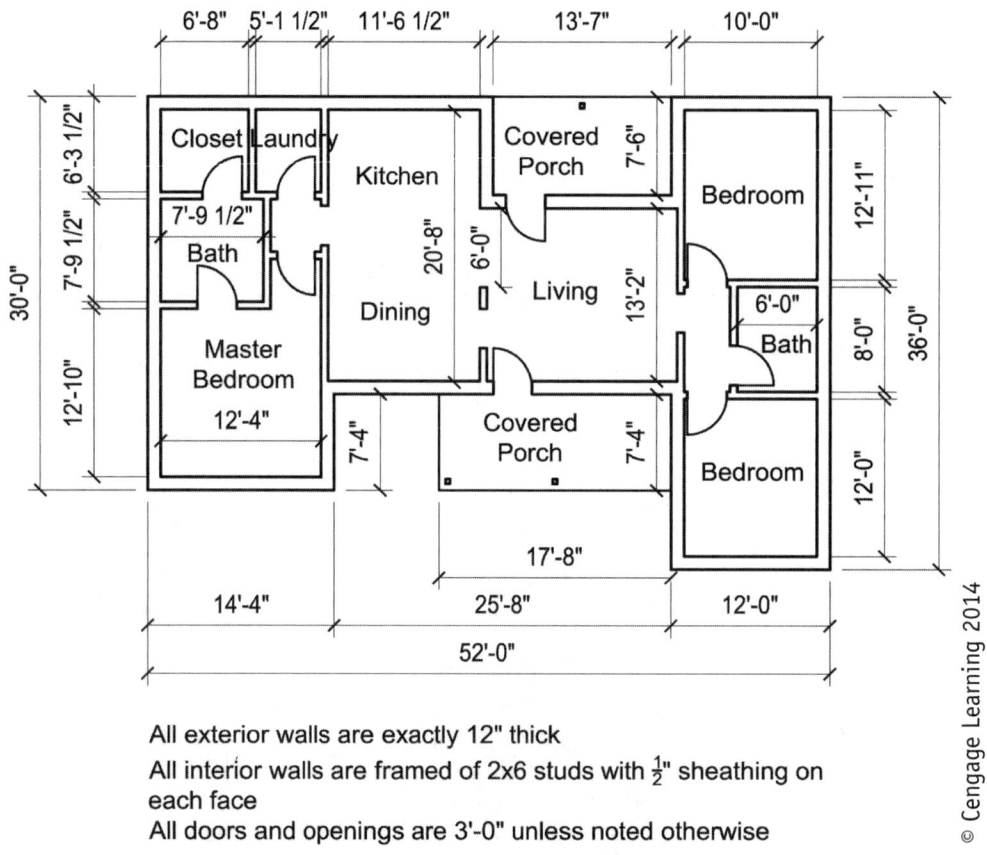

All exterior walls are exactly 12" thick
All interior walls are framed of 2x6 studs with $\frac{1}{2}$" sheathing on each face
All doors and openings are 3'-0" unless noted otherwise

FIGURE 9-3 *House plan for problems 9-18 and 9-19.*

Problem 9-20   Using figure 9-4, calculate the gross floor area for the house.

Problem 9-21   Using figure 9-4, calculate the net floor area for the house if all exterior walls are framed with 2 × 8 members, and interior walls are framed with 2 × 4 members. All framing will have 1/2" sheathing on each face.

Problem 9-22   Determine the overall area of a home with the following rooms:

- Living room – large
- Family room – medium
- Kitchen – large
- Dining room – small
- Bedrooms – two small and one large
- Full bathrooms – one medium and one large
- Garage – single car

176   Chapter 9: Residential Space Planning

Problem 9-23   Determine the overall area of a home with the following rooms:
- Living room – medium
- Family room – one small and one large
- Kitchen – large
- Dining room – large
- Bedrooms – four medium and two large
- Full bathrooms – one small, two medium, and one large
- Garage – large

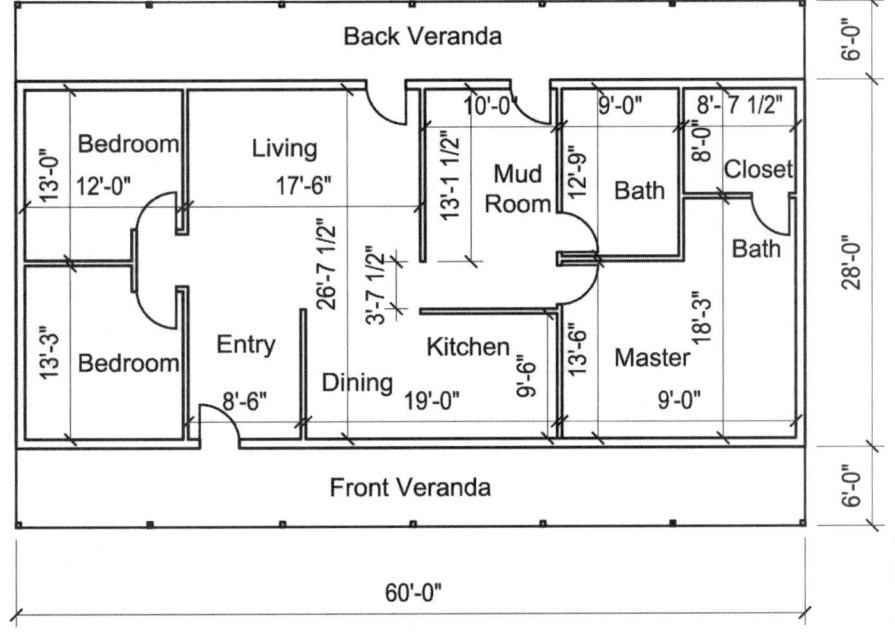

All exterior walls are framed of 2x8 studs with ½" sheathing on each face
All interior walls are framed of 2x4 studs with ½" sheathing on each face
All doors and openings are 3'-0" unless noted otherwise

FIGURE 9-4   *House plan for problems 9-20 and 9-21.*

Problem 9-24   Determine the overall area of a home with the following rooms:
- Living room – large
- Family room – large
- Kitchen – large
- Dining room – small
- Bedrooms – two large
- Full bathrooms – one small and two large
- Garage – double car

Analysis Leads to Optional Residential Space Planning   177

Problem 9-25 Develop an adjacency matrix for the residence in Problem 9-22.
Problem 9-26 Develop an adjacency matrix for the residence in Problem 9-23.
Problem 9-27 Develop an adjacency matrix for the residence in Problem 9-24.
Problem 9-28 Develop an adjacency matrix for the residence in which you live, and record it in your journal.

Major Development Considerations When Planning Functional and Appealing Spaces

Residential space planning must take many factors into consideration during the design of a home. For example, working, living, and sleeping spaces must be properly located such that they function well within the overall design. Some of these spaces require quiet conditions and thus should be located away from noisier spaces. In addition, within each room, the ceiling height must be considered along with window and door placement. Each of these affects the interior design of a home and also affects the home's exterior appearance. If the interior space planning and the exterior look of the home are not considered together, the home's beauty might not reach its full potential.

Planning within a Budget

Budgets are set based on how much a client wants to spend on a home, and on how much a lending institution will give to the clients as a mortgage, which is based partially on the amount of income the client makes. There are several cost-savings design issues that can be suggested to a client, including having multiple uses for rooms, centralizing and stacking stairs in the plan, and utilizing open floor plans, to name a few. There are many sources online for use in budgeting for residential design, as discussed at the start of this chapter. Budgets are often adjusted throughout the design and construction process as design changes are made, and final finish materials are selected. It is crucial that the designer keeps the clients apprised of changes in the budget to make sure they can afford the final cost of the home.

Planning Living Areas

When planning living areas, the designer must understand how homeowners will use the space. If the designer is familiar with the client's lifestyle, wants, and needs, the living areas can be planned and arranged within the home to successfully provide spaces that allow the client to relax, converse, and/or play. Rooms used as living areas include the living room, den, family room, great room, dining room, library, home theater, and sunroom. A typical home will have several of these living areas included, and the placement and access to each of the living areas with respect to the rest of the home becomes important so that access to the rooms can flow properly. The main entry to a home is often adjacent to a living area, and this entry is often used to move furniture and appliances in and out of the home, so the doors must be properly sized to allow for access.

Working and Service Areas

Working and service areas include the garage, utility room, kitchen, laundry room and bathrooms, which are often adjacent to each other to help reduce plumbing costs by sharing common access. Working and service areas are often noisy spaces that should be located away from quiet spaces within a home. Often, exterior entrances are associated with these spaces for convenient loading and unloading of supplies for the home. Bathrooms should be located in the home such that they can service the public and private spaces accordingly, and for two-story homes they should be stacked in plan to take advantage of plumbing in the walls of the home, with a one-half bath consisting of a sink and a toilet often provided for the public spaces.

Kitchen Planning

A Kitchen is typically the most used and most expensive room in a residence, so it becomes important that it be planned to function properly. If properly planned, the kitchen space will feel welcoming to the homeowner. Kitchen consultants are sometimes hired to provide a detailed design for the equipment and cabinets. At the beginning of a project, the client and architect decide on a budget allowance for the kitchen. Typically, this allowance includes the cost of appliances, cabinets, flooring, countertops, and backsplash. The final choices for the kitchen are often made well into construction, and it is important that the budget be properly developed to include the desired finished design. Proper counter space and placement of appliances within the kitchen will lead to a well-designed space appreciated by the clients. It should be noted that the kitchen is often the main selling point in a home, and a well-designed kitchen can increase the value of a home as well as decrease the time a home is on the market.

Kitchen Work Triangle

When planning the location of elements within a kitchen, the work triangle is a tool used to determine whether the kitchen plan is efficient. This triangle indicates the travel distances among the sink, stove and refrigerator. An efficient kitchen plan is achieved if the total straight-line distance among these three elements is from 15 to 22 feet with no distance between two of the elements being less than 4 feet. Additionally, there should be no foot traffic from another room through this triangle. Textbook figure 9-29 gives an example of the kitchen work triangle for a kitchen plan.

Kitchen Arrangements

The arrangements of kitchens can be designed in many configurations. Some of these, as shown in textbook figure 9-30, include:

- L-shape
- U-shape
- Peninsula
- One wall
- Corridor
- Island

With the exception of the one-wall kitchen configuration, all can utilize the work triangle mentioned in the previous section.

Cabinets and Pantries

Often, a kitchen supplier or kitchen planner is involved in helping plan the kitchen for a residence. Cabinets come in standard sizes, depending on their placement and use. Cabinets fall into the four main categories:

- Base cabinets – support the countertop and often incorporate drawers and appliances such as the stove, oven, dishwasher, and trash compactor
- Wall cabinets – hung from the walls and are used for storage of dishes, spices, and staples
- Utility cabinets – are often found in the laundry room and garage and are typically constructed of durable materials and can be constructed with or without shelves
- Specialty cabinets – can be base or wall cabinets that are designed for a specific residence, and are used to infill between standard cabinet sizes

Typical cabinet sizes include 24" depth for base cabinets, and 12" depth for wall cabinets. Utility cabinets often vary in depth, depending on their intended use.

Bathrooms

A typical bathroom may include the following elements:

- Water closet
- Vanity and lavatory
- Mirror
- Storage cabinet(s)
- Bathtub
- Shower

It is important as a cost issue to locate bathrooms adjacent to each other when possible. When two bathrooms can share a common wall, it can be used for efficient and economical placement of plumbing for both bathrooms within a single wall, which is typically wider than a normal wall in a home. Typical bathroom planning includes the following bathroom types:

- Half bath – contains a lavatory, and a water closet
- 3/4 bath – contains a lavatory, water closet, and a walk in shower
- Full bath – contains a lavatory, water closet, and a bathtub or bath tub/shower combination
- Bath and a half – contains a lavatory, water closet, and a separate bathtub and shower
- Master bath – contains a lavatory (often more than one), water closet, and a bathtub/shower (often separated) as a minimum. This room often has additional features, such as a makeup table, whirlpool, or sauna, and is often connected directly to the master bedroom closet for easy access when dressing.

When placing fixtures in bathrooms, the designer must ensure that minimum clearances are provided to allow proper movement within the space by the homeowners. Textbook figure 9-37 gives an example of these clearances.

Laundry Room

Laundry rooms can range in size from a hall-accessed closet with a stackable washer and dryer, to a full separate room that may include space for additional functions such as hobbies, sewing, or ironing. Similar to the front entry to the home, the laundry room doors must be wide enough to accommodate moving appliances in and out of the space. In addition, the laundry room should be located near other plumbing to reduce costs, should be planned with adequate storage space, and may be required to have direct access to the exterior so that a clothesline can be used to dry clothes.

Sleeping Area

Bedrooms are sleeping areas within a home, and the requirements for these rooms can vary greatly among clients. Planning for these spaces will depend on many factors, including the age and mobility of the clients and any children living in the home. For example, an older homeowner may chose to have the master bedroom on the first floor to accommodate their mobility, while a family with young children may choose to have the master bedroom adjacent to children's bedrooms located on the second floor. These factors must be considered by the designer when considering placement of these rooms within a residential design.

Mechanical Rooms

Mechanical rooms typically house equipment such as the hot water tank, the water filter/water softener and the heating, ventilating, and air-conditioning (HVAC) unit. For homes with basements, the mechanical rooms are usually located at that level. For homes without basements, the mechanical rooms are often accessed from the garage. Mechanical room size is dependent on conditions such as size and type of equipment as well as space needed for repairs.

Storage and Closets

Storage requirement will vary greatly between homeowners, but one thing is for sure: Storage is always needed and seldom seems to be supplied in ample amounts in a home design. When space planning, it is critical that proper storage spaces be provided throughout the home. Bedrooms, bathrooms, hallways, laundry rooms, and garages are locations where much of a home's storage space will be located, but it is important to the success of a residential design that storage be provided in each room of the home, and this storage space should be designed in a way as to be utilized efficiently by the occupants of the home.

Stairs and Hallways

Stairs and hallways can greatly affect both the quality and efficiency of a home design. Centrally located stairs and hallways can bring order to a residential design while making the plan efficient.

Dimensional requirements for use in stair design and construction have been set by the 2006 International Residential Code (IRC). For a non-circular stair, the code states that the minimum stair width including handrails is 36", the minimum clear height in a stair is to be 6'–8", the minimum tread depth (horizontal dimension) is 10", the maximum riser height (vertical dimension) is 7 3/4", and landings have a length in the direction of travel of at least the width of the stair. The building code requires that each stair tread and riser be essentially of equal dimensions. Stairs in homes often have landings at the top and/or bottom of the stair, and in some cases, intermediate landings are used when a stair changes direction. By utilizing the riser and tread dimensions, the plan area for a stair can be determined. Circular stairs also have dimensional requirements that vary slightly from those listed above. The procedure for determining the plan area of a straight-run staircase will require the designer to:

- Determine the type of stair configuration (refer to textbook figure 9-44) to be used and any requirements for stair landings.
- Determine the vertical finish floor-to-floor height in inches.
- Calculate the number of stair risers required, divide the floor-to-floor height by 7.5" per riser, and round up to arrive at the number of risers in the stair.
- The staircase run dimension can now be calculated by multiplying the length of the stair tread by the number of risers minus one for a straight-run staircase, and the number of risers minus two for a quarter-turn or half-turn staircase.
- The minimum stair width required by the IRC is 36".
- The plan area for a staircase is the stair width multiplied by the length of the staircase run plus any landing lengths.

Staircases that have intermediate landings are calculated in a similar manner. It is important to note that staircases can vary greatly in a home and that the building code (IRC) has various dimensional requirements to allow for a staircase that is functional and provides the occupants with adequate space for vertical circulation. Additionally, the IRC provides dimensional limits only, and the architect must take into consideration the client's needs when determining stair size and placement in a home.

Example 9.7 Determine the plan area for a straight-run staircase in a home with the following conditions:

- Stair width to meet minimum code requirements
- Finish floor to floor height = 10'-0"
- Riser height is to be 7 1/2"
- Tread depth is to be 11"
- Landings to match stair width are required at top and bottom of stair

Minimum stair width per IBC is 36". Number of risers required = 10'(12 in/ft)/(7.5 in/riser) = 16 risers. Length of stair treads = (16 − 1) 11" = 165". Landings at the top and bottom of the stair shall match the stair width of 36", thus the total length of staircase = 2(36") + 165" = 237" = 19'-9". thus, the staircase plan area = 3'-0" wide by 19'-9" long.

Example 9.8 Determine the dimensions, and draw a plan for a half-turn staircase in a home with the following conditions:

- Stair width = 48"
- Finish floor to finish floor height = 9'-0"
- A 4 1/2" wide guardrail is placed for interior support of the staircase
- Riser height not to exceed code maximum
- Tread depth is to be 12"
- Landings to match stair width, and are required at top and bottom of stair

Stair width is set at 48", hence the length of the landings will also be set at 48". Finish floor to floor = 9'-0" = 108". Using the maximum riser height of 7 3/4", the number of risers required = 108"/(7.75 in/riser) = 13.93 risers, therefore use 14 risers. For a half-turn staircase, it is preferred that the number of total risers be an even number so that the same number of risers can be placed above and below the intermediate landing. The length of the stair treads for this staircase = (14 − 2) 12" = 144" = 12'-0", thus 6'-0" for each run of stairs leading to the intermediate landing. The total width of the staircase = 2 (48" stair width) + 4.5" guardrail wall = 8'-4 1/2" wide. The total length of the staircase = 2 (48" landings) + 6"0" stair length = 14'-0" long. There for, the staircase plan area = 8'-4 1/2" wide by 14'-0" long. With the dimensions calculated, a plan can be drawn for the staircase, as shown in figure 9-5.

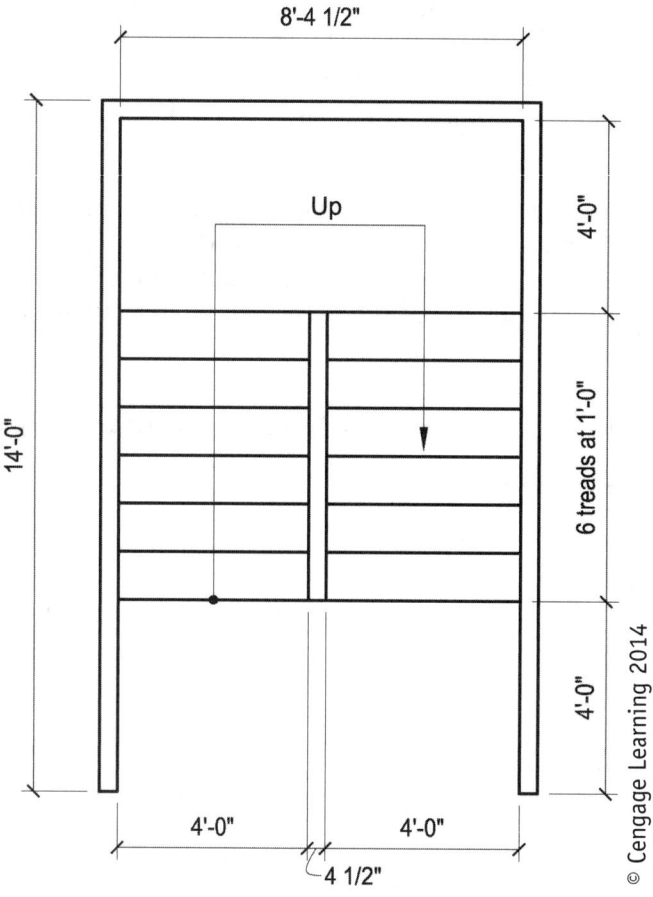

FIGURE 9-5 *Stair plan.*

Spiral Staircases

While spiral staircases take up little space, they may be difficult to use. In addition, when it is the only access to an upper level within a home, moving furniture to that level can be difficult, and a large operable window on the exterior of the home may be needed. Spiral stairs come in many different styles and materials, as can be seen on the Great Lakes Stair & Millwork Company website at www.stair.com.

Hallways

Hallway widths start at a minimum of 36", with wider hallways providing opportunities for incorporating better lighting and for providing a more comfortable space. Residential rooms accessed from a hallway typically have the doors swing into the room to avoid any kind of barrier protruding in the hallway space.

Planning a Garage

Vehicles vary greatly in size, and the designer should take into consideration that an owner may change the size of their vehicles over time. If a garage is not adequately planned, the owners might not be able to use the space properly if a large-scale vehicle is purchased. However, garage planning involves much more than determining the number of cars that will occupy the space. When planning a garage, several questions must be answered. These include:

- How many vehicles will be parked in the garage?
- What types and sizes of vehicles will be parked in the garage?
- Will the garage have an area dedicated to projects and hobbies? If so, will a utility sink and/or workbench be required?
- What are the storage requirements for the garage?
- Will there be an attic access located in the garage?
- What size of doors will be required to accommodate the vehicles parked in the garage?

Textbook figure 9-13 gives standard sizes for garages, but these sizes do not include extra space for many of the items listed above. It is critical when space planning for a garage that the intended use of the space is fully understood so that the appropriate areas, ceiling heights and door sizes can be properly included in the design.

Firewall

For multi-family residential building, a firewall may be required by the building code. A firewall is used to separate residential units and to protect the units in case of fire. A firewall is designed to impede the spread of a fire for a period of time that allows occupants to exit the building safely, and also to allow the fire to be contained. It is constructed of non-combustible materials of either a masonry, gypsum board on metal stud framing, or a combination of both.

Porches, Sunrooms, Decks, Patios, and Balconies

The architect can add value to a home design by extending the living space past the exterior wall of the home by adding porches, sunrooms, decks, patios, and balconies to act as additional spaces for the home. These can be utilized as exterior, covered, or seasonal spaces. It is important that building codes be consulted as there are often additional requirements pertaining to safety for these types of spaces.

Problem Set 9.3 Major Development Considerations When Planning Functional and Appealing Spaces

Problem 9-29: Draw a plan of the kitchen in your home, calculating and indicating the work triangle. Show dimensions, and determine whether the kitchen is efficiently planned.

Problem 9-30 Determine the dimensions, and draw a plan for a straight-run staircase in a home with the following conditions:

- Stair width = 42"
- Finish floor to floor height = 11'-6"
- Riser height not to exceed code maximum
- Tread depth = 10.5"
- Landings required at top and bottom of stair

Problem 9-31 Determine the dimensions, and draw a plan for a straight-run staircase in a home with the following conditions:

- Stair width to meet minimum code requirements
- Finish floor to floor height = 10'-6"

- Riser height is to be 7"
- Tread depth = 12"
- Landings to match stair width at top of stair only

Problem 9-32 Determine the dimensions, and draw a plan for a quarter-turn staircase in a home with the following conditions:

- Stair width to meet minimum code requirements
- Finish floor to floor height = 12'-0"
- Riser height not to exceed code maximum
- Tread depth = 10"
- Landings to match stair width are required at top and bottom of stair

Problem 9-33 Determine the dimensions, and draw a plan for a quarter-turn staircase in a home with the following conditions:

- Stair width = 44"
- Finish floor to floor height = 9'-8"
- Riser height not to exceed code maximum
- Tread depth = 11"
- Landings to match stair width are required at top and bottom of stair

Problem 9-34 Determine the dimensions, and draw a plan for a half-turn staircase in a home with the following conditions:

- Stair width = 40"
- Finish floor to finish floor height = 11'-8"
- A 6 1/2" wide guardrail is placed for interior support of the staircase
- Riser height = 7"
- Tread depth = 11"
- Landings required at top and bottom of stair to match stair width
- Intermediate landing to have a width = 5'-0"

Problem 9-35 Determine the dimensions, and draw a plan for a half-turn staircase in a home with the following conditions:

- Stair width = 54"
- Finish floor to finish floor height = 12'-6"
- A 4 1/2" wide guardrail is placed for interior support of the staircase
- Riser height not to exceed code maximum
- Tread depth is to be 12"
- Landings required at top and bottom of stair to match stair width
- Intermediate landing to have a width = 6'-6"

Problem 9-36 Sketch a plan for a home for the spaces shown in the bubble diagram given in figure 3-7. Assume that the floor areas given in the bubble diagram are gross floor areas.

CHAPTER 10
Commercial Space Planning

Skills List

After completing this chapter, you should be able to:

- Understand the differences between residential and commercial space planning

- Calculate the occupancy load for a building using the 2009 International Building Code (IBC)

- Calculate required minimum exit corridor and exit stair widths using the 2009 IBC

- Apply the American with Disabilities Act (ADA) Accessibility Guidelines

- Calculate the number of accessible parking spaces required for a parking lot

- Utilize resources to approximate floor areas required for office buildings

- Understand how to derive and use an adjacency matrix for a commercial building

- Understand standards and accessibility requirements for escalators and elevators in a commercial building

- Understand the basic concepts of an intelligent building

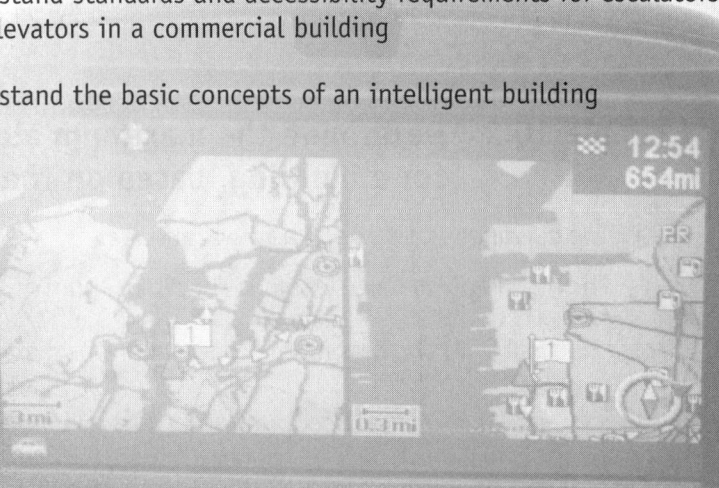

BACKGROUND

This chapter explores space planning for commercial buildings, which for the purposes of the workbook will include any non-residential building. Planning for a commercial space is different than planning for a residential space. One major difference is that commercial buildings must be accessible, meeting requirements outlined in the Americans with Disabilities Act (ADA) publication Accessibility Guidelines for Buildings and Facilities, available online at www.access-board.gov/adaag. There are other considerations that also need to be addressed in commercial space planning, such as specific restroom requirements, drinking fountains, fire stairs, and exit corridors. In addition, commercial buildings typically have public and private spaces, and the relationship of these spaces to each other could influence the placement of rooms within the building design. The inclusion of sustainability in commercial design to maximize natural daylighting and minimize energy use is commonplace in design. A building's use will often help in the decisions of where to place the functions within the building design. Recent studies have linked building occupant's work performance to the quality of the lighting and air quality, as well as to the aesthetics of the building's design. Each of these conditions, and many more, must be taken into consideration to arrive at a successful commercial building design.

Commercial Spaces are Designed for Safety, Accessibility, Flexibility, and Sustainability

Building codes are utilized to help achieve a building design that is both safe and accessible for building occupants. Many of the building code requirements are based on the occupancy load of a building. The occupancy load tells how many people a building can comfortably and safely accommodate, and is based on the use and size of the building. Calculation of a building's occupant load is covered in Section 1004 of the 2009 International Building Code (IBC). IBC Table 1004.1.1, shown in figure 10-1, gives the maximum floor area allowed per occupant for different building uses. For a known square footage of a building, the occupancy load can be determined based on its use. Or, if the number of occupants is known, the maximum square footage for the building can be determined.

Example 10.1   Determine the occupancy load for an office building with a gross square footage of 25,000 sq. ft.

Using figure 10-1, the maximum square footage per occupant for an office building would fall into the business areas category, and has a value of 100 gross sq. ft. per occupant. The occupancy load can be found from equation:

25,000 sq. ft. / (100 sq. ft. / occupant) = 250 occupancy load

Example 10.2   Determine the maximum allowable gross square footage for a building, based on the following conditions:

- The building is to be used as a warehouse.
- The building exits are designed for a maximum of 200 occupants in the building at any time.

Using figure 10-1, the maximum gross square footage allowed per the 2009 IBC is:

200 occupants * 500 sq. ft. / occupant = 100,000 sq. ft.

Once the occupant load for a building is determined, the required number of exits, and exit stair widths for multi-story buildings, can be calculated.

The 2009 IBC has set the required widths for exit corridors and exit stairs based on the number of occupants using the exit. For an exit corridor, the total required minimum width is 0.2 inches per occupant, and for an exit

MAXIMUM FLOOR AREA ALLOWANCES PER OCCUPANT			
FUNCTION OF SPACE	FLOOR AREA IN SQ. FT. PER OCCUPANT	FUNCTION OF SPACE	FLOOR AREA IN SQ. FT. PER OCCUPANT
Accessory storage areas, mechanical equipment room	300 gross	Exercise rooms	50 gross
Agricultural building	300 gross	H-5 Fabrication and manufacturing areas	200 gross
Aircraft hangars	500 gross	Industrial areas	100 gross
Airport terminal Baggage claim Baggage handling Concourse Waiting areas	20 gross 300 gross 100 gross 15 gross	Institutional areas Inpatient treatment areas Outpatient areas Sleeping areas	240 gross 100 gross 120 gross
Assembly Gaming floors (keno, slots, etc.)	11 gross	Kitchens, commercial	200 gross
Assembly with fixed seats	See Section 1004.7	Library Reading rooms Stack area	50 net 100 gross
Assembly without fixed seats Concentrated (chairs only—not fixed) Standing space Unconcentrated (tables and chairs)	7 net 5 net 15 net	Locker rooms	50 gross
Bowling centers, allow 5 persons for each lane including 15 feet of runway, and for additional areas	7 net	Mercantile Areas on other floors Basement and grade floor areas Storage, stock, shipping areas	60 gross 30 gross 300 gross
Business areas	100 gross	Parking garages	200 gross
Courtrooms—other than fixed seating areas	40 net	Residential	200 gross
Day care	35 net	Skating rinks, swimming pools Rink and poolDecks	50 gross 15 gross
Dormitories	50 gross	Stages and platforms	15 net
Educational Classroom area Shops and other vocational room areas	20 net 50 net	Warehouses	500 gross

FIGURE 10-1 "Portions of this publication reproduce excerpts from the 2009 International Building Code, International Code Council, Inc., Washington, D.C. Reproduced with permission. All rights reserved." www.iccsafe.org

For SI: 1 square foot = 0.0929 m².

stair, the total required minimum width is 0.3 inches per occupant. This total width is divided among the number of exit corridors and stairs within a building. Additionally, the building code has set minimum widths for exit corridors and exit stairs to be used in building design. For the exit corridor, the minimum width is set at 44", unless the corridor has an occupant load of less than 50, when the minimum width can be 36". For an exit stair, the width of stair required at each level is based on the number of occupants at that level. However, once a stair width is determined for a floor level, that width must be provided below the floor level down to the building exit level. Note that the minimum widths listed above are for the majority of commercial building exits, but there are exceptions to these minimum widths, such as school and hospital corridors.

Example 10.3 Using IBC requirements, determine the minimum width of exit corridors for the following conditions:

- One-story building with an occupancy load of 1,200 persons
- Five exit directions
- Assume one exit becomes blocked during an emergency

The IBC requires a width of 0.2 inches per person for an exit corridor on a given level of a building, thus the equation to determine the minimum width is:

1,200 persons * 0.2 inches/person / 4 exits = 60 inch minimum width for the exit corridors within this building

A more complex example in which the exit stair widths are calculated for a multi-story building with varying floor areas at each level in the building is shown below.

Example 10.4 Using IBC requirements, calculate the exit stair width for each level of a five story dormitory that utilizes 5 exit stairs with the following floor areas:

Building level	Gross square footage
5th	40,000
4th	28,000
3rd	46,000
2nd	50,000
Ground	50,000

TABLE 10-1 *Dormitory gross floor areas.*

First one needs to determine the occupancy load for each level. Using figure 10-1, it is determined that dormitories allow for a maximum of 50 square feet of gross floor area per occupant. With this value, the occupancy load can be calculated. Once the occupancy load is found for each level, the exit stair width can be calculated using the IBC required width of 0.3 inches per occupant for an exit stair, with a minimum exit stair width of 44 inches. Note that occupants on the ground floor will not use the exit stairs. This example will now perform the calculations for the 5th-floor level. The occupancy load is determined using the equation:

Occupancy load = Gross square footage / Floor area allowance per occupant (figure 10-1)
= 40,000 sq. ft. / (50 sq. ft. / occupant) = 800 occupants

With the occupancy load known, the width of each exit stair can be calculated from the equation:

Exit stair width = (occupancy load) * (0.3 inches/occupant) / number of exit stairs
= 800 occupants * (0.3 inches / occupant) / 5 exit stairs
= 48 inch minimum stair width

Note that if the minimum width had been less than the code-required 44 inches, then 44 inches would be required. Table 10-2 gives the calculations for all levels of the building.

Building level	Gross square footage	Occupancy load	Calculated stair width (inches)	Code-required stair width (inches)
5th	40,000	800	48	48
4th	28,000	560	33.6	48
3rd	50,000	1000	60	60
2nd	46,000	920	55.2	60
Ground	50,000	1000	—	—

TABLE 10-2 *Calculation of exit stair width at each level.*

Note that for each level, the exit stair width is for the stair directly below that level. For example, the 48 inch required width for the 5th floor of the building will be for the stair between the 5th and 4th floors. Also, note that once a width is required at a level, it cannot be reduced at lower levels within the building. For the 5th floor, a 48 inch width was calculated, and for the 4th floor, a 33.6 inch width was calculated. However, since the 5th level exit required a 48 inch width, that width must be provided as a minimum down the entire stair. Another location where this occurs is at the 2nd floor exit stair width. Finally, no values are calculated for the ground-floor occupancy load, because they would be using alternate building exits instead of the exit stairs.

Note that for certain building types, such as schools and hospitals, the minimum corridor width is greater than as required per the minimum width calculations given in this chapter. Additionally, ADA requirements often require additional minimum width for corridors.

For commercial buildings, facilities such as drinking fountains and restrooms must be provided. These requirements are based on the number of occupants in a building and can vary greatly based on the occupancy classification for the building. As a general rule, a drinking fountain is required for every 75 occupants, with the common requirement of locating a drinking fountain on each floor. Restrooms are required for most building types, and the number of toilets in a restroom is also based on the occupancy rate. One toilet per 75 to 125 occupants is required for a men's restroom, and one toilet per 65 to 75 occupants is required for a women's restroom. One sink is required for every 150 to 200 occupants, but this number can increase to one sink for every 80 occupants for office buildings. Table 2902.1 of the 2009 IBC indicates the number of drinking fountains, toilets, sinks, and showers required in buildings of different occupancies.

When designing a commercial building, the designer must consider both the owners and users of the building. As discussed in Chapter 3: Research, Documentation, and Communication, an extensive amount of research must be conducted before beginning the planning stage. This information provides the designer a starting point for meeting the needs and wants of the building owner. In commercial design, the research involves identifying public, private, and support spaces within the building and giving floor space allotments to each of the spaces. Typical floor areas, such as those shown in table 10-3, give ranges of different spaces within a commercial building. Websites such as the Officefinder.com website at www.officefinder.com/officespacecalc.html have space calculators that can be utilized in determining floor-area requirements.

President's or Chairman of the Board's Office	250 to 400 sq. ft. (4 to 5 windows in length)
Vice-President's Office	150 to 250 sq. ft. (3 to 4 windows in length)
Executive's Office	100 to 150 sq. ft. (2 widows in length)
Employee Office	100 to 125 sq. ft. (desk, lateral file, 1 visitor chair)
Partitioned Open Space	80 to 110 sq. ft.
Open Space	60 to 110 sq. ft.
Workstation Area	50 to 100 sq. ft. for clerical (depending on file and equipment needs) 64 to 80 sq. ft. for technical personnel
Conference Rooms	15 sq. ft. per person: theater style 25 to 30 sq. ft. per person: conference seating
Mail Room	8 to 9 ft. wide with 30" counters on either side. Length depends upon usage
Reception Area	125 to 200 sq. ft. Receptionist and 2 to 4 people 200 to 300 sq. ft. Receptionist and 6 to 8 people
File Room	7 sq. ft. per file with a 3' to 4' aisle width
Library	Allow 12" for bookshelf width 175 to 450 sq. ft. with seating for 4 to 6
Lunch Rooms	15 sq. ft. per person, not including kitchen. Kitchen should be 1/3 to 1/2 of seating area
Coat Closets	1 lineal ft. for 4 coats, 3 per person

TABLE 10-3 *Typical business space estimates.*

Example 10.5 Using Table 10-3, determine the approximate floor area for a business office with the following room requirements:

- One President office
- Two Vice-president offices
- Four executive offices
- Twenty employee offices
- Workstation area for 8 clerical employees
- Reception area (6 to 8 people)

- Library
- File room for 50 file cabinets
- Lunch room for 20 employees with a kitchen

Each of the floor areas for the spaces listed above are included in table 10-3, though the decision needs to be made on whether to consider the lower- or upper-limit floor areas when estimating the overall floor area requirements. For this example, the solution will be determined using the lower-limit areas of table 10-3.

Refer to table 10-4 for the area calculation using the lower limits:

Room	Area
President office	250 sq. ft.
Vice-President offices (2)	2 * 150 = 300 sq. ft.
Executive offices (4)	4 * 100 = 400 sq. ft.
Employee offices (20)	20 * 100 = 2,000 sq. ft.
Workstation area for eight clerical employees	8 * 50 = 400 sq. ft.
Reception area (6 to 8 people)	200 sq. ft.
Library	175 sq. ft.
File room for 50 file cabinets	50 * 7 = 350 sq. ft.
Lunch room for 20 employees with a kitchen	1.333*(20 * 15) = 400 sq. ft.
Total Floor Area for office building	**4475 sq. ft.**

TABLE 10-4 *Summation of office building floor areas.*

Note that in addition to the areas listed above, additional floor area may be required for such rooms as storage and restrooms if they are not provided outside the office space. Generally, these areas are used as a starting point to determine the amount of space needed.

Example 10.6 Using the online calculator at the Officefinder.com website at www.officefinder.com/officespacecalc.html, approximate the square footage of an office for the following spaces:

- 1 President office
- 2 Vice-President offices
- 6 Executive offices
- Partitioned open space for 18 employees
- 1 mail room
- 1 file room
- 1 library

Design for Safety, Accessibility, Flexibility, and Sustainability

- 1 conference room for 30 people
- 1 conference room for 10 people
- Lunch room for 20 people
- Reception area for 10 people

For this particular office space calculator, the information is input as shown in figure 10-2.

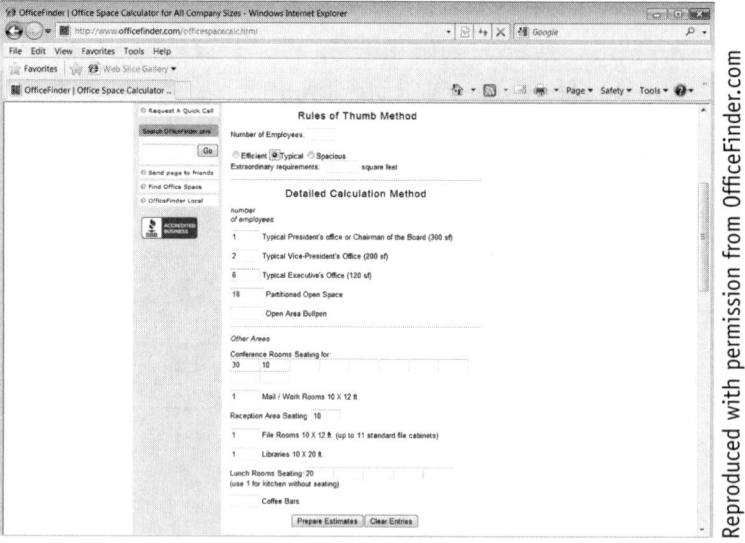

FIGURE 10-2 *Input screen from officefinder.com website.*

Note that in figure 10-2, the option exists with this space calculator to perform a calculation based on rules of thumb, or the option to perform a more detailed calculation which was chosen for this example. Once the data are entered, the space can be calculated, with results given as shown in figure 10-3.

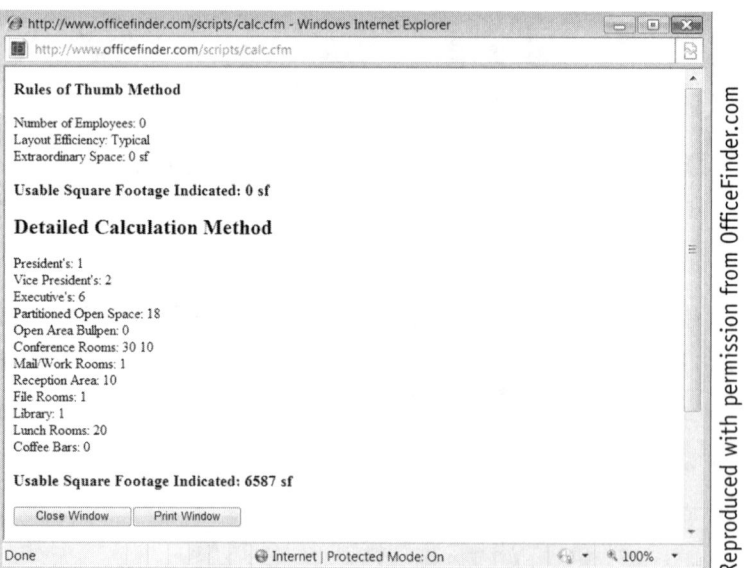

FIGURE 10-3 *output screen from officefinder.com website.*

Based on the conditions given in this example, an office space floor area of approximately 6,600 sq. ft. will be needed. Note that the floor areas from this example are only estimates, and the final design can vary greatly from what is shown.

192   Chapter 10: Commercial Space Planning

Once a list of spaces required for the building and the estimated floor area for each space has been determined, the designer can research the activities to occur in each space along with the furniture, equipment, and occupancy required to determine whether the floor area will be sufficient for each space within the building. The end result of this work is known as the program for space types, as shown in figure 10-8 of the textbook. In addition to furniture and equipment required in each space, the designer must also consider lighting requirements as well as whether the space needs to be a public, private, or secured space. This information is utilized in arriving at an adjacency matrix for the building.

Adjacency Matrix

Example 10.7   Derive an adjacency matrix for the building spaces listed in Example 10-5.

Each room is shown on the adjacency matrix and is indicated as a primary or secondary adjacency to all other spaces. In addition, each of the spaces are indicated as public, private, or secured. Refer to figure 10-4 for the adjacency matrix for this example.

		President Office	Vice President Office	Executive Office	Employee Office	Clerical Workstation	Reception	Library	File Room	Lunch room / Kitchen
■ : Primary adjacency □ : Secondary adjacency	**Commercial Adjacency Matrix**									
President Office			■	□	□			□		
Vice President Office				■	□	□	□	□		
Executive Office					■	■	□			
Employee Office						■	□	□	□	□
Clerical Workstation							■	□	■	□
Reception								□	■	□
Library									□	
File Room										
Lunch room / Kitchen										

FIGURE 10-4   *Adjacency matrix for example 10-7.*

Problem Set 10.1   Commercial Spaces are Designed for Safety, Accessibility, Flexibility, and Sustainability

Problem 10-1   Determine the occupancy load for a university dormitory with a gross square footage of 42,000 sq. ft.

Problem 10-2   Determine the occupancy load for an airport terminal with a total gross square footage of 240,000 sq. ft. Within the airport, the concourse has an area of 100,000 sq. ft., the waiting areas have an area of 70,000 sq. ft., the baggage-claim area has an area of 50,000 sq. ft., and the baggage-handling area has an area of 20,000 sq. ft.

Problem 10-3 Determine the occupancy load for a library with a gross square footage of 100,000 sq. ft. The stack area in the library will make up 75 percent of the gross area, and the reading rooms will make up 25 percent of the gross area, with the net-area to gross-area ratio equal to 0.9.

Problem 10-4 Calculate the required minimum width of each exit corridor for a building with an occupancy load of 2,400 persons and 8 building exits.

Problem 10-5 Calculate the required minimum width of each exit corridor for a building with an occupancy load of 1,500 persons and 10 building exits.

Problem 10-6 Using the ADA Accessibility Guidelines for Buildings and Facilities (ADAAG), determine the minimum number of accessible parking spaces required for a commercial parking lot with 325 parking spaces.

Problem 10-7 Using the ADA Accessibility Guidelines for Buildings and Facilities (ADAAG), determine the minimum number of accessible parking spaces required for a commercial parking lot with 850 parking spaces.

Problem 10-8 Using the ADA Accessibility Guidelines for Buildings and Facilities (ADAAG), determine the minimum number of accessible parking spaces required for a commercial parking lot with 1,850 parking spaces.

Problem 10-9 Using IBC requirements, calculate the exit stair width for each level of a four story library that utilizes 3 exit stairs. The upper floors of the building accommodate the book stacks, with the ground floor accommodating the reading rooms. The library has the gross floor areas at each level as shown in table 10-5.

Building level	Gross square footage
4th	40,000
3rd	50,000
2nd	32,000
Ground	50,000

TABLE 10-5 *Library gross floor areas.*

Problem 10-10 Rework example 10-5 to determine the approximate floor area for a business office using the upperlimit areas of table 10-3. What percentage difference exists between the lower and upper area limits for this office building?

Problem 10-11 Rework example 10-6 using the rules-of-thumb setting for an efficient setting. How does this vary from the total area determined in example 10-6?

Problem 10-12 Rework example 10-6 using the rules-of-thumb setting for both the typical and spacious settings and compare these floor areas with the efficient setting calculation.

The Space Plan Allows Safe and Efficient Circulation of Building Occupants

Previously in this chapter, the minimum requirements for exit corridors and stairs based on 2009 IBC requirements were discussed. In addition to these requirements, the Americans with Disabilities Act Accessibility Guidelines (ADAAG) must be consulted during the design process to ensure that people with disabilities will be able to use the building successfully. The ADAAG is available online at www.access-board.gov/adaag and is to be used as a resource for determining clearance requirements in a commercial building.

Example 10.8 Using the Americans with Disabilities Act Accessibility Guidelines (ADAAG), how many wheelchair locations must be provided for an assembly area with fixed seating for 1,250?

In section 4.1.3(19a) of the ADAAG, the required number of wheelchair locations for fixed seating greater than 500 is 6, plus 1 added location for every seating increase of 100. The calculation of the number of locations is 6 + (1,250 − 500)/100 = 13.5 locations, which can be rounded down to 13 wheelchair locations.

Example 10.9 Using the Americans with Disabilities Act Accessibility Guidelines (ADAAG), sketch a section through a library book stack area showing the required and preferred minimum aisle width and height.

In section 8 of the ADAAG, the minimum aisle width is 36". Per section 8.5, the preferred minimum aisle width is 42", and the stack height is unrestricted. Figure 10-5 shows a sketch of these conditions.

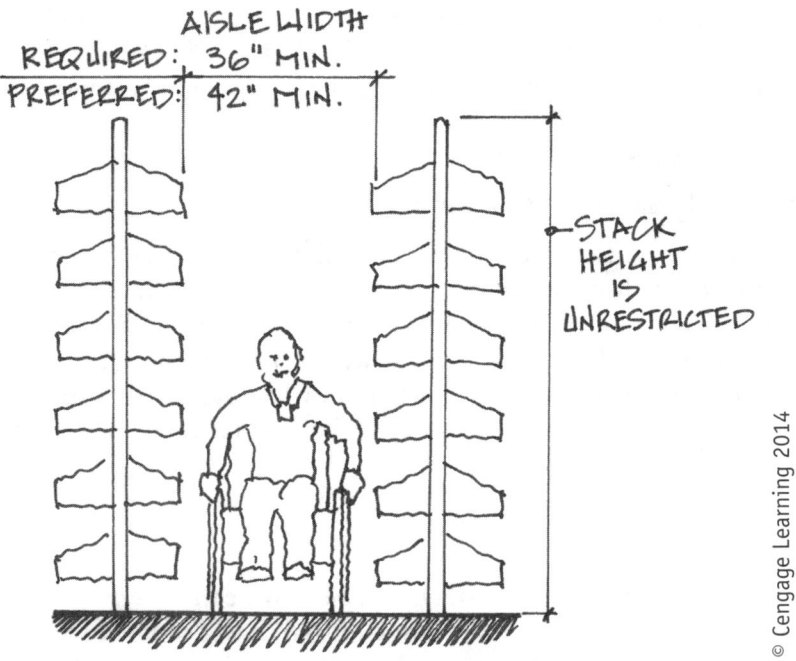

FIGURE 10-5 Aisle widths for library stack areas.

Problem Set 10.2 The Space Plan Allows Safe and Efficient Circulation of Building Occupants

For problems 10-13 through 10-22, use the Americans with Disabilities Act Accessibility Guidelines (ADAAG) available online at www.access-board.gov/adaag.

Problem 10-13   What are seven conditions when the ADA guidelines are not required to be met?

Problem 10-14   Calculate the number of wheelchair locations required for an auditorium with fixed seating for 2,500 people.

Problem 10-15   Calculate the number of wheelchair locations required for an auditorium with fixed seating for 3,750 people.

Problem 10-16   How many accessible boat slips are required in a boat dock with 325 total boat slips?

Problem 10-17   How many accessible boat slips are required in a marina with 2,600 total boat slips?

Problem 10-18   For a newly constructed grocery store, calculate the number of accessible check-out aisles required if there are to be 30 total check-out aisles.

Problem 10-19   Determine the number of accessible rooms required for a hotel with 800 rooms. In addition, determine how many of the accessible rooms are required to have a roll-in shower.

Problem 10-20   Determine the number of accessible rooms required for a hotel with 2,700 rooms. In addition, determine how many of the accessible rooms are required to have a roll-in shower.

Problem 10-21   Draw a sketch of an accessible bench, indicating the minimum dimensions of the bench, minimum and maximum heights above the floor, and back-support requirements.

Problem 10-22   Draw a sketch, and show dimensions for the location of grab bars for an accessible water closet.

Public Spaces Are Located for Visibility and Accessibility

Commercial building entries should be located and designed as easily identifiable and centrally located. Visitors to the building should find the lobby without trouble and should have direct access to the stairways, elevators, restrooms, and other public functions through the lobby. The lobby should also function to allow for proper monitoring of security.

Exits and Entrances Provide Safety and Accessibility

Building entries and exits are designed based on how they will be used. These can include emergency exits for building occupants, as previously discussed in this chapter, employee entrances that can have added security measures, and public entries. Public entries often feature multiple doors, vestibules, and revolving doors. If a revolving door is present, additional doors meeting ADA accessibility requirements must be provided. A vestibule and revolving door can be used as an airlock to help reduce heating and cooling losses within the building. The number and size of doors is based on the occupancy of the building and the number of users anticipated to utilize the public entries.

Access to Other Floors

Access to multiple levels within commercial buildings can be provided through the use of several systems, including stairs, ramps, escalators, and elevators. For sizing of stairs, refer to the discussion that occurred previously in this chapter. Ramps need a large amount of floor space within a building and are used sparingly in design. Some well-known buildings have interior ramps, such as the Guggenheim Museum, in New York City, designed by architect Frank Lloyd Wright, shown in figure 10-6.

FIGURE 10-6 *Guggenheim museum, New York City.*

The Americans with Disabilities Act Accessibility Guidelines (ADAAG) set limits for the amount of slope and length of slope in a ramp. A ramp shall be considered any part of a building that has a minimum slope of 1 to 20 (1:20) and a slope no greater than 1 to 12 (1:12). The maximum vertical rise for a ramp is 30", with the maximum horizontal run of a ramp being dependent on the slope, as illustrated in ADAAG figure 16, shown in figure 10-7.

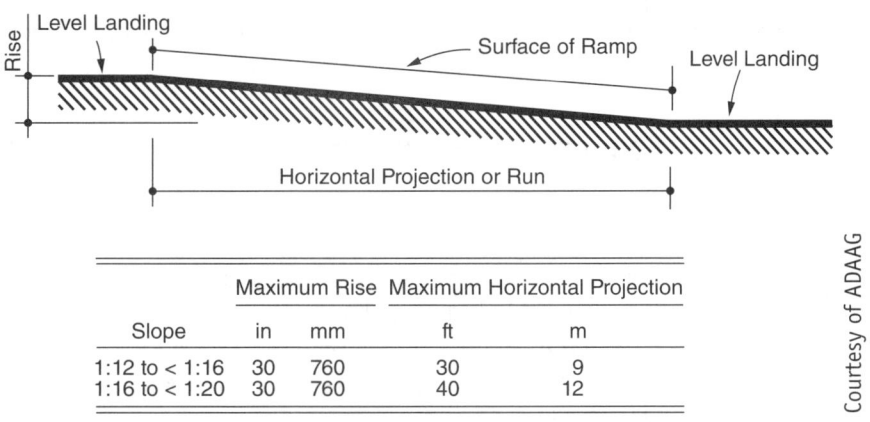

Slope	Maximum Rise		Maximum Horizontal Projection	
	in	mm	ft	m
1:12 to < 1:16	30	760	30	9
1:16 to < 1:20	30	760	40	12

FIGURE 10-7 *ADAAG ramp slope limits.*

For required ramps with vertical rises greater than 30", intermediate landings must be provided per ADAAG section 4.8. In this section of the ADAAG, additional requirements for ramps are given and should be reviewed by the student to understand accessible requirements for commercial design.

Escalators are used in buildings where large numbers of people need to move between floors. The sizing of escalators is based on the number of people who will move from floor to floor in an hour. A general rule of thumb for sizing of escalators is a 32" escalator width to move 3,000 people per hour between floors, and a 48" width to move 4,000 people per hour between floors. More information can be found online by searching the topic of escalators, such as found at http://architecture.cua.edu/res/docs/courses/arch457/report-1/10b-escalators-movingwalks.pdf.

Elevators are found in most multi-story commercial buildings, and they make vertical transport in these buildings possible. The number of elevators required in a building is based on its building type, occupancy, number of floors, traffic patterns, and speed and capacity of elevators. Average elevators have a weight capacity between 3,500 and 4,000 pounds, accommodating between 18 and 20 people. In addition, service elevators are often required in multi-story commercial buildings to allow for movement of equipment, furniture, and other large items. Rules of thumb for the number of elevators needed in a building vary greatly, based on the use of the building. Some criteria and constraints for elevators, as well as general information on elevators, can be found at www.elevatoradvisors.com/docs/BeyerSpecSeries.pdf. The ADA has minimum size requirements for elevators, as shown in figure 10-8.

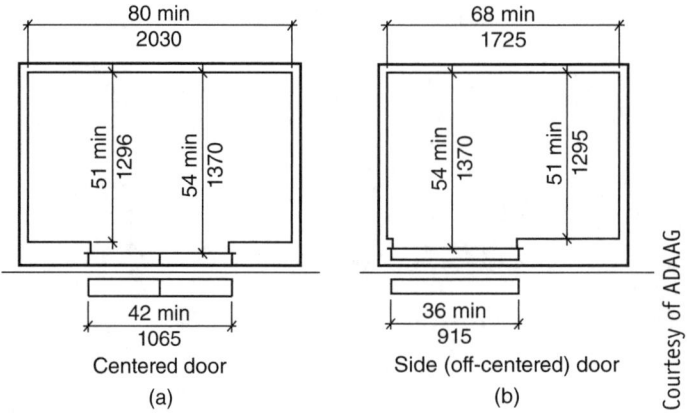

FIGURE 10-8 *ADAAG minimum elevator dimensions.*

Public Restrooms

Restrooms must be provided that can be used by the public, and in particular, they must be designed to be accessible. Previously in this chapter, requirements for the amount of restrooms in a commercial building were discussed. Additionally, the ADAAG provides space requirements for restrooms to ensure that the facilities can be used by all occupants in a building. Study the ADAAG, and become familiar with the requirements for accessible restrooms.

Problem Set 10.3 Public Spaces Are Located for Visibility and Accessibility

Problem 10-23 Research online websites for information on revolving door systems. Sketch a typical building entry plan that uses a revolving door.

Problem 10-24 Discuss conditions under which standard-duty and heavy-duty escalators are used. What is the difference between the two types of escalators?

Problem 10-25 For a standard escalator, what is the angle of inclination? What is the maximum linear speed allowed? What is the maximum load per tread (number of persons) allowed for a 48" model?

Problem 10-26 What is the relevant cost increase (percent increase) for a 48" model with a 30-foot rise, compared to a 32" model with a 14-foot rise?

Problem 10-27 List the three types of elevator systems, and indicate the building heights (number of stories) for which these systems are used.

Problem 10-28 Using rules of thumb, how many elevators are required for a 12-story office building with a net area of 180,000 sq. ft.? Should an additional service elevator be included for this building?

Problem 10-29 For the building in Problem 10-28, list the possible elevator systems that could be used to service the building, indicating the speed range that should be used for each system.

Problem 10-30 What are the minimum dimensions for a standard accessible toilet stall? Draw a sketch showing these dimensions.

Problem 10-31   What are the clear space dimensions for an accessible lavatory? Draw a sketch of this condition.

Problem 10-32   List the required minimum knee clearance for an accessible sink, as well as the two exceptions to these requirements.

A Sustainable Space Plan Provides Flexibly, Adaptability, and Affordability

Current design philosophy in commercial architecture tends to stress the term flexibility. This term can affect many aspects of the building, from flexibility of changing the physical spaces within a building, having the flexibility to adapt to future technologies, and perhaps having the flexibility to change the use of the building at a later date. Currently, flexibility in a commercial building can be achieved by utilizing planning based on grid systems, modular components or moveable walls, standard ceiling heights, and wireless technology. Flexibility tends to stray away from anything permanent. For example, implementing partition walls instead of permanent walls allows future flexibility in a building. Though these partition walls could be removed and relocated, modular systems exist that can be moved more easily without the need for demolition and reconstruction, which can greatly affect the operations of the building occupants. There are many different varieties of modular office furniture systems that can be used to furnish commercial spaces. Research these online to become familiar with the systems available.

Ceiling Heights

For most commercial spaces, the IBC 2009 requires a minimum 7'-6" ceiling height. However, this height often is too low for commercial spaces to feel comfortable. To achieve lighting and cross-ventilation requirements, and to make a space feel more inviting, the ceiling height often needs to extend several feet above this minimum. The designer must understand the volume of space the occupants require when designing the ceiling height for a commercial space. Individual room size must also be considered when designing ceiling heights, and it often occurs that there exist differing ceiling heights in different rooms within a commercial building.

Space Planning and Energy Consumption

In Chapter 8, design based on energy conservation was discussed. When incorporated properly, these design considerations can greatly reduce the lifecycle cost of a commercial building. A building's shape, size, number of floors, building materials, solar orientation, and energy supply are just a few of the categories that affect a building's energy efficiency, and thus its lifecycle cost.

Due to the projection of the increase in energy consumption by commercial buildings in the coming decades, the United States Department of Energy has developed a program to promote the development and implementation of technology in commercial buildings so that they can become net-zero energy buildings (ZEB). Definitions of different net-zero buildings have been developed by the National Renewable Energy Laboratory (NREL) and are given in figure 10-17 of the textbook. To learn more about these types of buildings, search for "zero energy buildings" online, and view the PowerPoint presentation titled "The Drive for Net-Zero Energy Commercial Buildings" by Drury B. Crawley, Ph.D., on the Department of Energy website at www.energy.gov/search/site/.

Problem Set 10.4   A Sustainable Space Plan Provides Flexibly, Adaptability, and Affordability

Problem 10-33   For a commercial office plan size of 14'-0" × 14'-0", draw a section with a ceiling height of 7'-6" and 10'-0" and compare the spaces. Which would you rather work in, and why?

Problem 10-34   Think about a typical multi-story office building. Why is the ceiling height in the lobby space often higher than the ceiling height in other parts of the building? What are positive and negative aspects of the higher ceiling in a lobby space?

Problem 10-35   Research online, and list five advantages of net-zero energy buildings.

Problem 10-36 Research online, and list five disadvantages of net-zero energy buildings.

Problem 10-37 According to the National Renewable Energy Laboratory (NREL), what percentage of energy consumption in the United States in 2006 was attributed to commercial buildings? By 2025, what percentage of energy consumption growth will be used to power commercial buildings?

Problem 10-38 What are the Department of Energy's goals for commercial buildings by the years 2020, 2040, and 2050?

Workplace Environments Influence Worker Performance

It has been shown through research that a person's performance can be affected by the physical environment in which they work and live. The architect must take the physical environment into consideration to help achieve the best performance by a building's occupants. Some design aspects that must be taken into consideration by the architect include:

- Sunlight and indirect daylight
- Air quality and natural ventilation
- Colors and materials
- Furnishings and decor
- Layout of spaces due to work patterns
- Open vs. closed office plans

Through considering the occupants and how they will utilize the building, the architect can achieve a design that will enhance the productivity of its users.

An Open Versus a Closed Office Plan

The design decision to use an open or closed office plan is often related directly to the function of the building. In researching the client's needs and wants, the architect will arrive at a plan that will best enhance the productivity of the occupants. In many buildings, the use of some closed offices along with open office planning can allow for privacy when needed. There are many advantages and disadvantages in using an open office plan, depending on the type of business and occupants using the space. Students should research online to understand these differences. One website that is beneficial in explaining aspects of the two plans is the Knoll website at www.knoll.com/research/downloads/OpenClosed_Offices_wp.pdf. Read through this document to further understand the debate about the systems.

Planning Space for Building Equipment and Utilities

When designing a building, space must be allocated for the mechanical, electrical, plumbing, and communications to be included in the design. It is important that access to these spaces be included in such a manner that allows easy access while remaining secured. In plan, the locations of these spaces are often aligned vertically in a multi-story building, and are typically located in the centrally located service area for cost-effectiveness and for ease of access. For horizontal distribution of equipment and utilities, space must be allocated either below or above the structure in the building. When a building is designed to have a raised-access flooring system, space above the floor structure must be provided to accommodate the wiring for electrical and communication systems that are often located in this space, as shown in figure 10-9.

FIGURE 10-9 *Raised-access flooring system.*

Additionally, the access floor space can be utilized as a plenum space, which is a sealed space used for the distribution of conditioned air through the building. If a plenum space is to be used in a building and is not provided above the floor structure, then it can be provided between the bottom of the structure and the finish ceiling of the floor below. Special requirements exist pertaining to fire and smoke issues when a plenum is to be used to supply conditioned air in a building, which can increase costs. A more common choice for supplying conditioned air throughout a building is the use of mechanical ductwork, as shown in figure 10-10.

FIGURE 10-10 *Ductwork to circulate conditioned air.*

When ductwork is used, space must be allocated for the placement of ductwork below the structure. As a general rule of thumb, 24" of clear space is required between the bottom of the structure and the ceiling construction to allow for proper placement of the ductwork. The thickness of ceiling systems can be as thin as 2" and can increase from there, depending on the system chosen. Thus, overall floor-to-floor height in a building is a function of the clear height of a space, plus the structure thickness, plus the ceiling thickness, plus the thickness of the system (plenum space or ductwork) used to supply air to the building.

Building Support Spaces

Additional spaces must be included by the architect that are utilized in supporting the function of a building. In a typical commercial building, these can include:

- Loading docks
- Staging areas
- Storage areas
- Security/fire control rooms
- Trash collection areas
- Maintenance and custodial rooms
- Equipment rooms

Each of these spaces should be efficiently located within a building, since most of the spaces are private and not open to the public. Issues pertaining to security, noise, views, and safety must be considered when locating support spaces.

With the focus on efficient building design, intelligent buildings are becoming more popular. An intelligent building is one that integrates both technology and process and arrives at a facility that is safe, comfortable and productive, and more efficient. This allows for a building that provides an environment that optimizes occupant comfort and productivity and reduces energy-consumption. For these buildings, a central control room is typically required for monitoring and control of the systems throughout the building. Information on intelligent buildings can be found at many websites, such as the Robert H. Lane and Associates, Inc. website at www.laneconsul.com/blog/building-intelligence-planning-program-survey. Review this website for general information on intelligent building systems.

Spaces are Identified by Their Name, Number, and Square Footage

Once the floor plan is finalized, each room is identified within the building. This provides an easy way to navigate through the building, both during construction and once occupied. Information provided on the construction documents for a building includes the room number, room name, and area of each room. The area is given in square footage and is typically the gross area of the room. Refer to Chapter 11 for more information on calculating the gross area of a room.

Room Numbers

Room numbers are formatted to allow for clarity within the building, and typically utilize similar patterns of numbering at each level in a building, as shown in figure 10-11. Though there is no set method for numbering rooms in a building, there exist common rules that are followed.

For example, the entry level of a building will typically be given the 100 series room numbers (101, 102, …), with above grade levels following the same format. Basement levels often have a prescript (B101, B102, …), and any rooms that are accessed directly from a given room will be

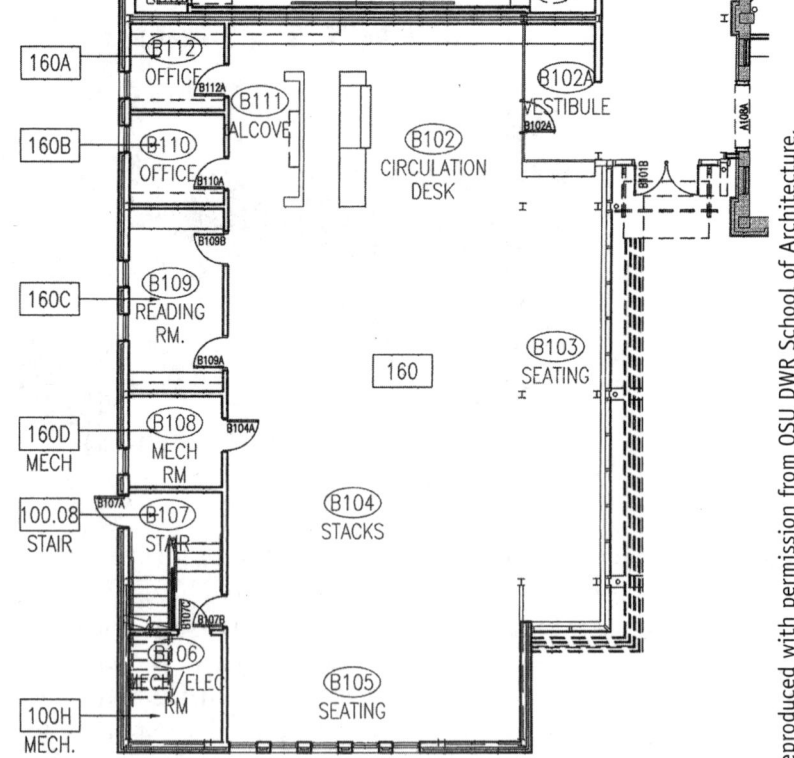

FIGURE 10-11 *Room numbering on architectural plan.*

given a postscript (e.g., 101A for a room that is accessed from Room 101). Mechanical rooms, corridors, and stairs often have specific indicators, as discussed in the textbook.

In some cases, two sets of room numbers are shown on the architectural plans, as shown in figure 10-11. The first number, shown in the box, is the actual room number. The second number, shown in the oval, is a number corresponding to the room finish schedule, shown in figure 10-12.

	NUMBER	ROOM NAME	FLOORS		WALLS				NOTES:
			FLOOR	BASE	NORTH	EAST	SOUTH	WEST	
SECTOR B	B101	LINK CORRIDOR	CT4	RB2	–	EXP	–	PT1	
	B102	CIRCULATION DESK	CPT1	RB1	PT1/TK1	GLASS	–	PT5	
	B102A	VESTIBULE	CT4	RB2	PT1/TK3	PT1	GLASS	GLASS	
	B103	SEATING	CPT1	–	GLASS	GLASS	GLASS	–	SOLAR SHADES ON E/S
	B103A	SEATING	CPT1	RB1	–	–	–	PT1	
	B104	STACKS	CPT1	RB1	–	–	–	PT1	
	B105	SEATING	CPT1	RB1	–	PT1	PT1	PT1	SOLAR SHADES ON S
	B106	MECH/ELEC ROOM	SC	RB2	PT1	PT1	PT1	PT1	
	B107	STAIR 4	SC	RB2	PT1	PT1	PT1	PT1	
	B108	MECHANICAL ROOM	SC	RB2	PT1	PT1	PT1	PT1	
	B109	READING ROOM	CPT1	RB1	PT1	WTK1	PT1	PT1	
	B110	OFFICE	CPT1	RB1	PT1	PT1	PT1	PT1	
	B111	ALCOVE	CPT1	RB1	PT/TK1	PT1	–	PT1	
	B112	OFFICE	CPT1	RB1	PT1	PT1	PT1	PT1	
	B113	STAGE VESTIBULE	–	–	–	–	–	–	CEILINGS TO BE PAINTED PT6.
	B114	SOUND & LIGHT LOCK	–	–	–	–	–	–	
	B115	VESTIBULE	–	–	–	–	–	–	
	B116	LECTURE STAGE	CT2	WVP1	PT3	WVP1	WVP1	WVP1	CEILINGS TO BE PAINTED PT6.
	B116A	LECTURE LWR SEATING	CPT2	CT2	–	PT3	PT3/WVP1	PT3	CEILINGS TO BE PAINTED PT6.
	B116B	LECTURE UPR SEATING	CPT2	CT2	FC1	PT3	PT3	PT3	CEILINGS TO BE PAINTED PT6.
	B116C	CROSS AISLE	CPT2	CT2	–	PT3	–	PT3	CEILINGS TO BE PAINTED PT6.
	B117	MECHANICAL ROOM	SC	RB2	PT1	PT1	PT9		
	B118	STAIR 6	SC	RB2	PT9	EXP	PT1	PT1	
	B119	CORRIDOR	SC	RB2	PT1	–	–	–	
	B120	AV ROOM	CPT1	RB2	WVP	BP/WVP	GLASS	GLASS/BP	CEILINGS TO BE PAINTED PT6.
	B121	ELECTRICAL	SC	RB2	PT9	PT1	PT1	PT1	CEILINGS TO BE PAINTED PT6.

FIGURE 10-12 *Room finish schedule.*

The room finish schedule gives information on the finishes to be constructed in each room and in some cases indicates changes in finish materials in the same room. This allows the design team and contractor to easily identify the finish requirements for a particular room or space in a building, for use in design, cost estimating, and construction. The schedule uses a series of abbreviations and notes to indicate the finishes to be used for floors, walls, and ceilings.

Problem Set 10.5 Workplace Environments Influence Worker Performance/ Planning Space for Building Equipment and Utilities/ Spaces Are Identified by Their Name, Number, and Square Footage

Problem 10-39 Give the definition of an *open office plan*, and discuss the benefits of this system.

Problem 10-40 Give the definition of a *closed office plan*, and discuss the benefits of this system.

Problem 10-41 Calculate the floor-to-floor height of a typical floor in an office building with a 27" depth of structure, and a 9'-6" clear height to the finish ceiling. The finish ceiling system has an overall depth of 3", and the required plenum depth is 20". Draw a section showing these dimensions.

Problem 10-42 Calculate the floor-to-floor height of ground floor in a commercial building with a 24" depth of structure at the second floor, and a 11'-6" clear height to the finish ceiling. The finish ceiling system has an overall depth of 6" and mechanical ductwork depth of 24". Draw a section showing these dimensions.

Problem 10-43 Calculate the total building height from ground floor up to roof level for a 4-story building with 20" of structural depth at each level, 16" of mechanical depth at each level, 2" of finish ceiling system depth at each level, and clear heights of 8'-6" at all other levels.

Problem 10-44 Calculate the total building height from ground floor up to roof level for a 6-story building with 24" of structural depth at each level, 28" of mechanical depth at each level, 6" of finish ceiling system depth at each level, and clear heights of 12'-0" at the ground floor and 9'-0" at all other levels.

Problem 10-45 Calculate the total building height from ground floor up to roof level for a 12-story building with 26" of structural depth at each level, 30" of mechanical depth at each level, 4" of finish ceiling system depth at each level, and clear heights of 13'-0" at the ground floor and 9'-6" at all other levels.

Problem 10-46 List the three characteristics that make up the basis of an intelligent building.

Problem 10-47 List and discuss the four steps included in the development of an intelligent building plan.

CHAPTER 11
Dimensioning and Specifications

Skills List

After completing this chapter, you should be able to:

- Understand rules and techniques for dimensioning and labeling construction drawings

- Recognize that construction drawings are organized from general to specific information

- Understand the function of the American National Standards Institute

- Understand the common symbols used on construction drawings

- Scale and dimension architectural plans

- Recognize abbreviations used on construction drawings

- Understand the use of elevations, sections, and details in construction drawings

- Read and utilize building project schedules

BACKGROUND

Upon completion of the design for a building, construction begins. The contractor, who will be in charge of constructing the building, must be given information on what and how to build. Working drawings, which are the end result of the design team's work on a building design, are used to communicate this information to the contractor and builders. These drawings graphically represent the building design and give information on what is to be built. They help ensure that the completed project is what the owner and design team have envisioned. Working drawings are broken into the different disciplines (e.g., Architectural, Civil, Structural, Mechanical) that are involved in a building design. By breaking up the drawings, each discipline can provide specific requirements and information pertaining to their area of expertise for the project. The drawings provide dimensions and specific notes and instructions on how the building is to be constructed. This chapter will explore working drawings, addressing the different types of information given on the drawings and how to understand the drawings to be used in the construction of a building.

Dimensioning and Annotations Follow Drawing Conventions for Universal Understanding

Drawing conventions are standards used to develop construction drawings that can be understood by all parties involved in a construction project, from client and design team to contractor and material suppliers. The drawing conventions have been accepted over time as the common method for communicating how a building is to be constructed. These conventions include standards for symbols, text, dimensions, notes, and abbreviations. Many of the standards used today fall under American National Standards Institute (ANSI) designations. ANSI is a non-profit organization that provides a clearing house for the development of voluntary standards in the United States for personnel, products, services, systems, and processes. More can be found out about ANSI at the website www.ansi.org.

Symbols

Symbols on construction drawings give a quick and easy method to identify common aspects included in the drawings. They are used to reduce the amount of text required to communicate information. Symbols are also used to refer to additional drawings or schedules that give more detailed information. A legend or key is typically included in the construction drawings to identify symbols used. Common symbols found on construction drawings include:

- Drawing title
- Building orientation
- Room name and number
- Floor levels
- Materials
- Components
- Section and detail cuts
- Elevation views

Without the use of symbols, the number of construction drawings and their complexity needed to construct a building would make the process very difficult. Through the use of accepted symbols, communication among all parties involved in the design and construction of a building has been simplified.

Example 11.1 For the floor plan shown in figure 11-1, identify the different symbol types used to communicate information.

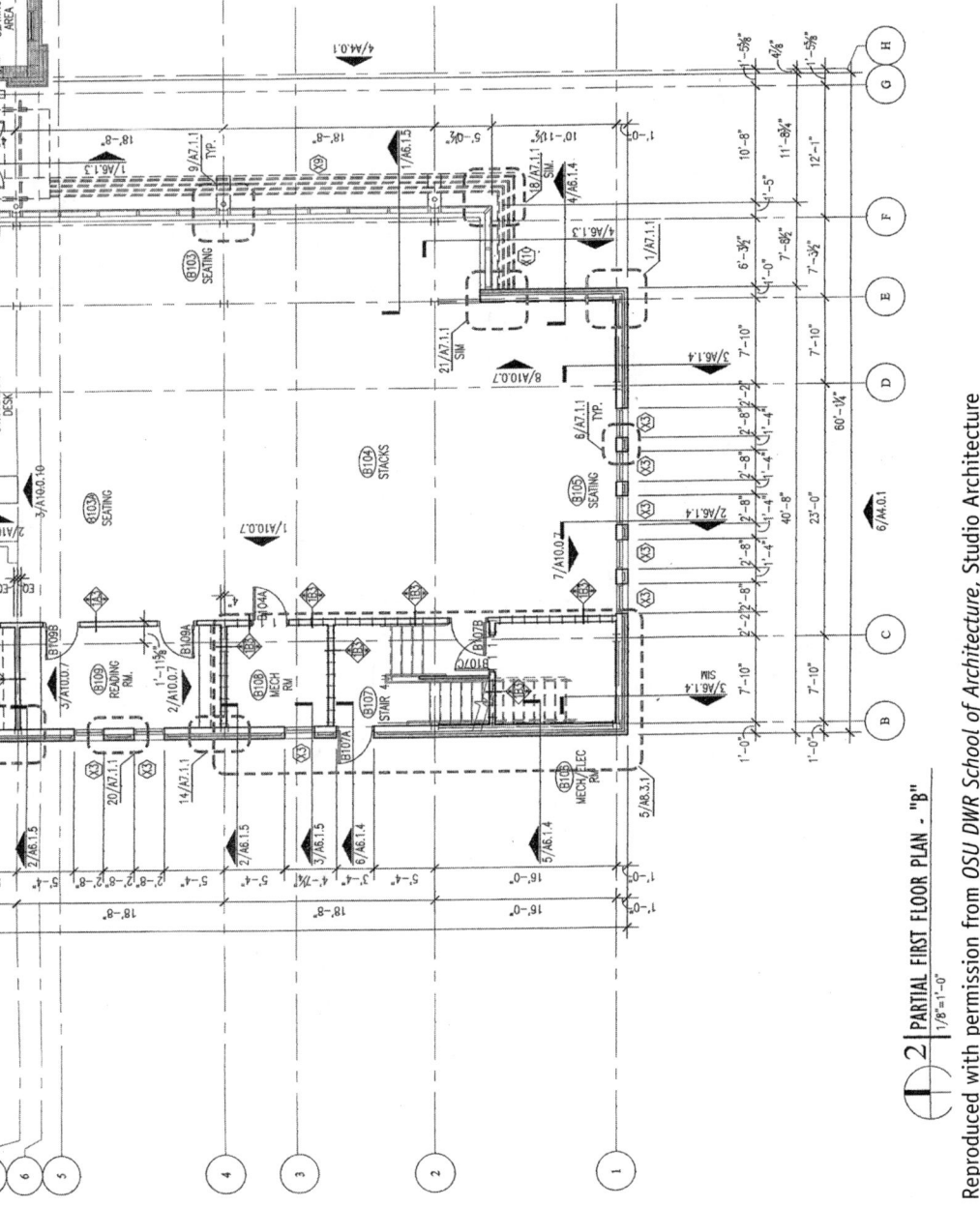

FIGURE 11-1 *Partial architectural floor plan.*

Several symbols exist on the plan shown in figure 11-1, including:

- Interior and exterior elevation callouts
- Window type designations
- Door-type designations
- Room number
- Partition wall type
- Exterior wall building section cuts
- Enlarged plan detail callouts

Refer to figure 11-1A for the location of each of these symbols shown on the plan.

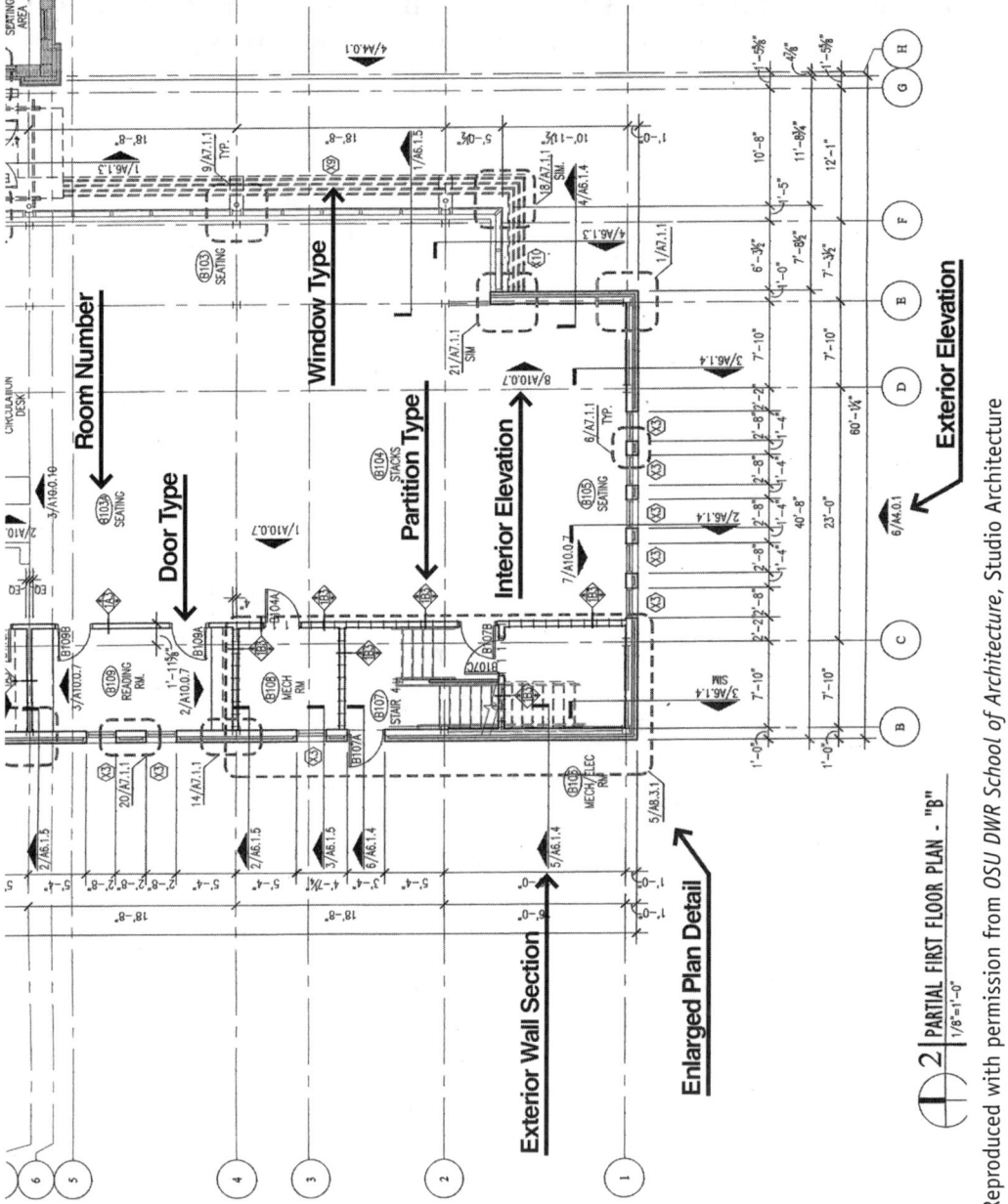

FIGURE 11-1A *Symbol locations on partial floor plan.*

Text and Dimensions

Text and dimensions on construction drawings are used to communicate what is to be built. When placing text and dimensions on construction drawings, it is important that it be sized such that is can be clearly read, and yet small enough to have enough room to include it on the drawing. Though the height of text and dimensions on construction drawings are often set by the individual firm, many have a standard height of 1/8", with drawing titles having a standard height of 1/4". The font used on drawings can vary based on the architectural firm, and even on the project, but it is important that the font size remain consistent throughout the set of drawings to avoid confusion.

Dimensions are used to give precise distances between specific points or locations in a building project. These dimensions are required to be able to construct a building that is secure, functional, and aesthetically pleasing. Dimensions are used to communicate precisely how big each component of a building needs to be to end up with a successful completed project. Dimension lines use ticks to indicate points between which measurements are given on the drawings. These ticks can be seen in figure 11-1, indicating dimensions between grid lines, to exterior edges of the building, and for window and door openings.

Note that the dimensions are positioned along several lines, with the smaller dimensions being located closer to the building. Using several lines, or strings, dimensions allow the different features of the building to

208   Chapter 11: Dimensioning and Specifications

be located with clarity. In the United States, dimensions are typically given in feet and inches on the architectural drawings, though for most government projects, metric dimensions are used. On other drawings, such as the civil engineering drawings, dimensions are often given in decimal form.

Example 11.2 The floor plan shown in figure 11-2 is drawn at a scale of 1/8" = 1'-0". For the dimension strings shown, measure and indicate the dimensions for the plan.

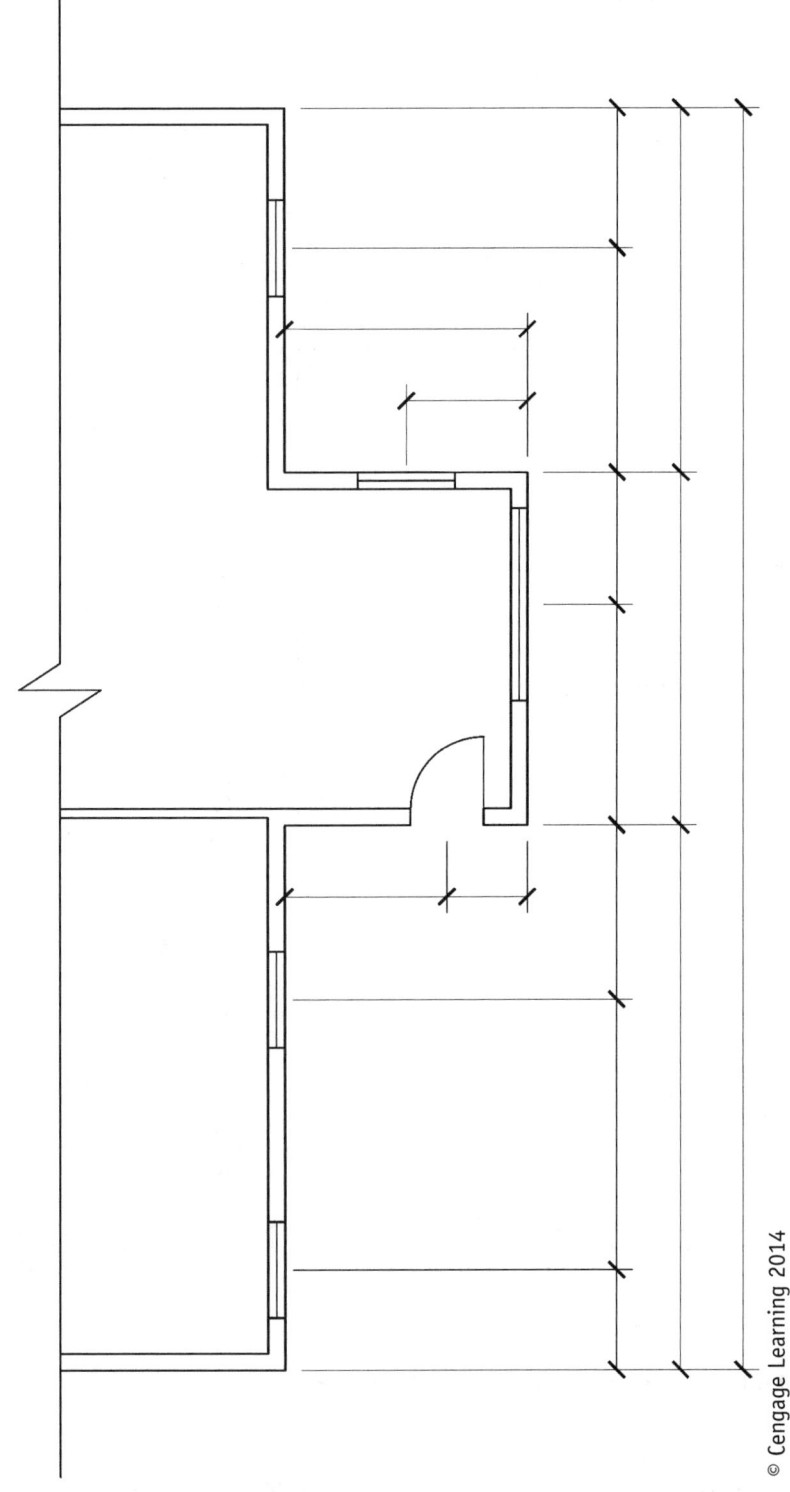

FIGURE 11-2 *Partial plan.*

Using an architecture scale, the 1/8" = 1'-0" side of the scale can be used to measure the distances in feet and inches. Note that when a measurement is in whole feet, it should still be given in feet and inches (i.e., 24'-0") to avoid confusion when reading a dimension. The floor plan with scaled dimensions is shown in figure 11-3.

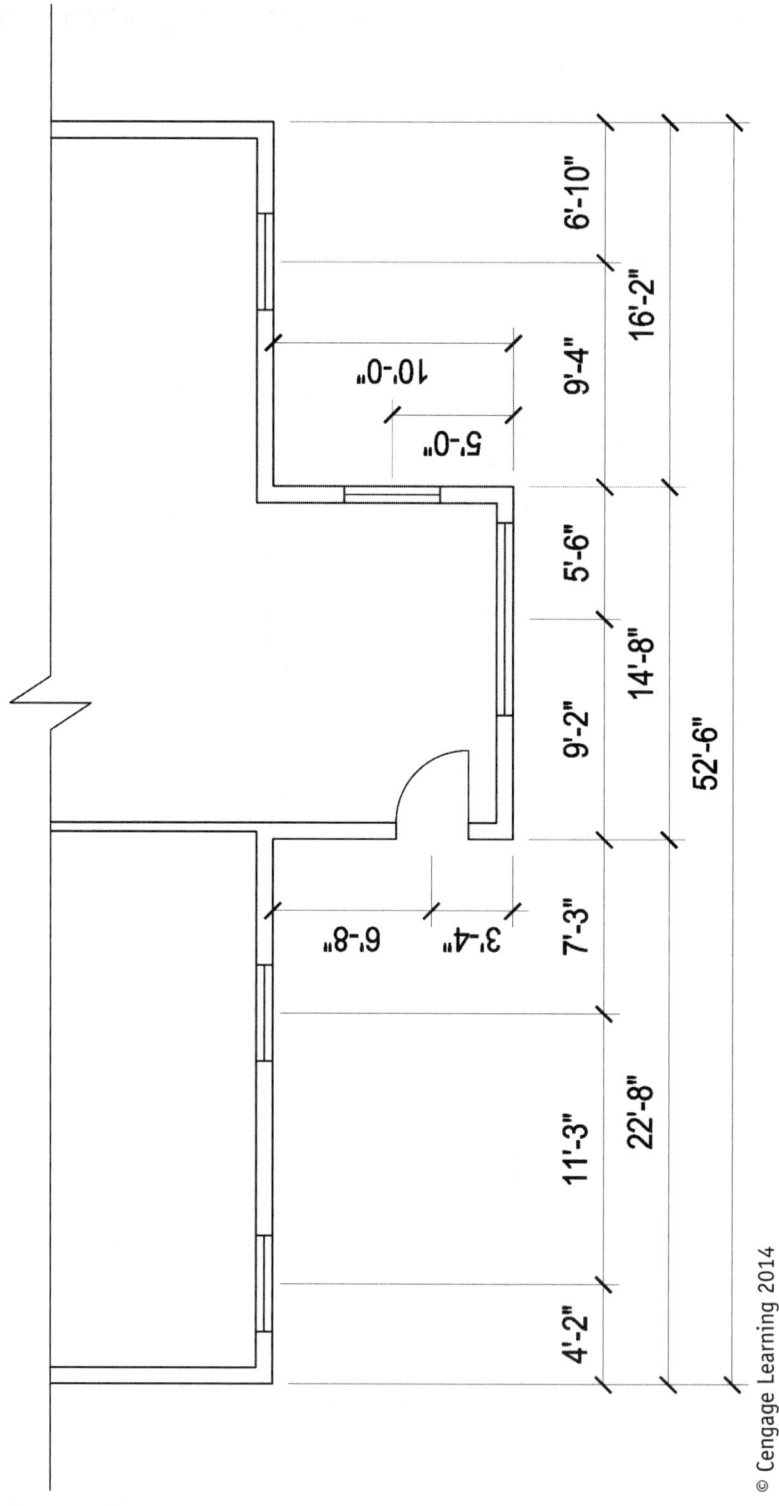

FIGURE 11-3 *Partial plan with dimensions.*

Notes

During any construction project, there will be conditions the builder will face for which explanations cannot be easily or clearly given through the use of symbols, text, and dimensions. In these cases, notes are used to further clarify the requirements for the project. Notes fall into the categories of general and specific. General notes give an overview of conditions that will exist throughout a project and how they are to be addressed. One example of a general note is information pertaining to the edition of the building code for which the building has been designed. General notes are typically placed at the beginning of a set of drawings so they will not be overlooked. Specific notes give detailed information on requirements found at isolated locations in a building. One example of a specific note is information on the thickness, strength and reinforcing for a concrete slab used for the ground floor of a building. Specific notes are found on plans, sections, and details and address conditions at specific locations. These notes use leaders to indicate location or general area to which the note pertains. Leaders are arrows that point toward the location or area of interest for a given specific note and are used throughout the set of construction drawings.

Abbreviations

Abbreviations are used to shorten the amount of text required on construction drawings. They are for a used word, or group of words, that is frequently repeated in the drawings. To give a clear understanding of the abbreviations, a list is often given on the title sheet of a construction drawing set. One is given in the textbook figure 11-9, and another is shown in figure 11-4.

ABBREVIATIONS

A.F.F.	ABOVE FINISHED FLOOR	ELEV.	ELEVATION	O.F.O.I.	OWNER FURNISHED OWNER INSTALLED
ALUM.	ALUMINUM	EXT.	EXTERIOR	O.H.	OPPOSITE HAND
BLK.	BLOCK	F.E.	FIRE EXTINGUISHER	PT	PAINT
B.O.M.	BOTTOM OF MASONRY	F.E.C.	FIRE EXTINGUISHER CABINET	R.O.	ROUGH OPENING
B.O.S.	BOTTOM OF STEEL	F.F.	FINISHED FLOOR	SCHED.	SCHEDULE
B.O.T.	BOTTOM OF TRUSS	F.D.	FLOOR DRAIN	SIM.	SIMILAR
CLNG.	CEILING	FLR.	FLOOR	STL.	STEEL
CFMF	COLD FORMED METAL FRAMING	GALV.	GALVANIZED	S.S.	STAINLESS STEEL
CONC.	CONCRETE	GA.	GAUGE	STRUCT.	STRUCTURE
C.J.	CONSTRUCTION JOINT	GWB	GYPSUM WALL BOARD	SUSP.	SUSPENDED
C.M.U.	CONCRETE MASONRY UNIT	H.M.	HOLLOW METAL	THK.	THICK
CONT.	CONTINUOUS	M.O.	MASONRY OPENING	T.O.C.	TOP OF CONCRETE
DET.	DETENTION	MECH.	MECHANICAL	T.O.M.	TOP OF MASONRY
DS	DOWNSPOUT	MTL.	METAL	T.O.P.	TOP OF PARAPET
E.J.	EXPANSION JOINT	N.T.S.	NOT TO SCALE	T.O.S.	TOP OF STEEL
ELEC.	ELECTRICAL	O.C.	ON CENTER	TYP.	TYPICAL
		O.F.C.I.	OWNER FURNISHED CONTRACTOR INSTALLED	W/	WITH

FIGURE 11-4 *Abbreviations used on architectural drawings.*

There is no set abbreviation for a given word, so the manner in which words are abbreviated, as well as the number of abbreviations given, can vary between projects.

Problem Set 11.1 Dimensioning and Annotations Follow Drawing Conventions for Universal Understanding

Problem 11-1 Define the mission statement for the American National Standards Institute(ANSI), and list when the institute was founded.

Problem 11-2 Write a brief overview of the history of ANSI.

Problem 11-3 List the code of ethics for volunteers and members of ANSI.

Problem 11-4 List and define the six types of membership available for the American National Standards Institute.

Problem 11-5 Who makes up the membership of ANSI, and what are the benefits of being a member of ANSI?

Problem 11-6 The floor plan shown in figure 11-5 is drawn at a scale of 1/8" = 1'-0". For the dimension strings shown, measure using an architectural scale, and indicate the dimensions on the plan.

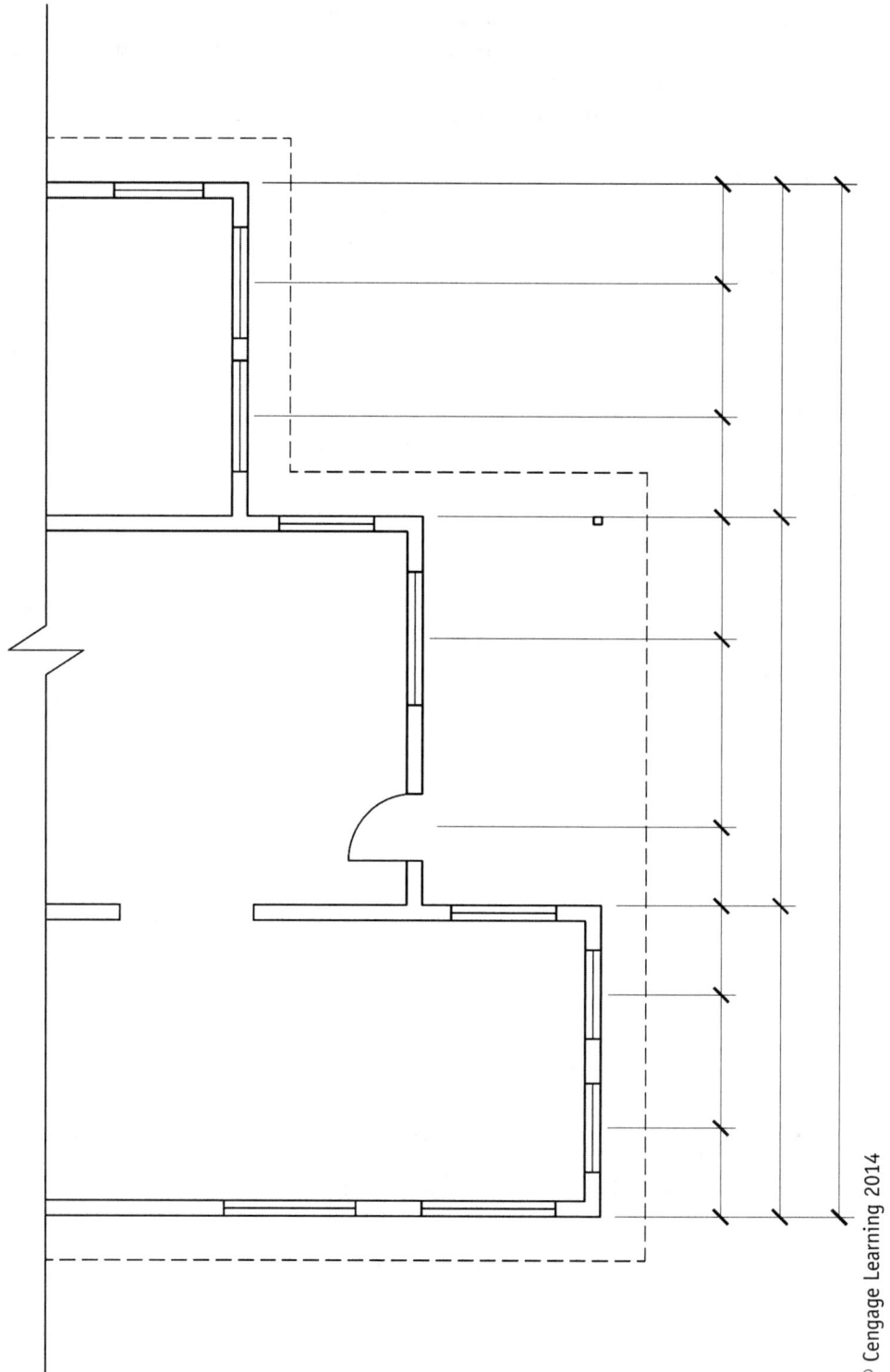

FIGURE 11-5 *Partial plan for use in problem 11-6.*

Problem 11-7 The floor plan shown in figure 11-6 is drawn at a scale of 1/8" = 1'-0". For the dimension strings shown, measure using an architectural scale, and indicate the dimensions on the plan.

212 Chapter 11: Dimensioning and Specifications

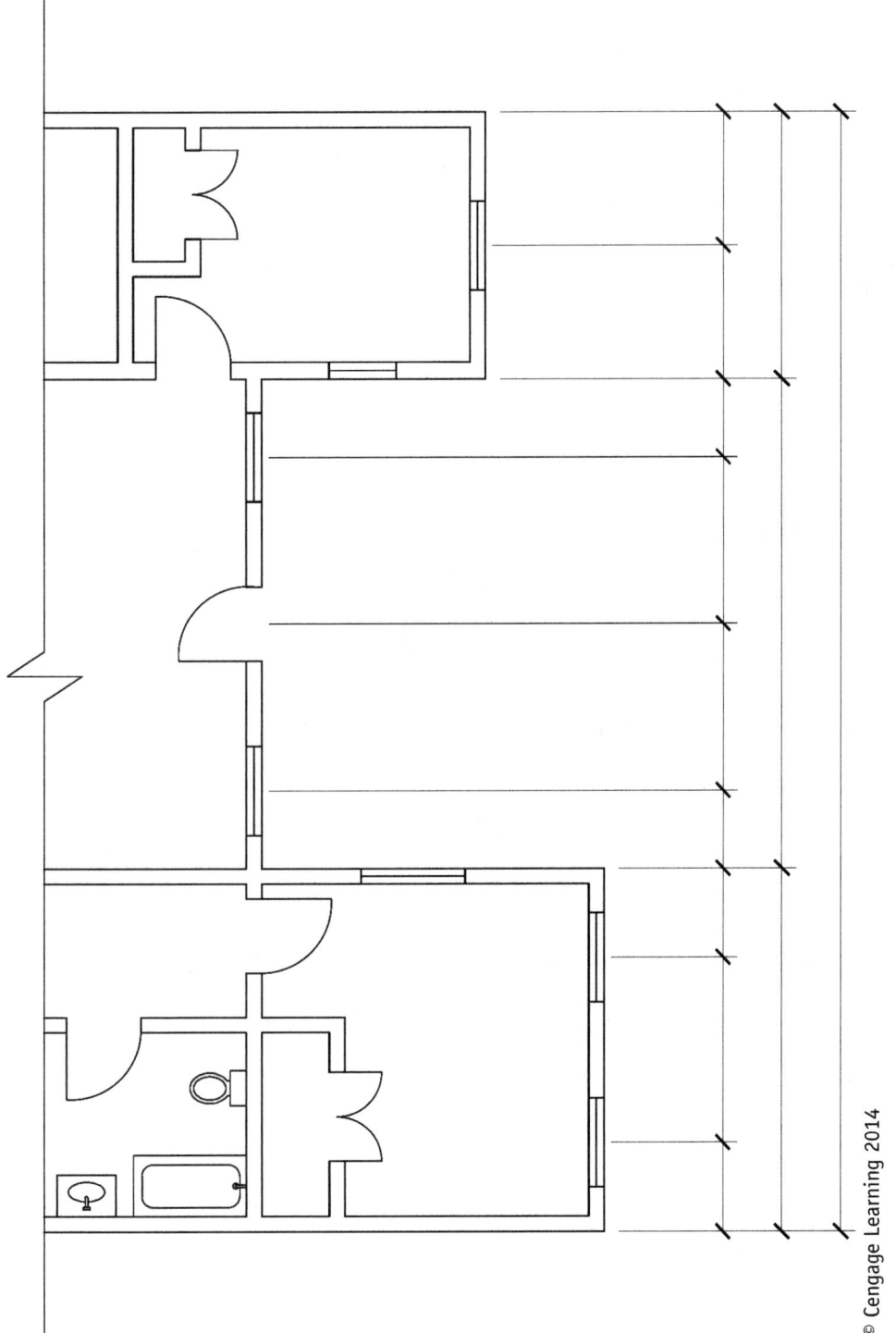

FIGURE 11-6 *Partial plan for use in problem 11-7.*

Problem 11-8   The floor plan shown in figure 11-7 is drawn at a scale of 3/16" = 1'-0". For the dimension strings shown, measure using an architectural scale, and indicate the dimensions on the plan.

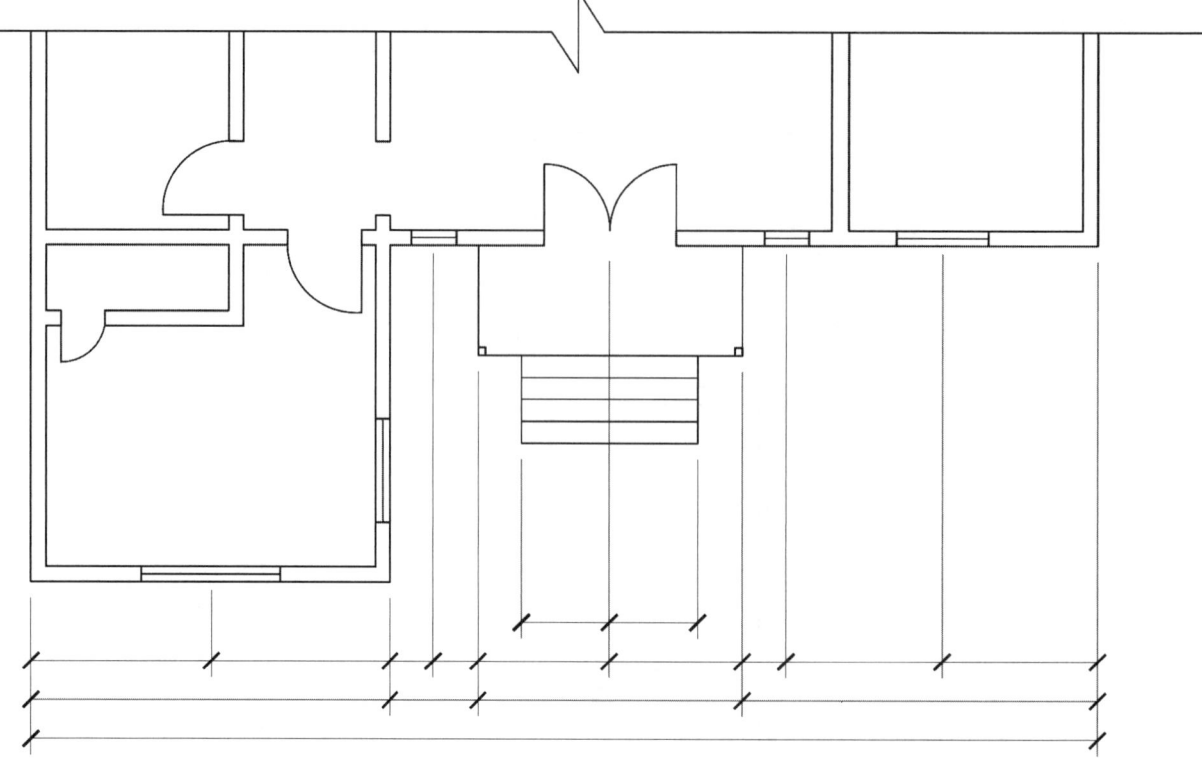

FIGURE 11-7 *Partial plan for use in problem 11-8.*

Problem 11-9 The floor plan shown in figure 11-8 is drawn at a scale of 3/16" = 1'-0". For the dimension strings shown, measure using an architectural scale, and indicate the dimensions on the plan.

Problem 11-10 The floor plan shown in figure 11-9 is drawn at a scale of 1/4" = 1'-0". For the dimension strings shown, measure using an architectural scale, and indicate the dimensions on the plan.

Problem 11-11 Using the textbook and workbook, define each of the terms in the following list:

- NTS
- CMU
- MFR
- FD
- HVAC
- W/O

Problem 11-12 Using the textbook and workbook, define each of the terms in the following list:

- TYP
- GWB
- EXT
- BRK
- FD
- CJ

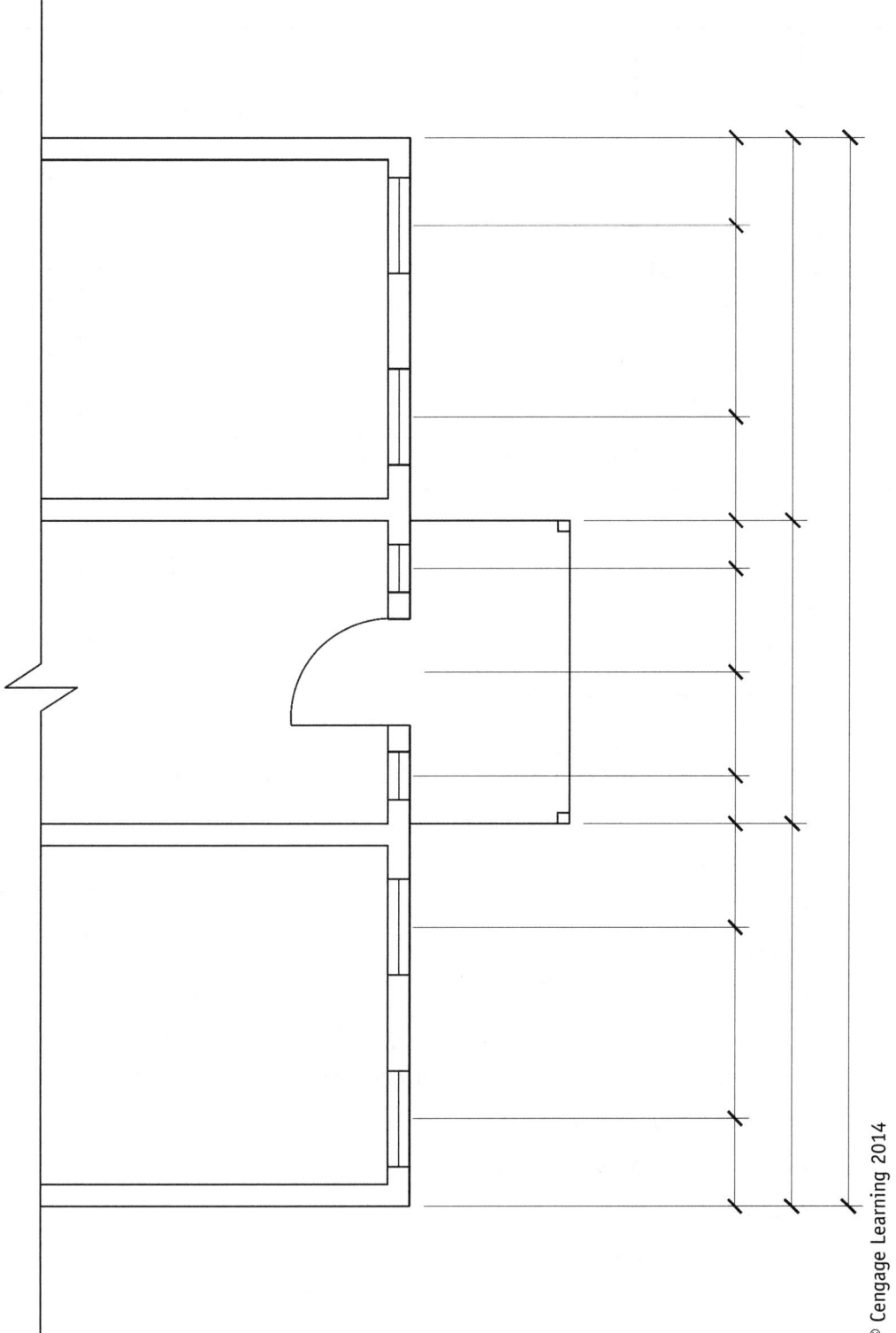

FIGURE 11-8 *Partial plan for use in problem 11-9.*

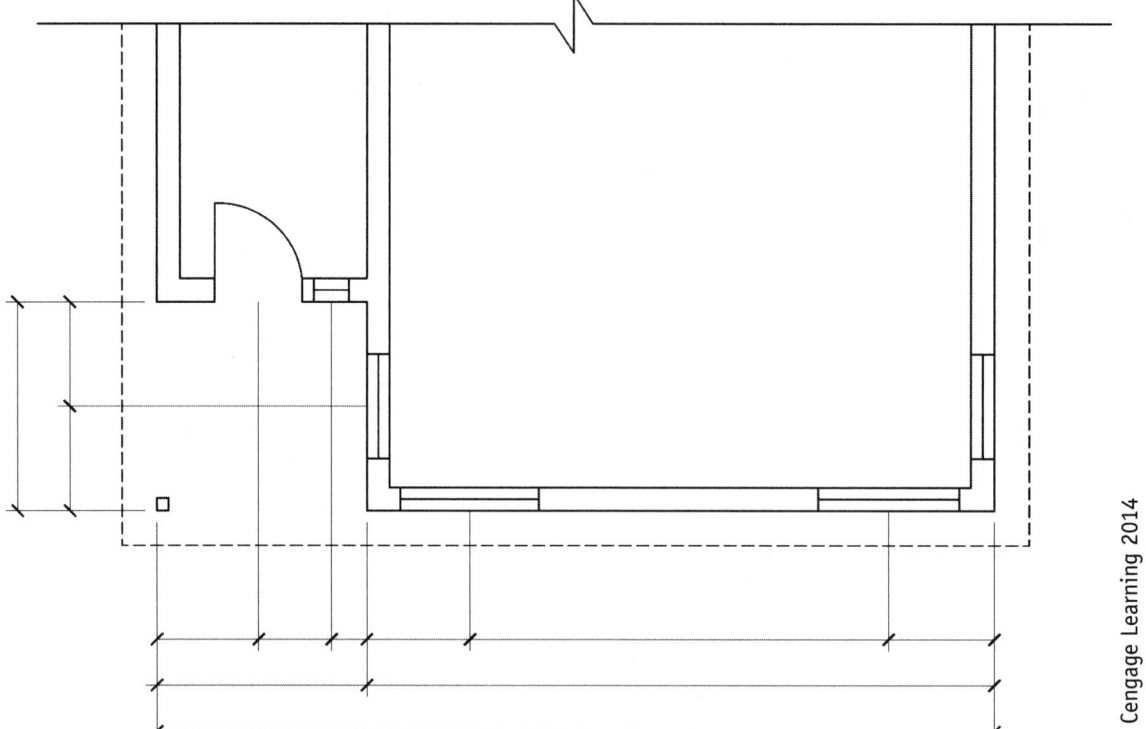

FIGURE 11-9 *Partial plan for use in problem 11-10.*

Problem 11-13   Using the textbook and workbook, define each of the terms in the following list:
- OFOI
- EXIST
- WC
- GALV
- AFF
- BSMT

Problem 11-14   Using the textbook and workbook, give the abbreviations for the following terms:
- Fire Extinguisher Cabinet
- Cold Formed Metal Framing
- Top of Parapet
- Tongue & Groove
- Double Hung
- Drawing

Problem 11-15   Using the textbook and workbook, give the abbreviations for the following terms:
- Owner Furnished, Contractor Installed
- Standard
- Rough Opening
- Bottom of Steel
- Concrete
- Bedroom

Dimensions and Annotations are Used to Communicate Specifications Necessary for Construction

Drawings such as plans, elevations, sections, and details are scaled graphical representations of what is to be built in a construction project. Dimensions and annotations are used to convey specific information, such as size, and to give the requirements for the finished project. For clarity in construction drawings, sections and details are typically drawn at the location where they occur. Floor plans are viewed from above at a distance of four feet above the floor plane, with objects above this plane being represented on the plan with a dashed line. Roof plans are viewed from above the highest peak of the roof.

Dimension Placement is Influenced by Wall Construction, and Material

Construction drawings are dimensioned using several different common techniques, and often the technique used is a function of the building components being dimensioned. For example, in floor plans, load-bearing masonry, wood stud, and metal stud walls are dimensioned from the face of the framing for the wall. For openings within these walls, wood or metal stud walls typically measure to the centerlines of the openings, while load-bearing masonry walls are dimensioned showing the actual opening sizes. For buildings with structural steel or reinforced concrete framing, dimensions to the centerline of columns, referred to as grid lines, provide a frame of reference to which all the horizontal dimensions in a building are tied.

The placement of dimensions on a plan is based on a hierarchy of smaller dimensions being located closest to the face of a building. Though the format can be adapted based on a project's size and complexity, there are typically three strings of dimensions used for each face of a building. The string placed nearest the building dimensions windows and doors. The second string dimensions shifts in the building façade, and the outermost string gives the overall dimension for the building.

Example 11.3 For the plan shown in figure 11-10, scale and dimension the exterior of the building for overall dimensions, steps in building exterior, and window and door locations for the wood stud framed construction. Use an architectural scale of 1/8" = 1'-0" for dimensioning the plan.

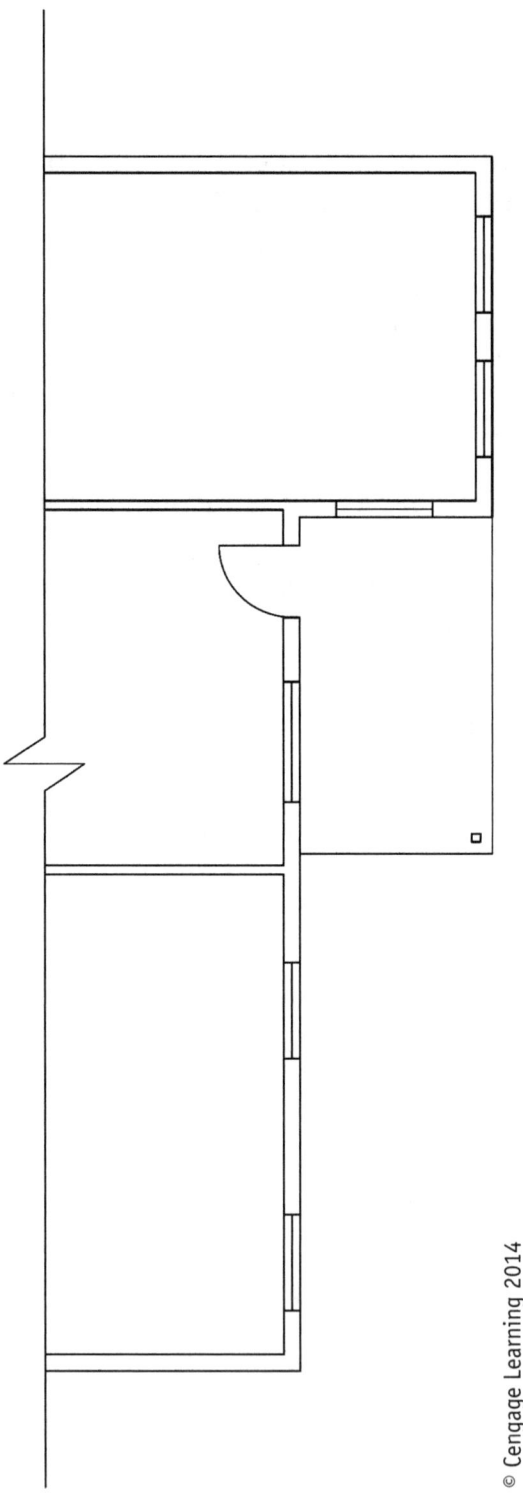

FIGURE 11-10 *Undimensioned partial plan.*

This plan should have three dimension strings. The interior string will dimension the centerlines of windows and doors. The middle string will dimension shifts in the exterior face of the building, and the exterior string will dimension the overall building size. The dimensioned plan is shown in figure 11-11.

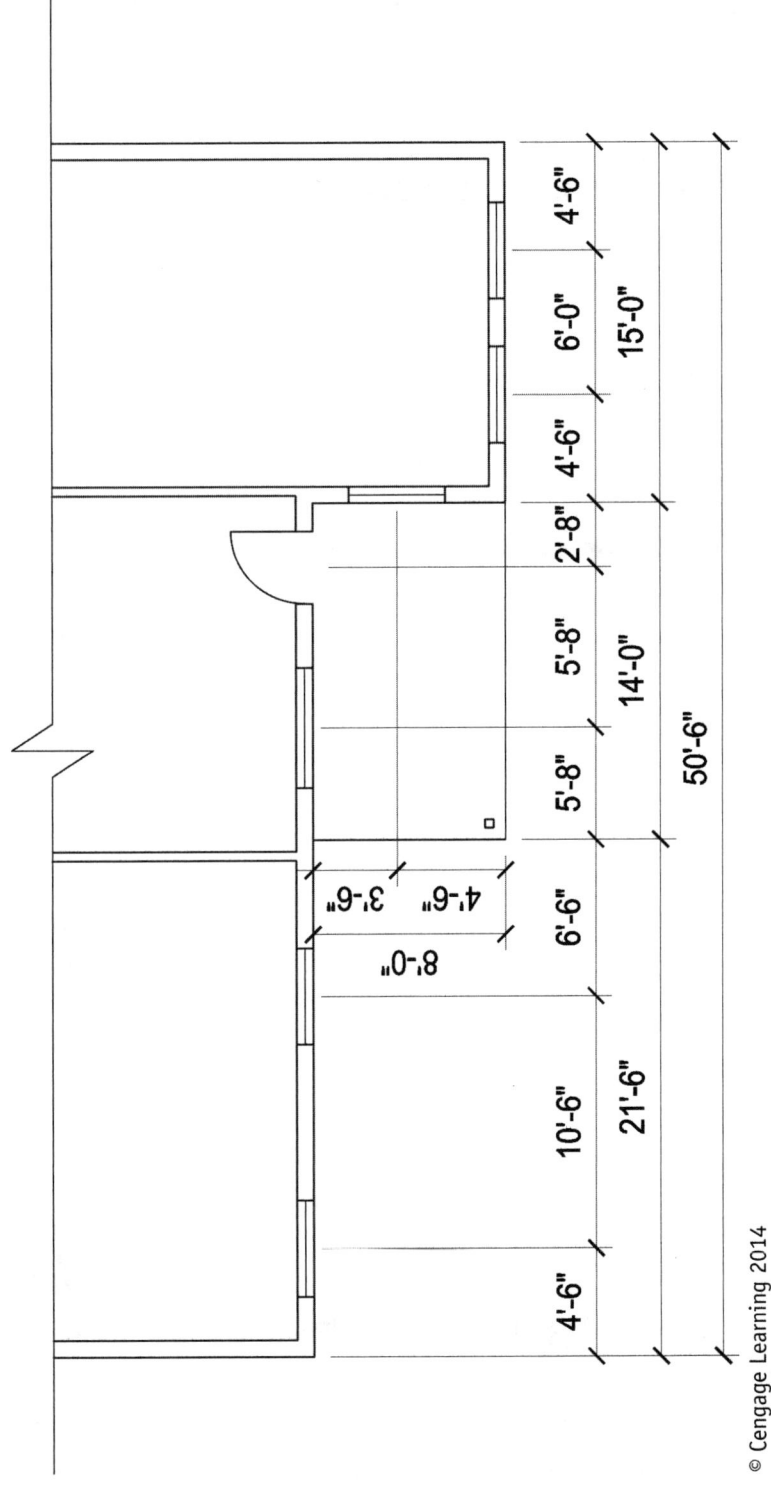

FIGURE 11-11 *Dimensioned partial plan.*

Example 11.4 For the plan shown in figure 11-12, scale and dimension the exterior of the building for overall dimensions, steps in building exterior, and window and door locations for the loadbearing masonry construction. Use an architectural scale of 3/16" = 1'-0" for dimensioning the plan.

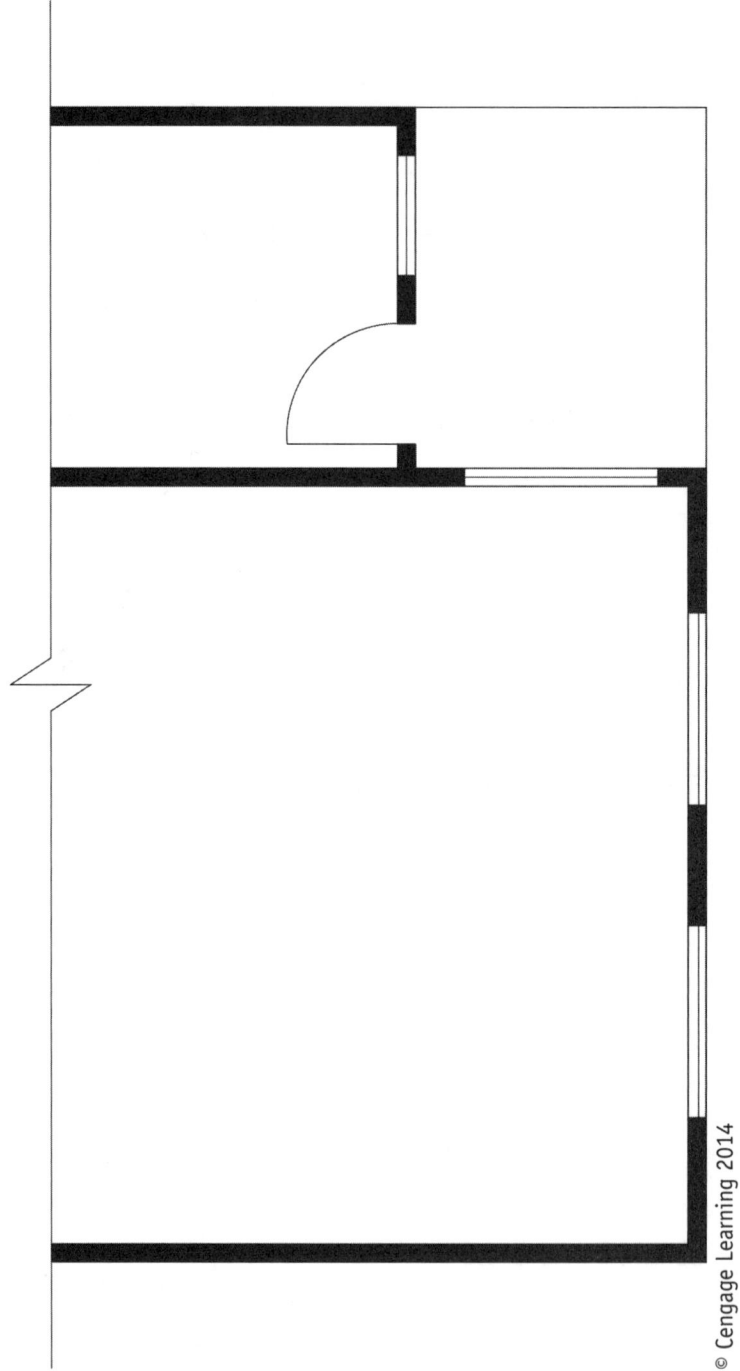

FIGURE 11-12 *Partial plan, undimensioned.*

Similar to example 11-3, this plan should have three dimension strings. The interior string will dimension the actual opening size in the load-bearing masonry walls. The middle string will dimension shifts in the exterior face of the building, and the exterior string will dimension the overall building size. The dimensioned plan is shown in figure 11-13.

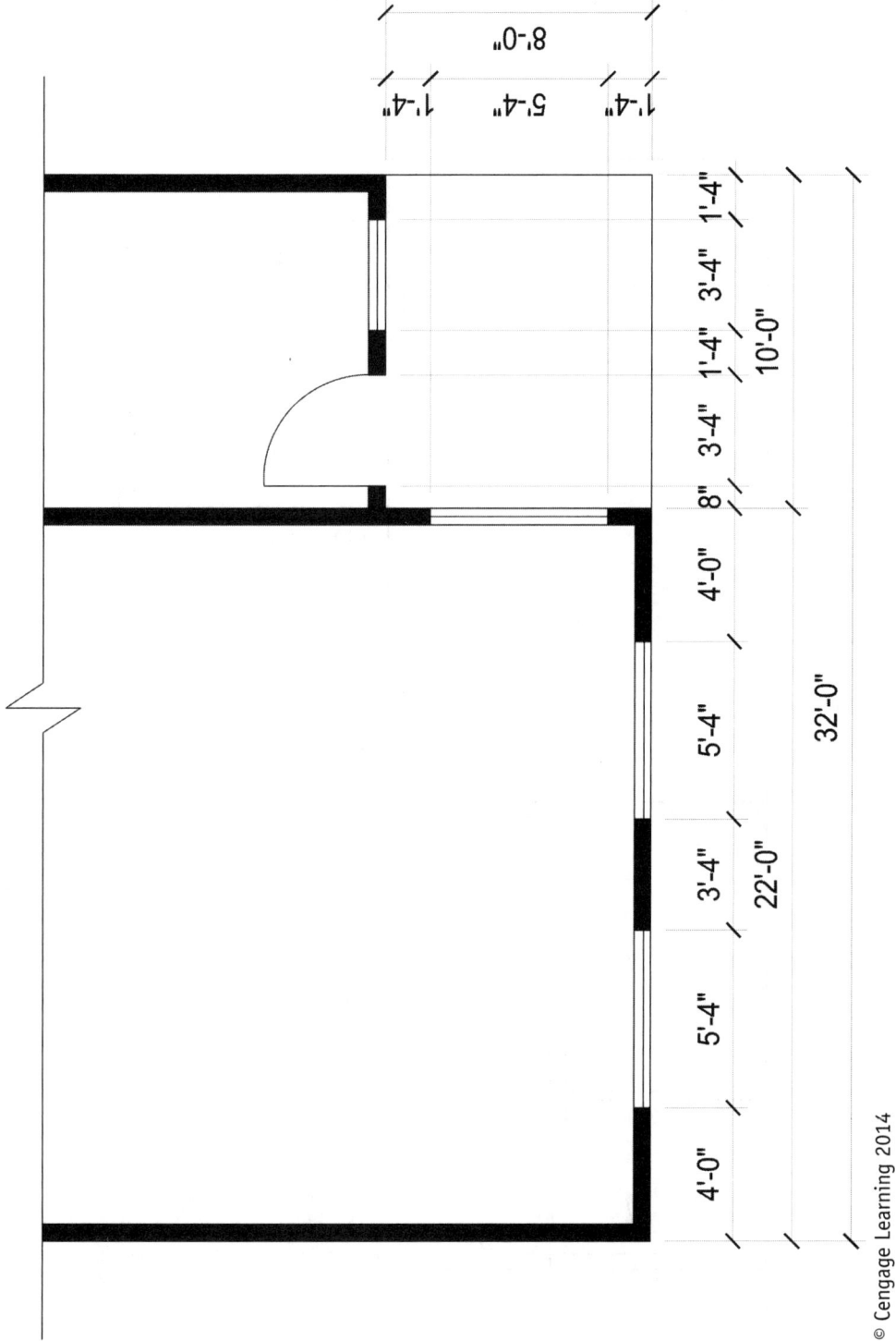

FIGURE 11-13 *Partial plan, dimensioned.*

Problem Set 11.2 Dimensions and Annotations are Used to Communicate Specifications Necessary for Construction

Problem 11-16 For the plan shown in figure 11-14, scale and dimension the exterior of the building for overall dimensions, steps in building exterior, and window and door locations for the wood stud framed construction. Use an architectural scale of 1/8" = 1'-0" for dimensioning the plan.

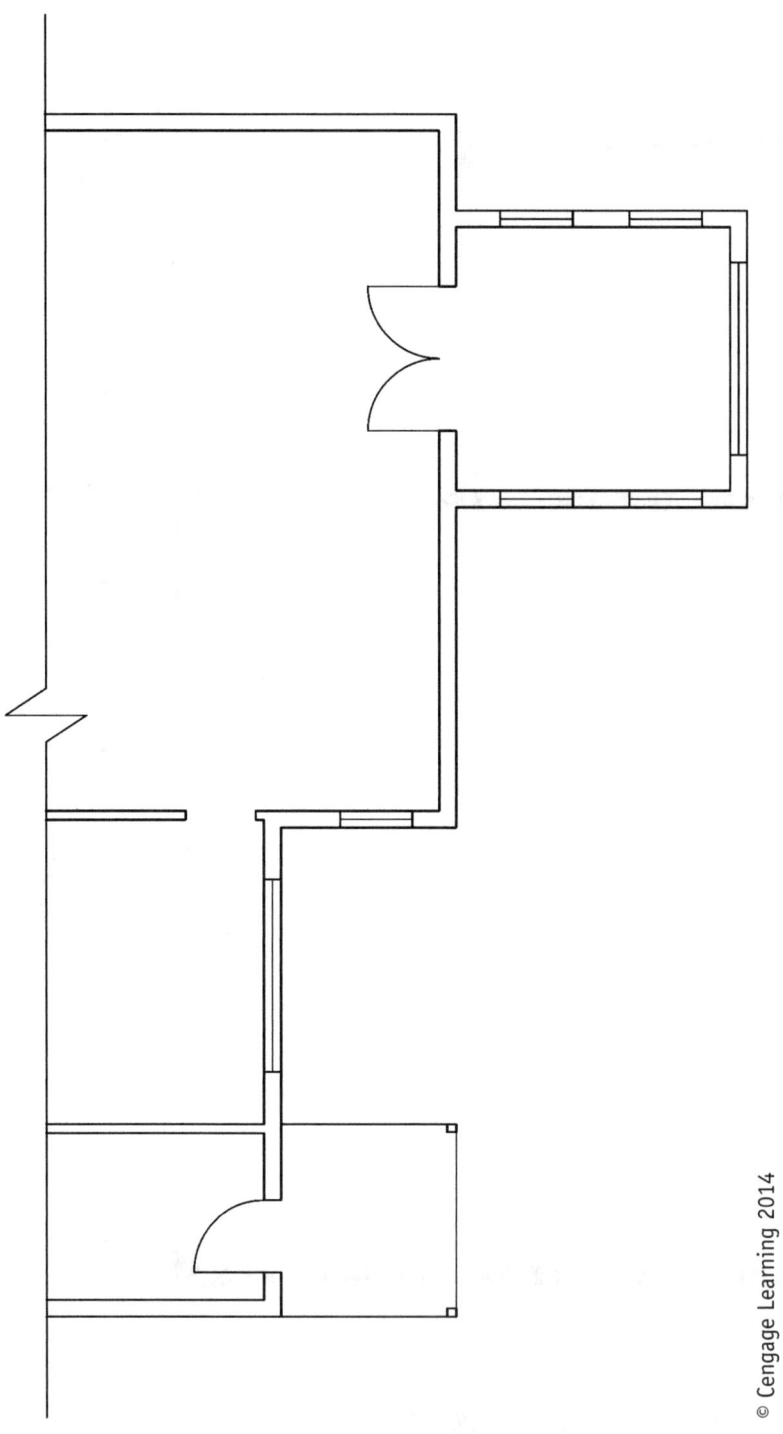

FIGURE 11-14 *Partial plan for use in problem 11-16.*

Problem 11-17 For the plan shown in figure 11-15, scale and dimension the exterior of the building for overall dimensions, steps in building exterior, and window and door locations for the load-bearing masonry construction. Use an architectural scale of 1/8" = 1'-0" for dimensioning the plan.

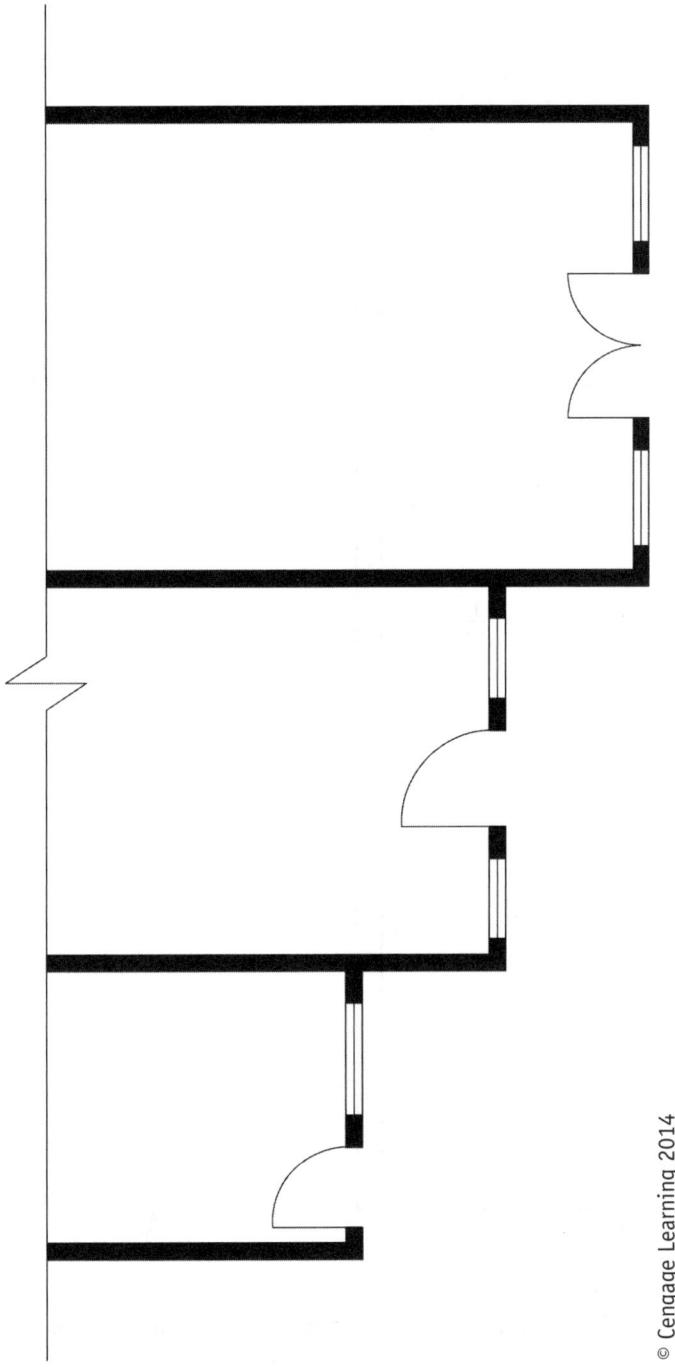

FIGURE 11-15 *Partial plan for use in problem 11-17.*

Problem 11-18 For the plan shown in figure 11-16, scale and dimension the exterior of the building for overall dimensions, steps in building exterior, and window and door locations for the wood stud framed construction. Use an architectural scale of 3/16" = 1'-0" for dimensioning the plan.

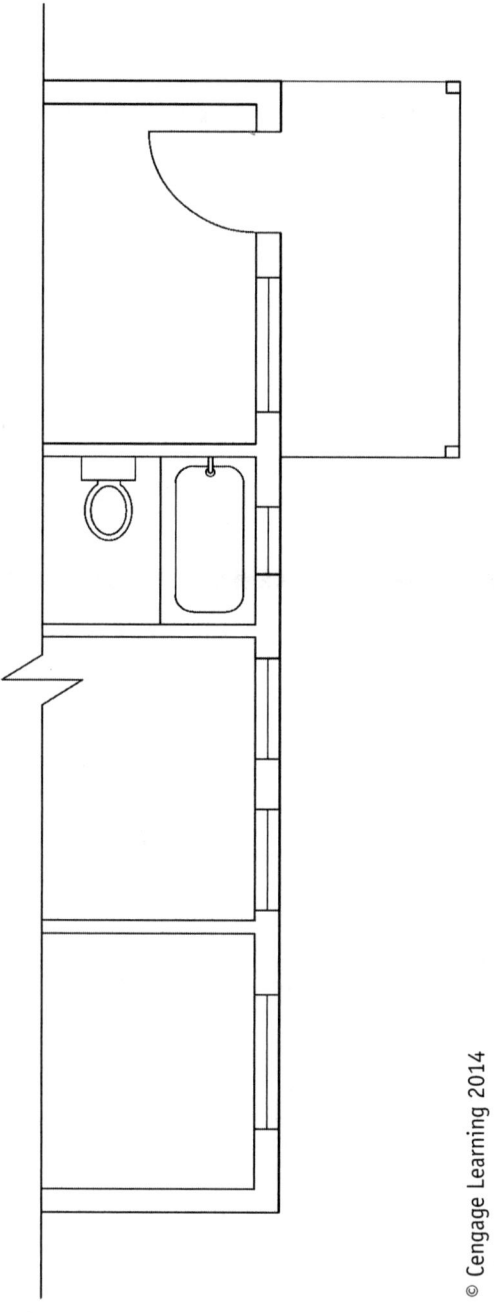

FIGURE 11-16 *Partial plan for use in problem 11-18.*

Problem 11-19 For the plan shown in figure 11-17, scale and dimension the exterior of the building for overall dimensions, steps in building exterior, and window and door locations for the load-bearing masonry construction. Use an architectural scale of 3/16" = 1'-0" for dimensioning the plan.

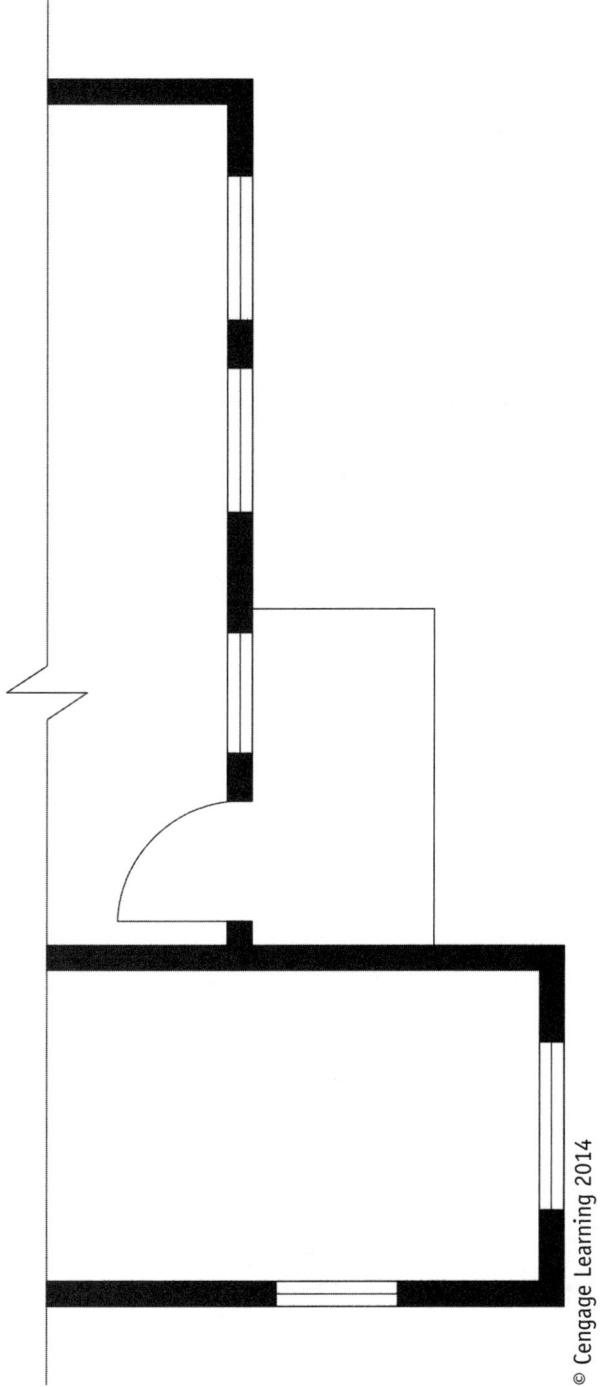

FIGURE 11-17 *Partial plan for use in problem 11-19.*

Problem 11-20   For the plan shown in figure 11-18, scale and dimension the exterior of the building for overall dimensions, steps in building exterior, and window and door locations for the wood stud framed construction. Use an architectural scale of 1/4" = 1'-0" for dimensioning the plan.

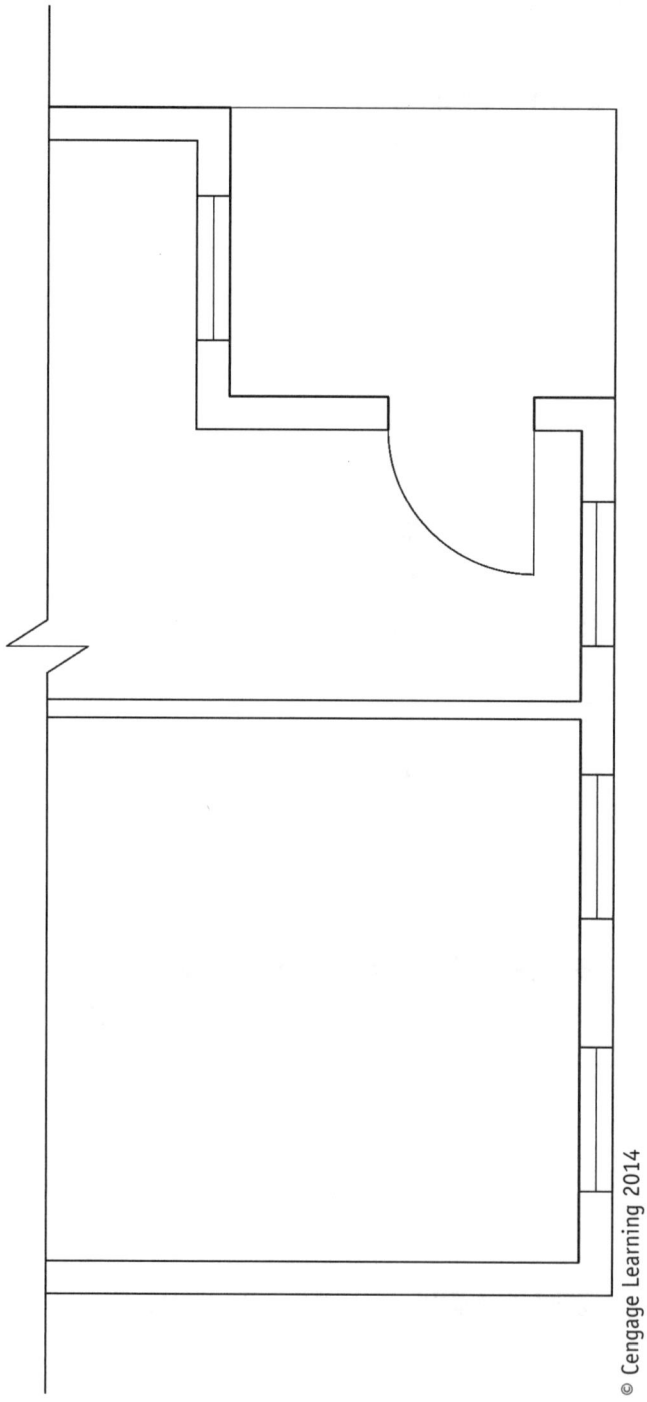

FIGURE 11-18   *Partial plan for use in problem 11-20.*

226   Chapter 11: Dimensioning and Specifications

Problem 11-21 For the plan shown in figure 11-19, scale and dimension the exterior of the building for overall dimensions, steps in building exterior, and window and door locations for the load-bearing masonry construction. Use an architectural scale of 1/4" = 1'-0" for dimensioning the plan.

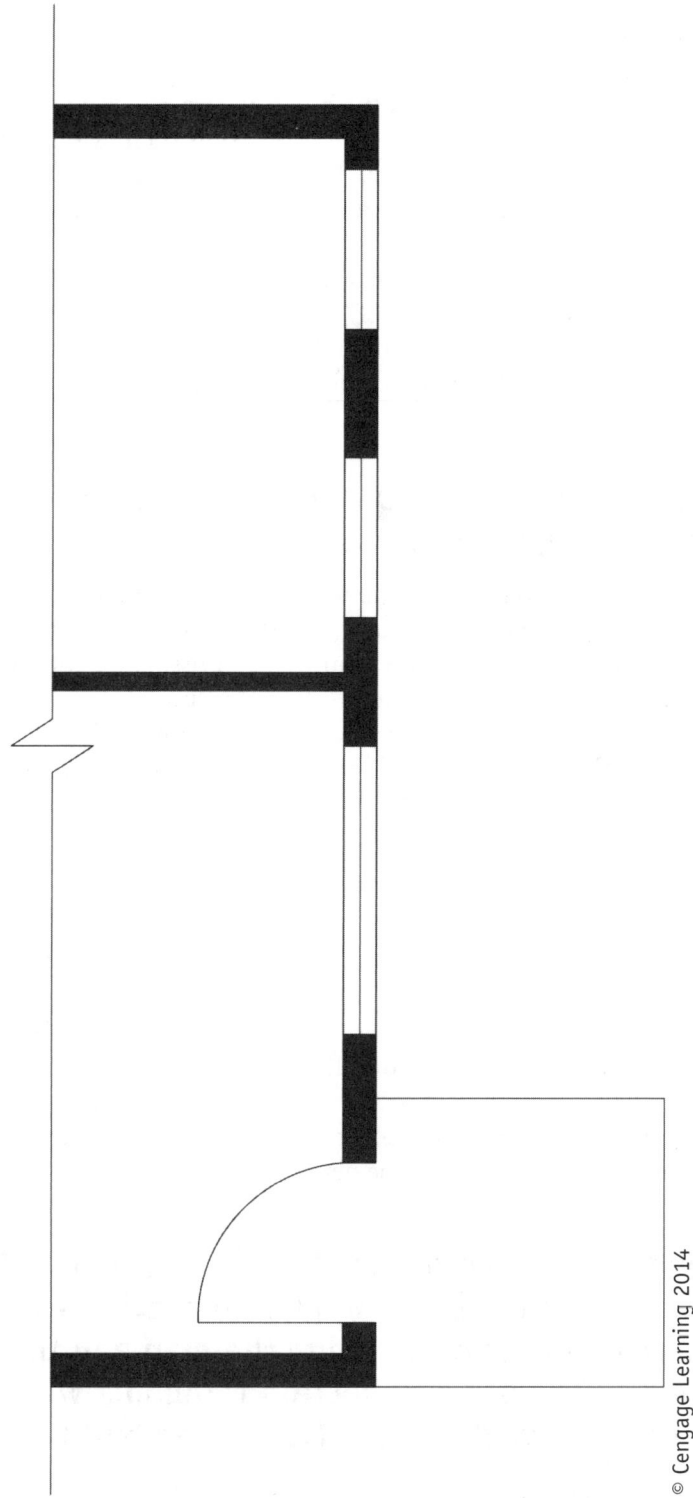

FIGURE 11-19 *Partial plan for use in problem 11-21.*

Elevation Views Provide a Preview of the Finished Construction

Elevations are used in construction drawings to provide information on aspects of a building that cannot be given in plans. This information will include vertical dimensions, materials, window and door styles, architectural features, and construction details. Elevation symbols are placed on plans indicating the direction of the view, the elevation designation (typically a letter or number), and the sheet number on which the elevation is placed. Elevations are drawn for a building's exterior, and for many projects, interior elevations exist as well.

Exterior elevations are provided for each face of the building and are drawn at the same scale as the floor plans, typically 1/8" = 1'-0" or 1/4" = 1'-0". They give information such as types and changes in materials; finish grade around the building; heights of windows, doors, roof eaves and ridges; and roof features such as chimneys or penthouses. In addition, exterior elevations provide information on roof pitch and callouts for sections and details.

Interior elevations are provided for locations within a building where detailed information needs to be given to the builders, and are typically drawn at a scale of 1/2" = 1'-0". They give information such as wall treatments and ceiling heights, placement of lighting fixtures, and built-in construction, such as cabinets and shelving. In addition, notes and construction specifications can be placed on the elevations. Interior elevations are typical for offices, bathrooms, and kitchens.

Section Views Allow Dimensioning and Annotation of Hidden Materials and Components

There are many components that go into the construction of a building that are not viewable once the building is completed. Sections allow the designer to show how the building is to be constructed, and how and where to place those components that are hidden within walls, floors, and roofs. Some of these components include structure, insulation, weather proofing, and electrical wiring. Sections allow vertical heights to be given for placement of openings and features at a wall or roof, along with the width of the wall or roof. Sections are typically drawn at a scale of 1/2" = 1'-0". For locations within a section where additional precise information needs to be provided, the user is referred to enlarged detail locations elsewhere within the construction drawings.

Details

Details are used to indicate how specific conditions in a building are to be constructed. Details can be drawn as enlarged portions of a building plan, elevation, or section, providing information that cannot be shown on the drawing from which they are referenced. Details are drawn at a larger scale than the plans, elevations, and sections to allow for more information about the condition to be provided. Typical scales used for details include 3/4" = 1'-0", 1" = 1'-0", 1 1/2" = 1'-0", and 3" = 1'-0". Details are usually placed on separate sheets, and callouts on plans, elevations, and sections refer to the detail location in the construction drawings by indicating the detail designation and the sheet number where the detail is located.

Example 11-5 From the plan shown in figure 11-11, draw a 1/8" scale elevation for the building plan indicating exterior finish materials. In addition, using the plan and the elevation, draw a 1/2" scale wall section through a window showing the wood-framed construction of the building.

When designing the exterior of a building, many aspects of the design must be considered. These include the style of architecture, the height of interior spaces and how they relate to the exterior elevation, the selection of materials, the roof type and pitch, and the style and size of doors and windows. There are many options in the design of an elevation, which can result in many solutions. One potential solution for the building plan shown in figure 11-11 is shown in figure 11-20.

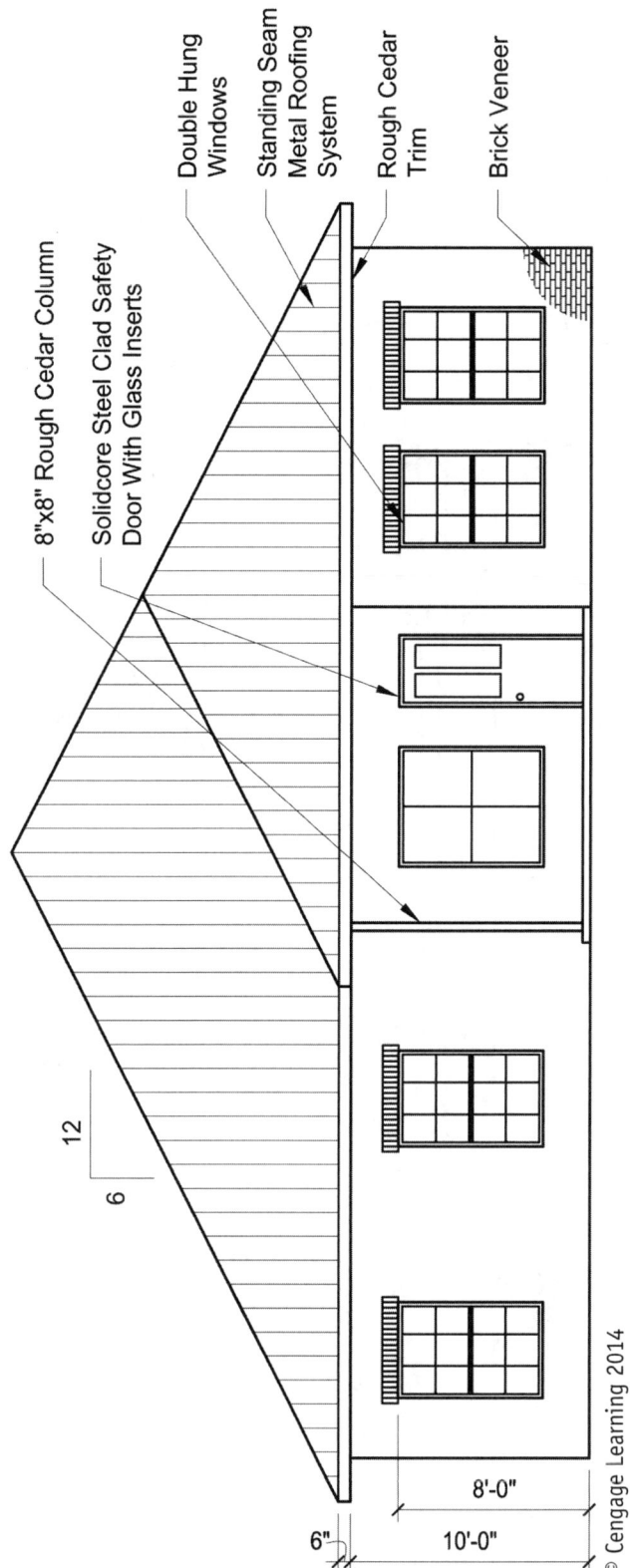

FIGURE 11-20 *Building elevation for plan given in figure 11-11.*

With the plan and elevation, the section can be drawn showing the conditions for construction of the building. Note that drawing of the plan, elevation, and section are often performed concurrently instead of one first and then the others. Revisions to each of these drawings are common throughout the design process to arrive at the final building design. The section for the building is shown in figure 11-21. Note that cut lines are shown at the window. This is done at times to reduce the size of the drawing while keeping the rest of the drawing to scale.

Section Views Allow Dimensioning and Annotation of Hidden Materials and Components 229

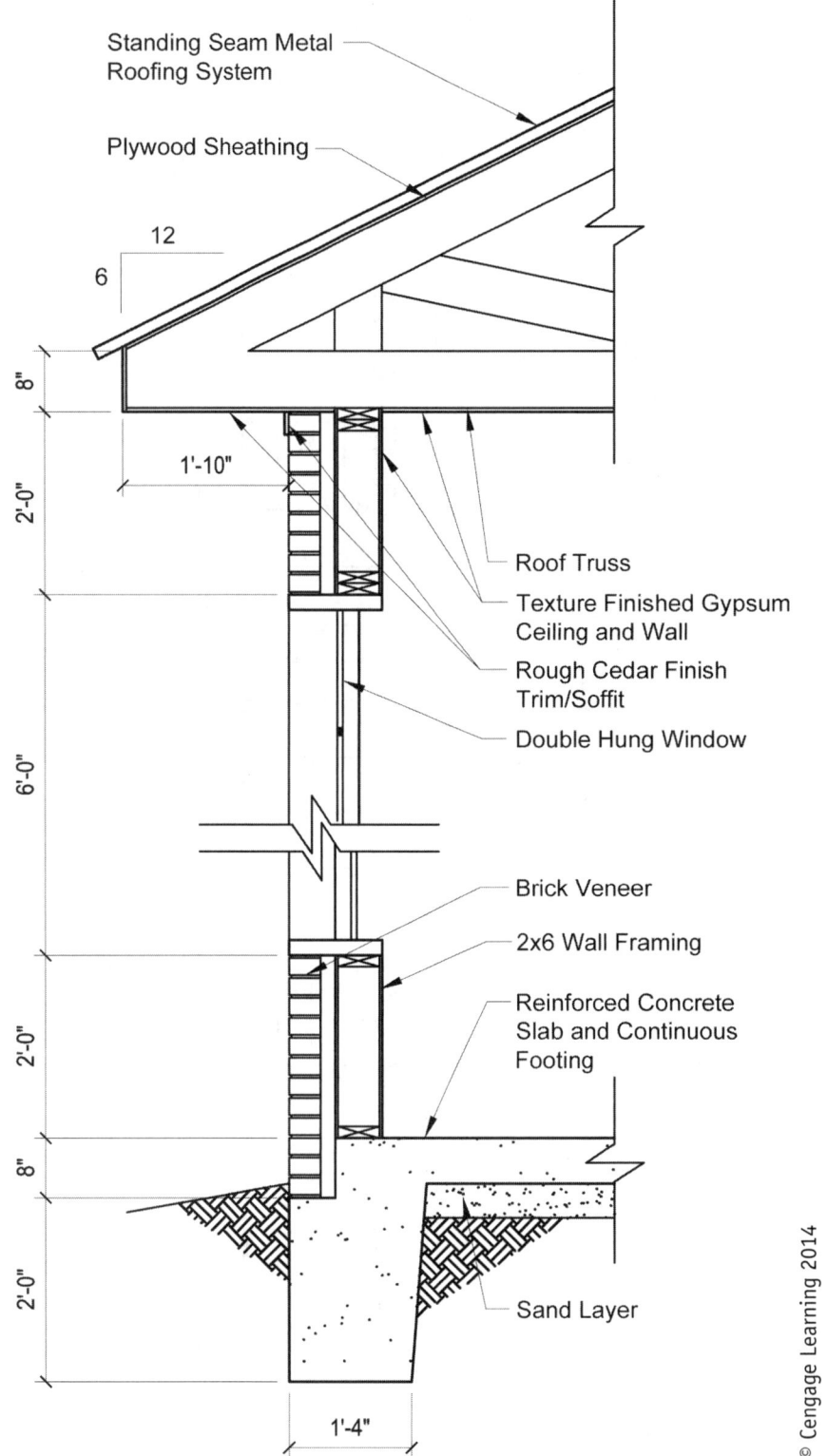

FIGURE 11-21 *Building section for building elevation given in figure 11-20.*

230 Chapter 11: Dimensioning and Specifications

Schedules Provide Numerous Specifications without Complicating a Drawing

Schedules give the design team for a project a convenient method to supply information to the builder that is easily understood. They are also used by the contractor during the bidding process to help in determining building costs more easily. Schedules are used for many components in a building design, including:

- Room finishes
- Windows and doors
- Lighting fixtures
- Mechanical and electrical components
- Structures including footings, columns, beams, slabs, and connections

Schedules are used in the construction drawings to give detailed information for components that are repeated through a building. For example, a window schedule will list the window type, size, finish material, rough opening size required, and, in some cases, the manufacturer and model number. Trying to place all of this information on the building plan would not be feasible. Symbols referring to these schedules are placed on the plans, and these vary based on what they are calling out on plan. Refer to figure 11-1 for a plan with room-finish symbols, and figure 11-22 for a room-finish schedule for the same building.

			FLOORS		WALLS				NOTES:
	NUMBER	ROOM NAME	FLOOR	BASE	NORTH	EAST	SOUTH	WEST	
SECTOR B	B101	LINK CORRIDOR	CT4	RB2	–	EXP	–	PT1	
	B102	CIRCULATION DESK	CPT1	RB1	PT1/TK1	GLASS	–	PT5	
	B102A	VESTIBULE	CT4	RB2	PT1/TK3	PT1	GLASS	GLASS	
	B103	SEATING	CPT1	–	GLASS	GLASS	GLASS	–	SOLAR SHADES ON E/S
	B103A	SEATING	CPT1	RB1	–	–	–	PT1	
	B104	STACKS	CPT1	RB1	–	–	–	PT1	
	B105	SEATING	CPT1	RB1	–	PT1	PT1	PT1	SOLAR SHADES ON S
	B106	MECH/ELEC ROOM	SC	RB2	PT1	PT1	PT1	PT1	
	B107	STAIR 4	SC	RB2	PT1	PT1	PT1	PT1	
	B108	MECHANICAL ROOM	SC	RB2	PT1	PT1	PT1	PT1	
	B109	READING ROOM	CPT1	RB1	PT1	WTK1	PT1	PT1	
	B110	OFFICE	CPT1	RB1	PT1	PT1	PT1	PT1	
	B111	ALCOVE	CPT1	RB1	PT/TK1	PT1	–	PT1	
	B112	OFFICE	CPT1	RB1	PT1	PT1	PT1	PT1	
	B113	STAGE VESTIBULE	–	–	–	–	–	–	CEILINGS TO BE PAINTED PT6.
	B114	SOUND & LIGHT LOCK	–	–	–	–	–	–	
	B115	VESTIBULE	–	–	–	–	–	–	
	B116	LECTURE STAGE	CT2	WVP1	PT3	WVP1	WVP1	WVP1	CEILINGS TO BE PAINTED PT6.
	B116A	LECTURE LWR SEATING	CPT2	CT2	–	PT3	PT3/WVP1	PT3	CEILINGS TO BE PAINTED PT6.
	B116B	LECTURE UPR SEATING	CPT2	CT2	FC1	PT3	PT3	PT3	CEILINGS TO BE PAINTED PT6.
	B116C	CROSS AISLE	CPT2	CT2	–	PT3	–	PT3	CEILINGS TO BE PAINTED PT6.
	B117	MECHANICAL ROOM	SC	RB2	PT1	PT1	PT9	–	
	B118	STAIR 6	SC	RB2	PT9	EXP	PT1	PT1	
	B119	CORRIDOR	SC	RB2	PT1	–	–	–	
	B120	AV ROOM	CPT1	RB2	WVP	BP/WVP	GLASS	GLASS/BP	CEILINGS TO BE PAINTED PT6.
	B121	ELECTRICAL	SC	RB2	PT9	PT1	PT1	PT1	CEILINGS TO BE PAINTED PT6.

FIGURE 11-22 *Room finish schedule.*

Reproduced with permission from OSU DWR School of Architecture, Studio Architecture

Problem Set 11.3 Elevation Views Provide a Preview of the Finished Construction/Section Views Allow dimensioning and Annotation of Hidden Materials and Components/ Schedules Provide Numerous Specifications without Complicating a Drawing

Problem 11-22 Draw an exterior elevation for the plan shown in figure 11-14 that matches the scale of the plan drawing.

Problem 11-23 Draw an exterior wall section through a window for the exterior elevation drawn in Problem 11-22. Use a scale of 1/2" = 1'-0" for the section.

Problem 11-24 Draw an exterior elevation for the plan shown in figure 11-15 that matches the scale of the plan drawing.

Problem 11-25 Draw an exterior wall section through a window for the exterior elevation drawn in Problem 11-22. Use a scale of 1/2" = 1'-0" for the section.

Problem 11-26   Draw an exterior elevation for the plan shown in figure 11-16 that matches the scale of the plan drawing.

Problem 11-27   Draw an exterior wall section through a window for the exterior elevation drawn in Problem 11-22. Use a scale of ½" = 1'-0" for the section.

Problem 11-28   For the plan shown in figure 11-7, place window symbols on the floor plan, and develop a window schedule indicating window mark, quantity, type, finish, size, rough opening size, and comments (pertaining to mullions, manufacturer, and/or manufacturer number).

Problem 11-29   For the plan shown in figure 11-14, place window symbols on the floor plan, and develop a window schedule indicating window mark, quantity, type, finish, size, rough opening size, and comments (pertaining to mullions, manufacturer, and/or manufacturer number).

Problem 11-30   For the plan shown in figure 11-15, place window symbols on the floor plan and develop a window schedule indicating window mark, quantity, type, finish, size, rough opening size, and comments (pertaining to mullions, manufacturer, and/or manufacturer number).

Problem 11-31   For the plan shown in figure 11-7, place window symbols on the floor plan and develop a door schedule indicating door mark, quantity, type, finish, size, rough opening size, and comments (pertaining to mullions, manufacturer, and/or manufacturer number).

Problem 11-32   For the plan shown in figure 11-14, place window symbols on the floor plan and develop a door schedule indicating door mark, quantity, type, finish, size, rough opening size, and comments (pertaining to mullions, manufacturer, and/or manufacturer number).

Problem 11-33   For the plan shown in figure 11-15, place window symbols on the floor plan and develop a door schedule indicating door mark, quantity, type, finish, size, rough opening size, and comments (pertaining to mullions, manufacturer, and/or manufacturer number).

CHAPTER 12
Building Materials and Components

Skills List

After completing the problems in this chapter, you should be able to:

- Classify and specify materials used in building construction
- Understand important material properties
- Be aware of building construction types and their limitations
- Understand criteria for material selection
- Understand how building materials are made
- Be aware of environmental impact of material selection

BACKGROUND

Architects and engineers must be able to select building materials that are safe, economical, durable, and sensitive to the environment. The construction process requires numerous materials and their components. The understanding of their production, fabrication, and construction are important to any successful building project.

The History of Building Materials

Natural materials that are readily available have historically been used for building construction. The vernacular architecture in most areas of the world evolves and is dependent on the environment, culture, context, and other local considerations. Ancient buildings were constructed from natural materials such as stone, sun-baked brick, and wood. They were designed and developed based upon their local environment.

The Industrial Revolution of the mid- to late nineteenth century gave rise to new manufacturing and production techniques, along with the use of modern building materials. The strength of bricks increased due to kiln firing, and large quantities could be made through the extrusion process. Timber production became standardized by mechanization and grading rules. The Bessemer process removed impurities from iron by oxidation and allowed for higher-strength steel and iron production. The design and construction of the braced steel frame allowed for high-rise construction to begin. Finally, the production of ordinary Portland cement and steel reinforcement were standardized, which lead to the development of reinforcement concrete construction.

Many new materials were developed in the twentieth century, but most were used as finishes and fixtures in buildings. Timbers, structural steel, and reinforced concrete were the primary material structural systems used throughout the modern age. In the late twentieth century, environmental concerns lead to the development of newer and clean manufacturing and production methods. Finally, rising transportation costs have redeveloped the financial need to use regional materials whenever possible.

Construction Materials Fundamentals

Architects and engineers should understand how materials will behave in buildings and the classification systems used for materials.

Material Families

Construction materials can be classified as metals, polymers, ceramics, composites, or organic/natural materials.

Table 12-1 lists materials under each material family.

Category	Material
Metals	• Ferrous metals and alloys (iron and steel) • Nonferrous metals and alloys (aluminum, copper, nickel, zinc, titanium, and precious metals)
Polymers	• Thermoplastic plastic (polyethylene, polypropylene, polystyrene, and poly chloride) • Thermosetting plastics (alkyds, amino and phenolic resins, epoxies, polyurethanes, and unsaturated polyesters)
Ceramics	• Structural clay products (brick, sewer pipe, roofing tiles, and flue linings) • Refractories (brick and monolithic products used in metal, glass, cements, ceramics, energy conversion, and petroleum and chemical industries) • Glasses (flat windows), containers (bottles), blown (dinnerware), and fiber (insulation) • Abrasives (natural and synthetic abrasives used for grinding, cutting, polishing, and blasting) • Cements (for roads, bridges, buildings, and dams) • Advanced ceramics (structural, electrical, coatings, chemical, and environmental)
Composites	• Reinforced plastics • Metal-matrix composites • Ceramic-matrix composites • Sandwich structures • Concrete
Organic/Natural	• Wood • Stone • Clay

TABLE 12-1 *Material families.*

Metals account for about two-thirds of all the elements and about one-fourth of the mass of the planet. They have useful properties, including strength, ductility, high melting points, thermal and electrical conductivity, and toughness.

Polymers contain many parts or units, which are chemically bonded together to form a solid. The word *polymer* literally means "many parts." Two important polymeric materials are plastics and elastomers. Plastics are a large and varied group of synthetic materials that are processed by forming or molding into shape. Elastomers or rubbers can be elastically deformed a large amount when a force is applied to them and can return to their original shape when the force is released. The term *thermoplastic* indicates that these materials melt on heating and may be processed by a variety of molding and extrusion techniques. Alternately, *thermosetting* polymers cannot be melted or re-melted.

Ceramics are an inorganic, nonmetallic solid that is prepared from powdered materials. They are fabricated into products through the application of heat, and display such characteristic properties as hardness, strength, low electrical conductivity, and brittleness. The word *ceramic* comes the from Greek word *keramikos*, which means "pottery."

A composite is commonly defined as a combination of two or more distinct materials, each of which retains its own distinctive properties, to create a new material with properties that cannot be achieved by any of the components acting alone. Using this definition, it can be determined that a wide range of engineering materials fall into this category. For example, concrete is a composite because it is a mixture of Portland cement and aggregate. Fiberglass sheet is a composite because it is made of glass fibers imbedded in a polymer. Composite materials are said to have two phases. The reinforcing phase is the fibers, sheets, or particles that are embedded in the matrix phase. The reinforcing material and the matrix material can be metal, ceramic, or polymer. Typically, reinforcing materials are strong with low densities, while the matrix is usually a ductile, tough material.

Example 12.1 In which family of materials do platinum, gypsum, and fiberglass belong?

Using table 12-1 platinum is a precious metal, so it belongs to the metal family. Gypsum is a naturally occurring compound, so it belongs to the organic/natural family. Fiberglass is fine glass fibers reinforced with polymers, so it belongs to the composite family.

Material Properties

Materials have many physical, mechanical, and chemical properties that influence their performance. Physical properties are those that can be observed without changing the identity of the substance. Mechanical properties are those that involve reaction to applied load. Chemical properties are those that describe how a substance changes into a completely different substance. Table 12-2 lists some of the important material properties.

Category	Material
Physical	Color, density, specific density, hardness, thermal conductivity, permeability, thermal expansion, electrical conductivity, and resistivity
Mechanical	Strength, ductility, elasticity, hardness, creep, impact resistance, and fracture toughness
Chemical	Flammability, corrosion resistance, and phase temperatures

TABLE 12-2 *Material properties.*

Most structural materials are anisotropic, which means that their material properties vary with orientation. The variation in properties can be due to production of the material, the controlled alignment of the material, naturally occurring material, and a variety of other causes. Isotropic materials have the same properties in all direction.

Mechanical properties are a measure of the material strength and stiffness in resisting different loadings. Figure 12-1 shows the five basic ways that a material can be loaded.

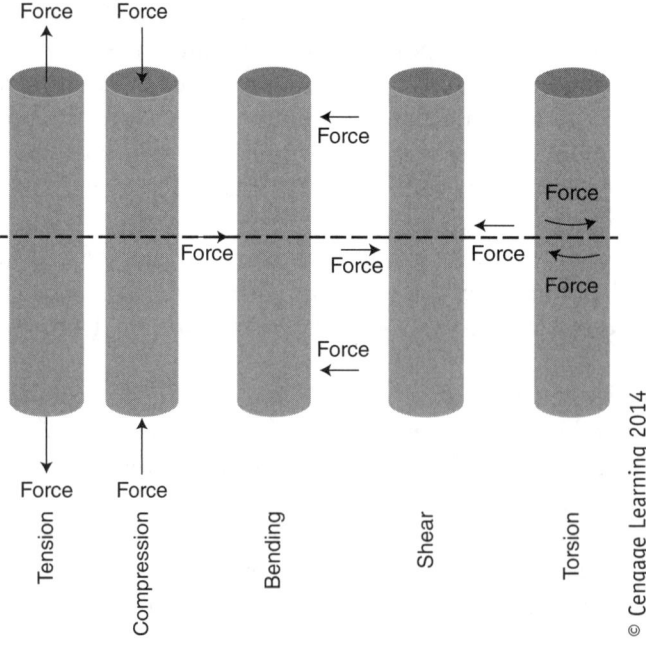

FIGURE 12-1 *Types of loading.*

Chapter 12: Building Materials and Components

Tensile strength and properties are important in structures. Tensile tests are used to determine the modulus of elasticity, elastic limit, elongation, proportional limit, reduction in area, tensile strength, yield point, yield strength, and other tensile properties. Figure 12-2 shows the typical stress–strain for an elastic and ductile material, such as steel. The elastic region of the curve is where stress and strain are linearly proportional. The ratio of the stress to the strain, σ/ϵ, is the modulus of elasticity, E.

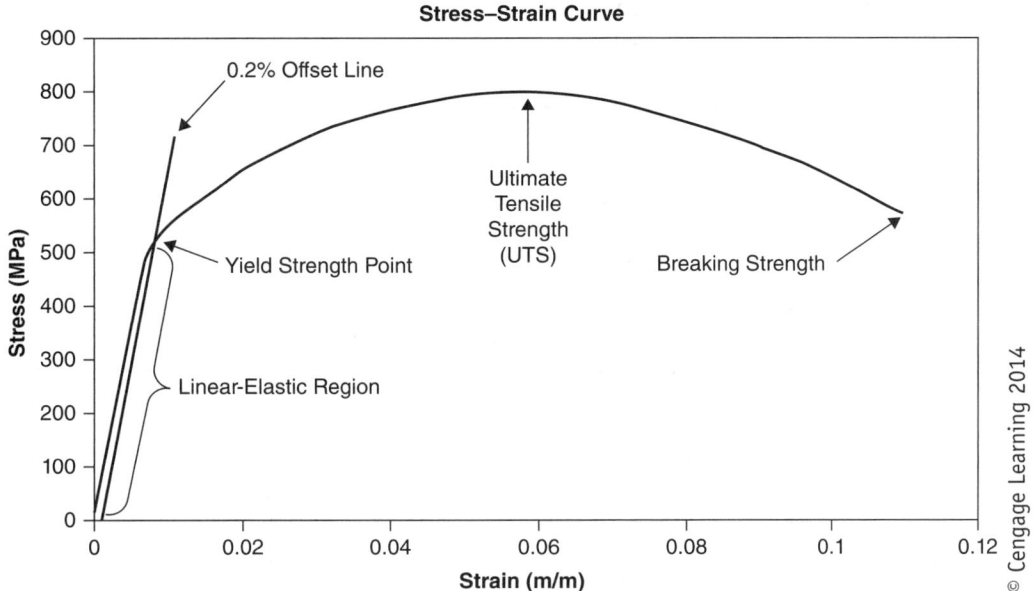

FIGURE 12-2 *Stress–strain curve for ductile material.*

The material can stretch many times its length beyond the initial yield point before it breaks. This is known as *ductility*. Brittle materials have little ductility and will break without deformation. Concrete is a brittle material and is strong in compression but weak in tension. Compression tests result in mechanical properties that include the compressive yield stress, compressive ultimate stress, and compressive modulus of elasticity. Compressive yield stress is measured in a manner identical to that done for tensile yield strength. Figure 12-3 shows the stress–strain curves for brittle and ductile materials.

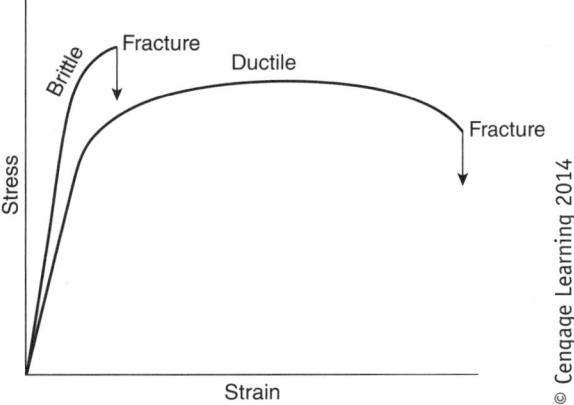

FIGURE 12-3 *Stress–strain curve for ductile vs. brittle materials.*

Other material properties that might influence selection are hardness, toughness, and creep. The textbook defines these terms as well as permeability and thermal properties. Table 12-3 lists some important material properties. For the brittle materials, there is no yield stress, and the modulus of elasticity is for compression only.

Material	Specific Weight γ, lb/ft³	Modulus of Elasticity E, k/in²	Yield Stress σ_y, k/in²	Ultimate Stress σ_u, k/in²	Coefficient of Thermal Expansion α, 10^{-6}/°F
Aluminum	169	10,000	3	10	13
Brick	110–140	1,500–3500 (C)	—	1–10	3–4
Normal Concrete	150	3,600 (C)	—	4	6
Glass	150–180	7,000–12,000	—	5–150	3–6
Steel A992	490	29,000	50	65	8
Granite	165	6,000–10,000 (C)	—	10–40	3–5
Limestone	125–180	3,000–10,000 (C)	—	3–30	3–5
Douglas Fir	30–35	1,600–1,900	5–8	8–12	—
Southern Pine	35–40	1,600–2,000	6–9	8–14	—

TABLE 12-3 *Mechanical and physical properties of materials.*

Additional thermal properties will be covered in chapter 17. The primary chemical property is flammability governed by the building codes in the next section.

Example 12.2 What is the weight of a slab of concrete 4 feet wide by 8 feet long by 4 inches thick?

Table 12-2 contains the specific weight of concrete $\gamma = 150$ lb/ft³. The total weight of the slab is the specific weight time the volume.

The weight is specific gravity time the total volume, which is the product of the width, length, and thickness.

$$W = \gamma(V) = 150 lb/ft^3 (4ft)(8ft)(4in)\left(\frac{1ft}{12in}\right) = 1600 lb$$

The weight of the slab is 1,600 lbs or 1.6 kips.

Materials and Building Codes

The primary code for buildings is the International Building Code (IBC). Specific materials and material construction systems are governed by the IBC. The most commonly used materials have their own codes and specification. The chapters in the IBC that are related to specific materials, along with the material specification, are listed in table 12-4.

Chapter	Material Content	Material Specification
19	Concrete	American Concrete Institute (ACI)
20	Aluminum	American Society of Metals (ASM)
21	Masonry	Masonry Standards Joint Committee (MSJC)
22	Steel	American Institute of Steel Construction (AISC)
23	Wood	American Wood Council (AWC)
24	Glass and Glazing	American Society for Testing and Materials (ASTM)
25	Gypsum Board and Plaster	ASTM
26	Plastic	ASTM

TABLE 12-4 *IBC material.*

Fire Resistance

The IBC classifies buildings into five basic building types based on the structural construction materials. The textbook defines these construction types, and they are summarized in table 12-5. The construction type will determine the type of fire resistance needed for the various components of the building.

Classification	Construction Materials (elements in table 12-6)
I and II	Noncombustible
III	Exterior wall noncombustible and interior elements of code approved materials
IV	Heavy Timber (see table 12-7)
V	Any code-approved material

TABLE 12-5 *Types of construction.*

The construction type along with the building occupancy classification group will determine the height and area limitations of the building. The fire-resistance ratings for building elements are listed in table 12-6.

Construction Materials Fundamentals

Building Element	Type I or II		Type III		Type III		Type IV	Type V	
	A	B	A	B	A	B	HT	A	B
Structural Frame	3	2	1	0	1	0	HT	1	0
Bearing Walls Exterior Interior	3 3	2 2	1 1	0 0	2 1	2 0	2 1/HT	1 1	0 0
Nonbearing walls and partitions exterior	See Table 602 in the IBC								
Nonbearing walls and partitions interior	0	0	0	0	0	0	See Section 602.4.6 IBC	0	0
Floor construction including supporting beams and joist	2	2	1	0	1	0	HT	1	0
Roof construction including supporting beams and joist	1 1/2	1	1	0	1	0	HT	1	0

TABLE 12-6 *Fire-resistance rating requirements for building elements (hours).*

In Type IV construction, also known as *heavy timber* (HT), the wood will burn on the exterior to a certain depth, known as char. Once the char occurs the wood will not burn rapidly and will maintain structural strength. In Type IV construction a minimum size of heavy timber is used to provide some of the fire rating. Table 12-7 lists the minimum sizes of wood members that are approved for Type IV construction.

Minimum Nominal Solid Sawn Size		Minimum Glued-Laminated Net Size	
Width, inch	Depth, inch	Width, inch	Depth, inch
8	8	6 3/4	8 1/4
6	10	5	10 1/2
6	8	5	8 1/4
6	6	5	6
4	6	3	6 7/8

TABLE 12-7 *Wood member size for type IV construction.*

Finally, the building's height and area are limited by the construction type and the occupancy use. Generally, the lower the construction type number, the taller and larger a building can be constructed.

Example 12.3 What is the required fire rating on steel column in Type IIIA construction?

A column is part of the structural frame, and from table 12-6 the required rating is 1 hour.

Example 12.4 What is the minimum depth for required for a 3-inch-wide glued-laminated roof beam in Type IV construction?

From table 12-6 we see that the roof construction, including beams, can be heavy timber to provide the fire rating. In table 12-7, the minimum depth for a 3-inch-wide glued-laminated member is 6 7/8 inches.

Specifying Materials – The Master Format

The textbook list the twenty-three divisions of the Construction Specification Institute (CSI) Master Format used in building material specification in table 12-1. The divisions that are material specific are 03 Concrete, 04 Masonry, 05 Metals, and 06 Wood, Plastics and Composites.

Problem Set 12.1   Construction Material Fundamentals

Problem 12-1	In which historical period were most of the manufacturing and production techniques for modern building materials developed?
Problem 12-2	Who is credited with developing the oxidation process in iron production that led to mass-production of steel?
Problem 12-3	In which family of materials do zinc, Portland cement, polyurethane insulation, and carbon fiber belong?
Problem 12-4	What is the primary difference between ferrous and nonferrous metals?
Problem 12-5	What is the primary difference between plastics and elastomers?
Problem 12-6	What is the primary difference between thermoplastic and thermosetting plastics?
Problem 12-7	What is meant by an *anisotropic material*? Use a couple of examples in your answers.
Problem 12-8	What are the weight and modulus of elasticity of a 3/4" × 14" × 1'-4" steel plate?
Problem 12-9	What publication regulates building structure in the United States?
Problem 12-10	What organization governs the specification of structural steel?
Problem 12-11	Under which construction type would you find a reinforced concrete frame?
Problem 12-12	What is the required fire rating on steel roof truss in Type IIIA construction?
Problem 12-13	What is the required fire rating on masonry exterior loadbearing wall in Type IIB construction?
Problem 12-14	What is the minimum depth for required for an 8-inch-wide solid sawn column in Type IV construction?
Problem 12-15	In what division of a CSI formatted document would you find information on concrete?

BACKGROUND

Material Selection Criteria

The selection of materials is seldom straightforward, and primarily standards are used as a guide. The American Society of Testing and Materials (ASTM) was developed to provide uniform characteristic and performance of materials. ASTM covers materials behavior and testing methods to ensure safety. Many other material organizations have specifications and standards for building. The primary materials in the International Building Code (ICC) are given in table 12-8 along with their trade organizations and their specifications.

Concrete	American Concrete Institute (ACI)	Building Code Requirements for Structural Concrete and Commentary
Aluminum	The Aluminum Association (AA)	Aluminum Design Manual
Masonry	Masonry Standards Joint Committee (MSJC)	Building Code Requirements and Specification for Masonry Structures
Structural Steel	American Institute of Steel Construction (AISC)	Steel Construction Manual
Cold-Formed Steel	American Iron and Steel Institute (AISI)	North American Specification for the Design of Cold-Formed Steel Structural Members
Wood	American Forest & Paper Association (AF&PA)	ASD/LRFD NDS National Design Specification for Wood Construction
Glass	Glass Association of North America (GANA)	GANA Glazing Manual
Gypsum	Gypsum Association (GA)	Fire Resistance Design Manual
Plastic	Society of Plastics Industry (SPI) and American Plastic Council (APC)	

TABLE 12-8 *Material trade organizations and specification.*

These trade organizations and their specifications contain a variety of technical data and information for material selection. The following sections contain some of the criteria used for material selection.

Strength

Basic material strength for the yield and ultimate conditions are listed in table 12-3, but a given material strength can vary widely. This variation can be due to the chemical composition of the material as in steel, the proportion of the components in the material as in concrete, or in the direction of the loading with respect to the material as in wood. The strength characteristics of each of these materials will be discussed later, but basic selection can be made based on how a material will be loaded.

Constructability

If a material is easy to manipulate and assemble, like wood products, then much of the fabrication and construction can be done at the construction site. On the other hand, structural steel normally is prefabricated in a plant and assembled with machinery on the site. The availability of workers to construct the systems should also be considered in the material selection process.

Durability

Durability can govern the selection and specification of a material. In the selection process a material should be able to perform under weathering or other changing conditions. Organic and other materials that change composition due to weathering should not be used in exterior exposures, like wood and steel. Some materials, like brick, have different grades specified for various weathering conditions. Other materials should have additives to increase durability, like air added to concrete. Finally, some materials, such as wood and concrete, will exhibit creep, which is a material deformation due to sustained load. If you have ever been in an old house with a wood floor structure, you have probably noticed the uneven floor due to creep.

Expense

The expense of a material selected is dependent on the cost of the basic material, fabrication and erection, maintenance, and possible replacement. All of these should be considered in a life-cycle cost analysis. The use of the R.S. Means Cost Data can be a great aid in selection of materials based on cost. This publication can be found in both print and online.

Availability

Materials should be readily available to a construction site, otherwise transportation cost needs to be considered. Most natural materials are available at many sites, like wood and concrete materials, but should be avoided in remote sites highly fabricated materials systems like precast concrete and steel trusses. Location of material producers can be found in *Sweets Catalog* for availability and selection. This resource is also available online.

Environmental Concerns

The Environmentally Preferable Products (EPPs) specified by the Environmental Protection Agency (EPA) list the following attributes to look for when selecting building and construction products.

- Biobased content
- Energy efficient
- Enhanced indoor environmental quality
- Low embodied energy
- Recyclable or reusable components
- Recycled content
- Reduced environmental impact over the life cycle
- Reduced or eliminated toxic substances
- Reduced waste
- Responsible stormwater management
- Sustainable development, smart growth
- Use of renewable energy
- Water efficient
- Water reuse and recycling

Other helpful resources for environmental concerns in buildings are Leadership in Energy and Environmental Design (LEED) from the U.S. Green Building Council (USGBC) and the Whole Building Design Guide (WBDG) from the National Institute of Building Sciences. LEED is a rating system for design, construction, and operation of green buildings and has classifications of silver, gold, and platinum. WBDG is an integrated design approach and an integrated team process to building design that strives to achieve a high-performance building. Environmental concerns will continue to grow in building design. The future trends in sustainable design are the living building and restorative design as shown in figure 12-4.

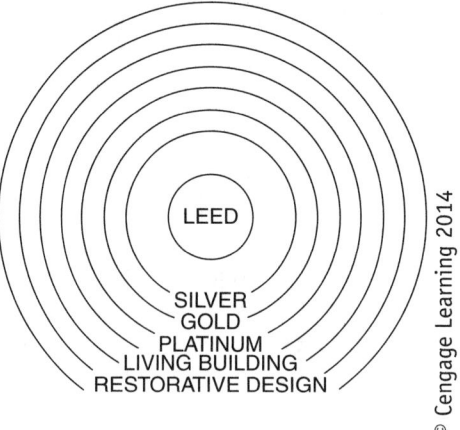

FIGURE 12-4 *Sustainable design.*

Embodied Energy

There are two forms of embodied energy in buildings. The initial embodied energy is that required to acquire, process, manufacture, transport, and construct the material. The initial embodied energy has two components. Direct energy is used to transport and construct the material. Indirect energy is used to acquire, process, and manufacture the product. The second type of embodied energy is recurring energy used to maintain, repair, restore, refurbish, or replace a material. Both of these energies are included in a Life Cycle Analysis (LCA) used to assess sustainable architecture. The four basic steps in LCA are shown in figure 12-5, and table 12-9 lists some construction materials and their embodied energy.

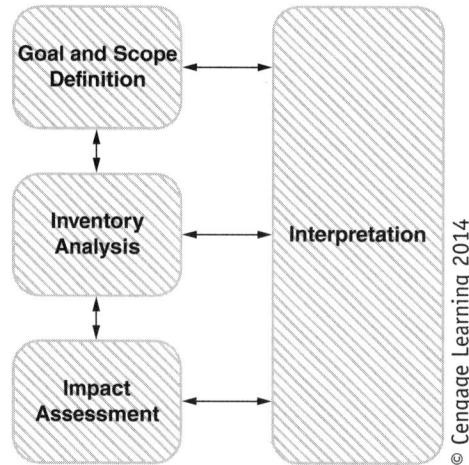

FIGURE 12-5 *Life cycle analysis.*

Material	MJ/kg	MJ/m³
Aluminum	227	515,700
Aluminum (recycled)	8.1	21,870
Brick	2.5	5,170
Concrete (normal)	1.3	3,180
Concrete (precast)	2.0	2,780
Glass	15.9	37,550
Gypsum board	6.1	5,890
Lumber	2.5	1,380
Masonry block	0.94	2,350
Plywood	10.4	5,720
Steel	32.0	251,200
Steel (recycled)	8.9	4,930
Stone (local)	0.79	2,030

TABLE 12-9 *Embodied energy of materials.*

In general, natural materials have a lower embodied energy. Recycled and reused materials can greatly decrease embodied energy. Transportation cost can cause a variation in embodied energy, as can local sources.

Natural Resources/Habitat Degradation

How and where a natural material is extracted should have an impact on selecting it for construction use. For instance, timber should be used from managed forests that replenish the natural environment and trees.

Materials and Human Health

There are many construction materials that are proven to be harmful to humans, such as asbestos, lead paint, arsenic, formaldehyde, and other volatile organic compounds. Another consideration for human health is how a material is produced. For instance, the production of Portland cement creates 8 percent of the global man-made carbon dioxide (CO_2). Also, global production of Portland cement is increasing about 5 percent per year. Therefore, ways to reduce the amount of cement in concrete, such as fly ash or blast slag replacement, can help reduce the health impact.

Summary of Material Selection Criteria

The selection of materials is complex, and whatever method is used, the selection should be evaluated with a life-cycle analysis and on environmental concerns.

Problem Set 12.2   Material Selection Criteria

Problem 12-16   What trade organization governs the specification of concrete in buildings?
Problem 12-17   What specification is used for the design of structural steel buildings?

Problem 12-18   What specification is used for the design of cold-formed steel members?
Problem 12-19   What publication can be used to aid in material cost?
Problem 12-20   What publication can be used to aid in locating material producers?
Problem 12-21   What are four attributes to look for when selecting Environmentally Preferred Products?
Problem 12-22   What are the three levels of LEED-certified buildings?
Problem 12-23   What are the four steps in a Life Cycle Analysis?
Problem 12-24   Why does normal concrete have a lower embodied energy than precast concrete?
Problem 12-25   Why is the embodied energy of recycles steel so much less than steel that is not recycled?

BACKGROUND

Major and Emerging Construction Materials

The following section covers the predominant materials used in building construction and some of their characteristics.

Concrete

Concrete is a widely available material that has good compressive strength and is very economical. It does lack in tension strength and must be reinforced with steel. The four basic components of concrete are Portland cement, coarse aggregate, fine aggregate, and water as shown in figure 12-14 of the textbook. Aggregate makes up 75 percent of the volume concrete, and fine aggregate is 3/16 in (5 mm) or smaller in size. The primary measure of strength in concrete is the water-to-cement (w/c) ratio. The general range is 0.45 to 0.6 for the w/c ratio. There are numerous types of cement, and the most common are shown in table 12-10.

Type	Use
I	Ordinary construction when special properties are not required
II	Ordinary construction when moderate sulfate resistance or moderate heat of hydration is desired
III	When high early strength is desired; has considerably higher heat of hydration then Type I cement
IV	When low heat of hydration is desired
V	When high-sulfate resistance is desired
K	Expansive or shrinkage compensating cement

TABLE 12-10   *Types of cement.*

The properties of the cement can be altered with admixtures. The most common admixture is air entraining, where minute bubbles are added to the concrete mix to increase durability. The addition of an "A" to the cement type indicates air entraining as in Type IA. Table 12-11 gives a list of other admixtures and their use per ASTM C494.

Type	Use
A	Water-reducing increases workability, sometimes called *plasticizers*
B	Retarding decrease time of curing for hot weather
C	Accelerating increase time of curing for cold weather
D	Water-reducing and retarding
E	Water-reducing and accelerating
F	Water-reducing, high-range, called *superplasticizers* (water reduced by 12% or more)
G	Water-reducing, high-range, and accelerating

TABLE 12-11 *Types of concrete admixtures.*

Common fine materials used for water-reducing include fly ash, slag, and silica fume. They not only will replace water but also can increase performance and provide for a reduction in cement. The term *High Performance Concrete* (HPC) is used for concrete that has improved properties over normally proportioned concrete. A new and emerging concrete called *Self-Consolidating Concrete* or flowable concrete is being used in precast plants and where increased viscosity is desired. Another emerging concrete is *Autoclave Aerated Concrete*, where aluminum powder is added to cement, sand, lime and other materials, then steam cured at higher temperature (360°F) and pressure (175 psi) for 8 to 12 hours. resulting in a very lightweight material with improved thermal properties.

Since concrete is a mixture of material produced in batches, it will have a variation of strength. The primary measure of strength is the specified compressive strength at 28 days after placement, denoted as f'_c. The batches must be sampled and tested to ensure quality. Test cylinders 6 inches in diameter and 12 inches high are tested in a compression or cylinder test. A field test, known as the *slump test*, can be performed on site to approximate the w/c ratio and the workability of the batch. A truncated cone-shaped 12-inch metal mold is filled with fresh concrete and then lifted off. The distance the top of the wet mass drops from its original position is the slump (3 to 4 inched is normally desired).

Concrete weight can also be reduced with the use of lightweight aggregates. This can slightly decrease the strength, but in areas of the country where earthquakes are prevalent, the benefit by the reduction of weight offsets the strength reduction. The modulus of elasticity will vary with both the weight and the strength of the concrete. The general equation for the modulus of elasticity is as follows:

$$E_c = 33 w_c^{1.5} \sqrt{f'_c}$$

E_c = modulus of elasticity of the concrete in pounds per square inch (psi)

w_s = weight of the concrete in pounds per cubic foot (pcf)

f'_c = specified compressive strength of the concrete in pounds per square inch (psi)

The weight of normally proportioned concrete is 145 lb/ft³, although 150 lb/ft³ is normally used to account for steel reinforcement. The modulus of elasticity of normal weight concrete becomes the following equation.

$$E_c = 57000 \sqrt{f'_c}$$

Example 12.5 Determine the modulus of elasticity of normal weight concrete with a specified compressive strength of 4000 psi using both equations.

$$E_c = 33(145)^{1.5}\sqrt{4000} = 3,644,147 psi$$

$$E_c = 57000\sqrt{4000} = 3,604,997 psi$$

The modulus of elasticity is approximately 3.6×10^6 psi or 3,600 ksi.

Since concrete has little to no tensile strength, it must be reinforced with steel. Steel reinforcement may consist of bars, welded wire fabric, or wires (usually deformed bars). Deformed bar sizes are given in table 12-12.

Bar Size Designation	Cross Sectional Area (sq. in.)	Weight (lb/ft)	Nominal Diameter (in.)
#3	0.11	0.376	0.375
#4	0.20	0.668	0.500
#5	0.31	1.043	0.625
#6	0.44	1.502	0.750
#7	0.60	2.044	0.875
#8	0.79	2.670	1.000
#9	1.00	3.400	1.128
#10	1.27	4.303	1.270
#11	1.56	5.313	1.410
#14	2.25	7.650	1.693
#18	4.00	13.600	2.257

TABLE 12-12 *Properties of deformed steel reinforcing bars.*

Masonry

Masonry can be made of stone, concrete, clay, or glass. The two primary types of masonry units are clay masonry (brick) and concrete masonry (CMU). The units are held together with mortar to form masonry construction. There are four types of mortar, designated as types M, S, N, and O in decreasing order of strength. The compressive strength of both the units and the mortar determine the strength of the construction. The tensile strength of masonry construction is minimal and usually ignored. The compressive strength of clay and concrete masonry are given in tables 12-13 and 12-14, respectively.

Net area compressive strength of clay masonry units, psi		Net area compressive strength of masonry, psi
Type M or S mortar	Type N mortar	
1,700	2,100	1,000
3,350	4,150	1,500
4,950	6,200	2,000
6,600	8,250	2,500
8,250	10,300	3,000
9,900		3,500
11,500		4,000

TABLE 12-13 *Compressive strength of clay masonry.*

Net area compressive strength of concrete masonry units, psi		Net area compressive strength of masonry, psi
Type M or S mortar	Type N mortar	
	1,900	1,350
1,900	2,150	1,500
2,800	3,050	2,000
3.750	4,050	2,500
4,800	5,250	3,000

TABLE 12-14 *Compressive strength of concrete masonry.*

Cement for mortar comes in three basic types, based on type of material used in the mixture. The three types of cementitious materials are cement-lime, mortar cement, and masonry cement. There are two methods for specifying mortar, by proportion and properties. Either of the methods may be used to specify mortar, but never use both methods. Tables 12-15 and 12-16 give the mortar specifications requirements by proportion and properties, respectively. In table 12-15, the ratio of the damp, loose aggregate to the sum of the separate volumes of cementitious materials should not be less than 2 1/4 and not more than 3. In table 12-16, the ratio of the damp, loose aggregate to the sum of the separate volumes of cementitious materials should not be less than 2 1/4 and not more than 3 1/2.

Major and Emerging Construction Materials

Mortar	Type	Portland cement or blended cement	Mortar cement			Masonry cement			Hydrated lime or lime putty
			M	S	N	M	S	N	
Cement-lime	M	1	—	—	—	—	—	—	1/4
	S	1	—	—	—	—	—	—	over 1/4 to 1/2
	N	1	—	—	—	—	—	—	over 1/2 to 1 1/4
	O	1	—	—	—	—	—	—	over 1 1/4 to 2 1/2
Mortar cement	M	1	—	—	1	—	—	—	—
	M	—	1	—	—	—	—	—	—
	S	1/2	—	—	1	—	—	—	—
	S	—	—	1	—	—	—	—	—
	N	—	—	—	1	—	—	—	—
	O	—	—	—	1	—	—	—	—
Masonry cement	M	1	—	—	—	—	—	1	—
	M	—	—	—	—	1	—	—	—
	S	1/2	—	—	—	—	—	1	—
	S	—	—	—	—	—	1	—	—
	N	—	—	—	—	—	—	1	—
	O	—	—	—	—	—	—	1	—

TABLE 12-15 *Mortar proportion specification requirements.*

Mortar	Type	Average compressive strength at 28 days, psi	Water retention min, percent	Air content max, percent
Cement-lime	M	2500	75	12
	S	1800	75	12
	N	750	75	14
	O	350	75	14
Mortar cement	M	2500	75	12
	S	1800	75	12
	N	750	75	14
	O	350	75	14
Masonry cement	M	2500	75	18
	S	1800	75	18
	N	750	75	20
	O	350	75	20

TABLE 12-16 *Mortar property specification requirements.*

The hollow cells in the masonry can be filled with grout to protect steel reinforcing and to provide added strength. Grout may be place in high (12.67 ft) or low (5 ft) maximum lifts. High lifts may be used if the masonry has cured at least 4 hours, and the grout has a slump between 10 and 11 inches. The maximum height of a grout pour is given in table 12-17, based on the minimum space that the grout has to fill. The space may be a continuous gap as in the case of two walls connected together or the cells within a hollow unit.

Grout type	Maximum grout pour height, ft	Minimum width of grout space, in	Minimum grout space for cells of hollow units, in x in
Fine	1 5 12 24	3/4 2 2 1/2 3	1 1/2 x 2 2 x 3 2 1/2 x 3 3 x 3
Coarse	1 5 12 24	1 1/2 2 2 1/2 3	1 1/2 x 3 2 1/2 x 3 3 x 3 3 x 4

TABLE 12-17 *Mortar property specification requirements.*

After the wall has been constructed, the net area is used to determine the wall strength. The net areas will be based on the unit width and the spacing of the grouted cells. Tables 12-18 and 12-19 give the net areas of clay and concrete masonry walls, respectively.

Nominal wall thickness (actual), in	Face-shell mortar bedding	Full mortar bedding	Solid or fully grouted
4 (3 1/2)	18.0	22.0	42.0
6 (5 1/2)	24.0	32.8	66.0
8 (7 1/2)	30.0	42.5	90.0
10 (9 1/2)	33.0	51.6	114
12 (11 1/2)	36.0	59.4	138

TABLE 12-18 *Net area of clay masonry walls, in^2/ft of wall.*

Major and Emerging Construction Materials

Nominal wall thickness (actual), in	Face-shell mortar bedding	Full mortar bedding	Solid or fully grouted
4 (3.625)	18.0	21.6	43.5
6 (5.625)	24.0	32.2	67.5
8 (7.625)	30.0	41.5	91.5
10 (9.625)	33.0	50.4	116.0
12 (11.625)	36.0	57.8	140.0

TABLE 12-19 *Net area of concrete masonry walls, in^2/ft of wall.*

Example 12.6 Determine compressive load capacity of an 8-inch hollow concrete masonry block wall with type S fully bedded mortar. The specified compressive strength of the masonry is 2,000 psi. Also specify the minimum unit strength of the block.

From table 12-14, the minimum unit strength for the concrete block is 2,800 psi for masonry strength of 2,000 psi. The net area for an 8-inch thick wall with fully bedded mortar given in table 12-19 is 41.5 in^2/ft. The strength of the wall is the product of the masonry strength and the net area.

$$P = f'_m A_{net} = 2,000 \, psi (41.5) in^2/ft = 83,000 \, lb/ft$$

f'_m = specified compressive strength of the masonry in pounds per square inch (psi)
A_{net} = net cross sectional area of the masonry wall in in^2 per foot (in^2/ft)
The capacity of the wall is 83,000 lb/ft of wall or 83.0 k/ft.

Other types of masonry construction include natural stone, glass block, and autoclaved aerated concrete (AAC) block. Stone walls are similar in design to block walls but are rarely reinforced; therefore, they should always remain in compression. Glass block is primarily used as a curtain wall that is subjected to wind pressure but not vertical load. AAC block can be made into many configurations and can be used the same as concrete and clay masonry construction.

Wood

Wood construction products come in many forms, including sawn lumber, plywood, oriented strand board (OSB), and other engineered wood product. Examples of engineered wood products include, trusses, glued laminated timber, laminated veneer lumber (LVL), parallel strand lumber (PSL), and even sheathing products like plywood and OSB. The most commonly used softwood species used in construction are Southern Yellow Pine (SYP) and Douglas-Fir (DF). Typical grade stamps are shown in figure 13.3, which indicates No. 2 grade SYP kiln dried to below 19 percent. Table 12-20 gives the size categories and subcategories for sawn lumber.

Category	Subcategory	Thickness, in	Width, in	Examples
Boards		1 to 1 1/2	2 and wider	
Dimension Lumber		2 to 4	2 and wider	
	Light Framing (LF) Structural Light and Framing (SLF)	2 to 4	2 to 4	2 x 2, 2 x 4, 4 x 4
	Structural Joist and Plank (SJ&P)	2 to 4	5 and wider	2 x 6, 2 x 14, 4 x 10
	Stud	2 to 4	2 to 6	2 x 4, 2 x 6, 4 x 6
	Decking	2 to 4	4 and wider	2 x 4, 2 x 8, 4 x 6
Timbers		5 and thicker	5 and wider	
	Beams and Stringers (B&S)	5 and thicker	More than 2 greater than thickness	6 x 10, 6 x 14, 12 x 16
	Post and Timbers (P&T)	5 and thicker	Not more than 2 greater than thickness	6 x 6, 6 x 8, 12 x 14

TABLE 12-20 *Sawn lumber size categories and subcategories.*

Note that decking is laid flat and loaded about its minor axis. Sawn lumber also varies in grading based on defects in the piece, as shown in figure 12-21 of the textbook, and the straightness of the grain, known as *slope to grain*. The slope of grain is the angle between the grain of the wood and its long direction axis. The stress grades are listed below in decreasing order of strength for dimensional lumber except as noted.

- Dense Select Structural (Timbers)
- Select Structural
- Dense No. 1 (Timbers)
- No. 1 and Better
- No. 1
- Construction (LF)
- Dense No. 2 (Timbers)
- No. 2
- Stud
- Standard (LF)
- No. 3
- Utility (LF)

Engineered wood products are used for a variety of uses, including sheathing product, such as plywood, oriented strand board, and other. Plywood is manufactured with thin layers of veneer cross laminated to provide strength in both directions as shown in figure 12-6. The *face ply* is the higher-quality outer ply and the *back* is the lower-quality outer ply. The *crossbands* are the inner plys perpendicular to the outer plys, and the *center* are the inner plys parallel to the outer plys.

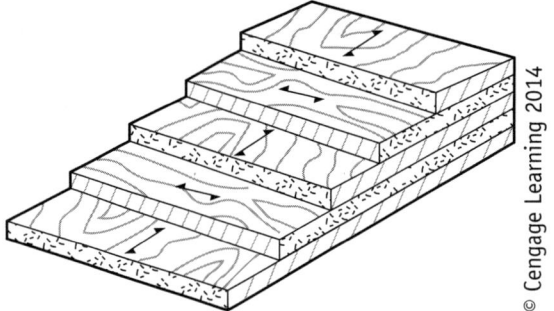

FIGURE 12-6 *Plywood layers.*

More information on plywood is given in chapter 13. Other sheathing products, like oriented strand board, must meet performance specifications and be stamped as shown in figure 13-2. Laminated veneered lumber and parallel strand lumbers are shown in figure 13-4. These are used to replace sawn lumber for heavier loading. Structural I-joists are shown in figure 13-7 and are used to replace dimensional lumber joists for longer spans. Wood trusses come in a variety of configurations for use for floors and roofs. They are covered in detail in chapter 13.

Steel and Other Metals

Steel used in building construction comes in two broad categories. The first is structural steel that is hot rolled into a variety of shapes as shown in figure 12-7. The most common shape is the I-shape, known as a *wide flange shape*.

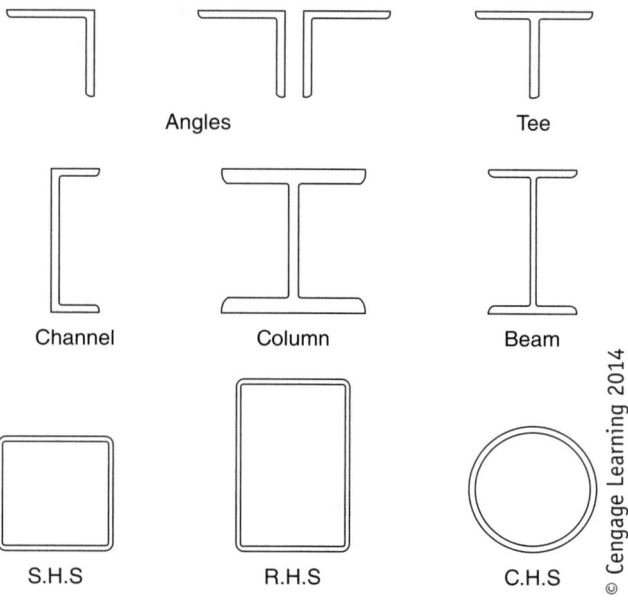

FIGURE 12-7 *Structural steel shapes.*

The common shapes and designations are listed below:

- W – Wide flange I beam
- M – Miscellaneous I beams
- S – American standard I beams
- HP – Bearing piles
- C – American standard channels
- MC – Miscellaneous channels
- L – Angles
- HSS – Hollow structural section
- P – Pipe

Other shapes include structural tees (WT, MT, and ST) cut from I shapes (W, M, & S) and combinations of shape. Common combinations are double angles, double channels, and W or S shapes with a channel cap. All of these are covered in the "Manual of Steel Construction" published by the American Institute of Steel Construction. Structural steel comes in a variety of strength grades, depending on the use. The strength grades are governed by ASTM, and a few common ones are listed table 12-21.

Steel Type	ASTM Designation	F_y Minimum Yield Stress, ksi	F_u Tensile Stress, ksi	Applicable Shapes	Preferred Shape
Carbon	A36	36	58–80	All but HSS & P	M, S, C, MC, L
	A53 Gr. B	35	60	P	P
	A500 Gr. B	42	58	HSS	HSS Round
		46	58	HSS	HSS Rectangle
High Strength Low-Alloy	A572 Gr. 50	50	65	All but HSS & P	HP
	A992	50–65	65	All but HSS & P	W

TABLE 12-21 *ASTM specification for structural steel shapes.*

The second category of steel used in construction is cold-formed steel. These are relatively thin structural shapes cold rolled from steel sheets. They are normally made from 50 ksi steel. Cold-formed steel is used in a C-shape for metal studs as shown in figure 13-13. Other uses include metal decking and forms for concrete floor. Aluminum is commonly used in construction for window and door frames, since it is lightweight and easily formed. Many other metals are used for finishes and systems in construction, as indicated in the textbook. Corrosion is always a concern with metals, and the galvanic table shown in chapter 12 of the textbook should be consulted when different metals are placed in contact.

Glass

Windows in building provide for daylighting to reduce energy needs, but they must block harmful ultraviolet rays, solar heat gain, and air and moisture infiltration. One way to select windows is to look for the ENERGY STAR® label. Figure 12-8 shows the four climate zones used by the U.S. Department of Energy and Environmental Protection Agency on the ENERGY STAR® label.

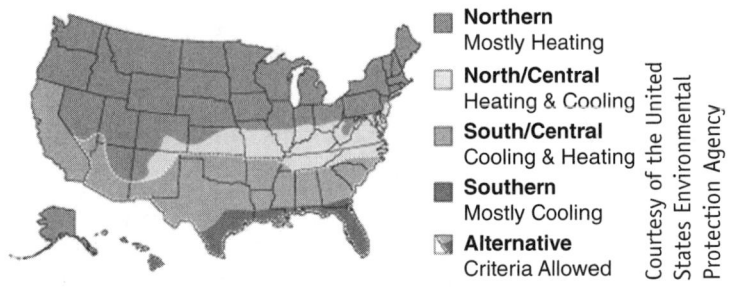

FIGURE 12-8 *Climate zones for ENERGY STAR® label.*

Another, more detailed, label has been developed by the National Fenestration Rating Council (NFRC). This label gives values for energy performance. The five values listed on the label are summarized as follows:

- U – Factor (U) measures how well a product prevents heat from escaping. The values generally fall between 0.20 and 1.20, where the lower the value, the better insulating provided.
- Solar Heat Gain Coefficient (SHGC) measures how well a product blocks heat caused by sunlight. The values fall between 0 and 1, where the lower the value, the less heat transmitted.
- Visible Transmittance (VT) measures how much light comes through a product. The values fall between 0 and 1, where the higher the value, the more light is transmitted.
- Air Leakage (AL) is indicated by an air leakage rating in ft^3 of air per ft^2 of window. The lower the value, the less air is transmitted.
- Condensation Resistance (CR) measures the ability of a product to resist the formation of condensation on the interior surface of the product. The values fall between 0 and 100, where the higher the value, the better.

These values can be used for a more detailed analysis for the windows performance. The NFRC label is shown in figure 12-9.

FIGURE 12-9 *NFRC fenestration label.*

Finish Materials

Interior finish materials must be durable and safe. Primary safety for fire and toxicity is regulated by the building code.

Flooring

Many flooring options exist for buildings, including wood, engineered laminate, linsleum, vinyl, rubber, tile, brick, stone, and carpet. With more brittle materials, it is important to provide joints to provide for controlled cracking. Many products have quality-control programs, such as the Carpet and Rug Institute's (CRI) "Green Label" for indoor air quality. A typical CRI label is shown in figure 12-10.

FIGURE 12-10 *CRI label "Green Label."*

Walls and Ceilings

The most common finish material used for walls is gypsum wallboard. This can be installed over wood or metal framing members and provides fire protection based on the type and thickness used. The GREENGUARD® Certification Program established by the GREENGUARD® Environmental Institute oversees air-quality standards. A typical label is shown in figure 12-11

FIGURE 12-11 *GREENGUARD® label.*

Problem Set 12.3   Major and Emerging Construction Materials

Problem 12-26   What type of Portland cement should be used when high sulfate resistance is desired?

Problem 12-27   What does an "A" added to the cement type indicate?

Problem 12-28   What type of concrete admixture should be used to increase workability?

Problem 12-29   What concrete field test is performed to approximate the water-to-cement ratio and workability?

Problem 12-30   Calculate the modulus of elasticity of concrete that weighs 120 pounds per cubic foot.

Problem 12-31   Determine the required compressive strength of a concrete masonry unit with type S mortar to achieve a net compressive strength of 2,500 psi.

Problem 12-32   Determine the material proportions required for a type S mortar using only mortar cement and aggregate.

Problem 12-33   What is maximum grout pour for fine grout in a 2-inch grout space?

Problem 12-34   What is net area of a solid 8-inch-thick concrete masonry wall?

Problem 12-35   Determine the compressive load capacity of a solid 8-inch concrete masonry wall with type S mortar. The specified compressive strength of the masonry is 2,500 psi.

Problem 12-36   What broad category of sawn lumber is 2 to 4 inches thick and 2 inches and wider?

Problem 12-37   What size range do post and timbers fall into?

Problem 12-38   What is the higher-quality outer layer of plywood called?

Problem 12-39   What are the inner layers perpendicular to the outer layer of plywood called?

Problem 12-40   What is most commonly used structural steel shape?

Problem 12-41   What is minimum yield stress for A500 Gr. B structural steel?

Problem 12-42   What is preferred grade of steel for a W shape used as a structural steel shape?

Problem 12-43   What climate zone does a window in Kentucky fall into?

Problem 12-44   What values are used to measure how well a product prevents heat from escaping?

Problem 12-45   Is a lower value of solar heat gain desired?

CHAPTER 13
Framing Systems Residential and Commercial

Skills List

After completing the problems in this chapter, you should be able to:

- Identify residential and commercial foundation systems
- Identify residential floor, wall, and roof framing components
- Select sizes for residential floor and roof framing components
- Determine approximate depths of commercial floor and roof systems

BACKGROUND

Architects and engineers must be able to select the various framing systems that make up the basic structure of a building. The following problems address the objectives stated in the textbook.

The structural framing systems are introduced and described in the textbook. Identification and preliminary sizes of the systems should be performed to facilitate the design process. Finally, all systems should be connected together and provide a safe load path through the building structure while meeting the appropriate building codes.

Residential Construction

The basic systems for residential construction are the foundation, floor framing, wall framing, and roof framing. The construction process starts with the foundation and proceeds upward to the roof framing.

Basic Foundations for Residential Construction

Most foundation systems for residential building construction are typically classified as shallow foundations. The depth of the foundation should be to a suitable soil and below the frost line and the zone of seasonal moisture variation. A map of the frost depth from the National Oceanic and Atmospheric Administration (NOAA) is shown in figure 13-1.

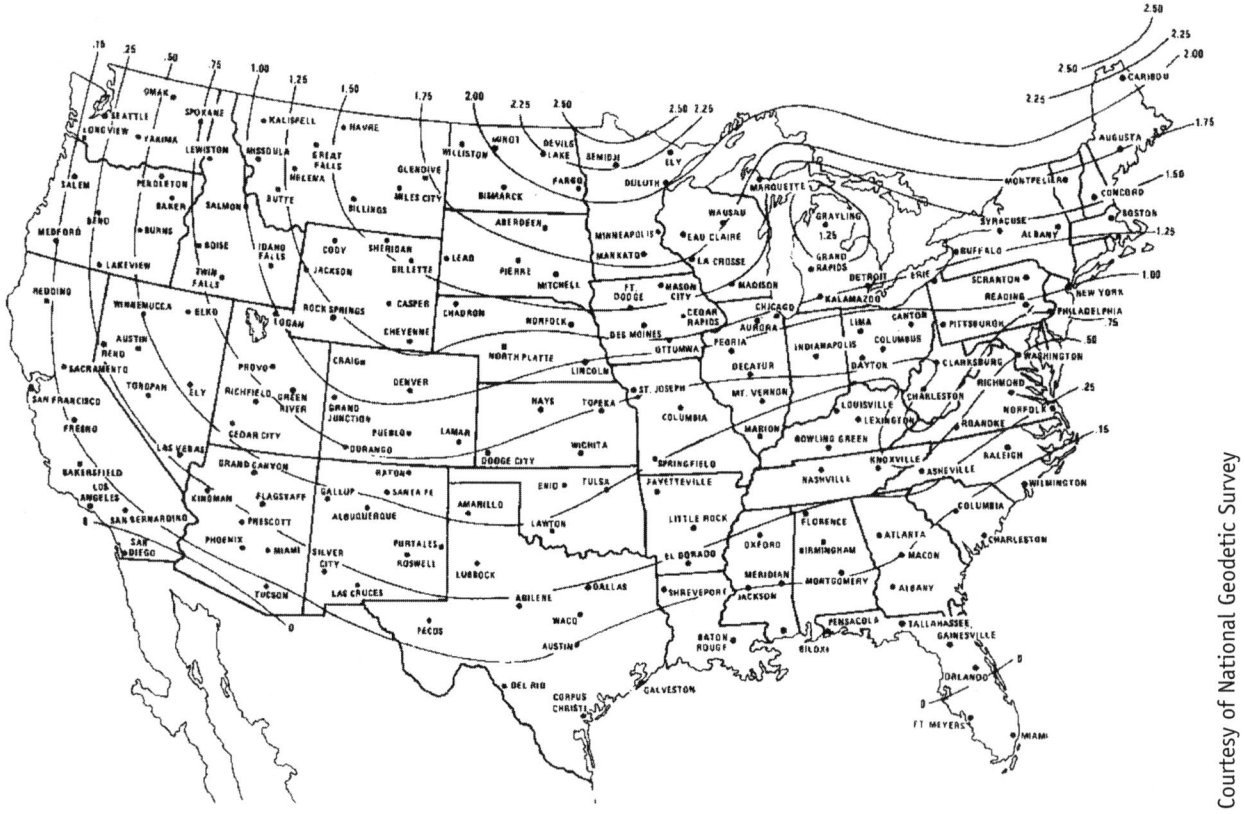

FIGURE 13-1 *Frost depth map for the United States.*

Soils with expansive properties are affected by changes in moisture. NOAA also publishes maps showing expansive soil materials. This is highly variable across the country, and your local code officials should be contacted for more information. Typical foundation depths in northern climates are governed by frost depth, and those in moderate, southern climates by the zone of seasonal moisture variation.

Spread footing foundations are generally used to distribute building loads uniformly over the soil. If the load is from a column, a spot or isolated footing is used with the column placed at the center of the footing. To determine the required area of an isolated spread footing, use the equation:

$$A_{required} \geq \frac{F}{q_{net}}$$

where $A_{required}$ = Required area for the shallow foundation
F = Total supported load = column load
q_{net} = Net allowable soil bearing pressure = $q - p$
with q = Allowable soil bearing pressure
p = Pressure due to weight of shallow foundation

Example 13.1 Calculate the required square footing size for a total load of 100 kips. The footing bears on a soil with an allowable bearing pressure of 2,500 psf. The building is in Kansas City, Kansas. Size all footing dimensions based on 3" increments.

The depth of the footing must be below the frost line shown in figure 13-1, which is located at 0.75 m in Kansas City, Kansas. The footing depth, d, is converted from meters to inches.

$$d = 0.75\,m = 0.75\,m\,\frac{39.37\,in}{1\,m} = 29.53\,in$$

The depth is rounded up to the nearest 3 inch increment of 30 inches. The area of the footing and square width can be found. The footing pressure: p = footing thickness × density of concrete = 30 in (1 ft/12 in) (150 lb/ft³) = 450 psf. The net bearing pressure is:

$$q_{net} = q - p = 2500\,psf - 450\,psf = 2050\,psf$$

The required minimum footing area is:

$$A_{required} = \frac{F}{q_{net}} = \frac{100{,}000\,lbs}{2050\,lb/ft^2} = 48.78\,ft^2$$

For a square footing, the actual required width, w, is:

$$w = \sqrt{A_{required}} = \sqrt{48.78\,ft^2} = 6.98\,ft$$

The width of the footing is rounded up to 7'–0", and the resulting footing size is 7'-0" × 7'-0" × 2'-6" deep.

If the load is from a wall, a continuous or wall footing is used with the wall placed along the centerline of the footing. The load from the wall would be expressed as a force per length of wall, and only the width of the footing is required to be found.

$$w_{required} = \frac{f}{q_{net}}$$

where $w_{required}$ = Required width for the shallow foundation
f = Total supported load = Wall load
q_{net} = Net allowable soil bearing pressure = $q - p$
with q = Allowable soil bearing pressure
p = Pressure due to weight of shallow foundation

Example 13.2 Calculate the required width of a wall footing total load of 7,500 lb/ft. The footing bears on a soil with an allowable bearing pressure of 3,500 psf. The depth of the footing is 24 inches.

The footing pressure: p = footing thickness × density of concrete = 24 in (1 ft/12 in) (150 lb/ft^3) = 300 psf. The net bearing pressure is:

$$q_{net} = q - p = 3500\,psf - 300\,psf = 3200\,psf$$

The required minimum footing width is:

$$w_{required} = \frac{f}{q_{net}} = \frac{7500\,lb/ft}{3200\,lb/ft^2} = 2.344\,ft$$

The width of the footing could be rounded up to 2' - 6", and the resulting continuous footing size is 2' - 6" × 2' - 0" deep.

Foundation walls need to be designed for both vertical loads from the structure and lateral loads due to soil pressure. The wall should be supported at both the top and bottom. The preliminary thickness of a concrete foundation wall can be approximately the vertical span divided by 16, and the minimum thickness is 7-1/2 inches. For a masonry foundation wall, use the thickness in table 13-1.

Wall Construction	Nominal Wall Thickness, in. (mm)	Maximum Depth of Unbalanced Backfill, ft (m)
Hollow unit masonry	8 (203)	5 (1.52)
	10 (254)	6 (1.83)
	12 (305)	7 (2.13)
Solid unit masonry	8 (203)	5 (1.52)
	10 (254)	7 (2.13)
	12 (305)	7 (2.13)
Fully grouted masonry	8 (203)	7 (2.13)
	10 (254)	8 (2.44)
	12 (305)	8 (2.44)

TABLE 13-1 *Foundation wall construction.*

Slabs on grade are sometimes used for the first floor in residential construction. When the slab is utilized on the first floor, it should be placed on a stable soil, and the exterior wall supported on foundation of an appropriate depth. In areas where expansive soils exist, the soil should be removed and replaced. Post-tensioned slabs on grade are sometimes used in lieu of replacing expansive soils.

When soil conditions are extremely poor, and replacing the soil is not possible, a mat or floating foundation may be placed under the entire building. If there is suitable soil for bearing deeper within the ground, a deep foundation may be used. The design of deep foundations is covered in chapter 14.

Basic Floor Framing for Residential Construction

The components of a residential floor framing system consist of the decking and support system. The decking can be plywood, oriented strand board, tongue-and-groove wood decking, or other approved sheathing. Typical sheathing will be used and is stamped with a span rating that indicates the allowable span in inches for a roof or floor. Sheathing normally comes in four foot by eight foot sheets, and the long direction is intended to be in the spanning direction. Other information, such as the sheathing grade, thickness, and exposure classification, are indicated in the stamp shown in figure 13-2.

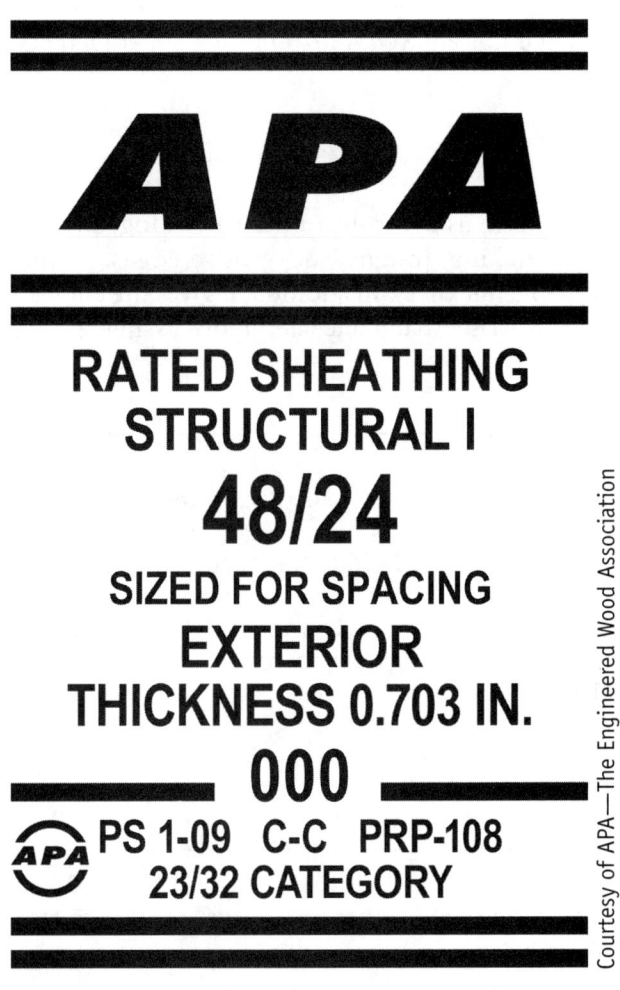

FIGURE 13-2 *APA sheathing stamp.*

This stamp shows a span rating of 48/24, indicating that it is designed to span floor members spaced 24 inches on center. The grade of the sheathing is Structural I and is 23/32 inches thick. The exposure classification is exterior. Finally, the exterior veneers of this sheathing are C and C. Veneer grades are listed in table 13-2. For structural purposes, A and C grades are similar, and B and C grades are similar. The grades are listed in order of finish.

Veneer Grade	Surface Condition and Defects
N	Special order "natural finish" (normally not for structural use)
A	Smooth paintable surface, repaired defects up to 2-1/4 inch wide
B	Solid surface veneer, repaired defects up to 4 inch wide
C	Minimum for exterior use, open defects up to 1 inch wide
D	Allows knotholes and not allowed in exterior use, open defects up to 2-1/4 inch wide

TABLE 13-2 *Sheathing veneer grades.*

Different thickness of sheathing may be used for the same span rating, and edge support may be required. Edge support may be provided by blocking, tongue-and-groove edges, panel clips, and other materials. Table 13-3 gives span rating and available thickness. In addition, it gives the allowable roof span with or without edge support. We will cover the use of this as a span table later in discussing residential roof systems.

Sheathing Span Rating	Sheathing Thickness, in	Maximum Roof Span		No. of Edge Supports
		With Edge Support	No Edge Support	1
24/0	3/8	24	20	1
24/16	7/16	24	24	1
32/16	15/32, 1/2	32	28	1
40/20	5/8, 19/32	40	32	1
48/24	3/4	46	36	2

TABLE 13-3 *Wood sheathing data.*

The structural framing that supports the floor is normally dimensioned lumber and referred to as *joists*. Dimension lumber is sawn and dressed wood 2 to 4 inches in thickness and 4 or more inches in depth. The most common species for dimensioned lumber are Southern Yellow Pine and Douglas Fir. They come from the southeastern and northwestern regions of the country, respectively. Timbers are sawn lumber 5 or more inches in thickness in each direction. Sawn lumber is marked with a grade stamp that indicates the species of wood, grading, surface and moisture conditons, and other informations as shown in figure 13-3. This stamp shows a species of Southern Yellow Pine (SYP) with a No. 2 grade that has been kiln dried below 19 percent moisture content. Other markings indicate the manufacturer, mill, and grading organization.

FIGURE 13-3 *SYP lumber stamp.*

The size of dimension lumber is given in whole inches, known as the *nominal dimension*, but the actual finished dimensions are less. Table 13-4 gives some dressed sizes of dimension lumber and timbers.

Nominal Size, b × d	Standard Dressed Size b × d, inches × inches
2 × 4	1-1/2 × 3-1/2
2 × 6	1-1/2 × 5-1/2
2 × 8	1-1/2 × 7-1/4
2 × 10	1-1/2 × 9-1/4
2 × 12	1-1/2 × 11-1/4
2 × 14	1-1/2 × 13-1/4
4 × 4	3-1/2 × 3-1/2
4 × 6	3-1/2 × 5-1/2
4 × 8	3-1/2 × 7-1/4
4 × 10	3-1/2 × 9-1/4
4 × 12	3-1/2 × 11-1/4
4 × 14	3-1/2 × 13-1/4
4 × 16	3-1/2 × 15-1/4
6 × 6	3-1/2 × 5-1/2
6 × 8	3-1/2 × 7-1/2
6 × 10	3-1/2 × 9-1/2
6 × 12	3-1/2 × 11-1/2
6 × 16	3-1/2 × 15-1/2
6 × 20	3-1/2 × 19-1/2
6 × 24	3-1/2 × 23-1/2

TABLE 13-4 *Standard dressed dimensions of sawn lumber.*

The joists need to have bridging, depending on their thickness-to-depth ratio. The bridging can be provided many ways, as indicated in table 13-5.

Depth/thickness, d/b	Lateral Support Requirement
d/b ≤ 2	No lateral support required
2 < d/b ≤ 4	End held in place by, e.g., blocking, bridging, hangers
4 < d/b ≤ 5	Top edge held in line over entire length by sheathing and ends held in place it prevent movement
5 < d/b ≤ 6	Blocking or bridging at 8 feet on center and ends held in place it prevent movement
6 < d/b ≤ 7	Both edges held in line over entire length and ends held in place it prevent movement

TABLE 13-5 *Wood beam bracing requirements.*

The design of wood joist can be simplifed with the use of span table for specific loads and joist spacings, as shown in the textbook table 13-1.

Example 13.3 Wood joists are used to span 14 feet with a desired spacing of 16 inches on center. What size of joist is required, assuming No. 1 and No. 2 grade Douglas Fir and bridging are provided? Use table 13-1 in the textbook for floor joist selection. Also determine the required bracing for the joist selected.

The first four rows of table 13-1 in the textbook are for Douglas Fir Larch (DFir-L), and the center three columns are for joists with bridging. The centermost of these three columns is for a spacing of 16 inches on center. Read down the column until the span is greater than 14 feet. From here. you will find 14'-4" as the maximum span for the 2 × 10 joist. Note that this table is limited to the loading and conditions listed in the footnotes.

The actual size of a 2 × 10 can be found in table 13-4, and the bracing requirements are given in table 13-5. For the 2 × 10, the d/b = 9.25 in/1.5 in = 6.17. This would indicate that both edges held in line over entire length, and ends held in place to prevent movement.

Many other structural members may be used to support a floor. Another common method for longer spans and heavier loads is the use of engineered wood floor sysytems. In the textbook, engineered wood I-joist are shown in the figures 13-13 and 13-14, and open web engineered wood framing in figure 13-15. Another type of engineered wood structural member is laminated veneered lumber (LVL). These are fabricated with thin veneers of wood glued together and arranged with their grain all in the same direction. This provides a member that is approximately 2 to 3 times stronger than an eqiuvalent dimension lumber member. Figure 13-4 shows a couple sizes and types of laminated veneer lumber.

FIGURE 13-4 *Laminated veneer lumber.*

The manufacturers of engineered wood members provide span tables for various uses. In the textbook, table 13-2 lists uniform load capacity for floor loading of 1-1/2 inch wide laminated veneer lumber in pounds per foot (plf). Values are given in the table for both live and total load. There are two values listed in the table for live load. These are the loads that produce deflections of L/480 and L/360.

Example 13.4   A header is needed to span a distance of 12 feet and supports a floor load. The total load is 200 plf, and the live load is 150 plf. The live load deflection should be limed to L/360. Select a laminated veneer member from table 13-2 in the textbook.

The first column lists the span in feet. Reading across the 12-foot row for a total load greater than 200 plf, we find that a 1-1/2" × 9-1/4" carries 204 plf, but the live load capacity at L/360 is only 139 plf. Continue across the row to find that a 1-1/2" × 9-1/2" carries a total load of 221 plf, and a live load of 150 plf at L/360. Therefore, select a 1-1/2" × 9-1/2" 1.75E LSL.

Before you try and design an entire residential floor system, read chapter 14 on loads. This will help in determining where loads come from and how loads are applied to the structural members.

Basic Wall Framing Systems for Residential Construction

Residential wall framing is usually constructed with dimensional wood framing and sheathing to provide lateral stability to the frame. In recent years, light-gauge metal studs and track have been used to frame residential construction. In some cases, masonry or concrete wall construction may be used in residential construction.

The sheathing used to cover wood-framed walls is normally plywood, oriented strand board, gypsum wall board, or other approved materials. The design and selection of the sheathing are similar to wall or roof sheathing. The basic components of a wood-framed wall system are shown in figure 13-5.

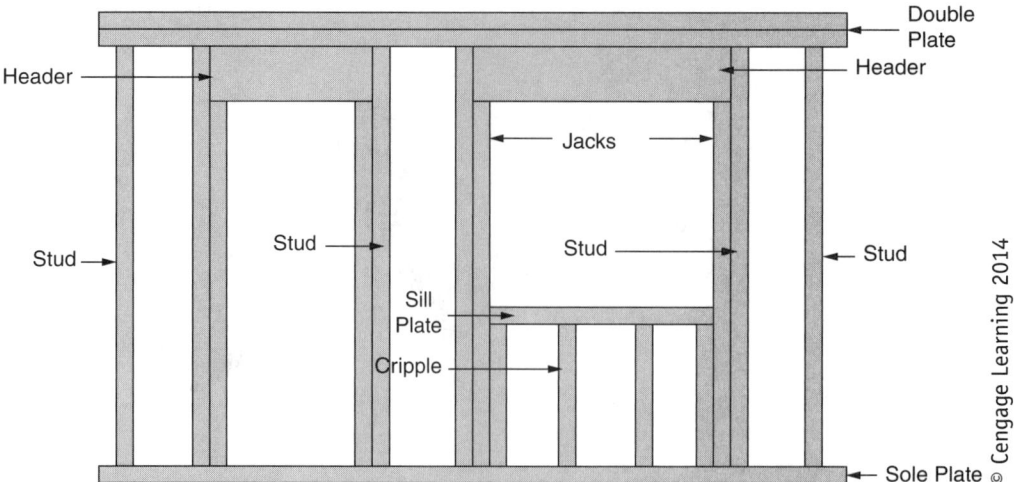

FIGURE 13-5 *Wood framed wall system.*

Vertical studs span the wall one level to another and are normally 2 × 4 or 2 × 6 in size with spacings of 12, 16, or 24 inches on center, based on the loading requirements. When an opening is need in the wall, a header is used to span the opening and is supported on additional studs called *jacks*. Under windows, shorter studs, called *cripples*, are used to infill vertically. The horizontal elements of the wall begin with the sole plate at the bottom. This is used to provide attachment to the floor or foundation and should be pressure treated if it is in contact with concrete or within 8 inches of the ground. The top plate consists of two members to provide continuity and for attachment to adjacent walls. Under wall openings like windows, a sill plate is supported by cripples and jacks.

The design of the wall studs depends on the magnitude and type of loading on the member. Approximate size and spacing can be found using table 13-6. This table should only be used for preliminary calculations.

Stud Size (inches)	Bearing Walls				Nonbearing Walls	
	Laterally Unsupported Height (feet)	Supporting Roof and Ceiling Only	Supporting One Floor, Roof and Ceiling	Supporting Two Floors, Roof and Ceiling	Laterally Unsupported Height (feet)	Spacing (inches)
		Spacing (inches)				
2 × 3	—	—	—	—	10	16
2 × 4	10	24	16	—	14	24
3 × 4	10	24	24	16	14	24
2 × 5	10	24	24	—	16	24
2 × 6	10	24	24	16	20	24

TABLE 13-6 *Size, height, and spacing of wood.*

Example 13.5 What size and spacing of wood stud should be used in an exterior, load-bearing wall supporting a two-story residence? Assume the unsupported height of the wood stud is less than 10 feet.

In table 13-6, the first column indicated the stud size, while the next four columns are for bearing walls. The last column under bearing walls is for studs supporting two floors, roof, and ceiling. Reading down this column, we see that 16-inch spacing will work for either 3 × 4 or 2 × 6 studs. The 2 × 6 would be preferred because they provide a thicker wall for increased strength, stiffness, and thermal insulation.

Light-gauge metal stud framing is similar to wood framing. Many manufactures provide tables with product specifications to aid in the design process. This construction method will be further explained in commercial wall framing.

Masonry can be used in residential wall construction as either the load-bearing structure or a finish material support by a wood or metal stud wall. Table 13-7 gives approximate height to thickness limitations for masonry walls.

Construction	Maximum Height/Thickness, h/t
Bearing walls	
Solid units or fully grouted	20
Other than solid units or fully grouted	18
Nonbearing walls	
Exterior	18
Interior	36

TABLE 13-7 *Masonry wall height limitations.*

Example 13.6 What thickness of partially grouted hollow concrete masonry block should be used for a 12-foot-high load-bearing wall?

In table 13-6, the first column indicated the construction, which in this case is a bearing wall other than solid units or fully grouted. The maximum height-to-thickness ratio is given as 18.

$$\frac{h}{t} = \frac{12 \text{ ft}(12 \text{ in/ft})}{t} = 18 \quad \therefore \quad t = \frac{144 \text{ in}}{18} = 8 \text{ in}$$

A nominal 8-inch thick hollow concrete masonry block can be used.

Openings in masonry wall should be spanned with a lintel made of masonry, concrete, or structural steel. The load on the lintel will vary, depending on the pattern of the masonry wall construction and the load above the opening. When a running bond pattern is used for the wall, load in the masonry can arch laterally over an opening to the walls on either side. This will occur as long as the wall height above the lintel is at least one-half the lintel span. Figure 13-6 shows the distribution of the load on a lintel with an ample amount of wall above. Only the weight of the wall within the triangle above the window and the lintel weight must be supported by the lintel.

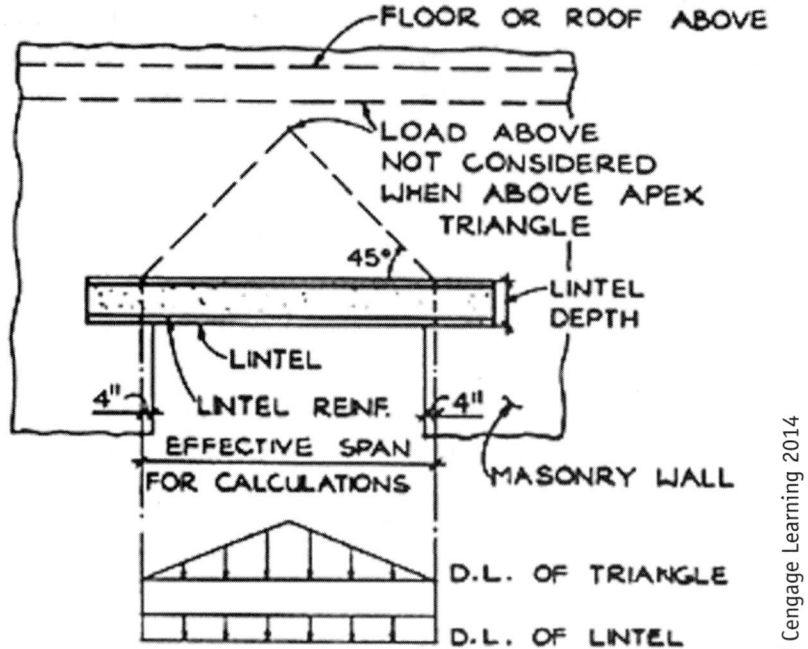

FIGURE 13-6 *Lintel loading.*

Basic Roof Framing for Residential Construction

The components of a residential roof framing system are similar to the floor framing with decking and support system. Some building materials used for decking may include plywood, oriented strand board, or composite materials. The thickness of the sheathing is governed by the roof loading, and the attachment is normally a function of the lateral loading on the building. A typical sheathing stamp with the span rating is shown in figure 13-2, and table 13-3 list maximum roof spans.

Example 13.7 What thickness of roof sheathing is required with roof trusses spaced at 24 inches on center? Assume that no edge support is provided for the sheathing.

The first column in table 13-3 lists the sheathing span rating. Recalling that the first number is the maximum roof span in inches, either the 24/0 or the 24/16 value may work. The fourth column lists the maximum roof span with no edge support. The 24/0 span rating has a maximum span of 20 inches, and the 24/16 rating has a maximum span of 24 inches. The required span is 24 inches; therefore, the 24/16 span rating selected. The second column lists the sheathing thickness as 7/16 inch.

The support system for residential roof framing can be dimensional lumber, engineered wood products, or engineered wood trusses. Dimensional lumber is usually 2 inches thick and spans between walls, beams, trusses, or other supporting members. There are two major differences in the design of roof members versus floor members. First is that the live load could be either roof load or snow. This will change the duration of the load and the corresponding design tables to be used. Second is that a roof is normally sloped. This will change the intensity of the dead loading on the roof and also require a change in the design table. Sloping load calculations are covered in Chapter 14. Many types of roof shape can be used, as shown in figure 13-24 of the textbook. To aid in the design of roof members, span tables are used. Table 13-3 in the textbook is for flat roof rafters made of Southern Yellow Pine and supporting snow load. The snow load for this table is 40 psf, and the dead load is 10 psf. In addition, the limitation of L/240 is for the live load only. Similar tables can be found for sloped roofs and other loading.

Example 13.8 It is desired to span flat roof rafters 16 feet, spaced at 16 inches on center. What size of No. 2 Southern Yellow Pine is required for a snow load of 40 psf and dead load of 10 psf?

Since the flat loading matches table 13-3 in the textbook, it may be used for selection. The third column under visually graded is for No. 2, and the second row for each size is for 16-inch spacing. Reading down the No. 2 column for 16-inch spacing, we find the first span larger than 16 feet if 17'-3" for a 2 × 10.

Engineered I-joist are becoming more popular as a roof framing system for a numer of reasons. Engineered I-joist are stronger and lighter than dimensioned lumber of equivalent depth. They are also straighter than dimensioned lumber. The basic components of the joist are top and bottom chords made of laminated veener lumber and a web made of oriented strand board. Some manufacturers may use dimensional lumber for the chords, and plywood for the web. The members can be made in various depth, and the chords also vary in size. Figure 13-7 shows three engineered I-joist. Design of these memebers is normally achieved using a table provided by the manufacturer.

FIGURE 13-7 *Engineered wood I-joist.*

Roof trusses can be used to span greater distance than dimensioned or engineered lumber. They can be made in many configuration as shown in figure 13-34 of the textbook. The members of the trusses are generally connected together with metal tooth plates as shown in figure 13-8.

FIGURE 13-8 *Metal truss plate connector.*

The basic compontent of a truss are shown in figure 13-9. For a simple span, the top chord is stressed in combined bending and compression. This chord must be braced by continuous sheathing to prevent lateral bucking. The bottom chord is in tension and should be braced at intervals sufficient to maintain stability and to keep the trusses in line during construction. The web could be in either tension or compression, based on the loading and the truss configuration.

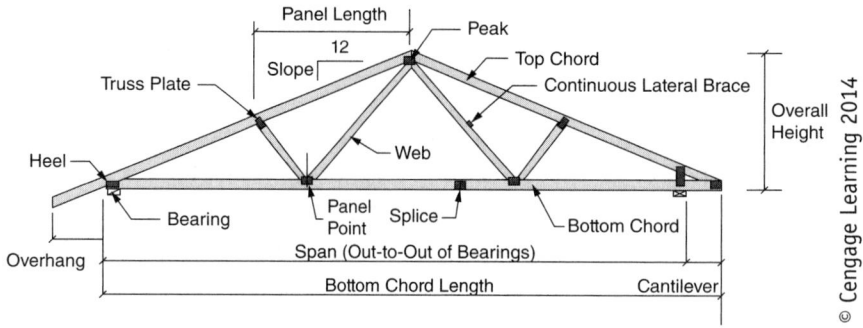

FIGURE 13-9 *Wood truss components.*

The design of wood trusses is normally performed by the manufacturer using their own proprietary software. Metal trusses are becoming more available but are primarily used in commercial construction.

Problem Set 13.1   Residential Construction

Problem 13-1   What should be the minimum depth for a shallow residential foundation in Chicago, Illinois?

Problem 13-2   Calculate the required square footing size for a total load of 125 kips. The footing bears on a soil with an allowable bearing pressure of 3,500 psf. The depth of the footing is 24 inches.

Problem 13-3   A footing carries a total load of 150 kips and bears on a soil with an allowable bearing pressure of 4,000 psf. If the width of the footing is limited to 5 feet, what is the minimum length of the footing? The depth of the footing is 24 inches.

Problem 13-4   What would be the allowable load, in pounds per foot, that a 24-inch-wide wall footing could support? The footing bears on a soil with an allowable bearing pressure of 2,500 psf. The depth of the footing is 24 inches.

Problem 13-5   What should be the minimum thickness of a hollow concrete masonry block foundation wall that supports 6 feet of backfill?

Problem 13-6   What grade of plywood veneer should not be used for exterior use?

Problem 13-7   What would be the maximum floor span for 1/2 inch plywood sheathing?

Problem 13-8   What type of beam bracing is required for a 2 × 8 floor joist?

Problem 13-9   Southern Pine 2 × 8 floor joists are used to span 12 feet. What are the possible combinations of joist spacing and bridging or strapping requirements that would be adequate?

Problem 13-10   What is the maximum span for a 1-1/2" × 11-1/4" − 1.75E LSL floor beam that supports a total load of 200 plf and a live load of 150 plf?

Problem 13-11   What wall framing members are used for support under door and window headers?

Problem 13-12   What wall framing members are used at the bottom to provide attachment to the floor?

Problem 13-13   What are the maximum height and spacing for 2 × 4s in a nonbearing wall?

Problem 13-14   What maximum span for 1/2 inch roof sheathing with edge support provided?

Problem 13-15   Southern Pine 2 × 10 roof rafters are used to span 18 feet. What are the possible combinations of rafter spacing and visual grade that would be adequate? Assume the snow load of 40 psf and dead load of 10 psf.

Problem 13-16   Why should the bottom chord of a simple span roof truss be laterally braced?

BACKGROUND

Commercial Construction

The basic systems for commercial construction are the same as residential construction, but they differ in type of materials for construction. Factors such as increased loads and span in commercial construction require more sophisticated engineered systems. The increased loads and span in commercial construction require more highly engineered systems.

Basic Foundations for Commercial Construction

Foundations for commerical buildings can be either shallow or deep, depending on the soil conditions and loads. If suitable stable soil is available near the ground surfaces, spread footings similar to residential construction may be appropriate. If the loads are large enough to require the footings to occupy more than approximately 25 percent of the plan area, or when the spread footings are fairly close to each other, a combined footing system may be appropriate. The three types of combined footings shown in figure 13-10 are rectangular, trapezoidal, and cantilever. The basic design concept for these footings is to size and configure the footing such that the soil pressure is uniformly distributed over each footing. This occurs when the location of the resultant applied loads, P_1 and P_2, corresponds with the resultant of the pressure load on the footings, R.

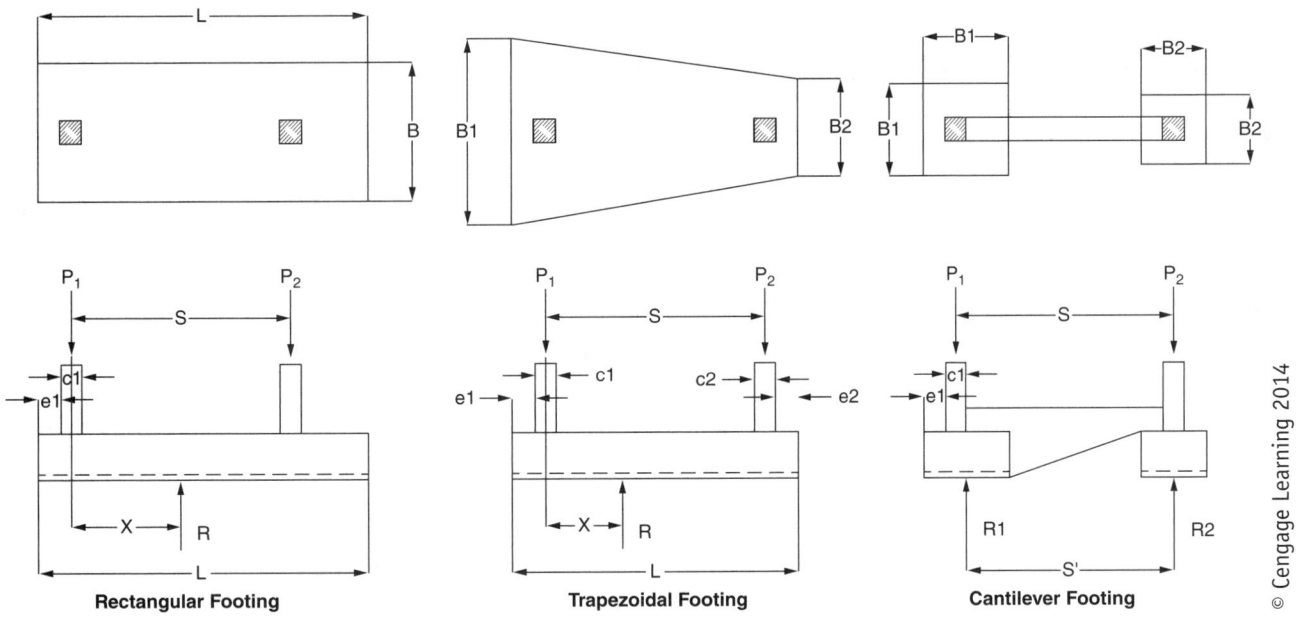

FIGURE 13-10 *Combined footings.*

The procedure for sizing a rectangular combined footing can be derived from the principles of static equilibrium and on the allowable soil bearing pressure. The three unknowns are the location of the resultant load, the length of the footing, and the width of the footing. The equations are as follows, with the varibles shown in figure 13-10:

$$x = \frac{P_2 s}{P_1 + P_2}$$

$$e_1 + \frac{c_1}{2} + x = \frac{L}{2} \quad \therefore \quad L = 2\left(e_1 + \frac{c_1}{2} + x\right)$$

$$B = \frac{P_1 + P_2}{L q}$$

where with q = Allowable soil bearing pressure

Example 13.9 Determine the length and width of a combined footing to support loads of $P_1 = 175$ kips and $P_2 = 350$ kips located 17 feet apart. The column width at load P_1 is $c_1 = 10$ inches and is located in $e_1 = 2$ inches from the edge of the footing. The allowable soil-bearing pressure is 2,500 psf. Round all sizes up to the nearest 3-inch increment.

$$x = \frac{P_2 s}{P_1 + P_2} = \frac{350 \text{ kips}(17 \text{ ft})}{175 \text{ kips} + 350 \text{ kips}} = 11.33 \text{ ft}$$

$$L = 2\left(e_1 + \frac{c_1}{2} + x\right) = 2\left(\frac{2 \text{ in}}{12 \text{ in/ft}} + \frac{10 \text{ in}}{2(12 \text{ in/ft})} + 11.33 \text{ ft}\right) = 23.83 \text{ ft}$$

Round the length of the footing up to the nearest 3-inch increment or 24 feet.

$$B = \frac{P_1 + P_2}{Lq} = \frac{175 \text{ kips} + 350 \text{ kips}}{24 \text{ ft}(2.5 \text{ ksf})} = 8.75 \text{ ft}$$

The size of the footing is then 24' - 0 long × 8' - 9" wide.

Other variations of shallow footing are given in figure 13-37 of the textbook. If the size of spread or combined footing occupys more than approximately one-half the plan area of the building, a raft or mat foundation may be used. This is shown in figure 13-7 of the textbook.

Deeper foundations are required when expansive or weak soils occur near the ground surface. Piers or piles are two types of deep foundations. These are shown in figure 13-38 of the texbook. Some design examples are covered in chapter 14.

Basic Floor Framing for Commercial Construction

The framing systems used for commerical floor construction typically span further and carry heavier loads than residential floor constructions. They are primarily steel and concrete framing systems. Design and analysis of these systems is sometimes complex, but basic rules of thumb can be used for approximate preliminary sizes. In lieu of other criteria, table 13-8 gives approximate depth for steel framing.

Member Type	Approximate Depth, d
Girders for typical framing	L/20
Beams for typical framing	L/24
Beams subjected to vibrating activities	L/16
Secondary purlins	L/32
Steel joist	L/24
Floor trusses	L/10

TABLE 13-8 *Approximate depth for steel floor framing.*

Another way to find approximate depths of a member is to find a size based on limiting the deflection of the member. The allowable deflection is typically based on the type of construction and loading that the member supports, as shown in table 13-9.

Construction	Live	Snow or Wind	Dead Plus Live
Roof Members			
Supporting plaster ceiling	L/360	L/360	L/240
Supporting non-plaster ceiling	L/240	L/240	L/180
Not supporting ceiling	L/180	L/180	L/120
Floor members	L/360	—	L/240

TABLE 13-9 *Deflection limitations.*

The design of structural steel framing is covered in chapter 14. One type of steel floor beam is a composite beam. This is where the concrete slab is connected to the steel beam to make them act as a single unit. The connection is normally made by a welded headed anchor stud attached to the top of the steel beam, as shown in figure 13-11. This can reduce the required depth and weight of the steel beam. Typically, the depth of a composite steel beam can be about 75 percent of the depth of a noncomposite steel beam.

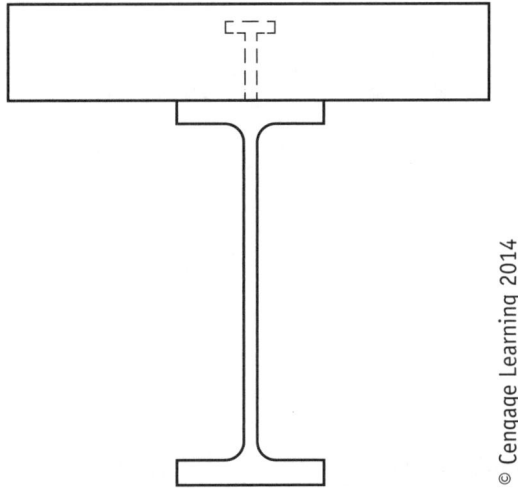

FIGURE 13-11 *Composite steel beam.*

Another common type of steel floor framing member is a steel joist. These are closely spaced prefabricated members that support a concrete floor system, normally with a metal deck. Steel joists are shown in figures 13-44 and 13-45 of the textbook. Figure 13-12 shows a standard load table for K-series steel joists. The standard K-series designation list the joist depth first in inches, followed by the joist series and a number indicating the chord size. The table lists the safe uniformly distributed total load and the live load that would produce a deflection in the joist of L/360. This table is for allowable stress design (ASD). If a deflection limit other than L/360 is needed, the second value can be factored to the desired value. For example, if a live load delfection limit of L/240 were needed, the second value in the table could be multiplied by 1.5 = 360/240. In addition, the approximate weight is given in the table if the most economical selection based on weight is desired.

Commercial Construction 275

STANDARD LOAD TABLE FOR OPEN WEB STEEL JOISTS, K-SERIES
Based on a 50 ksi Maximum Yield Strength - Loads Shown in Pounds per Linear Foot (plf)

Joist Designation	18K3	18K4	18K5	18K6	18K7	18K9	18K10	20K3	20K4	20K5	20K6	20K7	20K9	20K10	22K4	22K5	22K6	22K7	22K9	22K10	22K11
Depth (In.)	18	18	18	18	18	18	18	20	20	20	20	20	20	20	22	22	22	22	22	22	22
Approx. Wt. (lbs./ft.)	6.6	7.2	7.7	8.5	9	10.2	11.7	6.7	7.6	8.2	8.9	9.3	10.8	12.2	8	8.8	9.2	9.7	11.3	12.6	13.8
Span (ft.)																					
18	550 / 550	550 / 550	550 / 550	550 / 550	550 / 550	550 / 550	550 / 550														
19	514 / 494	550 / 523	550 / 523	550 / 523	550 / 523	550 / 523	550 / 523														
20	463 / 423	550 / 490	550 / 490	550 / 490	550 / 490	550 / 490	550 / 490	517 / 517	550 / 550	550 / 550	550 / 550	550 / 550	550 / 550	550 / 550							
21	420 / 364	506 / 426	550 / 460	550 / 460	550 / 460	550 / 460	550 / 460	468 / 453	550 / 520	550 / 520	550 / 520	550 / 520	550 / 520	550 / 520							
22	382 / 316	460 / 370	518 / 414	550 / 438	550 / 438	550 / 438	550 / 438	426 / 393	514 / 461	550 / 490	550 / 490	550 / 490	550 / 490	550 / 490	550 / 548	550 / 548	550 / 548	550 / 548	550 / 548	550 / 548	550 / 548
23	349 / 276	420 / 323	473 / 362	516 / 393	550 / 418	550 / 418	550 / 418	389 / 344	469 / 402	529 / 451	550 / 468	550 / 468	550 / 468	550 / 468	518 / 491	550 / 518	550 / 518	550 / 518	550 / 518	550 / 518	550 / 518
24	320 / 242	385 / 284	434 / 318	473 / 345	526 / 382	550 / 396	550 / 396	357 / 302	430 / 353	485 / 396	528 / 430	550 / 448	550 / 448	550 / 448	475 / 431	536 / 483	550 / 495	550 / 495	550 / 495	550 / 495	550 / 495
25	294 / 214	355 / 250	400 / 281	435 / 305	485 / 337	550 / 377	550 / 377	329 / 266	396 / 312	446 / 350	486 / 380	541 / 421	550 / 426	550 / 426	438 / 381	493 / 427	537 / 464	550 / 474	550 / 474	550 / 474	550 / 474
26	272 / 190	328 / 222	369 / 249	402 / 271	448 / 299	538 / 354	550 / 361	304 / 236	366 / 277	412 / 310	449 / 337	500 / 373	550 / 405	550 / 405	404 / 338	455 / 379	496 / 411	550 / 454	550 / 454	550 / 454	550 / 454
27	252 / 169	303 / 198	342 / 222	372 / 241	415 / 267	498 / 315	550 / 347	281 / 211	339 / 247	382 / 277	416 / 301	463 / 333	550 / 389	550 / 389	374 / 301	422 / 337	459 / 367	512 / 406	550 / 432	550 / 432	550 / 432
28	234 / 151	282 / 177	318 / 199	346 / 216	385 / 239	463 / 282	548 / 331	261 / 189	315 / 221	355 / 248	386 / 269	430 / 298	517 / 353	550 / 375	348 / 270	392 / 302	427 / 328	475 / 364	550 / 413	550 / 413	550 / 413
29	218 / 136	263 / 159	296 / 179	322 / 194	359 / 215	431 / 254	511 / 298	243 / 170	293 / 199	330 / 223	360 / 242	401 / 268	482 / 317	550 / 359	324 / 242	365 / 272	398 / 295	443 / 327	532 / 387	550 / 399	550 / 399
30	203 / 123	245 / 144	276 / 161	301 / 175	335 / 194	402 / 229	477 / 269	227 / 153	274 / 179	308 / 201	336 / 218	374 / 242	450 / 286	533 / 336	302 / 219	341 / 245	371 / 266	413 / 295	497 / 349	550 / 385	550 / 385
31	190 / 111	229 / 130	258 / 146	281 / 158	313 / 175	376 / 207	446 / 243	212 / 138	256 / 162	289 / 182	314 / 198	350 / 219	421 / 259	499 / 304	283 / 198	319 / 222	347 / 241	387 / 267	465 / 316	550 / 369	550 / 369
32	178 / 101	215 / 118	242 / 132	264 / 144	294 / 159	353 / 188	418 / 221	199 / 126	240 / 147	271 / 165	295 / 179	328 / 199	395 / 235	468 / 276	265 / 180	299 / 201	326 / 219	363 / 242	436 / 287	517 / 337	549 / 355
33	168 / 92	202 / 108	228 / 121	248 / 131	276 / 145	332 / 171	393 / 201	187 / 114	226 / 134	254 / 150	277 / 163	309 / 181	371 / 214	440 / 251	248 / 164	281 / 183	306 / 199	341 / 221	410 / 261	486 / 307	532 / 334
34	158 / 84	190 / 98	214 / 110	233 / 120	260 / 132	312 / 156	370 / 184	176 / 105	212 / 122	239 / 137	261 / 149	290 / 165	349 / 195	414 / 229	235 / 149	265 / 167	288 / 182	321 / 202	386 / 239	458 / 280	516 / 314
35	149 / 77	179 / 90	202 / 101	220 / 110	245 / 121	294 / 143	349 / 168	166 / 96	200 / 112	226 / 126	246 / 137	274 / 151	329 / 179	390 / 210	221 / 137	249 / 153	272 / 167	303 / 185	364 / 219	432 / 257	494 / 292
36	141 / 70	169 / 82	191 / 92	208 / 101	232 / 111	278 / 132	330 / 154	157 / 88	189 / 103	213 / 115	232 / 125	259 / 139	311 / 164	369 / 193	209 / 126	236 / 141	257 / 153	286 / 169	344 / 201	408 / 236	467 / 269
37								148 / 81	179 / 95	202 / 106	220 / 115	245 / 128	294 / 151	349 / 178	198 / 116	223 / 130	243 / 141	271 / 156	325 / 185	386 / 217	442 / 247
38								141 / 74	170 / 87	191 / 98	208 / 106	232 / 118	279 / 139	331 / 164	187 / 107	211 / 119	230 / 130	256 / 144	308 / 170	366 / 200	419 / 228
39								133 / 69	161 / 81	181 / 90	198 / 98	220 / 109	265 / 129	314 / 151	178 / 98	200 / 110	218 / 120	243 / 133	292 / 157	347 / 185	397 / 211
40								127 / 64	153 / 75	172 / 84	188 / 91	209 / 101	251 / 119	298 / 140	169 / 91	190 / 102	207 / 111	231 / 123	278 / 146	330 / 171	377 / 195
41															161 / 85	181 / 95	197 / 103	220 / 114	264 / 135	314 / 159	359 / 181
42															153 / 79	173 / 88	188 / 96	209 / 106	252 / 126	299 / 148	342 / 168
43															146 / 73	165 / 82	179 / 89	200 / 99	240 / 117	285 / 138	326 / 157
44															139 / 68	157 / 76	171 / 83	191 / 92	229 / 109	272 / 128	311 / 146

FIGURE 13-12 *Standard load table for open web steel joist, K-series.*

Example 13.10
Select the more economical 36 foot span steel floor joist to support a total load of 200 plf with a service live load of 100 plf. Limit the live load deflection to L/360, and use figure 13-12.

The left-hand column of the table gives the span in feet. Starting at 36 feet and proceeding to the right until a total load (top number) of 200 plf is exceeded and the live load at L/360 deflection (bottom number) of 100 plf is exceeded to find the best joist. The first joist that works is an 18K6, with values of 208 and 101. At the top of that column is the approximate weight of 8.5 lbs/ft. Continuing across the row, we find that both a 20K5 and 22K4 work. The table below summarizes the joists, their weight, and their load capacities. The 22K4 has the smallest weight and is therefore the most economical.

Joist Size	18K6	20K5	22K4
Weight	8.5	8.2	8.0
Total Load	208	213	209
L/360 Load	101	115	126

Concrete construction is more sensitive to deflection due to the brittle nature of the material. Table 13-10 lists approximate depths of concrete members that are not supporting construction likely to be damaged by a large deflection. The definition of a large deflection is the limit of the tolerable movement for fairly flexible or normal construction, such as metal stud wall construction with gypsum wall board. Any type of construction that is sensitive to deflection and will crack (such as masonry, glass, or concrete) is not considered flexible.

Member	Simple Span	Continuous End Span	Continuous Interior Span	Cantilever
Solid one-way slabs	L/20	L/24	L/28	L/10
Beams or ribbed one-way slabs	L/16	L/18.5	L/21	L/8

TABLE 13-10 *Minimum depth for concrete floor framing.*

Concrete, as well as timber, exhibits a behavior known as *creep*. Creep occurs when the material will continue to deform under a constant state of stress. The amount of creep deflection in concrete is dependent on many variables but can be as much as one to three times the initial elastic deflection. To account for this concrete deflection should not only be limited to a limit value based on the live load but should be based on a sustained load deflection. The sustained load is the dead load plus a certain percentage of the live load that is semi-permanent. The allowable deflection for concrete is shown in table 13-11.

Construction	Live Load	Sustained Load
Flat roofs not supporting construction not likely to be damaged by large deflections	L/180	L/240
Floors not supporting construction not likely to be damaged by large deflections	L/360	L/240
Roof or floors supporting construction likely to be damaged by large deflections	—	L/480

TABLE 13-11 *Maximum deflection for concrete.*

Another type of concrete construction is prestressed concrete construction. In prestressed concrete steel tendons are stretched and place the concrete in compression prior to loading. This increases the load capacity in the member by delaying the load at which tension in the concrete occurs. This results in shallower and lighter members than ordinary cast-in-place members. Typically, the approximate depth of a prestressed member is about 70 percent of an ordinary reinforced concrete member. A few types of prestressed members are shown in figure 13-59 of the textbook.

Basic Wall Framing for Commercial Construction

Some of the the framing systems used in commercial construction are the same as in residential construction. The major difference is that lightgauge metal stud framing is the predominant system in commercial construction. Wood and masonry wall construction was covered in residential construction. Here we will focus on light gauge metal framing. These stud can be used as either non-load-bearing or load-bearing or as a curtain wall. Curtain walls only resist the lateral load on the wall due to wind or seismic action.

The standard steel stud designation list the stud depth first in inches, multiplied by 100, followed by the stud type, then the stud width in inches multiplied by 100, and finally the material thickness in mils, which are thousandths of an inch. For example, a 600S162-54 is a 6-inch-deep stud shape that is 1.625 inches wide with a thickness of 0.054 inches. Figure 13-13 shows the basic dimensions of a metal stud with the depth A' and the width B'.

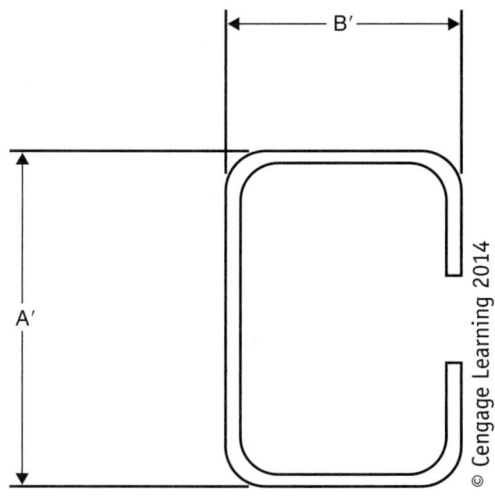

FIGURE 13-13   *Metal stud.*

Figure 13-14 is table for the allowable height of a curtain wall, based on the lateral pressure and the deflection limitation. In addition, it is further broken down into 12-, 16-, and 24- inch stud spacings. When the curtain wall supports brittle construction such as brick, the L/600 deflection limit should be used.

(S) Stud Member	Spacing (in) o.c.	5 psf L/120	5 psf L/240	5 psf L/360	15 psf L/240	15 psf L/360	15 psf L/600	20 psf L/240	20 psf L/360	20 psf L/600	25 psf L/240	25 psf L/360	25 psf L/600	30 psf L/240	30 psf L/360	30 psf L/600
600S137-33	12	34'7"	27'5"	24'0"	19'0"	16'7"	14'0"	17'3"	15'1"	12'9"	16'0"	14'0"	11'10"	15'1"	13'2"	11'1"
600S137-33	16	31'5"	24'11"	21'9"	17'3"	15'1"	12'9"	15'8"	13'8"	11'7"	14'7"	12'9"	10'9"	13'8"	12'0"	10'1"
600S137-33	24	27'5"	21'9"	19'0"	15'1"	13'2"	11'1"	13'8"	12'0"	10'1"	12'9"	11'1"	9'4"	12'0"	10'5"	8'10"
600S162-33	12	36'1"	28'7"	25'0"	19'10"	17'4"	14'7"	18'0"	15'9"	13'3"	16'9"	14'7"	12'4"	15'9"	13'9"	11'7"
600S162-33	16	32'9"	26'0"	22'8"	18'0"	15'9"	13'3"	16'4"	14'3"	12'0"	15'2"	13'3"	11'2"	14'3"	12'6"	10'6"
600S162-33	24	28'7"	22'8"	19'10"	15'9"	13'9"	11'7"	14'3"	12'6"	10'6"	13'3"	11'7"	9'9"	12'6"	10'11"	9'2"
600S200-33	12	37'9"	29'11"	26'2"	20'9"	18'2"	15'3"	18'10"	16'6"	13'11"	17'6"	15'3"	12'11"	16'6"	14'5"	12'1"
600S200-33	16	34'4"	27'3"	23'9"	18'10"	16'6"	13'11"	17'2"	14'11"	12'7"	15'11"	13'11"	11'8"	14'11"	13'1"	11'0"
600S200-33	24	29'11"	23'9"	20'9"	16'6"	14'5"	12'1"	14'11"	13'1"	11'0"	13'11"	12'1"	10'3"	13'1"	11'5"	9'7"
600S137-43	12	37'8"	29'10"	26'1"	20'8"	18'1"	15'3"	18'10"	16'5"	13'10"	17'5"	15'3"	12'10"	16'5"	14'4"	12'1"
600S137-43	16	34'2"	27'2"	23'8"	18'10"	16'5"	13'10"	17'1"	14'11"	12'7"	15'10"	13'10"	11'8"	14'11"	13'0"	11'0"
600S137-43	24	29'10"	23'8"	20'8"	16'5"	14'4"	12'1"	14'11"	13'0"	11'0"	13'10"	12'1"	10'2"	13'0"	11'4"	9'7"
600S162-43	12	39'3"	31'2"	27'3"	21'7"	18'10"	15'11"	19'7"	17'2"	14'5"	18'2"	15'11"	13'5"	17'2"	14'11"	12'7"
600S162-43	16	35'8"	28'4"	24'9"	19'7"	17'2"	14'5"	17'10"	15'7"	13'1"	16'6"	14'5"	12'2"	15'7"	13'7"	11'5"
600S162-43	24	31'2"	24'9"	21'7"	17'2"	14'11"	12'7"	15'7"	13'7"	11'5"	14'5"	12'7"	10'8"	13'7"	11'10"	10'0"
600S200-43	12	41'3"	32'9"	28'7"	22'8"	19'10"	16'8"	20'7"	18'0"	15'2"	19'1"	16'8"	14'1"	18'0"	15'9"	13'3"
600S200-43	16	37'6"	29'9"	26'0"	20'7"	18'0"	15'2"	18'9"	16'4"	13'9"	17'4"	15'2"	12'9"	16'4"	14'3"	12'0"
600S200-43	24	32'9"	26'0"	22'8"	18'0"	15'9"	13'3"	16'4"	14'3"	12'0"	15'2"	13'3"	11'2"	14'3"	12'6"	10'6"
600S137-54	12	40'4"	32'0"	28'0"	22'2"	19'5"	16'4"	20'2"	17'7"	14'10"	18'9"	16'4"	13'9"	17'7"	15'5"	13'0"
600S137-54	16	36'8"	29'1"	25'5"	20'2"	17'7"	14'10"	18'4"	16'0"	13'6"	17'0"	14'10"	12'6"	16'0"	14'0"	11'9"
600S137-54	24	32'0"	25'5"	22'2"	17'7"	15'5"	13'0"	16'0"	14'0"	11'9"	14'10"	13'0"	10'11"	14'0"	12'2"	10'3"
600S162-54	12	42'2"	33'5"	29'2"	23'2"	20'3"	17'1"	21'1"	18'5"	15'6"	19'6"	17'1"	14'5"	18'5"	16'1"	13'6"
600S162-54	16	38'3"	30'4"	26'6"	21'1"	18'5"	15'6"	19'1"	16'8"	14'1"	17'9"	15'6"	13'1"	16'8"	14'7"	12'3"
600S162-54	24	33'5"	26'6"	23'2"	18'5"	16'1"	13'6"	16'8"	14'7"	12'3"	15'6"	13'6"	11'5"	14'7"	12'9"	10'9"
600S200-54	12	44'3"	35'2"	30'8"	24'4"	21'3"	17'11"	22'1"	19'4"	16'3"	20'6"	17'11"	15'1"	19'4"	16'10"	14'3"
600S200-54	16	40'3"	31'11"	27'11"	22'1"	19'4"	16'3"	20'1"	17'7"	14'10"	18'8"	16'3"	13'9"	17'7"	15'4"	12'11"
600S200-54	24	35'2"	27'11"	24'4"	19'4"	16'10"	14'3"	17'7"	15'4"	12'11"	16'3"	14'3"	12'0"	15'4"	13'5"	11'3"
600S137-68	12	43'3"	34'4"	30'0"	23'9"	20'9"	17'6"	21'7"	18'10"	15'11"	20'1"	17'6"	14'9"	18'10"	16'6"	13'11"
600S137-68	16	39'4"	31'2"	27'3"	21'7"	18'10"	15'11"	19'8"	17'2"	14'5"	18'3"	15'11"	13'5"	17'2"	15'0"	12'7"
600S137-68	24	34'4"	27'3"	23'9"	18'10"	16'6"	13'11"	17'2"	15'0"	12'7"	15'11"	13'11"	11'9"	15'0"	13'1"	11'0"
600S162-68	12	45'2"	35'10"	31'4"	24'10"	21'8"	18'3"	22'7"	19'8"	16'7"	20'11"	18'3"	15'5"	19'8"	17'3"	14'6"
600S162-68	16	41'0"	32'7"	28'5"	22'7"	19'8"	16'7"	20'6"	17'11"	15'1"	19'0"	16'7"	14'0"	17'11"	15'8"	13'2"
600S162-68	24	35'10"	28'5"	24'10"	19'8"	17'3"	14'6"	17'11"	15'8"	13'2"	16'7"	14'6"	12'3"	15'8"	13'8"	11'6"
600S200-68	12	47'6"	37'8"	32'11"	26'2"	22'10"	19'3"	23'9"	20'9"	17'6"	22'0"	19'3"	16'3"	20'9"	18'1"	15'3"
600S200-68	16	43'2"	34'3"	29'11"	23'9"	20'9"	17'6"	21'7"	18'10"	15'10"	20'0"	17'6"	14'9"	18'10"	16'5"	13'10"
600S200-68	24	37'8"	29'11"	26'2"	20'9"	18'1"	15'3"	18'10"	16'5"	13'10"	17'6"	15'3"	12'10"	16'5"	14'4"	12'1"
600S137-97	12	47'10"	38'0"	33'2"	26'4"	23'0"	19'5"	23'11"	20'11"	17'7"	22'2"	19'5"	16'4"	20'11"	18'3"	15'4"
600S137-97	16	43'6"	34'6"	30'2"	23'11"	20'11"	17'7"	21'9"	19'0"	16'0"	20'2"	17'7"	14'11"	19'0"	16'7"	14'0"
600S137-97	24	38'0"	30'2"	26'4"	20'11"	18'3"	15'4"	19'0"	16'7"	14'0"	17'7"	15'4"	12'11"	16'7"	14'6"	12'2"
600S162-97	12	50'1"	39'9"	34'8"	27'6"	24'1"	20'3"	25'0"	21'10"	18'5"	23'3"	20'3"	17'1"	21'10"	19'1"	16'1"
600S162-97	16	45'6"	36'1"	31'6"	25'0"	21'10"	18'5"	22'9"	19'10"	16'9"	21'1"	18'5"	15'6"	19'10"	17'4"	14'7"
600S162-97	24	39'9"	31'6"	27'6"	21'10"	19'1"	16'1"	19'10"	17'4"	14'7"	18'5"	16'1"	13'7"	17'4"	15'2"	12'9"
600S200-97	12	52'9"	41'10"	36'7"	29'0"	25'4"	21'4"	26'4"	23'0"	19'5"	24'6"	21'4"	18'0"	23'0"	20'1"	16'11"
600S200-97	16	47'11"	38'0"	33'3"	26'4"	23'0"	19'5"	23'11"	20'11"	17'8"	22'3"	19'5"	16'4"	20'11"	18'3"	15'5"
600S200-97	24	41'10"	33'3"	29'0"	23'0"	20'1"	16'11"	20'11"	18'3"	15'5"	19'5"	16'11"	14'3"	18'3"	15'11"	13'5"

FIGURE 13-14   *Allowable height of curtain wall.*

Example 13.11 Select 6-inch metal studs placed at 16 inches on center for a curtain wall spanning 15 feet and resisting 30 psf of wind load. Use a deflection limit of L/360.

For 30 psf of wind pressure and a deflection limit of L/360, use the second to the last column on the right-hand side of the table in figure 13-14. For any given size of stud, there are three values listed for the allowable height. They correspond to 12-, 16-, and 24- inch stud spacing. The middle row is for a 16-inch stud spacing. Proceed down the column until you reach a height greater than or equal to 15 feet, which is 15'4". This is for a 600S200-54, and any other stud below that in the table would also work. This is a 6-inch-wide by 2-inch-wide stud with a thickness of 0.054 inches.

Finally, the minimum material thickness is required to be 95 percent of the design thickness, and the gauge number is often used to refer to the thickness. These values are listed in table 13-12.

Minimum Thickness, mils	Design Thickness, in	Reference Gauge
18	0.0188	25
27	0.0283	22
30	0.0312	20 – drywall
33	0.0346	20 – structural
43	0.0451	18
54	0.0566	16
68	0.0713	14
97	0.1017	12

TABLE 13-12 *Thickness of light gauge steel.*

Basic Roof Framing for Commercial Construction

The framing systems used for commerical roof construction typically span further and carry heavier loads than residential roof construction. They can be timber, steel, concrete, or other framing systems. They also can be divided into one-way and two-way system. Often, the span will be much longer than floor framing systems. This can lead to long span behavior, where the design of the system is not just dependent on the loads. Table 13-13 gives typical span range, typical depths, and approximate depth-to-span ratios for one-way systems.

System	Typical Spans ft	Typical Depths, ft	Approximate Depth-to-Span
Steel girders	10–72	2/3–3	L/20
Steel rigid frames	30–150	2–5	L/20–L/30
Glued laminated rigid frames	30–120	1 1/2 –4	L/20–L/30
Flat wood trusses	40–120	4–12	L/10
Pitched wood trusses	40–100	7–17	L/6
Flat steel trusses	40–300	4–30	L/10–L/12
Pitched steel trusses	40–150	5–20	L/6–L/8
Long span steel joist	25–96	1 1/2 –4	L/20–L/24
Deep long span steel joist	90–144	4–6	L/20–L/24
Steel joist girders	20–100	2–10	L/10–L/12
Glued laminated beams	10–60	1–4	L/24
Prestressed concrete single tee	20–120	1–4	L/20–L/30
Prestressed concrete double tee	20–60	1–2 1/2	L/20–L/30
Prestressed concrete girders	40–120	3–6	L/15–L/20
Steel arches	50–500	1–5	L/100
Concrete arches	40–320	1–7	L/50
Wood arches	50–240	1 1/2 –6	L/40

TABLE 13-13 *One-way systems.*

Table 13-14 gives typical span range, typical thickness, and approximate height-to-span ratios for two-way systems. These systems span in two perpendicular directions and are somewhat thinner than one-way systems.

System	Typical Spans ft	Typical Thickness, in	Approximate Height-to-span
Space Frames	80–220	—	L/15–L/25
Geodesic Domes	50–400	—	L/3–L/5
Thin Shell Domes	40–240	3–6	L/5–L/8
Hyperbolic Paraboloids	30–160	3–6	L/6–L/10
Barrel Vaults	30–180	3–5	L/10–L/15
Lamella Arches	40–150	—	L/4–L/6
Folded Plates	50–100	3–6	L/6–L/10
Suspended Cable Structures	50–450	—	L/8–L/15

TABLE 13-14 *Two-way systems.*

One of the most common one-way roof systems is the steel joist. In addition to the K-series joist discussed in commerical floor systems, they come in long span, LH-series and deep long span, DLH-series. These can span up to 96 feet and 144 feet, respectively. Some maunufacturers produce what are called *super long span SLH-series* that can span up to 240 feet. Figure 13-15 is a table for DLH-series joists.

STANDARD LOAD TABLE LONGSPAN STEEL JOISTS, DLH-SERIES
Based on a 50 ksi Maximum Yield Strength - Loads Shown in Pounds per Linear Foot (plf)

Joist Designation	Approx. Wt in Lbs. Per Linear Ft (Joists only)	Depth in inches	SAFELOAD* in Lbs. Between						CLEAR SPAN IN FEET								
			61-88	89	90	91	92	93	94	95	96	97	98	99	100	101	102	103	104
52DLH10	25	52	26700	298 / 171	291 / 165	285 / 159	279 / 154	273 / 150	267 / 145	261 / 140	256 / 136	251 / 132	246 / 128	241 / 124	236 / 120	231 / 116	227 / 114	223 / 110	218 / 107
52DLH11	26	52	29300	327 / 187	320 / 181	313 / 174	306 / 169	299 / 164	293 / 158	287 / 153	281 / 149	275 / 144	270 / 140	264 / 135	259 / 132	254 / 128	249 / 124	244 / 120	240 / 117
52DLH12	29	52	32700	365 / 204	357 / 197	349 / 191	342 / 185	334 / 179	327 / 173	320 / 168	314 / 163	307 / 158	301 / 153	295 / 149	289 / 144	284 / 140	278 / 135	273 / 132	268 / 128
52DLH13	34	52	39700	443 / 247	433 / 239	424 / 231	414 / 224	406 / 216	397 / 209	389 / 203	381 / 197	373 / 191	366 / 185	358 / 180	351 / 174	344 / 170	338 / 164	331 / 159	325 / 155
52DLH14	39	52	45400	507 / 276	497 / 266	486 / 258	476 / 249	466 / 242	457 / 234	447 / 227	438 / 220	430 / 213	421 / 207	413 / 201	405 / 194	397 / 189	390 / 184	382 / 178	375 / 173
52DLH15	42	52	51000	569 / 311	557 / 301	545 / 291	533 / 282	522 / 272	511 / 264	500 / 256	490 / 247	480 / 240	470 / 233	461 / 226	451 / 219	443 / 213	434 / 207	426 / 201	418 / 195
52DLH16	45	52	55000	614 / 346	601 / 335	588 / 324	575 / 314	563 / 304	551 / 294	540 / 285	528 / 276	518 / 267	507 / 260	497 / 252	487 / 245	478 / 237	468 / 230	459 / 224	451 / 217
52DLH17	52	52	63300	706 / 395	691 / 381	676 / 369	661 / 357	647 / 346	634 / 335	620 / 324	608 / 315	595 / 304	583 / 296	572 / 286	560 / 279	549 / 270	539 / 263	528 / 255	518 / 247
			66-96	97	98	99	100	101	102	103	104	105	106	107	108	109	110	111	112
56DLH11	26	56	28100	288 / 169	283 / 163	277 / 158	272 / 153	267 / 149	262 / 145	257 / 140	253 / 136	248 / 133	244 / 129	239 / 125	235 / 122	231 / 118	227 / 115	223 / 113	219 / 110
56DLH12	30	56	32300	331 / 184	324 / 178	318 / 173	312 / 168	306 / 163	300 / 158	295 / 153	289 / 150	284 / 145	278 / 141	273 / 137	268 / 133	263 / 130	259 / 126	254 / 123	249 / 119
56DLH13	34	56	39100	401 / 223	394 / 216	386 / 209	379 / 204	372 / 197	365 / 191	358 / 186	351 / 181	344 / 175	338 / 171	331 / 166	325 / 161	319 / 157	314 / 152	308 / 149	303 / 145
56DLH14	39	56	44200	453 / 249	444 / 242	435 / 234	427 / 228	419 / 221	411 / 214	403 / 209	396 / 202	388 / 196	381 / 190	375 / 186	368 / 181	361 / 175	355 / 171	349 / 167	343 / 162
56DLH15	42	56	50500	518 / 281	508 / 272	498 / 264	488 / 256	478 / 248	469 / 242	460 / 234	451 / 228	443 / 221	434 / 215	426 / 209	419 / 204	411 / 198	403 / 192	396 / 188	389 / 182
56DLH16	46	56	54500	559 / 313	548 / 304	537 / 294	526 / 285	516 / 277	506 / 269	496 / 262	487 / 254	478 / 247	469 / 240	460 / 233	452 / 227	444 / 214	436 / 209	428 / 204	420 / 204
56DLH17	51	56	62800	643 / 356	630 / 345	618 / 335	605 / 325	594 / 316	582 / 306	571 / 298	560 / 289	549 / 281	539 / 273	529 / 266	520 / 258	510 / 251	501 / 245	492 / 238	483 / 231

FIGURE 13-15 *Standard load table long span steel joist, DLH-series.*

Example 13.12 Select the more economical 100-foot-span steel roof joist to support a total load of 300 plf with a service live load of 150 plf. Limit the live load deflection to L/360, and use figure 13-15.

The top row of each section of the table gives the span in feet. Starting at 100 feet and proceeding downward in the 52-inch joist section until a total load (top number) of 300 plf is exceeded and the live load at L/360 deflection (bottom number) of 150 plf is exceed. The first joist that works is a 52DLH13 with values of 351 and 174. On the left side of that row by the size is the approximate weight of 34 lbs/ft. Moving to the 56-inch joist section and going down the 100-foot column, we find that a 56DLH12 has load capacities of 312 and 168, with an approximate weight of 30 plf. This would indicate that the 56DLH12 is more economical.

Problem Set 13.2   Commercial Construction

Problem 13-17 What are the three types of combined shallow footings that may be used when normal spot footings become close together or occupy more than 25 percent of the building plan area?

Problem 13-18 Determine the length and width of a rectangular combined footing to support loads of $P_1 = 200$ kips and $P_2 = 400$ kips located 20 feet apart. The column width at load P_1 is $c_1 = 12$ inches and is located $e_1 = 4$ inches from the edge of the footing. The allowable soil-bearing pressure is 2,500 psf. Round all sizes up to the nearest 3-inch increment. Refer to figure 13-10 for dimension of a rectangular combined footing.

Problem 13-19 What type of footing may be used when normal spot footings occupy more than 50 percent of the building plan area?

Problem 13-20 What would be the approximate depth needed for steel floor beams spanning 40 feet?

Problem 13-21 Select the more economical 32 foot span steel floor joist to support a total load of 320 plf with a service live load of 150 plf. Limit the live load deflection to L/360, and use figure 13-12.

Problem 13-22 What would be the approximate depth needed for interior continuous concrete floor beams spanning 40 feet?

Problem 13-23 What is the difference in the allowable sustained load deflection when it is supporting construction likely to be damaged by large deflection, versus when it is supporting construction not likely to be damaged by large deflection?

Problem 13-24 Select 6-inch metal studs paced at 16 inches on center for a curtain wall spanning 16 feet and resisting 20 psf of wind load. Use a deflection limit of L/360.

Problem 13-25 What are the design and minimum thicknesses of a 20 gauge structural steel stud?

Problem 13-26 What would be the approximate depth needed for flat steel roof trusses spanning 120 feet?

Problem 13-27 What would be the approximate height needed for a geodesic dome spanning 240 feet?

Problem 13-28 What size 56-inch DLH-series steel joist is required for a 110-foot-span steel roof joist to support a total load of 350 plf with a service live load of 200 plf? Limit the live load deflection to L/360, and use figure 13-15.

CHAPTER 14
Structural Systems: What Makes a Building Stand?

Skills List

After completing the problems in this chapter, you should be able to:

- Research and calculate gravity loads on structural members
- Solve static equilibrium problems on structural members
- Construct shear and bending moment diagrams on beams
- Calculate axial stresses and deformations on structural members
- Understand allowable strength design of structural steel
- Understand how the earth supports building structures

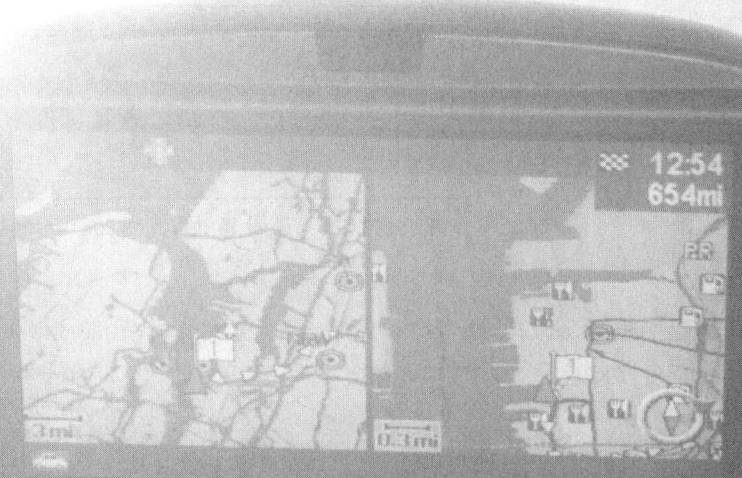

BACKGROUND

Architects and engineers must understand where loads come from and how a building reacts when loaded. The types of structural elements and the material they are made of will have an impact on how a building transfers forces. The following problems address the objectives stated in the textbook.

The structural design process introduced in the textbook required knowledge of how to determine loads on a building and configure members to safely transfer the forces created. This is not always a straightforward process and may require investigation of alternate solutions for a successful design.

Loads

Loads due to gravity normally act vertically. Dead, live, and snow loads are three examples of gravity loads. Loads due to wind and earthquake act in both the horizontal and the vertical direction. They primarily act horizontally and are called *lateral loads*. In general, a load has both a magnitude and a direction, but sometimes the direction is implied.

Dead and Live Loads

Dead loads are composed of materials fixed to the building at construction. These loads may be given by the manufacturer of a given component but are more commonly tabulated in referenced sources. Figure 14-5 in the textbook is a source for some dead loads and is used in the following examples.

Example 14.1 Calculate the total dead load in pounds per square foot that the floor trusses must support in figure 14-1.

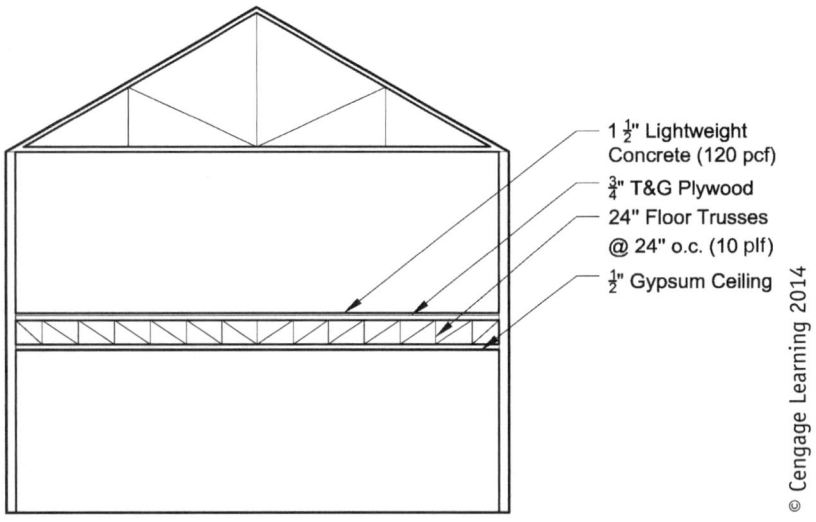

FIGURE 14-1 *Floor dead loads for example 14.1.*

Referring to the weights of common building materials in figure 14-5 in the textbook, materials and weights can be found for the system. Table 14-1 is an organized way to do the calculation. The plywood is tongue-and-groove (T&G), and the weight can be found using either plywood or sheathing.

284 Chapter 14: Structural Systems: What Makes a Building Stand?

Material	Calculation	Weight (psf)	Reference
1-1/2" Light wt. concrete	1.5 in (1 ft/12 in)(120 pcf)	15.0	given
3/4" T&G plywood	(3/4 in)(3 psf/in)	2.25	Text figure 14-5
24" Floor trusses @ 24" o.c.	(1 truss/2 ft)(10 lb/ft)	5.0	given
1/2" Gypsum ceiling	(2 psf/1/2 in)(1/2 in)	2.0	Text figure 14-5
Total floor dead load		24.25 psf	

TABLE 14-1 *Floor dead loads.*

To simplify calculations of forces on the structure, the load would commonly be rounded up to 25 psf.

Example 14.2 Calculate the total dead load in pounds per square foot that the roof trusses must support in figure 14-2. The load should be expressed on a horizontal plane.

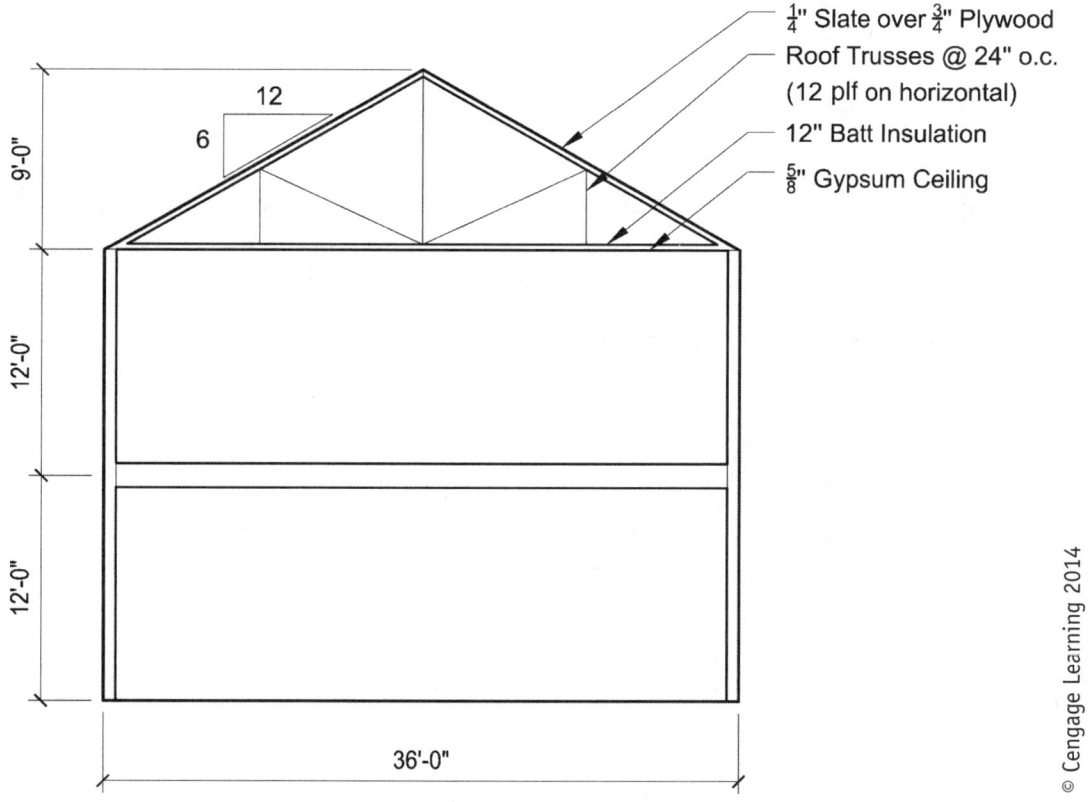

FIGURE 14-2 *Roof dead loads for example 14.2.*

Referring to the weights of common building materials in figure 14-5 material weights can be organized in table 14-2 for the system. First, the material that is on the slope surface must be determined, and then the load is converted to a horizontal plane.

Loads 285

Material	Calculation	Weight (psf)	Reference
1/4" Slate		9.5	Text figure 14-5
3/4" Plywood	(3/4 in)(3 psf/in)	2.25	Text figure 14-5
Total on sloped roof		11.75 psf	

TABLE 14-2 *Roof dead loads on slope.*

From trigonometry, the load on the horizontal plane is the load on the sloped plane divided the cosine of the angle between them. The rise over the run of the sloped plane is the tangent of the plane from the horizontal plane.

$$\tan\theta = \frac{rise}{run} = \frac{opposite}{adjacent} = \frac{9}{18} = \frac{6}{12} = slope$$

$$\theta = \tan^{-1}(9/18) = \tan^{-1}(6/12) = 26.57°$$

The 11.75 psf distributed load acts along the sloped surface and must be converted to a load on the horizontal plane. The horizontal length of one-half of the roof divided by the sloped length is the cosine of the angle; therefore, the load on the horizontal plane is the load on the sloped plane divided by the cosine of the angle. Another way to find this load is to multiply the sloped length by the distributed load, and then divide it by the horizontal length.

The converted loads can be added to the loads that exist in the horizontal plane in table 14-3.

Material	Calculation	Weight (psf)	Reference
Sloped dead load	(11.75 psf)/(cos 26.57°)	13.1	
Roof trusses @ 24" o.c.	(1 truss/2 ft)(12 lb/ft)	6.0	given
12" Batt insulation	(12 in)(0.4 psf/in)	4.8	Text figure 14-5
5/8" Gypsum ceiling	(5/8 in)(2 psf/ 1/2 in)	2.5	Text figure 14-5
Total Floor Dead Load		26.4 psf	

TABLE 14-3 *Roof dead loads on horizontal plane.*

To simplify calculations of forces on the structure, the load would commonly be rounded up to 27 psf.

As illustrated in the examples, the dead load may be expressed in many different ways, and sources for finding dead loads are numerous.

Live loads are due to occupants of a building and materials that are not fixed to the building during construction. The live loads are specified in codes like the International Building Code, The IBC live load requirements are given in figure 14-6 of the textbook and are used in the following examples. This live load may be reduced according to the type of structural member, and the area supported by the structural member. This will be illustrated in the following example:

Example 14.3 Calculate the design floor live load for the interior column supporting the second floor of the office building in figure 14-3. The column supports only the second floor, and no partitions are planned in the space.

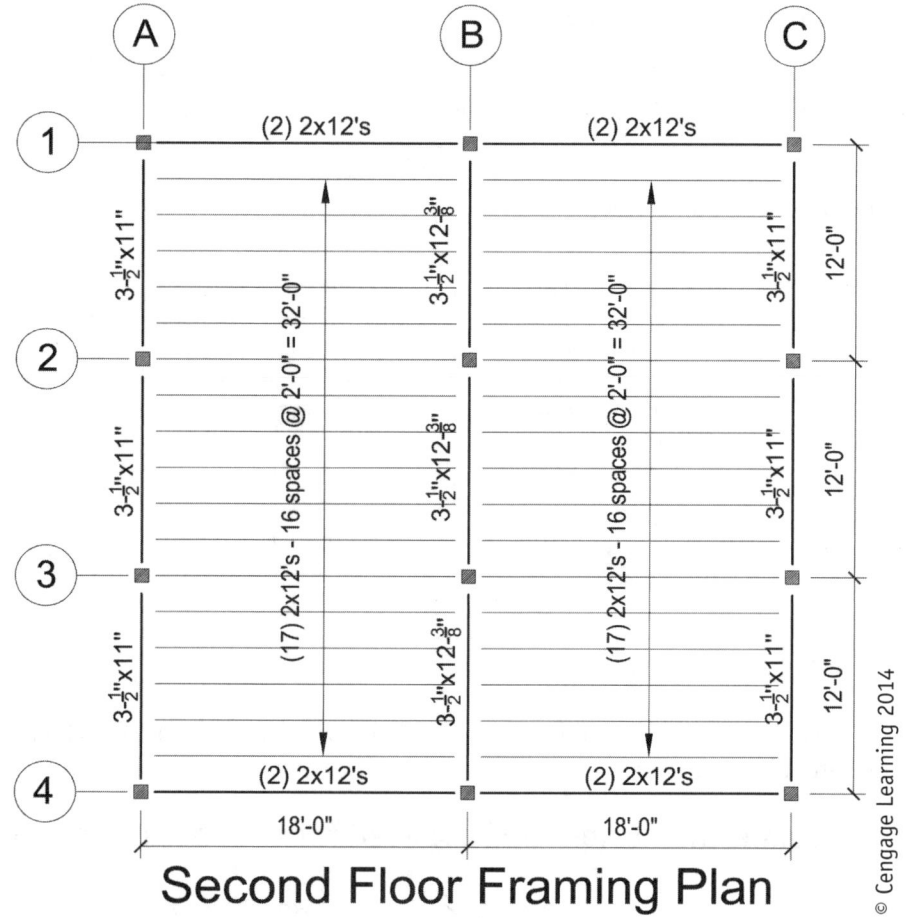

FIGURE 14-3 *Floor framing for example 14.3.*

The basic live load, L_o, is equal to 50 psf as given in figure 14-6 of the textbook. This can be reduced according to the following equation:

$$L = L_o\left(0.25 + \frac{15}{\sqrt{A_I}}\right)$$

Where A_I is equal to the tributary area, A_T supported by the member times K_{LL}. K_{LL} is specified in the IBC code as 1.0 for slabs, 2.0 for beams, and 4.0 for columns. The influence area must be greater than 400 ft² for the live load to be reduced. In addition, the minimum reduced live load is 0.5 L_o for members supporting only one floor, and 0.4 L_o for members supporting more than one floor.

$$A_T = (18 \text{ ft})(12 \text{ ft}) = 216 \text{ ft}^2$$

$$A_I = 4 A_T = (4)(216 \text{ ft}^2) = 864 \text{ ft}^2 > 400 \text{ ft}^2$$

Since the influence area is larger than 400 ft², live load reduction may be performed.

$$L = (50 \text{ psf})\left(0.25 + \frac{15}{\sqrt{864}}\right) = (50 \text{ psf})(0.7603) = 38.02 \text{ psf}$$

The maximum reduced live load is 0.5 L_o = (0.5)(50 psf) = 25 psf for members supporting one floor. This is less than the calculated live load reduction of 38.02 psf and therefore does not control in this case. The load on the column is then calculated as the design load multiplied by the tributary area.

$$P = LA_T = (38.02 \text{ psf})(216 \text{ ft}^2) = 8{,}212 \text{ lb} = 8.212 \text{ k}$$

The design live load on the column is 8.21 kips.

Live loading can also exist on a roof. The basic roof live load is 20 psf, for ordinary flat, pitched, and curved roofs per textbook table 14-6, and can be reduced as follows:

$$L_r = 20 \, R_1 \, R_2$$

The values of R_1 and R_2 are listed in table 14-4.

	1.0	$A_T \leq 200 \text{ ft}^2$
R_1	$1.2 - 0.001 A_T$	$200 < A_T < 600 \text{ ft}^2$
	0.6	$600 \leq A_T$
	1.0	$F \leq 4$
R_2	$1.2 - 0.05 F$	$4 < F < 12$
	0.6	$12 \leq F$

TABLE 14-4 *Roof live load reduction factors.*

A_T is the tributary area for the member, and F is the rise in inches per foot of run of the roof. The minimum roof live load is 12 psf after all reductions.

Example 14.4 Calculate the design roof live load for the roof trusses in figure 14-4.

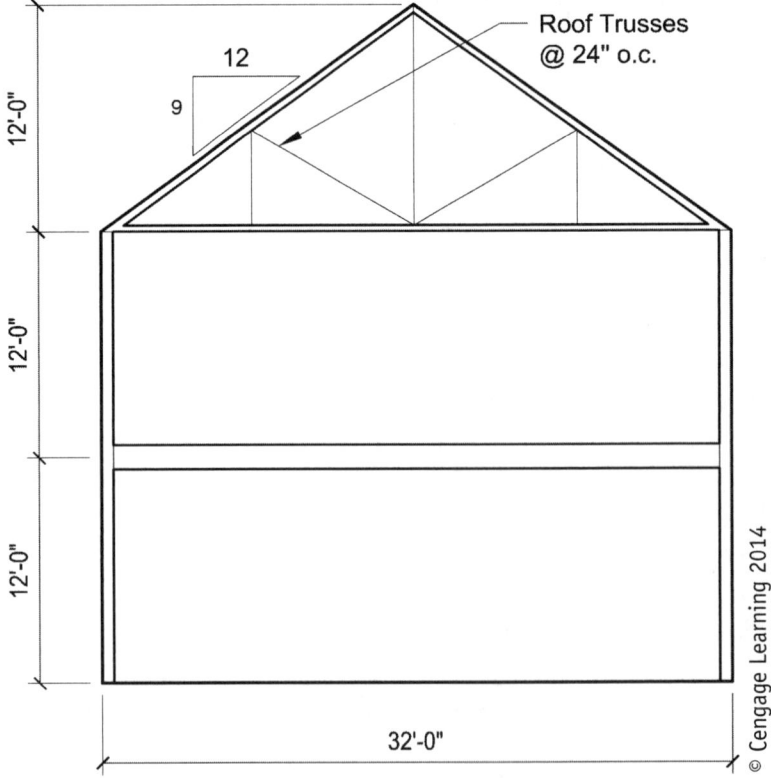

FIGURE 14-4 *Roof framing for example 14.4.*

The tributary area for the member is the span multiplied by the spacing, and F is 9 inches of rise per foot of run.

$R_1 = 1.0$ since $A_T = (2\text{ ft})(32\text{ ft}) = 64\text{ ft}^2 \leq 200\text{ ft}^2$

$R_2 = 1.2 - 0.05F = 1.2 - (0.05)(9) = 0.75$ for $4 < F < 12$

$$L_r = 20R_1R_2 = (20\text{ psf})(1.0)(0.75) = 15\text{ psf} > 12\text{ psf}$$

The design load, L_r is 15 psf, which is greater than the minimum of 12 psf. The distributed roof load on the truss would be design load multiplied by the 2 foot truss spacing, $w_r = (15\text{ psf})(2\text{ ft}) = 30\text{ plf}$.

Snow Load

The ground snow load is dependent on location as indicated in figure 14-9 of the textbook. The actual design snow load is also dependent on roof slope, wind exposure, thermal factors and the importance of the structure. The ASCE-7 formula for design snow load is as follows:

$$p_s = 0.7 C_s C_e C_t I_s p_g$$

p_s = design snow load
C_s = roof slope factor based on the roof slope and roofing materials (use 1.0 for flat roofs)
C_e = exposure factor based on the building's exposure to wind (table 14-5)

Terrain Category	Fully Exposed	Partially Exposed	Sheltered
B (urban)	0.9	1.0	1.2
C (suburban and rural)	0.9	1.0	1.1
D (coastal non-hurricane)	0.8	0.9	1.0
Above the mountain tree line	0.7	0.8	N/A
Alaska with no trees for 2 miles	0.7	0.8	N/A

TABLE 14-5 *Snow exposure factor, C_e.*

C_t = thermal factor (table 14-6)

Thermal Condition	C_t
Except noted below	1.0
Heated structures with cold vented attics	1.1
Unheated structures	1.2
Heated greenhouses	0.85

TABLE 14-6 *Snow thermal factor, C_t.*

I_s = occupancy importance factor (figures 14-10 and 14-11 of the textbook)
p_g = ground snow load (figure 14-9 of the textbook)

The minimum design snow load on a flat roof is $p_s = I_s p_g$ and where p_g exceeds 20 psf, $p_s = 20 I_s$.

Example 14.5 Calculate the design snow load for a flat roof on a police station in Kansas City, Missouri. Assume partial exposure to the wind.

The following factors apply to the design snow load equation.

C_s = 1.0 for flat roof

C_e = 1.0 for urban terrain and partially exposed (table 14-5)

C_t = 1.0 for not noted otherwise (table 14-6)

I_s = 1.2 for essential facility (police station, textbook figures 14-9 and 14-10)

p_g = 20 psf for Kansas City, Missouri (textbook figure 14-9)

$$P_s = 0.7\, C_s C_e C_t I_s p_g = (0.7)(1.0)(1.0)(1.0)(1.2)(20 \text{ psf}) = 16.8 \text{ psf}$$

The minimum design snow load on a flat roof is $p_s = I_s p_g = (1.2)(20 \text{ psf}) = 24$ psf, which controls in this case; therefore, the design snow load is 24 psf.

Lateral Loads

The building codes specify how to determine lateral forces due to wind and earthquakes, but the methods are complicated and beyond the scope of this workbook.

Load Combinations

Load combinations and the factors associated with the loads are a function of the probability that loads will exist together under full design values. The IBC code contains several sets of load-combination factors, depending on the design methodology used. One such set is as follows for what is known as *allowable stress design*.
Alternate Basic Load Combinations

$D + L + (L_r \text{ or } S \text{ or } R)$

$D + L + (\omega W)$

$D + L + \omega W + S/2$

$D + L + S + \omega W/2$

$D + L + S + E/1.4$

$0.9 D + E/1.4$

Where the basic loads are as follows:

D = dead load

L = live load

L_r = roof live load

W = wind load

ω = wind coefficient (if wind loads are calculated using Chapter 6 in ASCE 7, $\omega = 1.3$, else $\omega = 1$)

S = snow load

E = earthquake load

R = rain load

Example 14.6 Calculate the maximum and minimum design load resulting from each of the load combinations for a column with the following individual axial loads. The wind is per ASCE 7.

D = 20 kips

L = 30 kips

L_r = 10 kips

W = 10 kips

S = 8 kips

E = 15 kips

$D + L + (L_r \text{ or } S \text{ or } R) = 20 + 30 + 10 = 60$ kips OR $20 + 30 + 8 = 58$ kips

$D + L + (\omega W) = 20 + 30 + 1.3(10) = 63$ kips

$D + L + \omega W + S/2 = 20 + 30 + 1.3(10) + 8/2 = 67$ kips

$D + L + S + \omega W/2 = 20 + 30 + 8 + 1.3(10)/2 = 64.5$ kips

$D + L + S + E/1.4 = 20 + 30 + 8 + 15/1.4 = 68.7$ kips

$0.9D + E/1.4 = 0.9(20) + 15/1.4 = 28.7$ kips

The maximum load is 68.7 kips, and the minimum is 28.7 kips.

Tributary Width and Tributary Area

In example 14-3 on live load reduction, you were introduced to the concept of *tributary area*, that is, the area supported by an individual member. Tributary width is the dimension perpendicular to a horizontal member that has its load attributed to that member. One way to think of it is halfway to the next adjacent member. The uniformly distributed load on a horizontal member is the tributary width multiplied by the design load.

Example 14.7 Determine the uniformly distributed load on the 2 × 12 joists and the 3-1/2" × 12-3/4" beams in figure 14-3. The floor load is 75 psf, and no live load reduction will be included.

The tributary width of the joist is 2'-0", and the tributary width of the beam is 18'-0", as shown in figure 14-5.

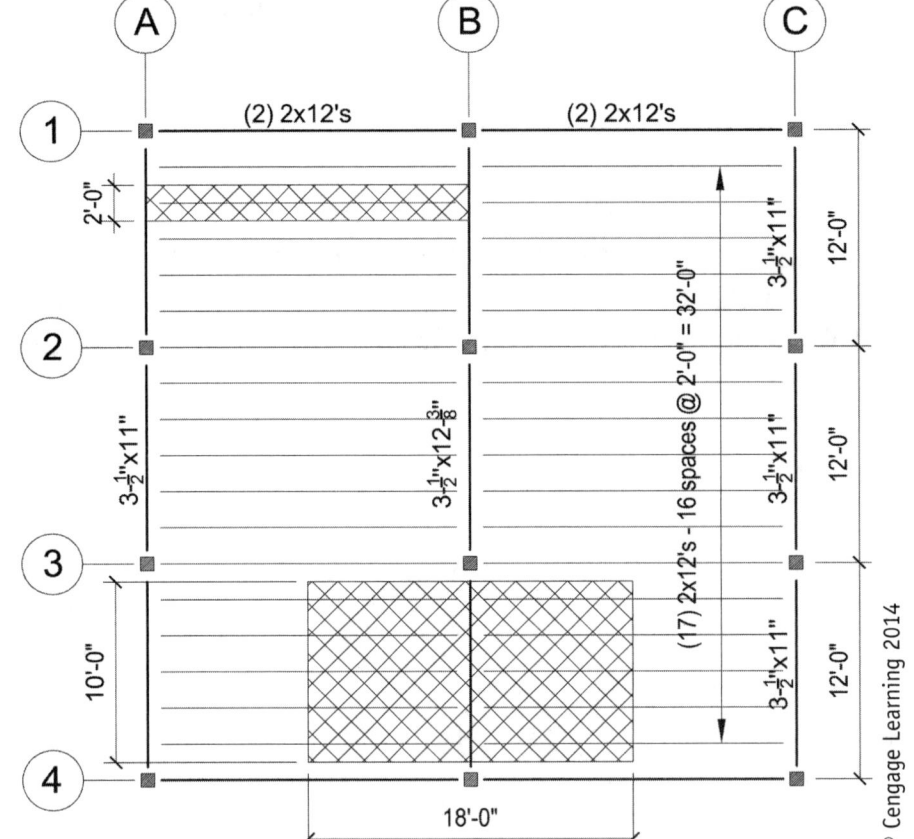

FIGURE 14-5 *Floor framing for example 14.7.*

The uniform load can be found from the floor load multiplied by the tributary width for each of the members.

$$w_{joist} = (75 \text{ psf})(2 \text{ ft}) = 150 \text{ lb/ft}$$
$$w_{beam} = (75 \text{ psf})(18 \text{ ft}) = 1350 \text{ lb/ft}$$

Load Path

Tributary width and tributary area are direct applications of load path for simple joist and beams. In more complicated structures, the load path may not be as direct and may required a more detailed method of analyzing the structure. One simple way to create a load path is to think of where the strongest part of the structure is, and the load will go in that direction.

Problem Set 14.1 Loads

Problem 14-1   Calculate the total dead load in pounds per square foot that the floor joists must support in figure 14-6.

Problem 14-2   Calculate the total dead load in pounds per square foot that the roof trusses must support in figure 14-6. The answer should be expressed on the horizontal plane.

Problem 14-3   Calculate the design floor live load in pounds per square foot for the 3-1/2" × 12-3/8" interior beam supporting the second floor of the office building in figure 14-3. There are no partitions planned in the space.

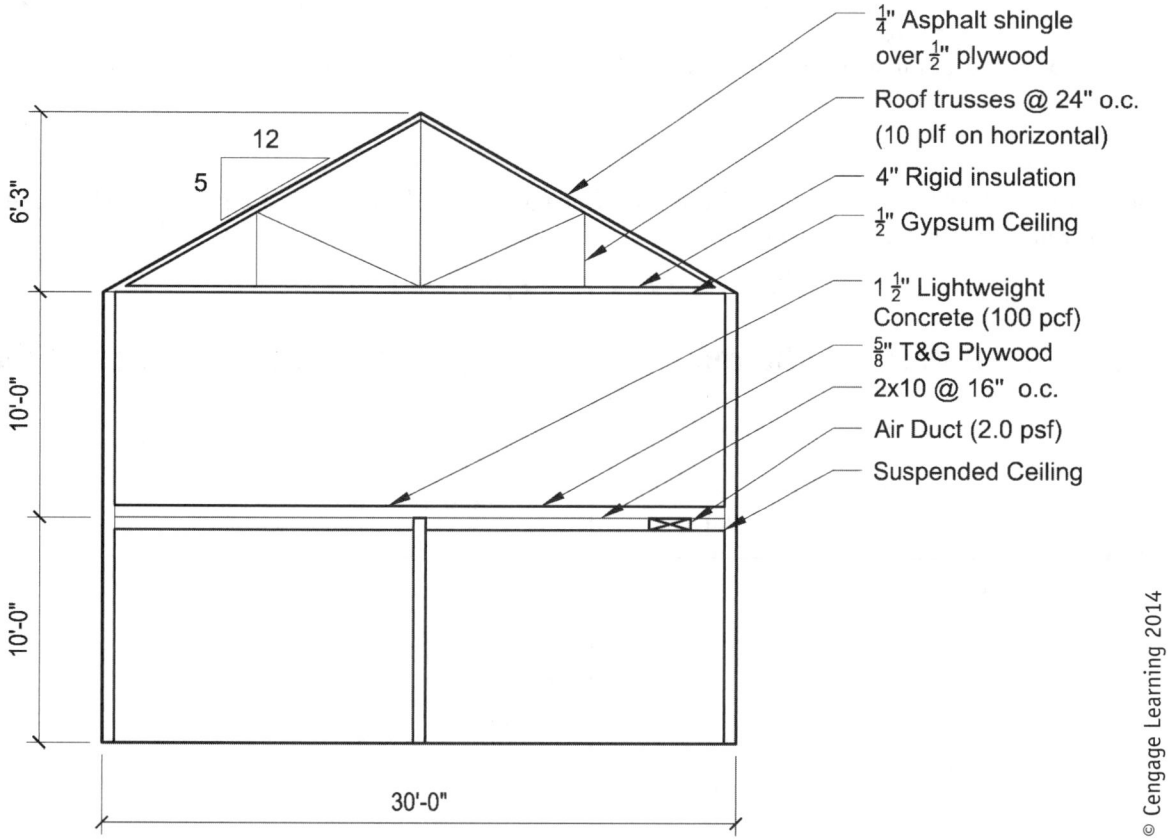

FIGURE 14-6 *Roof and floor framing for problems 14-1, 14-2, 14-7, 14-8, and 14-9.*

Problem 14-4   Calculate the design floor live load in pounds per square foot for the 2 × 12 joists supporting the second floor of the office building in figure 14-3. There are no partitions planned in the space.

Problem 14-5   How would the answer change in problem 14-3, if movable partitions were included?

Problem 14-6   What would be the minimum concentrated live load for the building in problem 14-3?

Problem 14-7   Calculate the design live load for the roof trusses in figure 14-6.

Problem 14-8   How would the answer in problem 14-7 change if the roof slope were changed to 4 inches of rise per foot of run?

Problem 14-9   Calculate the design snow load for the roof trusses in figure 14-6. The building is a residence in Omaha, Nebraska, with a ventilated attic insulated as shown.

Problem 14-10 Calculate the maximum and minimum design loads on a beam with the following individual uniform loads. $D = 25$ psf, $L_r = 20$ psf, $S = 25$ psf and $W = -10$ psf. Note that a negative load indicates an upward force.

Problem 14-11 Determine the uniformly distributed load in pounds per foot on the 3 1/2" × 11" beams in figure 14-3. The floor load is 75 psf, and no live load reduction will be included.

Problem 14-12 Would the answer to problem 14-11 change if live load reduction could be included? Why or why not?

BACKGROUND

Statics

Statics is the study of objects at rest and is commonly viewed as the cornerstone for all structural engineering analysis and design. One must first understand static (at rest) forces before comprehending dynamic (in motion) forces. External (applied) forces as well as internal forces within the structural member are considered in statics.

Forces

The sum of all of the forces in a system is called the *force resultant*. A resultant force has not only a magnitude, but also a direction.

$$F_R = F_1 + F_2 + F_3 + \cdots + F_n$$

For an object or particle to remain at rest, all of the externally applied forces must equal zero when added together. This is referred to as *static equilibrium*.

$$\Sigma F = 0 = F_1 + F_2 + F_3 + \cdots + F_n$$

Example 14.8 Determine the required force, F_4, for particle static equilibrium.

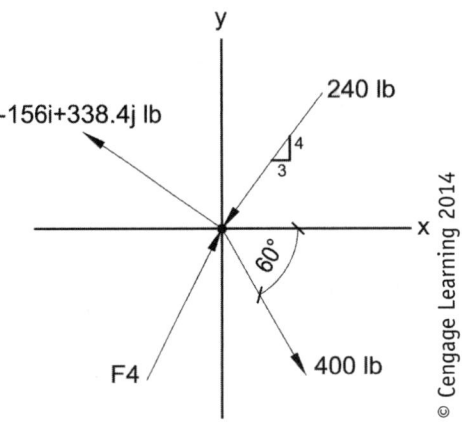

FIGURE 14-7 Particle FBD for example 14.8.

$$\rightarrow \Sigma F_x = 0 = -\frac{3}{5}(240) + 400 \cos 60° - 156 + F_{4x}$$

$$0 = -100 + F_{4x} \Rightarrow F_{4x} = 100 \text{ lb} \rightarrow$$

$$\uparrow \Sigma F_y = 0 = -\frac{4}{5}(240) - 400 \sin 60° + 338.4 + F_{4y}$$

$$0 = -200 + F_{4y} \Rightarrow F_{4y} = 200 \text{ lb} \uparrow$$

The magnitude of force F_4 is:

$$F_4 = \sqrt{F_{4x}^2 + F_{4y}^2} = \sqrt{100^2 + 200^2} = 223.61 \text{ lb}$$

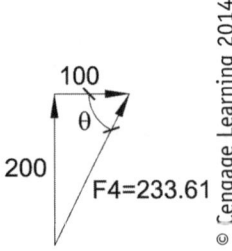

FIGURE 14-8 Force triangle for example 14.8.

$$\theta = \tan^{-1}\left(\frac{200}{100}\right) = 63.43°$$

Example 14.9 Determine the tension in each cable for equilibrium. Note that a cable can only carry a tension load.

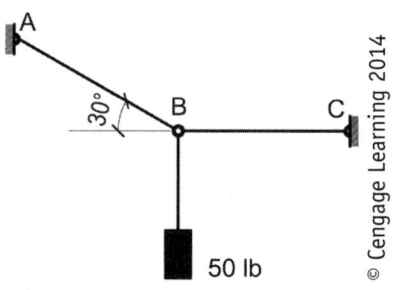

FIGURE 14-9 Cable structure for example 14.9.

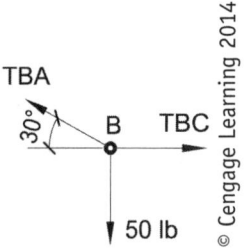

FIGURE 14-10 Particle FBD for example 14.9.

$$\uparrow \Sigma F_y = 0 = T_{BA} \sin 30 - 50$$

$$\frac{T_{BA} \sin 30}{\sin 30} = \frac{50}{\sin 30}$$

$$T_{BA} = 100 \text{ lb}$$

$$\rightarrow \Sigma F_x = 0 = -T_{BA} \cos 30° + T_{BC}$$

Substitute the value for T_{BA} into the equation to solve directly for T_{BC}

$$\frac{100 \cos 30}{\cos 30} = \frac{T_{BC}}{\cos 30}$$

$$T_{BC} = 86.60 \text{ lb}$$

Moments, or rotations, can be found about any point in space by taking the sum of all of the forces multiplied by their perpendicular distances from that point. These forces will create a positive or negative rotation, depending upon their direction in relation to the point of interest.

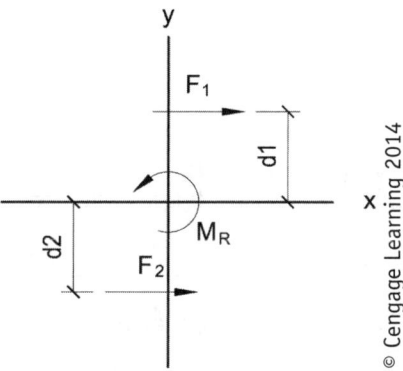

FIGURE 14-11 Moments and forces.

$$\circlearrowright M_R = -F_1 d_1 + F_2 d_2$$

Any applied moments are added directly to the moment equation.

Statics 295

Example 14.10 Determine the magnitude of moment required to place the system of forces in equilibrium about point O.

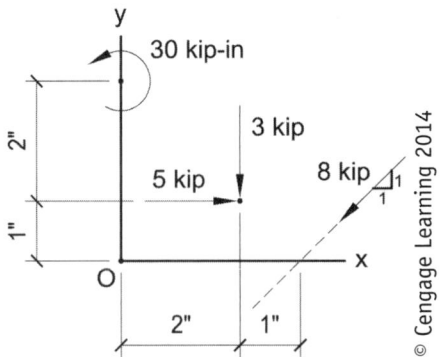

FIGURE 14-12 Moments and forces for example 14.10.

Sum moments about point O assuming a counter clockwise positive direction

$$\circlearrowleft \Sigma M_O = 30 - 5(1) - 3(2) - \left(\frac{1}{\sqrt{2}}\right)(8)(3) = 2.03 \text{ k-in}$$

An applied negative moment of 2.03 k-in (clockwise) is required for equilibrium of the system.

Free body diagrams

The free-body diagram is the most critical step in a statics problem solution. It is normally abbreviated as FBD. This diagram is a graphical representation of all forces acting on the structure and often contains the reactive support forces. Applied forces may consist of concentrated forces, distributed loadings, and rotational moments. A support reaction exists where the support is resisting a displacement or rotation. A list of these reactions and their corresponding support conditions are listed in figure 14-30 in the textbook.

Example 14.11 Draw the free-body diagram of the applied forces and support reactions for the simply supported beam.

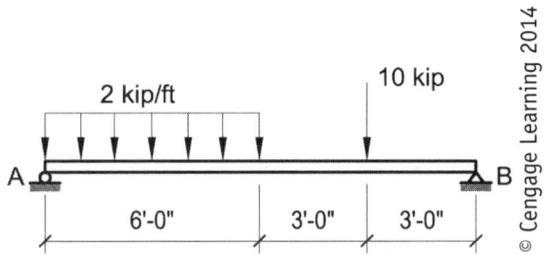

FIGURE 14-13 Beam for example 14.11.

Support A is a roller that yields one support reaction in the y direction, and support B is a pin with reactive forces in the x and y directions. Resolve the distributed load into an equivalent concentrated force. The total force is equal to area of the distributed load, and the resultant load should be located at the loading's centroid. For a rectangle, the centroid is at the midpoint.

$$F = \frac{2 \text{ kip}}{\text{ft}}(6 \text{ ft}) = 12 \text{ kip}$$

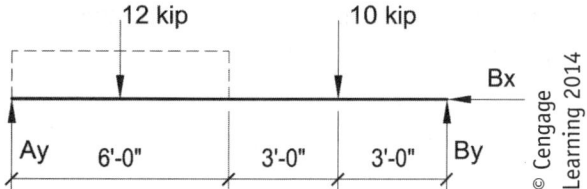

FIGURE 14-14 *Beam FBD for example 14.11.*

Example 14.12 Calculate the support reactions for the beam.

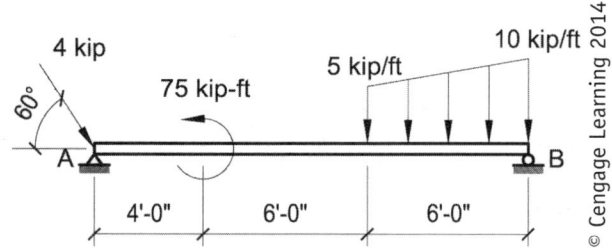

FIGURE 14-15 *Beam for example 14.12.*

Resolve the distributed load into an equivalent concentrated force by solving for the area of the load and applying a single force located at the loading's centroid. This is done separately for the rectangular and triangular portions of the distributed loads. The rectangular portion is done the same as example 14-11.

$$F = \frac{5 \text{ kip}}{\text{ft}}(6 \text{ ft}) = 30 \text{ kip}$$

The triangular portion is done in a similar manner to the rectangular portion. The area of a triangle is one-half the base multiplied by the height, and the centroid is one-third the distance from the wide end.

$$F = \left(\frac{1}{2}\right)\left(\frac{5 \text{ kip}}{\text{ft}}\right)(6 \text{ ft}) = 15 \text{ kip}$$

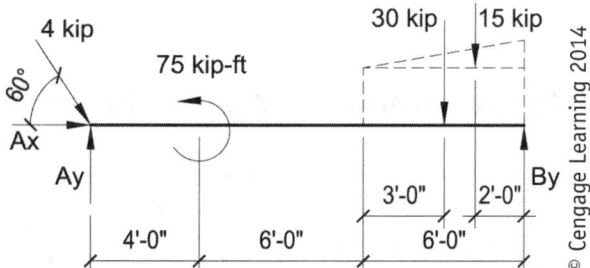

FIGURE 14-16 *Beam FBD for example 14.12.*

Apply equilibrium equations to solve three equations and three unknowns.

Solve the moment equation about point A to eliminate two unknowns, and solve directly for B_y. Assume counter clockwise as positive rotation.

$$\circlearrowleft \Sigma M_A = 0 = 75 - 30(13) - 15(14) + B_y(16)$$

$$\frac{16B_y}{16} = \frac{525}{16} \Rightarrow B_y = 32.81 \text{ kip} \uparrow$$

$$\uparrow \Sigma F_y = 0 = -4 \sin 60 - 30 - 15 + 32.81 + A_y$$

$$A_y = 15.65 \text{ kip} \uparrow$$

$$\rightarrow \Sigma F_x = 0 = 4 \cos 60 + A_x$$

$$A_x = -2 \text{ kip} \Rightarrow A_x = 2 \text{ kip} \leftarrow$$

Shear Force and Bending Moment

When designing a structure, it is important to understand the internal forces that are created from the externally applied loads. Internal forces can be calculated at any point along a structural member by taking a "cut" at the desired point of interest. A free-body diagram, FBD, of the structure up to the cut is then constructed. These forces consist of normal (axial) force, shear force, and bending moment. The positive sign conventions are shown in figure 14-17 and are dependent on which side of the member is considered. One critical point to keep in mind is that, unlike reactive forces, internal forces have a value but no direction.

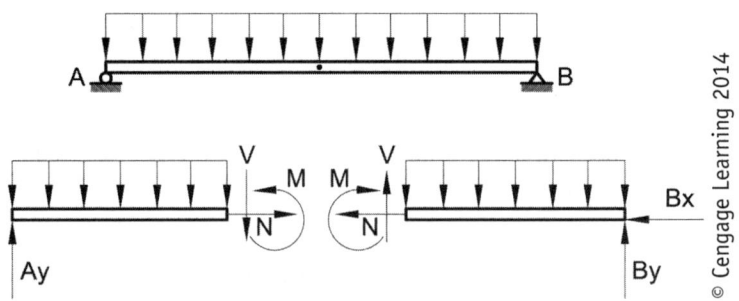

FIGURE 14-17 *Internal beam forces.*

Example 14.13 Calculate the internal forces for the beam at point C.

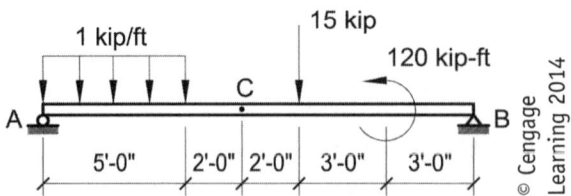

FIGURE 14-18 *Beam for example 14.13.*

Draw the free-body diagram of the entire beam, and then separate free-body diagrams of the left and right sides of the internal "cut."

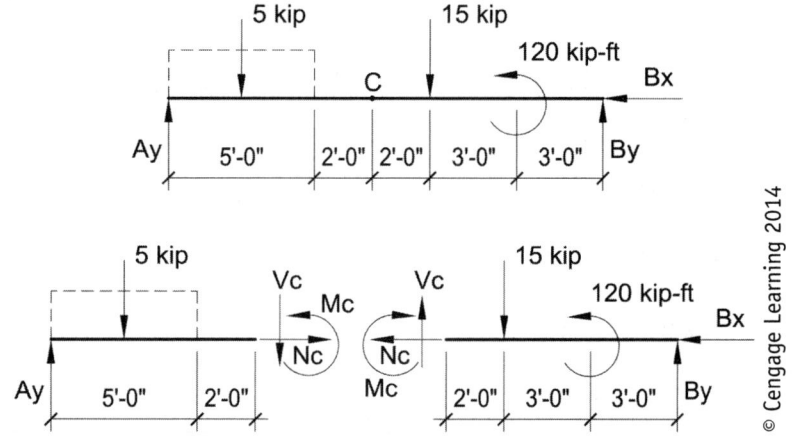

FIGURE 14-19 Beam FBDs for example 14.13.

Calculate the support reactions from the overall FBD by applying equations of equilibrium.

Assume counter clockwise rotation as positive:

$$\circlearrowleft \Sigma M_B = 0 = -A_y(15) + 5(12.5) + 15(6) + 120$$

$$A_y = 18.17 \text{ kip} \uparrow$$

$$\uparrow \Sigma F_y = 0 = 18.17 - 5 - 15 + B_y$$

$$B_y = 1.83 \text{ kip} \uparrow$$

$$\rightarrow \Sigma F_x = 0 = B_x \Rightarrow B_x = 0$$

Internal forces are now calculated from either the left or the right FBD. Using the left side FBD and applying equilibrium, utilize the moment equation about point C:

$$\circlearrowleft \Sigma M_C = 0 = -18.17(7) + 5(4.5) + M_c$$

$$M_c = 104.69 \text{ kip-ft}$$

$$\uparrow \Sigma F_y = 0 = 18.17 - 5 - V_c$$

$$V_c = 13.17 \text{ kip}$$

$$\rightarrow \Sigma F_x = 0 = N_c \Rightarrow N_c = 0$$

It should be noted that replacing distributed loads with equivalent concentrated force is acceptable for writing the statics equations only. This should not be done while constructing the shear and bending moment diagrams.

Shear and bending moment diagrams

Shear and bending moment diagrams are a graphical representation of the internal forces within a member. Once these graphs are plotted, the internal shear and moment are known at any point along the beam.

Example 14.14 Draw the shear and bending moment diagrams for the loaded beam below.

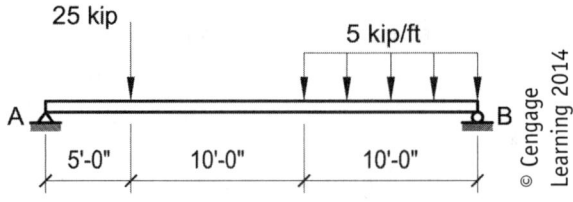

FIGURE 14-20 Beam for example 14.14.

Calculate the support reactions for the pin at A and roller at B.

Assume counter clockwise rotation as positive:

$$\circlearrowleft \Sigma M_A = 0 = -25(5) - 5(10)(20) + B_y(25)$$

$$B_y = 45 \text{ kip} \uparrow$$

$$\uparrow \Sigma F_y = 0 = A_y - 25 - 50 + 45$$

$$A_y = 30 \text{ kip} \uparrow$$

$$\rightarrow \Sigma F_x = 0 = A_x \Rightarrow A_x = 0$$

Now that the support reactions are known, the shear diagram may be drawn from the loading diagram. Two key points to remember are that the change in shear between two points is equal to the area of the load and that the slope of the shear diagram is equal to the value of the load. Beginning at zero, the shear diagram will "jump" at point A by an amount equal to the value of the vertical reaction, 30 kips. Once the 25 kip applied load is encountered, the shear diagram will decrease or jump down by the magnitude of the 25 kip load to the value of 5 kips. The shear diagram will remain constant here without a change in loading for the next ten feet. At the distributed load, the shear diagram will decrease linearly by an amount equal to the area of the distributed load, 50 kips, to 45 kips. At point B, the reaction is equal to 45 kips, which takes the shear value back to zero. Figure 14-21 shows the shear diagram in relation to the loading diagram.

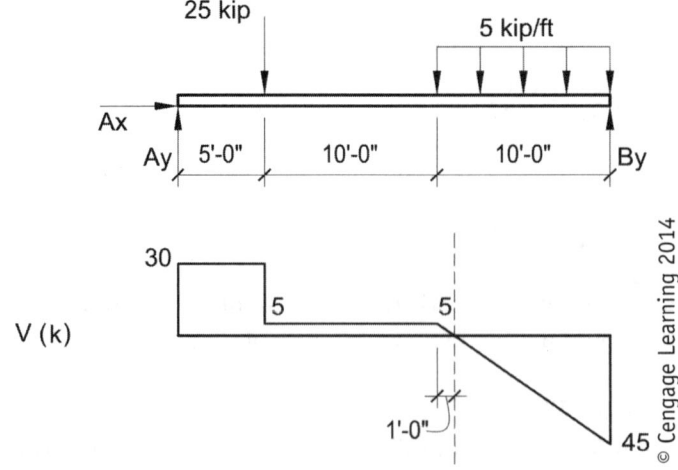

FIGURE 14-21 Loading and shear diagram for example 14.14.

The moment diagram can be constructed in a similar manner from the shear diagram with the change in the moment between two points equaling the area of the shear diagram, and the slope of the moment diagram is equal to the value of the shear. Since there are no applied moments, no "jumps" exist in the moment diagram.

Beginning at zero, the moment diagram will increase the value of the area of the shear diagram. Between zero and five feet, the shear diagram is constant at positive 30 kips. Therefore, the moment diagram will increase linearly at 30 kips/ft from zero to 150 kip-ft. The shear diagram is a constant 5 kip value from five to fifteen feet. The change in the moment is equal to the area of the shear. The moment will begin at 150 kip-ft and increase linearly at 5 kips/ft for a total increase of 50 kip-ft to 200 kip-ft. From here until the end of the beam, the shear diagram decreases linearly due to the distributed load. This shape in the shear diagram yields a curved shape in the moment diagram. The moment diagram will increase with the positive shear and then decrease with the negative portion of the shear diagram. The inflection point (point of zero shear) must be calculated in order to determine the areas of the shear diagram. This can be achieved by solving similar triangles.

$$\frac{50 \text{ kip}}{10 \text{ ft}} = \frac{5 \text{ kip}}{x} \Rightarrow x = 1 \text{ ft}$$

With the distance measured to the inflection point, the area can be calculated as 2.5 kip-ft. Beginning at 200 kip-ft, the moment diagram will curve up to a maximum point of 202.5 kip-ft. Now that the shear has become negative, the moment diagram decreases accordingly. The value of change is the area of the shear and is calculated to be 202.5 kip-ft. The moment diagram is now drawn back to zero. The direction of the curve in the moment diagram is directly related to the value of the shear diagram. The slope of the moment diagram is equal to the value of the shear diagram. If the shear diagram is zero, the moment diagram has a corresponding zero slope as shown in figure 14-22.

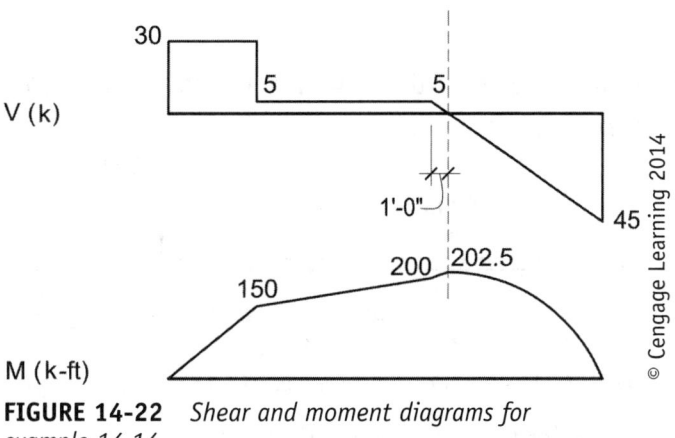

FIGURE 14-22 *Shear and moment diagrams for example 14.14.*

The use of software to aide in structural analysis and check results is prevalent in engineering practice. MDSolids© will solve statically determinate beam and plot shear and moment diagrams. The following figure shows the shear and moment diagrams from this software. You will only need to input the beam type, length, and loading.

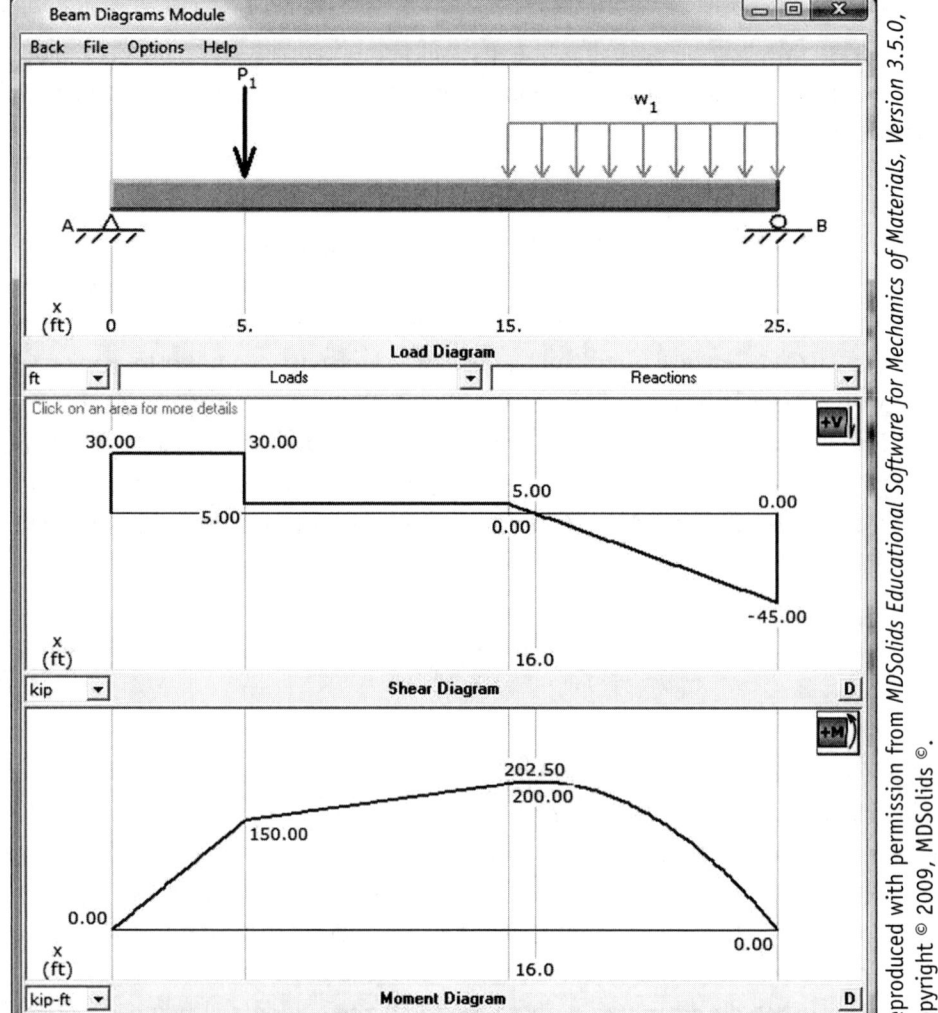

FIGURE 14-23 MDSolids© solution for example 14.14. Reproduced with permission from MDSolids©.

Example 14.15 Draw the shear and bending moment diagrams for the loaded beam below.

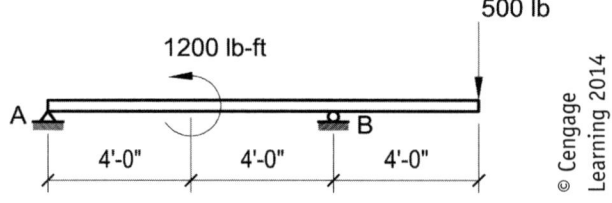

FIGURE 14-24 Beam for example 14.15.

Calculate the support reactions for the pin at A and roller at B.

Assuming counter clockwise rotation as positive:

$$\circlearrowleft \Sigma M_A = 0 = 1200 + B_y(8) - 500(12)$$

$$B_y = 600 \text{ lb} \uparrow$$

$$\uparrow \Sigma F_y = 0 = A_y + 600 - 500$$

$$A_y = -100 \Rightarrow 100 \text{ lb} \downarrow$$

$$\rightarrow \Sigma F_x = 0 = A_x \Rightarrow A_x = 0$$

With the reactive forces determined, the shear diagram begins at zero and drops to negative 100 lbs. The only change occurring in the loading diagram between the support reactions is an applied moment. This moment will not have any effect on the shear diagram but will play an important role in the moment diagram. With no other loadings present, the shear diagram remains constant to point B where the reaction takes the value from negative 100 lbs to a positive 500 lbs. The diagram will stay here to the end of the beam where the 500 lb concentrated load takes the diagram back to zero.

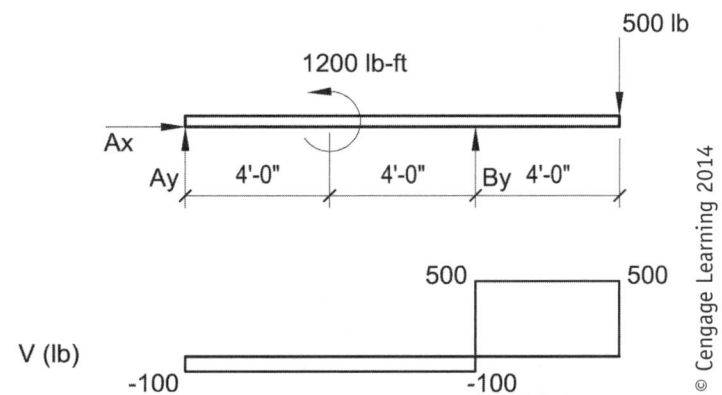

FIGURE 14-25 *Loading and shear diagrams for example 14.15.*

Beginning at zero, the moment diagram will decrease the value of the area of the shear diagram. Over the first eight feet, the shear diagram remains constant at negative 100 lbs. However, an applied moment of 1,200 lb-ft occurs at the midpoint of this load region and will result in a "jump" in the moment diagram. Plotting the moment based on the shear between 0 and 4 feet, a value of negative 400 lb-ft is obtained. Looking back on the loading diagram, the applied moment of 1,200 lb-ft is shown counter clockwise. When a moment is applied to a beam, it causes an opposite internal effect. Therefore, a "positive" or counter clockwise moment will create a negative drop in the moment diagram. The value of the diagram is now negative 1,600 lb-ft when the negative 1,200 lb-ft moment is added to the negative 400 value. Over the next 4 feet, the shear diagram remains constant at negative 100 lb, which results in a negative 400 lb-ft change to negative 2,000 lb-ft. From here to the end of the beam, the shear diagram is a constant 500 lb. The moment diagram will increase linearly at 500 lb/ft due to the positive shear diagram. The change is equal to the area that is 500 lb over a length of 4 feet. This results in an increase of 2,000 lb-ft that takes the moment diagram back to zero.

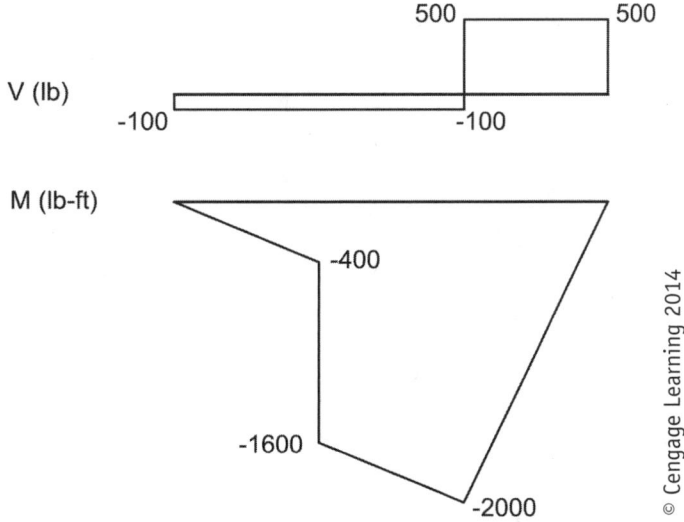

FIGURE 14-26 *Shear and moment diagrams for example 14.15.*

Statics 303

Problem Set 14.2 Statics

Problem 14-13 Determine the resultant of the force system and its direction measured clockwise from the positive x-axis.

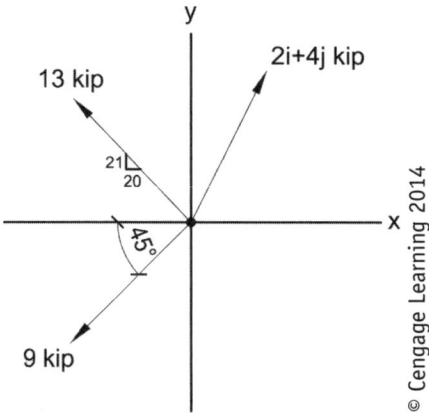

FIGURE 14-27 Particle FBD for problem 14-13.

Problem 14-14 Calculate the tensile force in each cable for equilibrium. Hint: You begin with a FBD at joint A.

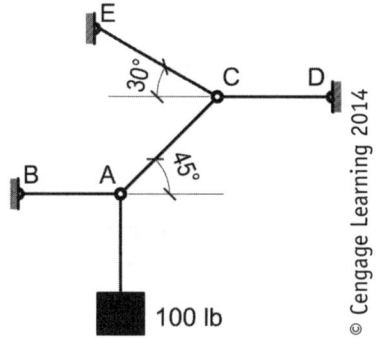

FIGURE 14-28 Cable structure for problem 14-14.

Problem 14-15 Resolve the forces into a resultant moment about point C.

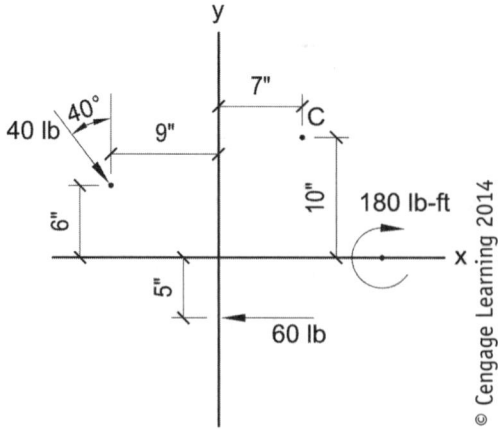

FIGURE 14-29 Force system for problem 14-15.

Problem 14-16   Draw the free-body diagram for the loading system below.

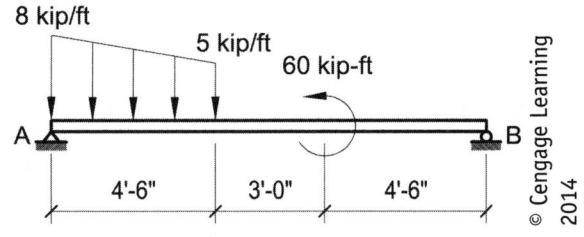

FIGURE 14-30   *Beam for problem 14-16.*

Problem 14-17   Draw the free-body diagram for the loading system below.

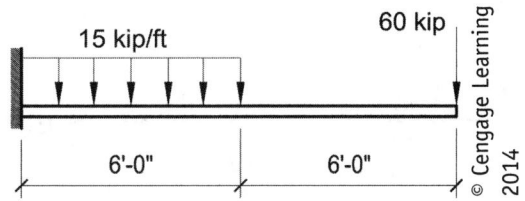

FIGURE 14-31   *Beam for problem 14-17.*

Problem 14-18   Draw the free-body diagram for the loading system below.

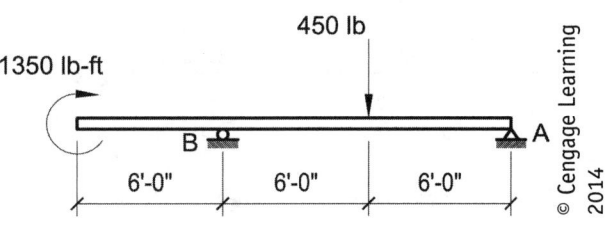

FIGURE 14-32   *Beam for problem 14-18.*

Problem 14-19   Calculate the support reactions for the beam.

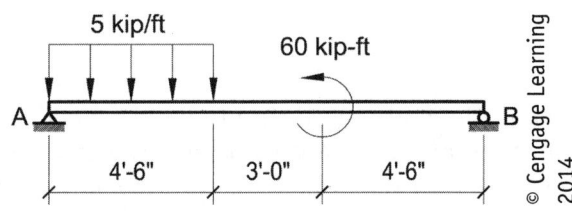

FIGURE 14-33   *Beam for problem 14.19.*

Problem 14-20   Calculate the internal shear, moment, and normal forces measured 8 feet from the left end of the beam.

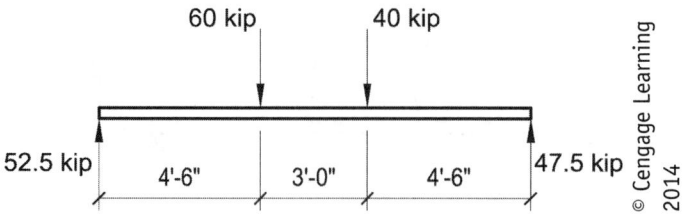

FIGURE 14-34   Beam for problem 14-20.

Problem 14-21   Draw the shear and bending moment diagrams for the loaded beam in figure 14-32.
Problem 14-22   Draw the shear and bending moment diagrams for the loaded beam in figure 14-31.
Problem 14-23   Draw the shear and bending moment diagrams for the loaded beam below.

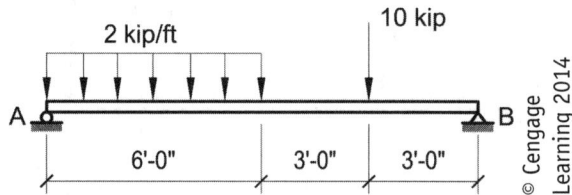

FIGURE 14-35   Beam for problem 14-23.

Problem 14-24   Draw the shear and bending moment diagrams for the loaded beam below.

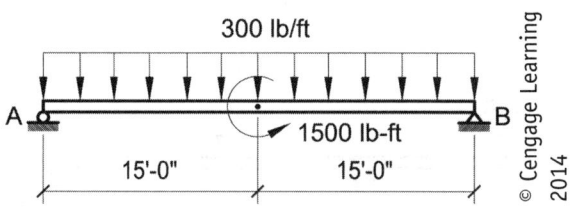

FIGURE 14-36   Beam for problem 14-24.

BACKGROUND

Mechanics

Stress

When a structural member is loaded along its length, and the force acts through the centroid of the cross-sectional area of the member, a uniform normal stress will result. This may be either a tension or a compression stress causing the member to elongate or shorten. The magnitude of the normal stress is equal to the force divided by the cross-sectional area of the member. The centroid is the center of gravity of a uniform material.

$$\sigma = \frac{P}{A}$$

where

σ = axial stress
P = applied force
A = resisting area

306   Chapter 14: Structural Systems: What Makes a Building Stand?

Example 14.16 A hollow pipe has an inside diameter of 4.0 in. and an outside diameter of 4.5 in. Determine the average compressive stress due to an axial force of 55 kips.

The area of a hollow pipe is π times the outer radius squared minus the inner radius squared. The inner and outer radii are shown in the figure below.

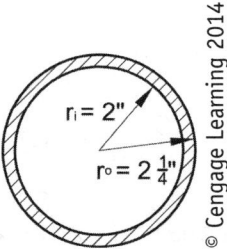

FIGURE 14-37 Pipe column for example 14.16.

$$A = \pi r_o^2 - \pi r_i^2 = \pi(r_o^2 - r_i^2) = \pi((2.25 \text{ in})^2 - (2.00 \text{ in})^2) = 3.338 \text{ in}^2$$

The compressive stress is the force divided by the area.

$$\sigma = \frac{P}{A} = \frac{55{,}000 \text{ lbs}}{3.338 \text{ in}^2} = 16{,}480 \text{ lb/in}^2$$

The resulting stress is normally given in kips/in² or ksi; therefore, the compressive stress is 16.5 ksi.

Example 14.17 A two-story column in a building is constructed of a hollow, square box section and shown in figure 14-38. The roof load is $P_1 = 80$ kip and the floor load at mid-height is $P_2 = 100$ kip. Calculate the compressive stresses in each story of the column.

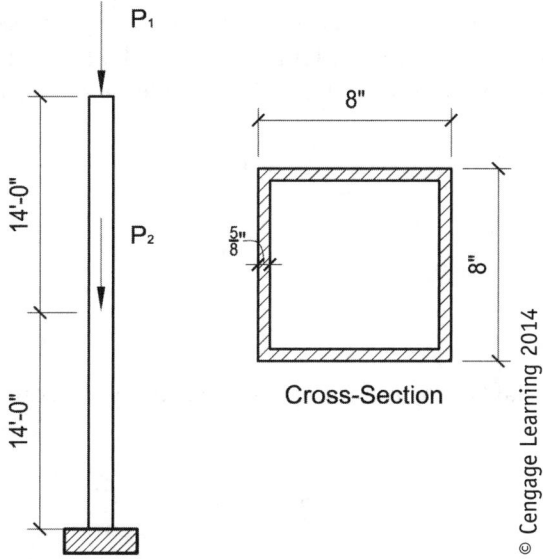

FIGURE 14-38 Tube column for example 14.17.

Mechanics 307

The area of a hollow tube is the outer square area minus the inner square.

$$A = (b_o^2 - b_i^2) = ((8.00 \text{ in})^2 - (6.75 \text{ in})^2) = 18.44 \text{ in}^2$$

Where b_o = width of outer edges of the square tube

Where b_i = width of inner edges of the square tube

The compressive stress in the top story is the roof load divided by the area.

$$\sigma_1 = \frac{P_1}{A} = \frac{80 \text{ k}}{18.44 \text{ in}^2} = 4.339 \text{ ksi}$$

The compressive stress in the bottom story is the roof load and the floor load divided by the area.

$$\sigma_2 = \frac{P_1 + P_2}{A} = \frac{80 \text{ k} + 100 \text{ k}}{18.44 \text{ in}^2} = 9.763 \text{ ksi}$$

The final stresses are 4.34 ksi in the top story and 9.76 ksi in the bottom story.

Strain

The strain is the measure of axial deformation of a member. It is the change in a member's length after loading, divided by the original member length before loading. Members in tension will elongate, while members in compression will shorten.

$$\varepsilon = \frac{\Delta L}{L} = \frac{\delta}{L}$$

where

ε = unit strain (in/in or unitless)
$\delta = \Delta L$ = total deformation, elongation (inches)
L = original length (inches)

Elastic and Plastic Behavior

An elastic material will deform under loading, by either stretching or compressing. As long as the yield stress is not reached, the member will return to its original length once the load is removed. This is called *linear elastic behavior*. The ratio of the stress to strain in the elastic region is called *modulus of elasticity* or *Young's modulus*.

$$E = \frac{\sigma}{\varepsilon}$$

Where

E = elastic modulus
σ = axial stress
ε = strain (elongation)

If the equations for axial stress, axial strain, and modulus of elasticity are combined, the axial deformation due to the load can be found as follows:

$$E = \frac{\sigma}{\varepsilon} = \frac{P/A}{\sigma/L} = \frac{PL}{\delta A} \therefore \delta = \frac{PL}{AE}$$

Example 14.18 Determine the total axial deformation of the two-story column in figure 14-38. The roof load is $P_1 = 80$ kips and the floor load at mid-height is $P_2 = 100$ kips. The column is steel, with a modulus of elasticity of 29,000,000 psi, and the cross-sectional area was calculated in example 14-17.

The axial deformation will be composed of two parts, since the load for each story is different. The total deformation will be the sum of the deformations of each story.

$$\delta = \sum \frac{PL}{AE} = \frac{80\,k(14\,ft)12\,in/ft}{18.44\,in^2\,(29{,}000\,k/in^2)} + \frac{180\,k(14\,ft)12\,in/ft}{18.44\,in^2\,(29{,}000\,k/in^2)} = 0.08168\,in$$

If the elastic limit is exceeded, the member will deform greatly under little added load. This is known as *plastic behavior*. Furthermore, if a member is loaded into the plastic range and then unloaded, a permanent deformation will exist.

Problem Set 14.3 Mechanics

Problem 14-25 A two-story column in a building is constructed of an I-shaped section and shown in figure 14-39. The roof load is $P_1 = 180$ kips and the floor load at mid-height is $P_2 = 400$ kips. Calculate the compressive stresses in each story of the column.

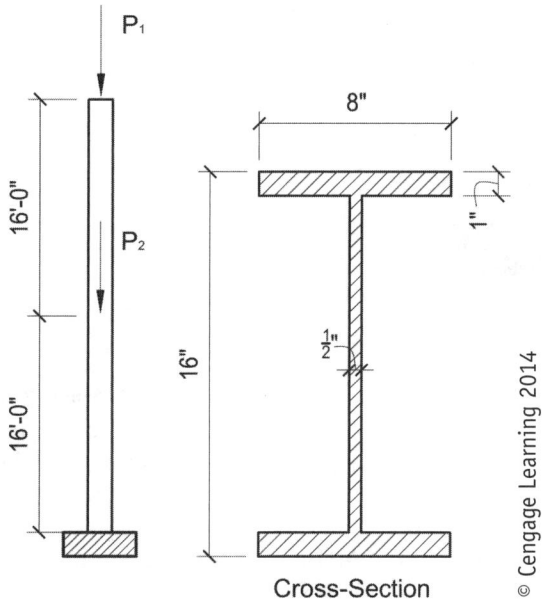

FIGURE 14-39 W-shape column for problem 14-25.

Problem 14-26 Determine the compressive stress in a 2-inch-diameter rod under a load of 100,000 pounds.

Problem 14-27 Determine the total axial deformation of the two-story column in figure 14-39. The roof load is $P_1 = 180$ kips and the floor load at mid-height is $P_2 = 400$ kips. The column is steel, with a modulus of elasticity of 29,000,000 psi and the cross-sectional area of the I-shape is 23.0 in^2.

Problem 14-28 Determine the amount of stretch in a 1-inch-diameter steel rod under a tension load of 100,000 pounds.

Problem 14-29 Given the following data from a tension test on aluminum alloy, plot the stress-strain curve and determine the yield stress. Also, estimate the modulus of elasticity. You may want to use a computer spreadsheet program to aid in your solution.

Stress (ksi)	Strain
8	0.0006
17	0.0015
27	0.0024
35	0.0032
43	0.0040
50	0.0046
58	0.0052
62	0.0058
64	0.0062
65	0.0065
67	0.0073
68	0.0081

BACKGROUND

Section Properties

The axial stress, strain, and deformations are only dependent on the cross-sectional area of the structural member. Other types of loading on a member, such as bending, shearing, or twisting, are dependent on other cross-sectional properties. These properties are more difficult to find but are normally tabulated for specific shapes.

- A_g, Gross area of the cross section.
- I, Area Moment of Inertia. The area moment of inertia is used to predict deflection in a bending member.
- r, Radius of Gyration. The radius of gyration is a used to predict the susceptibility of a compression member to buckling.
- Z, Plastic Section Modulus. The plastic section modulus is used to predict stress in a bending member.

We will not calculate these properties but instead will use published values for the sections we utilize.

Structural Design

In order to design the size of a structural member based on its strength, the forces to be carried by the member must be determined. Once these are found, the member can be sized based on the type of load it is to carry. Three broad categories for the types of members carrying these forces include: tension members, compression members and bending (or flexural) members. In addition to strength requirements, a member size can also be dependent on serviceability, fire resistance, and aesthetics as well as other conditions. The method by which the members are designed is based on building code requirements, which vary for each of the three main categories of structural materials: concrete, timber, and steel. For this workbook, strength and serviceability in the sizing of members, will be considered in the design of structural steel. Many strength requirements will be considered, and deflection will be the only serviceability requirement checked.

Allowable Strength Design (ASD) vs. Load and Resistance Factor Design (LRFD) in Steel Structures

The American Institute of Steel Construction (AISC) Steel Construction Manual gives the designer the option of using either ASD or LRFD for steel structures. This workbook will utilize ASD for the design of structural members.

Structural Steel

Structural steel is available in many shapes and strengths. Figure 14-47 in the textbook gives the most commonly used structural steel shapes along with the yield stress, F_y (in ksi), for each type of shape. The AISC manual uses a general form of the strength equation that is used on all types of loading conditions:

$$F_a \leq \frac{F_n}{\Omega} = \text{or } F_n \geq \Omega F_a$$

where F_a = applied load
F_n = nominal strength
Ω = safety factor

Tension Members

When a member supports only axial tension force, it is called a *tension member*. Hanger rods, cables, certain truss members and diagonal braces are tension members. The design of a tension member must satisfy the equation:

$$P_a \leq \frac{P_n}{\Omega_t} \text{ or } P_n \geq \Omega_t P_a$$

where P_a = applied tension force in member
P_n = nominal tension strength of member
Ω_t = 1.67 = safety factor for tension

The nominal tension strength, P_n, for a tension member is determined from the equation:

$$P_n = F_y A_g$$

where F_y = yield stress of the steel (ksi)

A_g = gross cross-sectional area of the member (in²)

Example 14.19 Determine the allowable tension force, in kips, of a L4 × 4 × 1/4 angle with a cross-sectional area of 1.94 in², fabricated of A36 steel.

From figure 14-47 from the textbook, the yield strength of an angle fabricated of A36 steel is F_y = 36 ksi. With the area of the angle given, the allowable tension force is calculated directly.

The nominal strength of the angle in tension is:

$$P_n = F_y A_g = 36 \text{ ksi}(1.94 \text{ in}^2) = 69.84 \text{ kips}$$

The allowable tension force for the steel angle is:

$$P_a = \frac{P_n}{\Omega_t} = \frac{69.84 \text{ k}}{1.67} = 41.82 \text{ kips}$$

Example 14.20 Design the most economical (lightest weight = least area) solid round tension hanger rod to support a load of 35 kips dead load and 30 kips live load using F_y = 50 ksi steel. Design the diameter of the hanger rod rounded up to the *nearest* 1/4 inch.

The total tension force to be supported by the hanger rod is: $P_a = 35 + 30 = 65$ kips.

The nominal required strength of the hanger rod is calculated with equation:

$$P_n \geq \Omega_t P_a = (1.67)(65 \text{ k}) = 108.55 \text{ kips}$$

With $P_n = F_y A_g$, the equation can be rewritten as:

$$F_y A_g \geq P_n = 108.55 \text{ kips}$$

With a known value of F_y = 50 ksi, the required area, A_g may be calculated from the equation:

$$A_g \geq \frac{P_n}{F_y} = \frac{108.55 \text{ k}}{50 \text{ ksi}} = 2.171 \text{ in}^2$$

For a round section, determine the required radius, and then multiply by 2 to arrive at the minimum diameter:

$$A_g \leq \pi r^2 \quad \therefore \quad r \geq \sqrt{\frac{A_g}{\pi}}$$

$$\text{Diameter}, d \geq 2\sqrt{\frac{A_g}{\pi}} = 2\sqrt{\frac{2.171 \text{ in}^2}{\pi}} = 1.663 \text{ in}$$

Rounding up to the next ¼ inch increment gives the final answer:

Use 1 3/4" round hanger rod of F_y = 50 ksi steel.

Compression Members

When a member supports only axial compression force, it is called a *compression member*. Certain truss members, diagonal braces, and columns are compression members. The design process for compression members follows the process used for tension members:

$$P_a \leq \frac{P_n}{\Omega_c} \text{ or } P_n \geq \Omega_c P_a$$

where P_a = applied compression force in member

P_n = nominal compression strength of member with:

$$P_n = F_{cr} A_g \text{ with } F_{cr} = \text{critical stress (explained below)}$$

$\Omega_c = 1.67$ = safety factor for compression

One distinct difference when compared to tension members is that the effects of buckling must be taken into account in the design process for columns. Refer to figure 14-53 in the textbook for buckling of a column. The buckling failure in a column is accounted for in the design process by including the slenderness ratio of the column in the strength equation. The AISC code limits the slenderness ratio to:

$$\frac{KL}{r} \leq 200$$

where K = effective length factor of column based on support conditions
L = unbraced height of column
r = least radius of gyration of column:

$$r = \sqrt{\frac{I}{A}} \text{ with } I = \text{moment of inertia (in}^4\text{), and } A = \text{cross-sectional area (in}^2\text{)}$$

As the slenderness ratio decreases, the strength capacity of a column increases. This ratio is taken into consideration when calculating the effective stress F_e:

$$F_e = \frac{\pi^2 E}{\left(\frac{KL}{r}\right)^2}$$

The effective stress is utilized in the equation to calculate the critical stress, F_{cr}:

For $F_e \geq 0.44 F_y$:

$$F_{cr} = \left[0.658^{\frac{F_y}{F_e}}\right] F_y$$

For $F_e < 0.44 F_y$:

$$F_{cr} = 0.877 F_e$$

Example 14.21 Calculate the compression capacity in kips of a W14 × 74 column of A 992 steel (F_y = 50 ksi) with an unbraced height of 20'-0" with pinned supports top and bottom (K = 1.0). The section properties include A_g = 21.8 in², r_x = 6.04 in. and r_y = 2.48 in.

To calculate the critical stress, F_{cr}, the maximum slenderness ratio is used, which corresponds to the least radius of gyration, r_y. The slenderness ratio is:

$$\frac{KL}{r} = \frac{1.0[20 \text{ ft}(12 \text{ in/ft})]}{2.48} = 96.77 \leq 200 \therefore \text{Okay}$$

The next step is to calculate the elastic critical buckling stress, F_e from equation:

$$F_e = \frac{\pi^2 E}{\left(\frac{KL}{r}\right)^2} = \frac{\pi^2 (29,000 \text{ ksi})}{(96.77)^2} = 30.56 \text{ ksi}$$

Since F_e = 30.56 ksi > 0.44 F_y = 0.44(50 ksi) = 22.0 ksi, the critical stress equation, F_{cr} is:

$$F_{cr} = \left[0.658^{\frac{50 \text{ ksi}}{30.56 \text{ ksi}}}\right] 50 \text{ ksi} = 25.21 \text{ ksi}$$

Structural Steel

Once the critical stress is calculated, the capacity of the column can be found from equation:

$$P_a = \frac{P_n}{\Omega_c} = \frac{F_{cr}A_g}{\Omega_c} = \frac{25.21 \text{ ksi} (21.8 \text{ in}^2)}{1.67} = 329.1 \text{ kips}$$

Compression capacity = 329.1 kips.

Example 14.22

Select the most economical (lightest) 5 in. square HSS (Hollow Structural Section) column with an unbraced height of 15'-0" in each direction, axial loads of 25 kips dead and 40 kips live with the base and top of the column pinned. Design for $F_y = 46$ ksi and use the *Steel Construction Manual* column design tables.

The *Steel Construction Manual* design tables give capacities of various column types and sizes based on effective length KL and the column capacity. For HSS shapes, columns can have the same exterior dimensions, but the thickness of the wall of the shape can increase, thus increasing the capacity of a column. Refer to figure 14-40 for a section through an HSS shape. Solve for P_a and for KL, and use the design tables to select the most economical HSS shape.

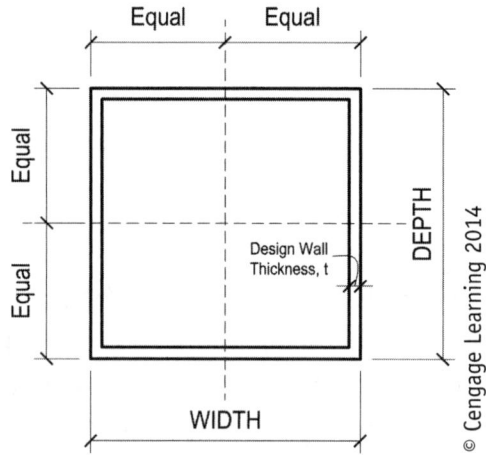

FIGURE 14-40 HSS shape section.

$$P_a = 25 k + 40 k = 65 \text{ kips}$$

The effective column height, $KL = 1.0(15.0 \text{ ft}) = 15.0$ ft

Steel Construction Manual table 4-4 lists capacity values for HSS square columns. For a $KL = 15.0$ ft and $P_a = 65$ kips, use the design table to select the lightest 5 in. square HSS column that works.

Refer to figure 14-41 for a portion of table 4-4 from the *Steel Construction Manual*. This table can be used for Allowable Stress Design (ASD) and for Load and Resistance Factor Design (LRFD). The values in the shaded columns are for ASD, and the other columns are for LRFD. It is important to differentiate between the values.

Table 4-4 (continued)
Available Strength in Axial Compression, kips
Square HSS
HSS5–HSS4½

F_y = 46 ksi

Shape	HSS5×5×										HSS4½×4½×	
	3/8		5/16		1/4		3/16		1/8 c		1/2	
t_{design}, in.	0.349		0.291		0.233		0.174		0.116		0.465	
Wt/ft	22.3		19.0		15.6		12.0		8.15		24.9	
Design	P_n/Ω_c	$\phi_c P_n$	P_n/Ω_c	$\phi_c P_n$	P_n/Ω_c	$\phi_c P_n$	P_n/Ω_c	$\phi_c P_n$	P_n/Ω_c	$\phi_c P_n$	P_n/Ω_c	$\phi_c P_n$
	ASD	LRFD	ASD	LRFD	ASD	LRFD	ASD	LRFD	ASD	LRFD	ASD	LRFD
0	170	256	145	218	119	178	90.4	136	56.5	84.9	191	288
1	170	255	145	217	118	178	90.1	135	56.4	84.8	191	287
2	168	253	143	216	117	176	89.5	134	56.1	84.4	189	283
3	166	250	142	213	116	174	88.3	133	55.7	83.7	185	278
4	163	245	139	209	114	171	86.8	130	55.1	82.8	180	271
5	159	239	136	204	111	167	84.8	128	54.3	81.7	174	262
6	154	232	132	198	108	162	82.5	124	53.4	80.3	167	252
7	149	223	127	191	104	157	79.9	120	52.3	78.6	159	240
8	143	214	122	184	100	151	76.9	116	51.0	76.6	151	227
9	136	205	117	175	96.0	144	73.7	111	49.5	74.4	142	213
10	129	194	111	167	91.4	137	70.2	106	47.8	71.9	132	198
11	122	183	105	158	86.5	130	66.6	100	45.7	68.6	122	183
12	114	172	98.6	148	81.5	122	62.8	94.4	43.2	64.9	112	168
13	107	160	92.2	139	76.4	115	59.0	88.6	40.6	61.0	102	153
14	99.1	149	85.8	129	71.2	107	55.1	82.8	38.0	57.1	92.3	139
15	91.5	138	79.4	119	66.0	99.2	51.2	76.9	35.4	53.2	82.8	124
16	84.0	126	73.0	110	60.9	91.5	47.3	71.1	32.8	49.3	73.7	111
17	76.7	115	66.9	100	55.9	84.0	43.6	65.5	30.2	45.4	65.3	98.1
18	69.6	105	60.9	91.5	51.0	76.7	39.9	59.9	27.7	41.7	58.2	87.5
19	62.7	94.3	55.0	82.7	46.4	69.7	36.3	54.6	25.3	38.1	52.3	78.6
20	56.6	85.1	49.7	74.7	41.8	62.9	32.8	49.3	23.0	34.5	47.2	70.9
21	51.4	77.2	45.1	67.7	37.9	57.0	29.8	44.8	20.8	31.3	42.8	64.3
22	46.8	70.3	41.1	61.7	34.6	51.9	27.1	40.8	19.0	28.5	39.0	58.6
23	42.8	64.3	37.6	56.5	31.6	47.5	24.8	37.3	17.4	26.1	35.7	53.6
24	39.3	59.1	34.5	51.9	29.0	43.7	22.8	34.3	16.0	24.0	32.8	49.2
25	36.2	54.5	31.8	47.8	26.8	40.2	21.0	31.6	14.7	22.1	30.2	45.4
26	33.5	50.4	29.4	44.2	24.7	37.2	19.4	29.2	13.6	20.4	27.9	42.0
27	31.1	46.7	27.3	41.0	22.9	34.5	18.0	27.1	12.6	18.9		
28	28.9	43.4	25.3	38.1	21.3	32.1	16.8	25.2	11.7	17.6		
29	26.9	40.5	23.6	35.5	19.9	29.9	15.6	23.5	10.9	16.4		
30	25.2	37.8	22.1	33.2	18.6	27.9	14.6	21.9	10.2	15.3		

Effective length KL (ft) with respect to least radius of gyration r_y

FIGURE 14-41 *Allowable Strength in Axial Compression Copyright © American Institute of Steel Construction. Reprinted with permission. All rights reserved.*

Find the effective length value, KL = 15 ft along the left column of the table, and then move horizontally to the right until a value is found in the shaded columns that is close to, but larger than or equal to, the value of P_a = 65 kips.

For this problem, the value for P_n/Ω = 66 kips.

Once this value is found, go to the top of the column, and the wall thickness size for the HSS 5 × 5 column is indicated as 1/4 inch.

Selection: HSS 5 × 5 × 1/4

Structural Steel 315

Bending Members

When a member supports a transverse load and transfers it to supports, it is called a *bending member*. A bending member, often referred to as a *beam*, typically spans horizontally and carries vertical loads, though a beam can also be oriented in other directions. The design process for a beam includes strength design, which takes into account the shear and bending moments supported by the beam, as well as the serviceability issue of deflection of the beam under load. Each of these components will be explored within this section of the workbook.

Moment strength

The moment strength of a beam must satisfy the equation:

$$M_a \leq \frac{M_n}{\Omega_b} \quad \text{or} \quad M_n \geq \Omega_b M_a$$

where M_a = applied maximum moment in the beam
$\Omega_b = 1.67$ = safety factor for bending
M_n = nominal moment strength of the beam: $M_n = F_y Z_x$
with Z_x = plastic section modulus of the beam

For the equations given above to be valid for bending, the compression flange must be braced properly against lateral buckling. The following examples are based on lateral bracing for the compression flange of the beam being provided.

Example 14.23 Calculate the moment strength for a beam that is continuously laterally braced. The beam is a W21 × 55 that has a plastic section modulus, $Z_x = 126$ in³, and the yield stress, $F_y = 50$ ksi.

To solve, use the equation:

$$M_{allowable} = \frac{M_n}{\Omega_b}$$

with $M_n = F_y Z_x = 50 \text{ ksi}(126 \text{ in}^3) = 6{,}300 \text{ kip-in}$

Moment is typically expressed in units of kip-feet, thus make the conversion:

$$M_n = \frac{6300 \text{ k-in}}{12 \text{ in/ft}} = 525.0 \text{ k-ft}$$

$$\Omega_b = 1.67$$

Solving for the moment:

$$M_a = \frac{M_n}{\Omega_b} = \frac{525.0 \text{ k-ft}}{1.67} = 314.4 \text{ k-ft}$$

Thus the allowable moment capacity, $M_{allowable} = 314.4$ k-ft

Example 14.24 For the loaded beam shown in figure 14-42, determine whether a W24 × 84 (Z_x = 224 in³) with F_y = 50 ksi is adequate for moment strength.

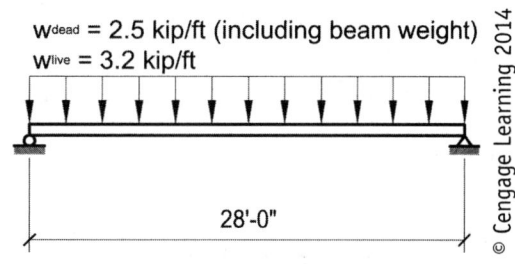

FIGURE 14-42 Beam for example 14.24.

To solve this problem, determine the maximum moment for the loaded beam condition, and calculate the allowable maximum moment, M_a for the given beam size and then compare the values to determine whether the beam can support the moment due to the applied loading. In equation form, with maximum moment of the loaded beam represented by M_{max}, this procedure becomes:

$$M_{max} = M_a \leq \frac{M_n}{\Omega_b}$$

As discussed in the previous example, moment is typically expressed in the units of kip-ft, and the conversion to these units will be made in the equations.

Allowable moment:

$$M_{max} = M_a \leq \frac{M_n}{\Omega_b} = \frac{F_y Z_x}{1.67} = \frac{50 \text{ ksi}(224 \text{ in}^3)}{1.67} = \frac{6706.6 \text{ k-in}}{12 \text{ in/ft}} = 558.9 \text{ k-ft}$$

For a uniformly loaded, simple span beam, the maximum moment occurs at mid-span and can be calculated from the following equation:

$$M_{max} = \frac{wl^2}{8}$$

where, w = total uniform load in units of kips/feet and l = beam span:

$w = w_{dead} + w_{live} = 2.5 + 3.2 = 5.7$ kips/ft

$l = 28.0$ ft

The maximum moment is:

$$M_{max} = \frac{5.7 \text{ k/ft}(28 \text{ ft})^2}{8} = 558.6 \text{ k-ft}$$

Compare the maximum moment with the allowable moment to determine whether the beam is adequate for loading:

$$M_{max} = 558.6 \text{ k-ft} < \frac{M_n}{\Omega} = 558.9 \text{ k-ft}$$

Thus the beam is adequate (just barely) for the applied loading condition.

Shear strength

The shear strength of a beam must satisfy the equation:

$$V_a \leq \frac{V_n}{\Omega_v} \quad \text{or} \quad V_n \leq \Omega_v V_a$$

where V_a = applied maximum shear in the beam
V_n = nominal shear strength of the beam: $V_n = 0.6 F_y A_w$
with A_w = area of the beam web: $A_w = d\, t_w$
$\Omega_v = 1.5$ = safety factor for shear

Example 14.25 Based on shear strength only, calculate the maximum load, P that can be supported by the beam in figure 14-43, if the beam is a W18 × 50, with $F_y = 50$ ksi, $d = 18.0$ in and $t_w = 0.355$ in.

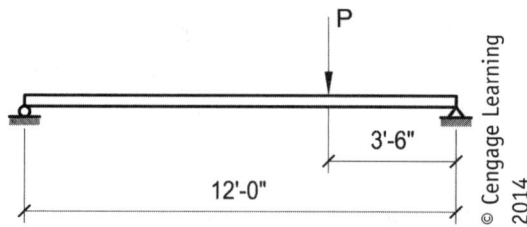

FIGURE 14-43 Beam for example 14.25.

First, calculate the maximum shear force that can be supported by the given beam:

$$V_a \leq \frac{V_n}{\Omega_v} = \frac{0.6 F_y A_w}{1.5} = \frac{0.6 F_y d\, t_w}{1.5} = \frac{0.6 (50 \text{ ksi})(18.0 \text{ in})(0.355 \text{ in})}{1.5} = 127.8 \text{ kips}$$

Next, solve for the maximum shear force along the length of the beam in terms of P:

For this loading condition the equation for shear is shown below where a is the largest distance from a support to the applied load.

$$V_{max} = \frac{Pa}{L} = \frac{P(8.5 \text{ ft})}{12 \text{ ft}} = 0.7083\, P$$

To solve for the maximum load P, set $V_a = V_{max}$ and solve for P:

$$0.7083\, P \leq 127.8 \text{ kips} = \therefore P \leq 180.4 \text{ kips}$$

The maximum load $P = 180.4$ kips

Example 14.26 Determine whether the W 12 × 72 beam shown in figure 14-44 is adequate in shear strength for $F_y = 50$ ksi.

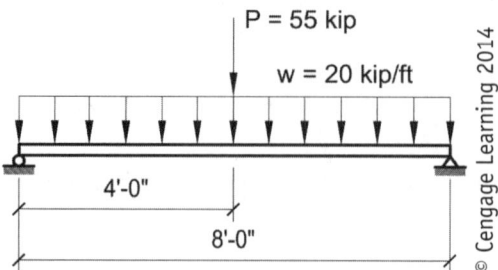

FIGURE 14-44 Beam for example 14.26.

For the beam to be adequate for shear strength, the maximum shear force must be such that:

$$V_a \leq \frac{V_n}{\Omega_v} = \frac{0.6 F_y A_w}{1.5}$$

Solving for V_{max} gives the equation:

$$V_a = V_{max} = \frac{wl}{2} + \frac{P}{2} = \frac{20.0 \text{ k/ft}(8.0 \text{ ft})}{2} + \frac{55.0 \text{ k}}{2} = 107.5 \text{ kips}$$

For the beam shown in figure 14-44, the depth of the section, $d = 12.3$ inches and the thickness of the web, $t_w = 0.43$ inches.

Solving for V_a gives the equation:

$$V_a \leq \frac{V_n}{\Omega_v} = \frac{0.6 F_y A_w}{1.5} = \frac{0.6 F_y d t_w}{1.5} = \frac{0.6(50 \text{ ksi})(12.3 \text{ in})(0.43 \text{ in})}{1.5} = 105.8 \text{ kips}$$

Since 107.5 kips > 105.8 kips, the beam is not quite adequate for shear. One way to solve this inadequacy is to select the next heaviest section in this size group, which is a W12 × 79.

Deflection

Deflection in a beam is a condition caused primarily by bending in a member, due to applied loads, as shown in figure 14-57 in the textbook. The deflection of a beam is dependent on several factors:

- Beam span length
- Type of load carried by the beam
- The type of support conditions for the beam
- The modulus of elasticity for the beam material
- The moment of inertia of the beam

For beams with typical load and support conditions, deflection equations have been derived and can be used to determine the maximum deflection along a beam span. Some of these are given in figure 14-59 in the textbook.

Example 14.27 Calculate the maximum deflection for a simple span beam with a span of 20'-0" and a uniform load, $w = 3.75$ kips/ft. The beam is a W 21×44 with a moment of inertia, $I = 843$ in⁴.

For a simple span, uniformly load beam, the maximum deflection occurs at mid-span and can be calculated by the equation:

$$\Delta_{max} = \frac{5 wL^4}{384 EI}$$

It is important that proper units are used in the equation. It is common to use the units of kips and inches when calculating deflections.

$$w = \left(3.75 \frac{k}{ft}\right)\left(\frac{1 \text{ ft}}{12 \text{ in}}\right) = 0.3125 \text{ k/in}$$

$$L = 20'\text{-}0" = 240"$$

$$E = 29{,}000{,}000 \text{ psi} = 29{,}000 \text{ ksi}$$

Structural Steel

Thus, the deflection equation is:

$$\Delta_{max} = \frac{5wL^4}{384EI} = \frac{5(0.3125 \text{ k/in})(240 \text{ in})^4}{384(29{,}000 \text{ ksi})(843 \text{ in}^4)} = 0.552 \text{ inches}$$

The maximum deflection for the beam, $\Delta_{max} = 0.552$ inches

Many building codes place deflection limitations on members, based on the type of load and the materials supported by the member. As a general rule of thumb, many engineers use a limit of L/240 for the total load and L/360 for the live load supported by the member.

Example 14.28 Determine whether the loaded beam in figure 14-45 is adequate, based on deflection only, for a total load deflection limit of L/240. The beam is a W24 × 55 with a moment of inertia, $I = 1{,}350 \text{ in}^4$.

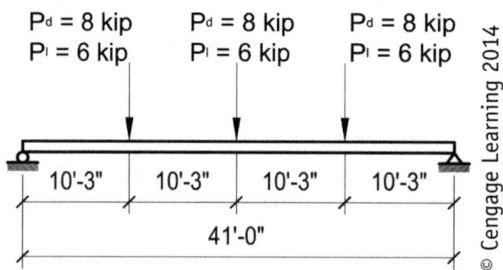

FIGURE 14-45 Beam for example 14.28.

Often, deflection is limited based on a span ratio, L/240 in this case, with L being the span of the beam in inches. In this example, the maximum total load deflection must satisfy the equation:

$$\Delta_{max} \leq \frac{L}{240} = \frac{41 \text{ ft}(12 \text{ in/ft})}{240} = 2.05 \text{ inches}$$

To calculate the maximum deflection for identical concentrated loads at the quarter points of a simple span, the equation from the *Steel Construction Manual* may be used.

$$\Delta_{max} = \frac{0.05 PL^3}{EI}$$

with P = total load at one of the concentrated load locations, thus $P = 8.0 \text{ k} + 6.0 \text{ k} = 14.0$ kips
$L = 41 \text{ ft }(12 \text{ in/ft}) = 492$ inches
E = modulus of elasticity of steel = 29,000 ksi
$I = 1{,}350 \text{ in}^4$ from the problem statement.

Thus the maximum deflection for this beam is:

$$\Delta_{max} = \frac{0.05 PL^3}{EI} = \frac{0.05(14 \text{ k})(492 \text{ in})^3}{29{,}000 \text{ ksi}(1{,}350 \text{ in}^4)} = 2.129 \text{ inches}$$

Comparing the maximum deflection with the deflection limits reveals:

$$\Delta_{max} = 2.129 \text{ inches} > \frac{L}{240} = 2.05 \text{ inches}$$

Thus the beam is not adequate for the stated deflection criteria. To satisfy the deflection criteria, a W24 × 62, with an $I = 1{,}550 \text{ in}^4$, could be used and the main deflection would be 1.85 inches.

Problem Set 14.4 Structural Design

Problem 14-30 Determine the allowable tension force, in kips, of an HSS4 × 4 × 3/8 tension member with a cross-sectional area of 4.78 in². The yield stress, F_y = 46 ksi for this member.

Problem 14-31 Determine whether a W6 × 9 (A = 2.68 in²), acting as a tension member, is capable of supporting a dead load of 45 kips and a live load of 30 kips, if F_y = 50 ksi for the W-shape.

Problem 14-32 For the W-shape given in problem 14-31, determine its capacity as a tension member, if the minimum yield strength is F_y = 65 ksi.

Problem 14-33 For a column with a total axial compression load of 50 kips and KL = 14 ft, select the most economical 5" square HSS column shape with F_y = 42 ksi. Use the table in figure 14-41.

Problem 14-34 Design for the most economical W8 column for an axial compression load of 140 kips and an effective length of 18'-0". The yield strength is F_y = 50 ksi and use figure 14-46 for design.

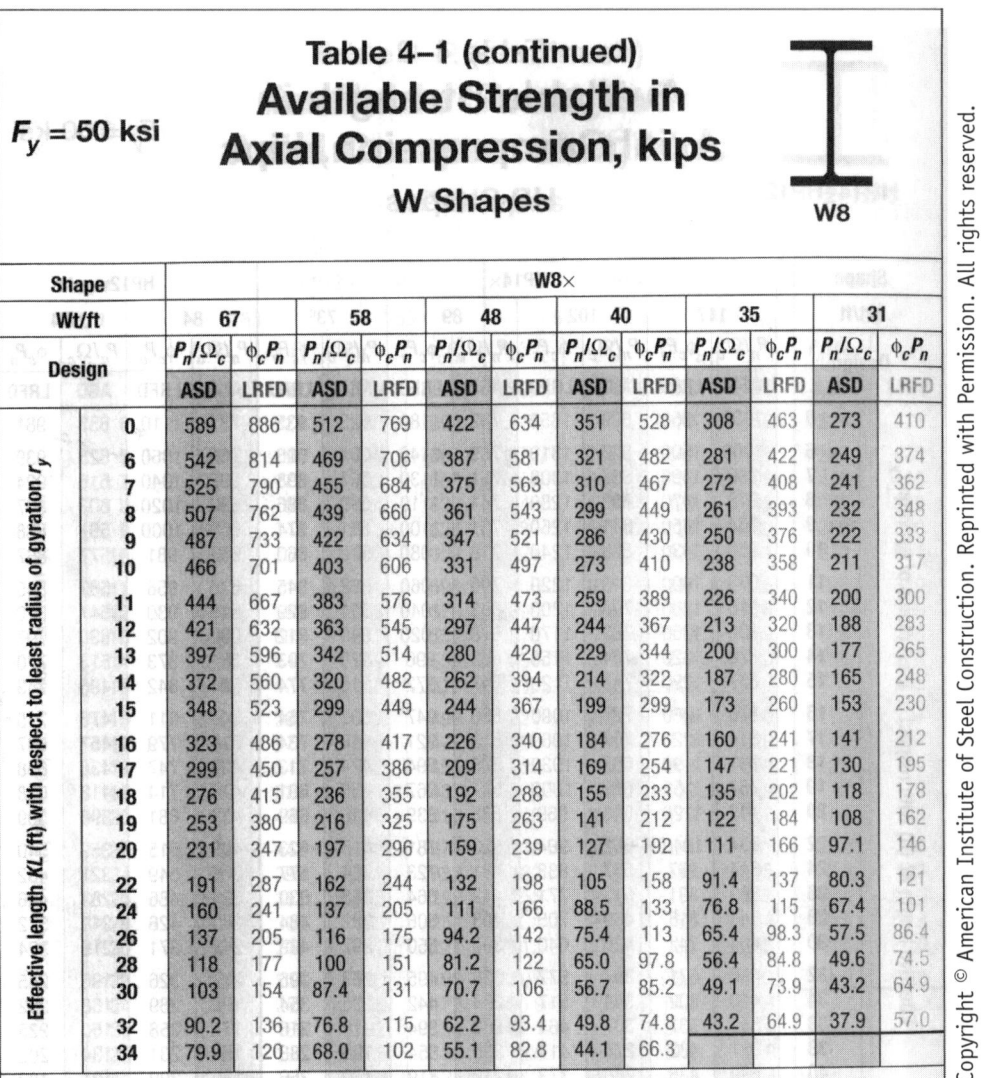

FIGURE 14-46 *Allowable Strength in Axial Compression for problem 14-34.*

Table 4–1 (continued) Available Strength in Axial Compression, kips — W Shapes (F_y = 50 ksi)

Shape Wt/ft	W8×67 ASD P_n/Ω_c	W8×67 LRFD $\phi_c P_n$	W8×58 ASD	W8×58 LRFD	W8×48 ASD	W8×48 LRFD	W8×40 ASD	W8×40 LRFD	W8×35 ASD	W8×35 LRFD	W8×31 ASD	W8×31 LRFD
0	589	886	512	769	422	634	351	528	308	463	273	410
6	542	814	469	706	387	581	321	482	281	422	249	374
7	525	790	455	684	375	563	310	467	272	408	241	362
8	507	762	439	660	361	543	299	449	261	393	232	348
9	487	733	422	634	347	521	286	430	250	376	222	333
10	466	701	403	606	331	497	273	410	238	358	211	317
11	444	667	383	576	314	473	259	389	226	340	200	300
12	421	632	363	545	297	447	244	367	213	320	188	283
13	397	596	342	514	280	420	229	344	200	300	177	265
14	372	560	320	482	262	394	214	322	187	280	165	248
15	348	523	299	449	244	367	199	299	173	260	153	230
16	323	486	278	417	226	340	184	276	160	241	141	212
17	299	450	257	386	209	314	169	254	147	221	130	195
18	276	415	236	355	192	288	155	233	135	202	118	178
19	253	380	216	325	175	263	141	212	122	184	108	162
20	231	347	197	296	159	239	127	192	111	166	97.1	146
22	191	287	162	244	132	198	105	158	91.4	137	80.3	121
24	160	241	137	205	111	166	88.5	133	76.8	115	67.4	101
26	137	205	116	175	94.2	142	75.4	113	65.4	98.3	57.5	86.4
28	118	177	100	151	81.2	122	65.0	97.8	56.4	84.8	49.6	74.5
30	103	154	87.4	131	70.7	106	56.7	85.2	49.1	73.9	43.2	64.9
32	90.2	136	76.8	115	62.2	93.4	49.8	74.8	43.2	64.9	37.9	57.0
34	79.9	120	68.0	102	55.1	82.8	44.1	66.3				

Effective length KL (ft) with respect to least radius of gyration r_y

Problem 14-35   Calculate the axial compression capacity for W14 × 90 of A992 steel that has lateral bracing and end supports as indicated in figure 14-47. Note the effective length factor K = 0.8 if one end of a member is attached to a fixed support, and the other end to a pinned support. The W14 × 90 has an area of A = 25.6 in² and radii of gyration r_x = 6.14 in. and r_y = 3.70 in. (weak axis).

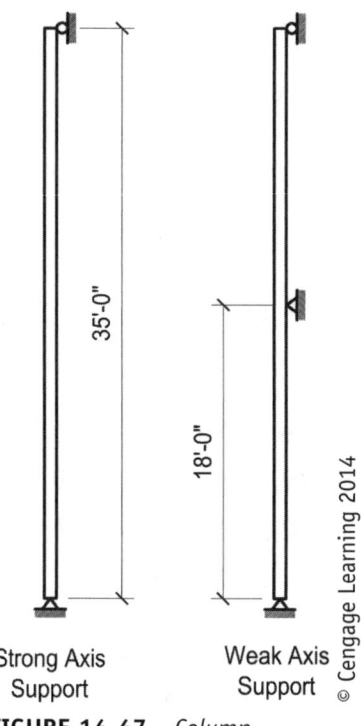

FIGURE 14-47   Column for problem 14-35.

Problem 14-36   Determine whether the W21 × 44 simple span beam (F_y = 50 ksi) shown in figure 14-48 is adequate for the loading shown based on moment strength. The plastic section modulus is Z_x = 95.4 in³.

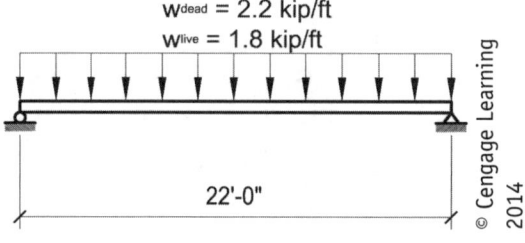

FIGURE 14-48   Beam for problem 14-36.

Problem 14-37   Determine the maximum uniform load that can be supported (in addition to the weight of the beam) by the beam in figure 14-49 for a W30 × 99 beam of A992 steel, based on moment strength. The plastic section modulus is Z_x = 312 in³.

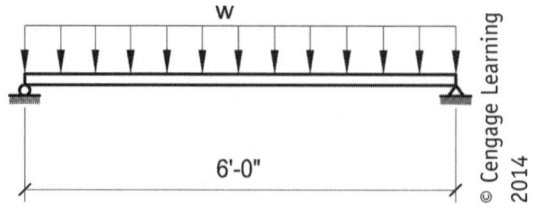

FIGURE 14-49   Beam for problem 14-37.

Problem 14-38 Design the most economical section based on moment strength for the loaded beam shown in figure 14-50 for a W-shape member with $F_y = 50$ ksi. Use the beam selection table shown, and ignore beam weight.

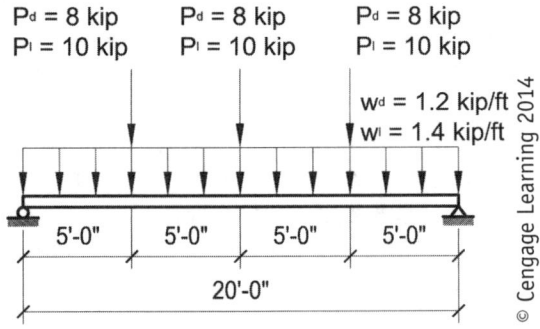

FIGURE 14-50 Beam for problem 14-38.

Table 3–2 (continued)
W Shapes
Selection by Z_x

$F_y = 50$ ksi

Shape	Z_x in.³	M_{px}/Ω_b kip-ft ASD	$\phi_b M_{px}$ kip-ft LRFD	M_{rx}/Ω_b kip-ft ASD	$\phi_b M_{rx}$ kip-ft LRFD	BF kips ASD	BF kips LRFD	L_p ft	L_r ft	I_x in.⁴	V_{nx}/Ω_v kips ASD	$\phi_v V_{nx}$ kips LRFD
W21×55	126	314	473	192	289	10.8	16.3	6.11	17.4	1140	156	234
W14×74	126	314	473	196	294	5.34	8.03	8.76	31.0	795	128	191
W18×60	123	307	461	189	284	9.64	14.5	5.93	18.2	984	151	227
W12×79	119	297	446	187	281	3.77	5.67	10.8	39.9	662	116	175
W14×68	115	287	431	180	270	5.20	7.81	8.69	29.3	722	117	175
W10×88	113	282	424	172	259	2.63	3.95	9.29	51.1	534	131	197
W18×55	112	279	420	172	258	9.26	13.9	5.90	17.5	890	141	212
W21×50	110	274	413	165	248	12.2	18.3	4.59	13.6	984	158	237
W12×72	108	269	405	170	256	3.72	5.59	10.7	37.4	597	105	158
W21×48ᶠ	107	265	398	162	244	9.78	14.7	6.09	16.6	959	144	217
W16×57	105	262	394	161	242	7.98	12.0	5.65	18.3	758	141	212
W14×61	102	254	383	161	242	4.96	7.46	8.65	27.5	640	104	156
W18×50	101	252	379	155	233	8.69	13.1	5.83	17.0	800	128	192
W10×77	97.6	244	366	150	225	2.59	3.90	9.18	45.2	455	112	169
W12×65ᶠ	96.8	237	356	154	231	3.60	5.41	11.9	35.1	533	94.5	142
W21×44	95.4	238	358	143	214	11.2	16.8	4.45	13.0	843	145	217
W16×50	92.0	230	345	141	213	7.59	11.4	5.62	17.2	659	124	185
W18×46	90.7	226	340	138	207	9.71	14.6	4.56	13.7	712	130	195
W14×53	87.1	217	327	136	204	5.27	7.93	6.78	22.2	541	103	155
W12×58	86.4	216	324	136	205	3.76	5.66	8.87	29.9	475	87.8	132
W10×68	85.3	213	320	132	199	2.57	3.86	9.15	40.6	394	97.8	147
W16×45	82.3	205	309	127	191	7.16	10.8	5.55	16.5	586	111	167
W18×40	78.4	196	294	119	180	8.86	13.3	4.49	13.1	612	113	169
W14×48	78.4	196	294	123	184	5.10	7.66	6.75	21.1	484	93.8	141
W12×53	77.9	194	292	123	185	3.65	5.48	8.76	28.2	425	83.2	125
W10×60	74.6	186	280	116	175	2.53	3.80	9.08	36.6	341	85.8	129
W16×40	73.0	182	274	113	170	6.69	10.1	5.55	15.9	518	97.7	146
W12×50	71.9	179	270	112	169	3.97	5.97	6.92	23.9	391	90.2	135
W8×67	70.1	175	263	105	159	1.73	2.60	7.49	47.7	272	103	154
W14×43	69.6	174	261	109	164	4.82	7.24	6.68	20.0	428	83.3	125
W10×54	66.6	166	250	105	158	2.49	3.74	9.04	33.7	303	74.7	112

FIGURE 14-51 W shapes selection by Z_x for problem 14-38.

Structural Steel

Problem 14-39 Determine if the W21 × 44 simple span beam (F_y = 50 ksi) shown in figure 14-48 is adequate for the loading shown based on shear strength. The W21 × 44 has d = 20.7 inches and t_w = 0.350 inches.

Problem 14-40 Determine the maximum uniform load that can be supported (in addition to the weight of the beam) by the beam in figure 14-49 for a W30 × 99 beam of A992 steel, based on shear strength. The W30 × 99 has d = 29.7 inches and t_w = 0.520 inches.

Problem 14-41 Determine the total load deflection of a W16 × 36 beam shown for the loading shown at the mid-span. At each location, the concentrated load P = 25 kips, and the uniform load w = 5.8 kip/ft (including allowance for weight of steel beam to be designed). The moment of inertia is I_x = 448 in^4 for a W16 × 36.

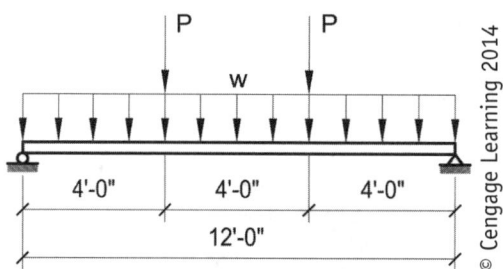

FIGURE 14-52 Beam for problem 14-41.

Problem 14-42 Determine whether the W21 × 44 simple span beam (F_y = 50 ksi) shown in figure 14-48 is adequate for the loading shown, based on a total load deflection limitation of L/180. The moment of inertia is I_x = 843 in^4 for a W21 × 44.

Problem 14-43 For the loaded beam shown in figure 14-48, calculate the minimum moment of inertia required satisfying a total load deflection limitation of L/240.

Problem 14-44 Calculate the maximum total load deflection for the loaded beam shown for a W24 × 55 with a moment of inertia I_x = 1,350 in^4.

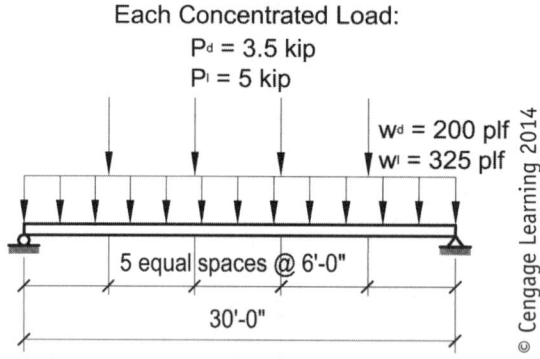

FIGURE 14-53 Beam for problem 14-44.

BACKGROUND

Foundations

A building's structural members carry the various types of loads (dead, live, wind ...) through the building and down to the foundation, where the loads are transferred to the supporting earth. Two categories of foundations are shallow foundations and deep foundations. Each of these will be explored in this workbook.

Shallow Foundations

Shallow foundations include spread footings, continuous footings, and mat foundations. For each of these, the bottom of the foundation is located near the surface of the surrounding ground, though it is required to be placed below the frost line, as indicated in figure 14-67 in the textbook. The required area of a shallow foundation is based on the amount of load being supported and the allowable bearing pressure of the soil. The allowable bearing pressure is typically given in a geotechnical report, but if one is not available, the building code provides bearing pressures to be used for design, as shown in figure 14-68 in the textbook. To determine the required area of a spread footing, use the equation:

$$A_{required} \geq \frac{F}{q_{net}}$$

where $A_{required}$ = required area for the shallow foundation
F = total supported load = column load + weight of foundation
q_{net} = net allowable soil bearing pressure = $q - p$
with q = allowable soil bearing pressure
p = pressure due to weight of shallow foundation

Example 14.29 Calculate the required square footing size for a total load (including column weight) of 108.0 kips. The column sits directly on top of the footing, and the footing bears on sedimentary rock. The frost depth for the location of this building is 22 inch below finish grade. Size all footing dimensions based on 6 inch increments.

The total load on the column footing, F = 108,000 lbs.

The allowable bearing pressure for sedimentary rock as given in figure 14-68 in the textbook is 4,000 psf.

The depth of the footing must be below the frost line, which is located 22 inches below finish grade, thus set the thickness of the footing at 24 inches based on the 6 inch increment requirement for footing dimensions.

Footing pressure: p = footing thickness × density of concrete = 24 inch(1 ft/12 in)(150 lb/ft^3) = 300 psf.

The net bearing pressure is:

$$q_{net} = q - p = 4,000 \text{ psf} - 300 \text{ psf} = 3,700 \text{ psf}$$

The required minimum footing area is:

$$A_{required} = \frac{F}{q_{net}} = \frac{108{,}000 \text{ lbs}}{3{,}700 \text{ lbs}} = 29.19 \text{ ft}^2$$

For a square footing, the actual required width, w is:

$$w = \sqrt{A_{required}} = \sqrt{29.19 \text{ ft}^2} = 5.403 \text{ ft} \cong 5'\text{-}5''$$

Based on the 6 inch increment requirement given in the problem statement, round the width up to 5'-6", resulting in a footing size of 5'-6" × 5'-6" × 2'-0" deep.

Example 14.30 Calculate the width of a continuous footing for a 10'-0" masonry wall that supports roof framing, as shown in figure 14-54. The dead load of the masonry wall is 75 psf. The roof dead load is 30 psf and live load is 40 psf. The tributary width that is supported by the masonry wall is shown on the figure, and the allowable bearing pressure from a geotechnical report is 1,100 psf. Use 3 inch increments in determining the width of footing.

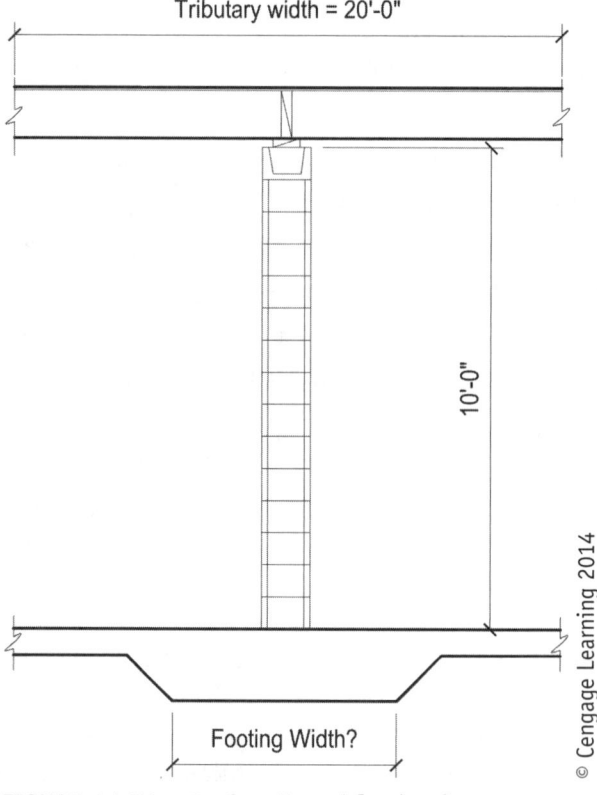

FIGURE 14-54 Roof, wall, and footing for example 14.30.

To determine the width of footing, use the equation:

$$w_{required} = \frac{F}{q_{net}} = \frac{F}{q - p}$$

For this form of the equation, calculate the loads and weights based on a 1-foot length of wall:

F = weight supported by the continuous footing = roof dead and live load + weight of masonry wall

F = 20.0 ft(30 psf + 40 psf) + 10.0 ft(75 psf) = 2,150 lb/ft

q = 1,100 psf, from the problem statement

p = weight of concrete footing = 1.5 ft(150 psf) = 225 psf

Thus, the required width for the continuous footing is:

$$w_{required} = \frac{F}{q-p} = \frac{2,150 \text{ lb/ft}}{1,100 \text{ psf} - 225 \text{ psf}} = 2.46 \text{ ft}$$

Increase the footing width to the next larger 3 inch increment:

Required footing width: 2'-6"

Deep Foundations

Deep foundations include bearing piles and friction piles. Bearing piles support vertical loads based on bearing at the end of the piles only. Friction piles utilize the friction that occurs between the surrounding soil and the face of the pile to transfer the vertical loads. These types of deep foundation systems are often circular in cross section and can be seen in figures 14-74 and 14-75 in the textbook. Piles can be fabricated of timber, steel, or concrete. The examples in this workbook will be based on concrete piles.

For a bearing pile, the equation to determine the required minimum area is similar to that for a spread footing:

$$A_{required} = \frac{F}{q_{net}}$$

where $A_{required}$ = required area for the shallow foundation
F = total supported load + Weight of pile
q_{net} = net allowable soil bearing pressure = $q - p$
with q = allowable soil bearing pressure
p = pressure due to weight of pile

The diameter of the pile can then be easily determined by using the equation for the area of a circle.

For a friction pile, the equation for the capacity is based on the frictional value, the circumference of the pile, and the length that the pile is embedded into the soil. The equation for the capacity of a friction pile is:

$$F_{capacity} = \pi d L q_{friction}$$

where $F_{capacity}$ = capacity of a single friction pile
d = diameter of pile
L = length that pile is embedded into soil
$q_{friction}$ = allowable frictional force in force/unit area

It is typical to set the diameter of piles at 2 inch increments.

Example 14.31 Calculate the diameter of a single 20'-0" deep concrete bearing pile for vertical load of 170 kips (including the weight of the pile) if the allowable bearing pressure at the bearing depth of the pile is 25 ksf. Use 2 inch increments in sizing the pile diameter.

qnet = q − p = allowable bearing pressure − weight of pile = 25 ksf − 20 ft(150 lb/ft^3)/(1,000 lb/kip) = 22.0 ksf

The minimum area of the pile is obtained from equation:

$$A_{required} = \frac{F}{q_{net}} = \frac{170.0 \text{ k}}{22.0 \text{ ksf}} = 7.73 \text{ ft}^2$$

Calculate the minimum diameter, d from equation:

$$d_{minimum} = 2\sqrt{\frac{A_{required}}{\pi}} = 2\sqrt{\frac{7.73 \text{ ft}^2}{\pi}} = 3.14 \text{ ft}$$

Rounding up to the next 2 inch increment gives the answer:

Required pile diameter: 3'-2"

Example 14.32 Calculate the capacity of a single concrete friction pile that is embedded 38'-9" into the soil and has a diameter of 2'-4". The allowable friction force, $q_{friction}$ = 800 psf.

The capacity of the friction pile is found by the equation:

$$F_{capacity} = \pi d L q_{friction} = \pi(2.333')(38.75')(0.8 \text{ ksf}) = 227.2 \text{ kips}$$

Friction pile capacity is 227.2 kips

Problem Set 14.5 Foundations

Problem 14-45   Design the diameter, using 2 inch increments, of a single concrete bearing pile for a total load of 50 kips that bears on crystalline bedrock.

Problem 14-46   For a 10'-0" long, 20 inch diameter concrete bearing pile that sets on a stratum with q = 25.0 kips/ft^2, determine whether it is adequate to support a column load consisting of 25 kips dead load (not including pile weight) and 30 kips live load.

Problem 14-47   Four bearing piles of equal diameter are to support a vertical total load of 360 kips (including allowance for weight of piles). Using 2 inch increments, determine the diameter of these piles if the stratum on which they are to bear has an allowable bearing pressure of 30 ksf.

Problem 14-48   Design the diameter, using 2 inch increments, of a single concrete friction pile for a total load of 85 kips that is embedded 26'-0" into the soil with an allowable friction force, $q_{friction}$ = 650 psf.

Problem 14-49   A 24 inch diameter friction pile is to be used to support a total column load of 50 kips (including allowance for weight of pile). If the allowable friction force is 400 psf, determine the length of embedment the pile must extend into the stratum, using 6 inch increments.

Problem 14-50   A single friction pile is to support 125 kips (including pile weight). The allowable friction capacity for the surrounding soil is 900 psf. Using diameter increments of 2 inch and length increments of 6 inch, design five possible solutions for the diameter and embedment length of the pile.

CHAPTER 15
Planning Electric Codes

Skills List

After completing the problems in this chapter, you should be able to:

- Understand common terms and properties of electricity
- Determine a building's electrical needs
- Calculate electrical loads on buildings
- Understand common electrical code requirements
- Identify electrical symbols used in drawings
- Be aware of the requirements for lighting design

BACKGROUND

Architects and engineers must be able to have a basic understanding as well as communicate basic electrical terminology to clients and other stakeholders. The various electrical systems and components need to be clearly defined and assembled properly.

Common Terms and Properties of Electricity

Electrical engineers are responsible for the design of the building's electrical system, but all of the design principles principals should understand basic electrical terminology. The following are definitions for fundamental electrical terms from the textbook:

- Alternating current (AC) – A signal or power source that varies with time, switching polarities
- Current (I) – The amount of charge that flows past a given point, per unit of time
- Direct current (DC) – Power that flows in one direction but can vary in intensity
- Energy – The ability to perform work, watts per hour for electricity
- Frequency – The number of occurrences per unit of time, cycles per second
- Ground – Reference point from which electrical voltage is measured
- Ohm's law – Current is directly proportional to the voltage and inversely proportional to the resistance
- Power – The amount of work per unit of time, watts for electricity
- Voltage (V or E) – Measure of electrical force

There are various types of electrical systems in buildings, as well as many components of those systems. The textbook lists many of the systems and components in table 15-1. There are many units of measure for electrical systems listed in table 15-2 of the textbook. The following are definitions of those units:

- Ampere (A) – Same as current, the amount of charge that flows past a given point, per unit of time
- British thermal units (Btu) – Energy needed to heat 1 pound of water 1°F, for electricity 3.41214 Btu/hr equals 1 watt.
- Capacitance (C) – Ability to store electrical charge of 1 farad or 1 coulomb per volt
- Horsepower (hp) – Power of 500 foot-pounds per second or approximately 745.7 watts
- Kilovolt-ampere (KVA) – Electrical load of 1,000 volt-amperes
- Kilowatt (KW) – 1,000 watts
- Volt-ampere (VA) – Voltage times current of an electrical load
- Volt (V) – Measure of electrical force, 1 volt will push 1 ampere through a 1-ohm resistive load
- Watt (W) – Power of volt-amperes
- Ohm – Same as impedance, opposition to electrical flow
- Megawatt (MW) – 1,000,000 watts of power
- Watt-hour (Wh) – Energy in watts times hours
- Kilowatt-hour (KWh) – 1,000 watt-hours

Direct and Alternating Current

The two types of electrical current are used differently. Direct current (DC) is normally used to power simple devices with energy stored in batteries. Alternating currents (AC) are used to power various devices that require different load. Nearly all of the power in buildings is in the form of alternating current. Typical commercial electrical loads are 480/277 or 120/208 volts. Residential loads are 120/240 volts. Refer back to figure 15-2 in the textbook to see how power is distributed to buildings.

Problem Set 15.1   Common Terms and Properties of Electricity

Problem 15-1   Describe the difference between AC and DC electricity.
Problem 15-2   What is known as the *flow of electricity* over a given period of time?
Problem 15-3   What is the basic measure of electrical power?
Problem 15-4   What is the basic measure of electrical force?
Problem 15-5   Describe Ohm's law and the units of each component.
Problem 15-6   What is a Btu, and how does it equate to electricity?
Problem 15-7   How does horsepower equate to electricity?
Problem 15-8   What are the two types of electrical loads used in homes?

BACKGROUND

Architects and engineers must be able to design electrical systems in buildings in an economical and efficient manner. The cost of energy to power buildings is one of the major factors in the success of a building's performance.

Power and Energy

Energy comes from many sources and is used in many ways. The U.S. Energy Information Administration (EIA) keeps track of energy statistics. The primary energy sources for each of the major sectors of the economy for 2008 are shown in figure 15-1.

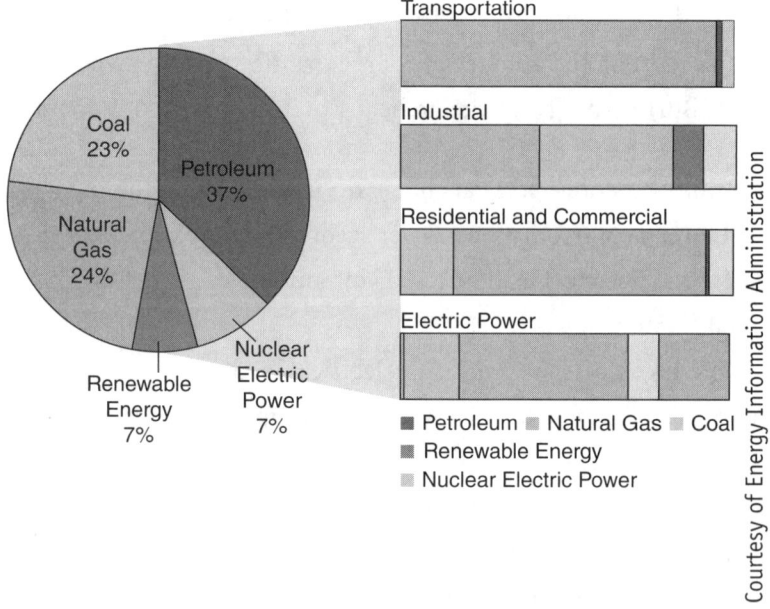

FIGURE 15-1   *U.S. primary energy computation by source and sector, 2008.*

The consumption of energy in 2009 for the major sectors of the economy is shown in figure 15-2. One should note that buildings are the major part of three of the four sectors shown.

332   Chapter 15: Planning Electric Codes

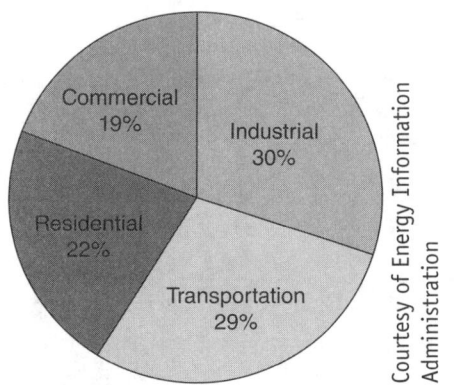

FIGURE 15-2 *Share of energy consumed by major sectors or the economy, 2009.*

There has been an increased focus on energy conservation over the last few decades in many sectors of the economy, but consumption has continued to increase. Consumption for the last five decades is shown in figure 15-3.

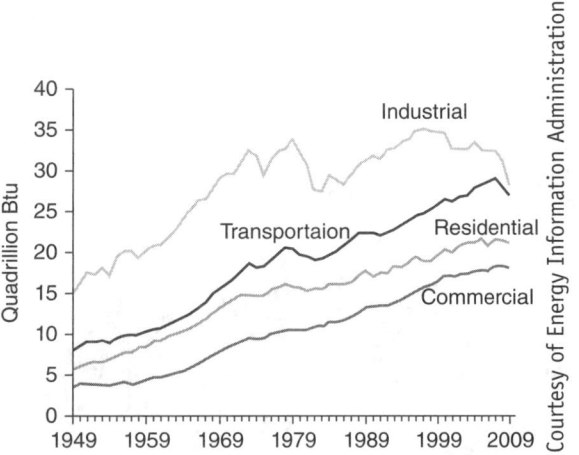

FIGURE 15-3 *Energy consumptions by sector, 1949–2009.*

Proper design of building electrical systems can help reduce the amount of energy consumed. The five basic steps in the electrical design procedure are covered in the following sections.

Determining the Building Needs

The source of the energy used in buildings and the way that it is used are highly variable. The source of energy used in commercial buildings for 2003 is shown in figure 15-4. Electrical makes up a majority of consumption, more than all other sources combined.

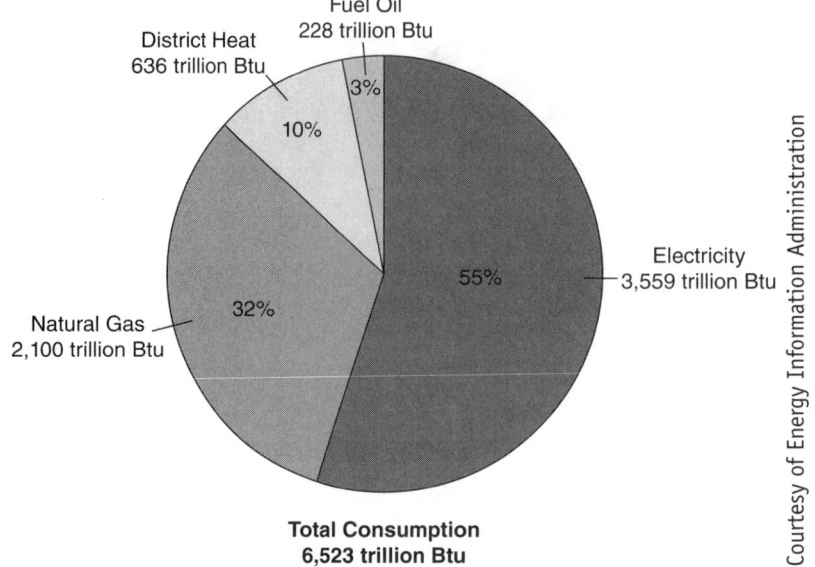

FIGURE 15-4 *Energy consumed by commercial buildings.*

The energy usage for the systems within commercial buildings in 2003 is shown in figure 15-5. The major systems of heating, cooling, and electricity make up the majority of the consumption.

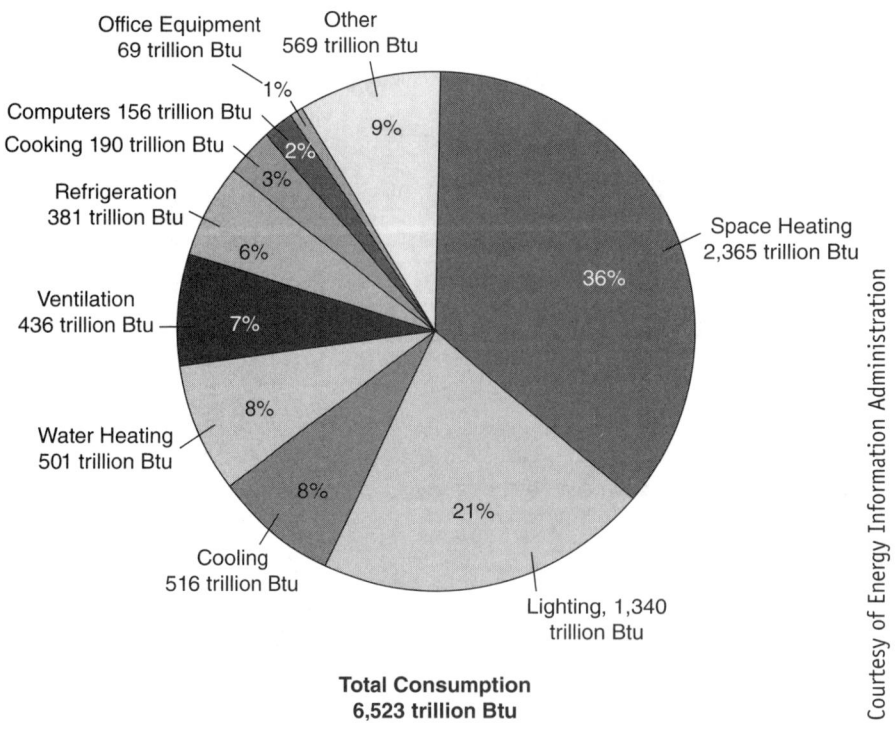

FIGURE 15-5 *Energy consumed by commercial buildings.*

334   Chapter 15: Planning Electric Codes

The energy needs for a given project will depend on the building occupancy use, location, and construction. The selection of the mechanical and electrical systems has an impact on the energy consumption. Careful consideration must be taken in these selections.

Calculating the Electrical Loads

The electrical loads for each of the components for the various building systems must be determined to evaluate the system. Various alternatives could be explored to find a more efficient system. The approximate power consumption for various systems, appliances, and components is given in table 15-1.

Appliance	Watts	Appliance	Watts	Appliance	Watts
Air Conditioner 1 ton	1,900	Fan (Ceiling)	90	Range	12,200
Air Conditioner 3 1/2 ton	6,500	Fan (Furnace)	500	Refrigerator 16 cu. ft.	600
Air Conditioner 5 ton	9,200	Freezer 18 cu. ft.	700	Refrigerator 20 cu. ft.	800
Portable heater	1,500	Garbage Disposal	450	Swimming Pool Pump	2,000
Boiler	1,100	Hair Dryer	1,000	Television	180
Clothes Dryer	2,800	Jacuzzi/Spa Pump	1,300	Vacuum Cleaner	630
Coffee Maker	1,200	Light, Daily use	3,000	Washing Machine	510
Computers	240	Light Incandescent	60	Water Heater	2,500
Dishwasher	1,200	Light Florescent	15	Toaster	1,150
Fan (Attic)	370	Microwave Oven	1,500	Toaster Oven	1,550

TABLE 15-1 *Typical power consumption of residential equipment.*

Most major appliances that operate continuously will run from four to six hours a day. Smaller appliances with high use will run two to four hours a day. Single use appliances will run from 10 to 60 minutes a day, depending on their timer.

Example 15.1 Calculate the daily electrical use for the three-bedroom residence with the appliances and daily usages in table 15-2.

Appliance	Watts	Hours/day	Watts/day
Air Conditioner 5 ton		6.00	
Water Heater		2.00	
Living Room w/150 watt lighting		1.00	
Dining Room w/160 watt lighting		1.00	
Master Bedroom w/ceiling fan		6.00	
Master Bath w/320 watt lighting		0.50	
Bedroom 2 w/ceiling fan		6.00	
Bedroom 3 w/ceiling fan		3.00	
Office w/computer		3.00	
Bedroom/Office lighting 300 watts		3.00	
Refrigerator 20 cu. ft		6.00	
Range		1.00	
Dishwasher		1.00	
Microwave Oven		0.25	
Coffee Maker		0.25	
Toaster Oven		0.25	
Garbage Disposal		0.10	
Washing Machine		1.00	
Clothes Dryer		1.00	
Kitchen/Den Lighting 300 watts		3.00	
Television 1		6.00	
Television 2		3.00	
Vacuum Cleaner		0.50	
Hair dryer		0.25	

TABLE 15-2 *Power consumption for a three-bedroom residence.*

From table 15-1, the power consumption for each appliance can be found and recorded in the second column of table 15-3. The total consumption for each appliance can be found from the product of the watts and hours/day and recorded in the last column of table 15-3. The sum of the last column is the total daily consumption.

Appliance	Watts	Hours/day	Watts/day
Air Conditioner 5 ton	9,200	6.00	55,200
Water Heater	2,500	2.00	5,000
Living Room w/150 watt lighting	150	1.00	150
Dining Room w/160 watt lighting	160	1.00	160
Master Bedroom w/ceiling fan	90	6.00	540
Master Bath w/320 watt lighting	320	0.50	160
Bedroom 2 w/ceiling fan	90	6.00	540
Bedroom 3 w/ceiling fan	90	3.00	270
Office w/computer	240	3.00	720
Bedroom/Office lighting 300 watts	300	3.00	900
Refrigerator 20 cu. ft	800	6.00	4,800
Range	12,200	1.00	12,200
Dishwasher	1,200	1.00	1,200
Microwave Oven	1,500	0.25	375
Coffee Maker	1,200	0.25	300
Toaster Oven	1,550	0.25	388
Garbage Disposal	450	0.10	45
Washing Machine	510	1.00	510
Clothes Dryer	2,800	1.00	2,800
Kitchen/Den Lighting 300 watts	300	3.00	900
Television 1	180	6.00	1,080
Television 2	180	3.00	540
Vacuum Cleaner	630	0.50	315
Hair dryer	1,000	0.25	250
Total consumption			89,343

TABLE 15-3 *Power consumption calculation for a three-bedroom residence.*

The cost of electricity varies among states and figure 15-6 is a map showing cost in 2008 according to the U.S. Energy Information Administration.

Power and Energy

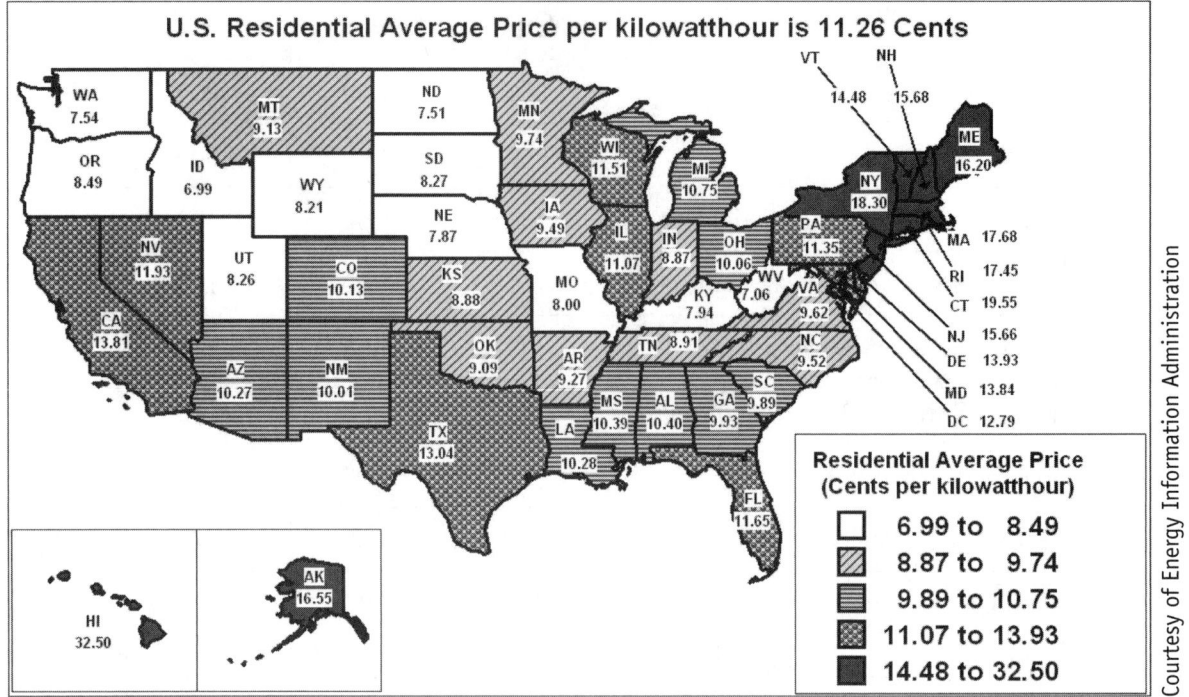

FIGURE 15-6 *U.S. residential average price per kilowatthour.*

The average energy bill for a typical single family home in 2009 was approximately $2,200. The percentage of energy cost for different uses according to the U.S. Environmental Protection Agency's (EPA) Energy Star program is shown in figure 15-7.

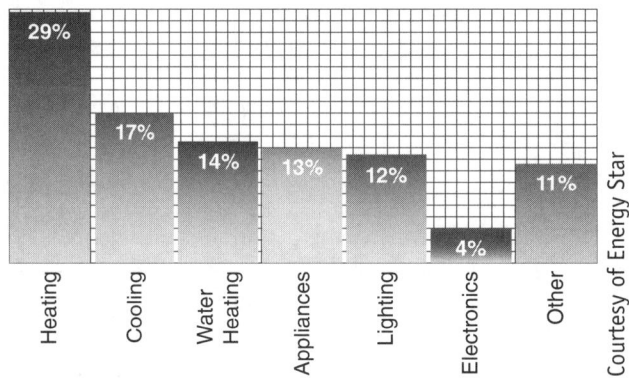

FIGURE 15-7 *Residential energy usage.*

Determining the Types of Electrical Service

The two basic types of electrical service are single-phase and three-phase systems. Single-phase systems supply all the same voltage and are normally used in residential construction with 120/240 volts. Single-phase systems contain two hot wires and a ground (three-wire). Three-phase systems have at least three hot conductors and can carry different voltages simultaneously (four-wire). Three-phase systems are used primarily in commercial buildings with 120/208, 277/480, and 480 volts.

Coordination of Other Design Aspects in the Building

The electrical space planning is controlled primarily by the systems selected. In buildings that are relatively small, a single-phase source can be used and is typically an electrical panel, which is common for residential and small commercial buildings. With larger commercial and multi-use buildings, electrical spaces should be provided throughout the building. Typically, a main electrical space is placed on the ground floor and small satellite rooms on each of the other floors. If a floor plan is fairly large, multiple electrical rooms may be needed on each floor where each room serves spaces approximately 125 feet away.

Preparation of Electrical Plans and Specifications

Several types of electrical drawings can be used to communicate designs. The following is a summary of drawing types included in the textbook:

- Floor plan for overall system layout (see textbook figure 15-5)
- Schematic diagrams to show circuitry in a system (see textbook figure 15-6)
- Connection diagrams to show wiring layouts of a system (see textbook figure 15-7)
- One-line diagrams for major equipment (see textbook figure 15-8)
- Riser diagrams that show vertical system layout (see textbook figure 15-9)

In most cases, the electrical engineer will design the systems and provide the drawings. On smaller projects, the architects may prepare floor plans that also include a layout with electrical symbols. The symbols fall into the three broad categories of switches (figure 15-9), outlets (figure 15-8), and devices (figure 15-10).

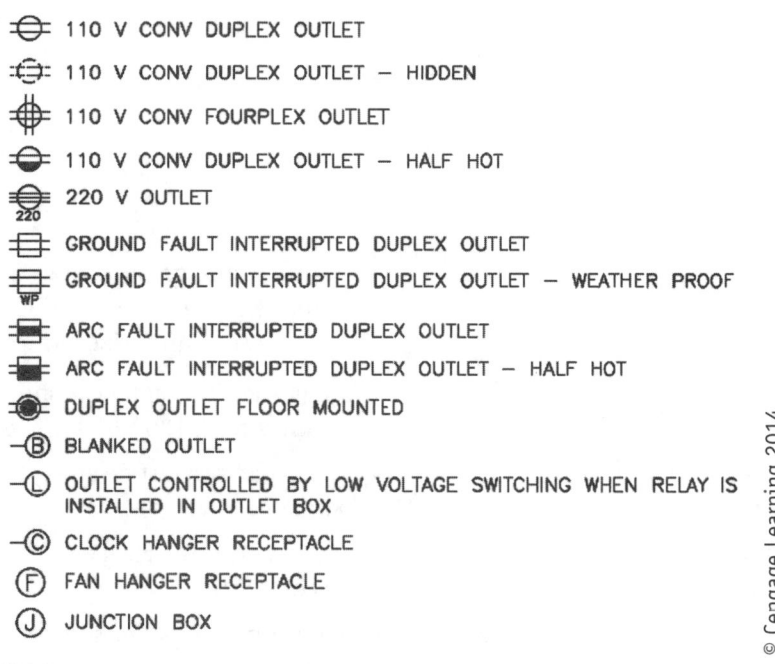

FIGURE 15-8 *Electric outlet symbols.*

Power and Energy

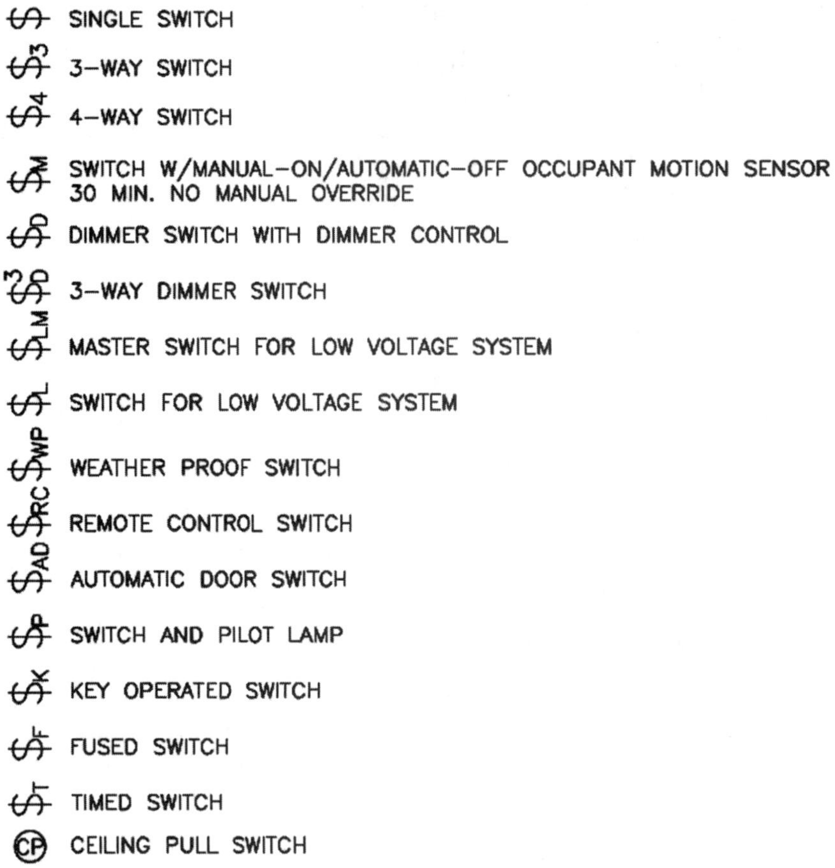

- SINGLE SWITCH
- 3-WAY SWITCH
- 4-WAY SWITCH
- SWITCH W/MANUAL-ON/AUTOMATIC-OFF OCCUPANT MOTION SENSOR 30 MIN. NO MANUAL OVERRIDE
- DIMMER SWITCH WITH DIMMER CONTROL
- 3-WAY DIMMER SWITCH
- MASTER SWITCH FOR LOW VOLTAGE SYSTEM
- SWITCH FOR LOW VOLTAGE SYSTEM
- WEATHER PROOF SWITCH
- REMOTE CONTROL SWITCH
- AUTOMATIC DOOR SWITCH
- SWITCH AND PILOT LAMP
- KEY OPERATED SWITCH
- FUSED SWITCH
- TIMED SWITCH
- CEILING PULL SWITCH

FIGURE 15-9 *Electric outlet symbols.*

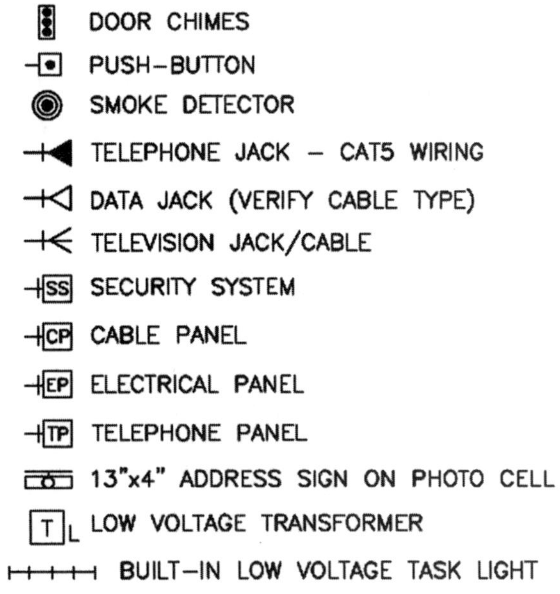

- DOOR CHIMES
- PUSH-BUTTON
- SMOKE DETECTOR
- TELEPHONE JACK – CAT5 WIRING
- DATA JACK (VERIFY CABLE TYPE)
- TELEVISION JACK/CABLE
- SECURITY SYSTEM
- CABLE PANEL
- ELECTRICAL PANEL
- TELEPHONE PANEL
- 13"x4" ADDRESS SIGN ON PHOTO CELL
- LOW VOLTAGE TRANSFORMER
- BUILT-IN LOW VOLTAGE TASK LIGHT

FIGURE 15-10 *Electric outlet symbols.*

Some of the general rules and guidelines from the National Electric Code (NEC) are given in the textbook and are summarized as follows:

- Single-family residences should have a minimum of 60 amps for a 120/240 service.
- The capacity must be at least the actual or calculated demand.
- Motors and electrical equipment must have a quick disconnect.
- Minimum 14-gauge wire for residential and 12-gauge wire for commercial buildings.
- Maximum 1,200 watts for 15 amp, and 1,500 watts for 20 amp in Two-wire 120 V circuits.
- Branch circuits shall not exceed 80 percent of load capacity and 70 percent for inductive fluorescent and HID lighting.
- Fixed-wire appliances shall not exceed 50 percent of branch capacity.

Problem Set 15.2 Power and Energy

Problem 15-9   What is the largest primary energy source in the United States?

Problem 15-10 What percentage of energy is consumed by the residential sector in the United States?

Problem 15-11 How much has energy consumption increased over the last 50 years in the residential sector of the United States?

Problem 15-12 What is the largest source of energy for commercial buildings in the United States?

Problem 15-13 What percentage of energy consumed by commercial buildings in the United States is for lighting?

Problem 15-14 Calculate the daily electrical use for the two-bedroom apartment with the appliances and daily usages in table 15-4.

Appliance	Watts	Hours/day	Watts/day
Air Conditioner 3.1/2 ton		5.00	
Water Heater		2.00	
Living Room w/150 watt lighting		1.00	
Master Bedroom w/ceiling fan		5.00	
Bathroom w/160 watt lighting		0.50	
Bedroom 2 w/150 watt lighting		2.00	
Computer		2.00	
Refrigerator 16 cu ft		5.00	
Range		0.50	
Microwave Oven		0.25	
Coffee Maker		0.25	
Toaster Oven		0.25	
Garbage Disposal		0.10	
Kitchen Lighting 160 watts		1.50	
Television		3.00	
Vacuum Cleaner		0.25	

TABLE 15-4 *Power consumption for a two-bedroom apartment.*

Problem 15-15   What is the cost of electricity in Nebraska?
Problem 15-16   What type of electrical service is used primarily in residential construction?
Problem 15-17   How many wires are needed for a three-phase electrical service?
Problem 15-18   What is the electrical symbol for a 220-volt outlet?
Problem 15-19   What is the electrical symbol for a three-way switch?
Problem 15-20   What is the electrical symbol for an electrical panel?
Problem 15-21   What is the minimum 120/240 V service amperage for single-family dwellings?
Problem 15-22   Does a blower motor on a furnace require a quick disconnect?
Problem 15-23   What is the minimum branch wire gauge required in residential construction?
Problem 15-24   What is the maximum branch wattage for a 15 amp, 120 V circuit?

BACKGROUND

Architects and engineers must understand the electrical distribution system within buildings. The layout of the electrical distribution system may affect the cost of the system itself, along with the cost based on the system's performance.

Electrical Panel Location

Each branch circuit should be controlled by an electrical panel. In residential construction, only one panel may be needed to feed all appliances, receptacles, and fixtures. There may be a separate panel to feed larger equipment, such as, the furnace and condensing unit. Typically, in commercial buildings, a sub panel would supply all electrical devices within about 100 feet. The main panels should be centrally located within 125 feet of the utility service provider. Electrical panels for single-phase, three-wire and three-phase, four-wire circuits are shown in figures 15-11 through 15-14 of the textbook. The minimum working space in front of a panel is shown in figure 15-15 of the textbook. In commercial buildings, a main electrical closet approximately 7 ft by 12 ft should be provided, and satellite closets should be approximately 5 ft by 3 ft. In commercial buildings, a telecommunications distribution closet should be provided on every floor. This closet should be approximately 10 ft by 12 ft for data-intensive buildings like offices and hospitals, and 4 ft by 7 ft for less-data-intensive occupancies. In large buildings, switchgear and transformer rooms should be provided. The size of these rooms varies with the floor area served, but for up to approximately 150,000 ft^2 served, a 30 ft by 30 ft space should be provided.

Problem Set 15.3   Electrical Panel Location

Problem 15-25   What is the minimum ceiling height above an electrical panel?
Problem 15-26   What is the minimum clearance in front of an electrical panel?
Problem 15-27   What is the approximate size of a main electrical closet in a commercial building?
Problem 15-28   What is the approximate size of a telecommunication closet in an office building?

BACKGROUND

Architects and engineers must be able to design and understand lighting systems in buildings to provide the proper level of illumination and be energy efficient. The choices in lighting types and layout are almost endless, and a good design procedure is very important in the success of the system's performance.

Lighting

The energy consumed for lighting in buildings is a major portion of all sources. Lighting consumes 20 percent of all energy in commercial buildings, as shown in 15-5. If you consider only the electrical energy consumption in commercial buildings, figure 15-11 shows that lighting consumes 38 percent, which is the largest portion by far of all sources.

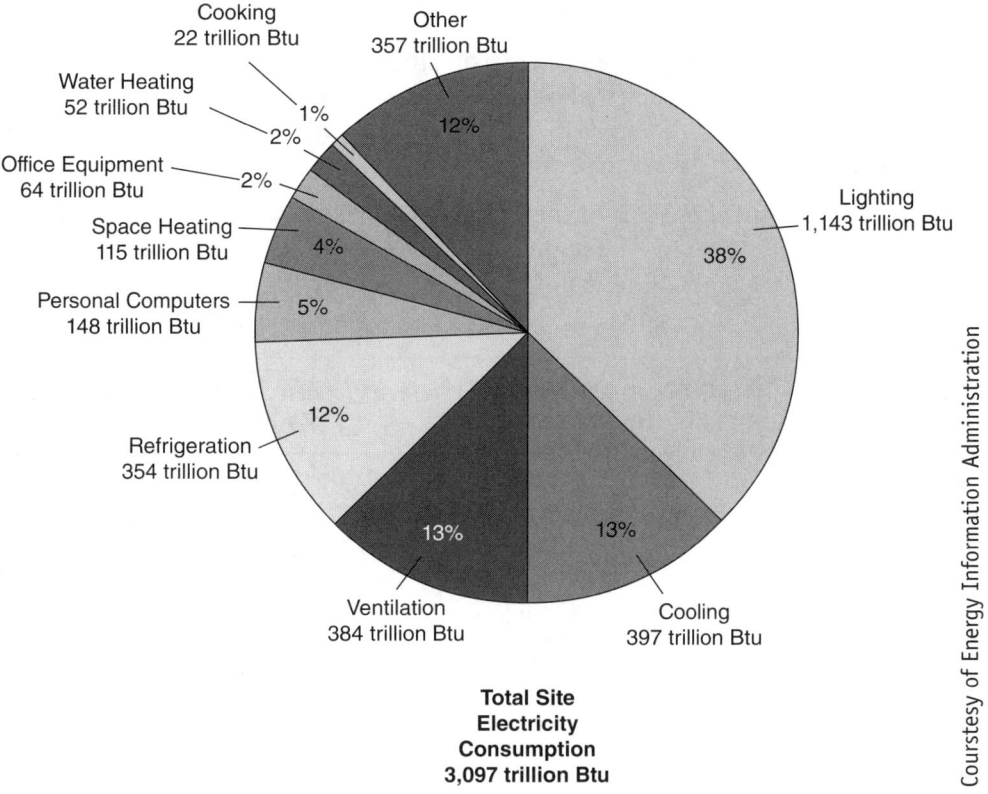

FIGURE 15-11 *Site electricity consumption.*

If you consider all forms of energy and uses, lighting consumes approximately 10 percent of the total energy in the United States.

The Design Process

The design process most commonly used in buildings has many steps and can be summarized in three basic stages:

- Programming, which involves research and development of design criteria
- Design development, where design selections are made and tested against the criteria
- Construction documentation, where the drawings, details, schedules, and specifications necessary to facilitate construction are prepared

Programming

The two major steps in the programming phase are data collection and establishing goals.

Data Collection

The information to be collected in the programming phase is detailed in the textbook and summarized here:

- Space dimensions involving room sizes and required tasks
- Furnishings to establish lighting layout and levels
- Surface reflectance of materials to determine amount of light needed (see textbook table 15-3)
- Space activities to classify the rooms and spaces

- Visual task to establish types of lighting
- Occupant type to develop illumination levels
- Special geometries, architectural features, existing conditions, image, budget, and other issues

The goals should be established and written toward specific lighting design criteria for each of the major tasks or functions within a building and reviewed often in the design development phase.

Design Development

In most cases, the design development phase is further subdivided into the schematic design phase, where various designs and selections are made, and design development, where the better solution is refined. The three primary needs in lighting design, according to the Illuminating Engineers Society (IES), are outlined in table 15-5.

Human Needs	Economics, Energy Efficiency, and the Environment	Architecture and Other Building- or Site-related Issues
• Visibility • Task performance • Visual comfort • Social communication • Mood and atmosphere • Health, safety, well-being • Aesthetic judgment	• Installation • Maintenance • Operation • Energy • Environment	• Form • Composition • Style • Codes and standards • Safety and security • Daylighting

TABLE 15-5 *Lighting quality needs and goals.*

Construction Documents

The construction documents necessary to communicate the design normally consist of drawings, details, schedules, and specifications. The primary drawings for lighting are the reflected ceiling plan and the electrical plan. Detail drawings may be individual portions of the lighting system or may be included in wall and ceiling sections. Schedules are used to quantify the fixtures and other elements of the lighting. Specifications are written to establish the criteria and standards that the design must meet.

Lighting Types

The four basic types of lighting are incandescent, low-pressure discharge (fluorescent), high-pressure discharge (HID), and miscellaneous. The following are the most widely used equipment according to the EIA:

- Incandescent lamps produce light by applying electrical current to a tungsten filament. The filament is mounted in a glass bulb that is filled with an inert gas, such as nitrogen and/or argon. The filament, when heated by the current, emits visible light. Incandescent lamps are energy inefficient and only convert 5 to 10 percent of the input energy into visible light, with the rest converted to heat.
- Halogen lamps are a type of incandescent lamp, but they last longer than standard incandescent lamps. These lamps also use a tungsten filament; however, it is enclosed in a halogen gas-filled (iodine or bromine) quartz capsule and is operated at a higher temperature than an incandescent lamp.
- Standard fluorescent lamps consist of a sealed gas-filled tube. The gas in the tube consists of a mixture of low-pressure mercury vapor and an inert gas, such as argon. The inner surface of the tube has a coating of phosphor powder. When an electrical current is applied to electrodes in the tube, the mercury vapor emits ultraviolet, radiation which then causes the phosphor coating to emit visible light. (The process is termed *fluorescence*.) A ballast is necessary to regulate and control the current and voltage. Two types of ballasts are used: magnetic and electronic. Electronic ballasts have several characteristics that make them attractive. They are more energy efficient, permit lamps to start quickly, and eliminate lamp flicker.

- Compact fluorescent lamps (CFL) produce light the same way as a standard fluorescent lamp. A CFL has a small diameter tube that is arranged in a shape and style that is more compact than a standard fluorescent lamp. A CFL with built-in ballast may have a screw-base and can replace a screw-in incandescent bulb. Other CFLs have a pin base that fits into fixtures with built-in ballasts.
- High-intensity discharge (HID) lamps produce light by an electrical arc discharge between two tungsten electrodes in a sealed arc tube mounted within an outer bulb. This lamp requires a ballast to start and regulate the arc. There are several types of HID lamps, including mercury vapor, metal halide, and high-pressure sodium lamps.

Many of these light sources are shown in table 15-4 of the textbook.

Selecting Light Sources and Equipment

There are many ways to measure lighting and light quality. These are covered in the textbook, and some are summarized here:

- A lumen is the basic measure of output for a light source that indicates the energy flow.
- Footcandles indicate the density of light that falls on a surface. One footcandle is equal to one lumen per square foot.
- Lux are a newer measure of light density. One footcandle is equal to 10.76 lux, but many designers use 10 as the conversion. A handheld lux meter is shown in figure 15-12.
- Efficacy is the amount of light produced per unit of energy consumed, expressed in lumens per watt (lm/W). Lamps with a higher efficacy value are more energy efficient.
- Average rated life of a particular type of lamp is defined by the number of hours when 50 percent of a large sample of that type of lamp has failed.
- Color rendering index (CRI) is a measurement of a light source's accuracy in rendering different colors when compared to a reference light source. The highest attainable CRI is 100. Lamps with a CRI above 70 are typically used in office and living environments.
- Correlated color temperature (CCT) is an indicator of the "warmth" or "coolness" of the color appearance of the lamp's light. The CCT is given in the Kelvin (K) temperature scale, and the higher the color temperature, the cooler the appearance of the light. Below 3,200 K the light has a "warm" appearance, and above 4,000 K the light has a "cool" appearance.

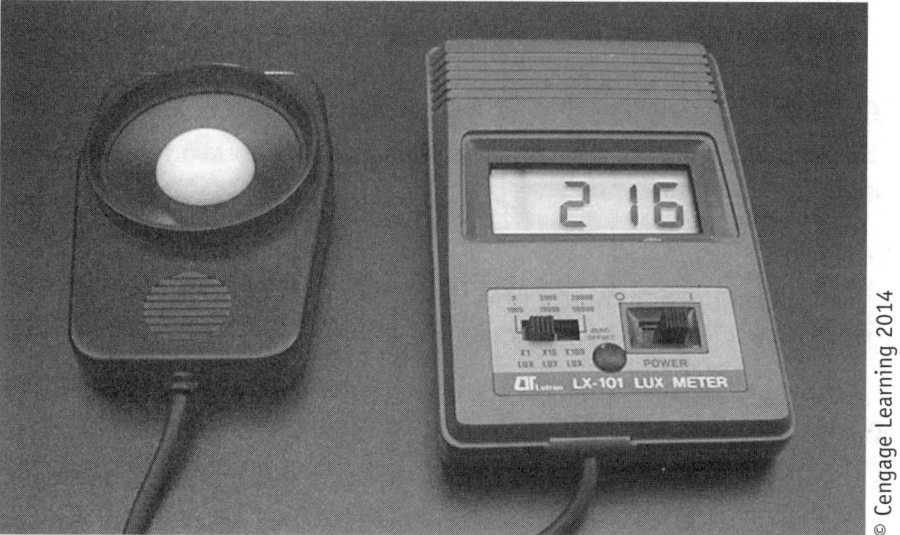

FIGURE 15-12 *Lux meter photo.*

When selecting lighting equipment efficacy, rated life, CRI, and CCT should all be considered. These values are shown in table 15-6 for the most common light sources according to the Office of Energy Efficiency and Renewable Energy, U.S. Department of Energy.

Type of Lighting	Efficacy (lumens/watt)	Typical Rated Lifetime (hours)	Color Rendering Index	Correlated Color Temperature (K)
Incandescent	10–19	750–2,500	97	2,500–3,000
Halogen	14–20	2,000–3,500	99	2,800–3,000
Standard Fluorescent				
T5	25–55	6,000–7,500	52–75	3,000–6,500
T8	35–87	7,500–20,000	52–90	3,000–6,500
T12	35–92	7,500–20,00	50–92	3,000–7,500
Compact Fluorescent	40–70	10,000	82	2,700–6,500
High Intensity Discharge				
Mercury Vapor	25–50	29,00	15–50	4,000–7,000
Metal Halide	50–115	3,000–20,000	65–90	3,000–4,400
High Pressure Sodium	50–124	29,000	22	1,900–2,200

TABLE 15-6 *Typical values of major lighting characteristics.*

Lighting Controls

Historically, the primary control of lighting has been manual switches, but to improve lighting efficiency and conserve energy, electronic switching is now standard in commercial buildings.

Manual Switching

Manual switches are used to open or close an electrical circuit. The two most common forms of manual switches are single-pole or single-throw and three-way.

Single-Pole, Single-Throw Switches

Single-pole switches are used to provide control from one location to one or more lighting sources by opening and closing a basic electric circuit. The wiring diagram for a single-pole switch is shown in figure 15-28 of the textbook.

Three-Way Switching

Three-way switches are used to provide control from more than one location to one or more lighting sources by opening and closing a continuous circuit. This requires a pair of wires between each switch location. The wiring diagram for a three-way switch is shown in figure 15-29 of the textbook, but a simple circuit diagram is shown in figure 15-13. This shows all four possible positions of two three-way switches that produce two ON and two OFF configurations.

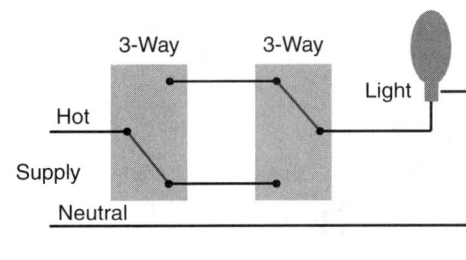 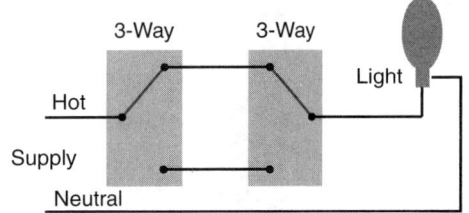

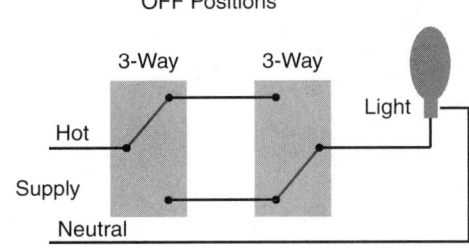 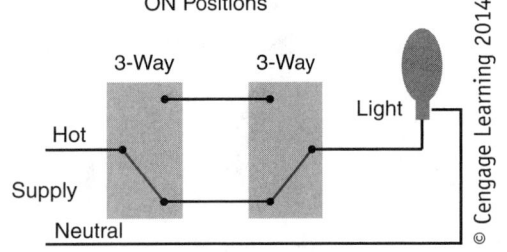

FIGURE 15-13 *Three-way switch.*

Multi-Level Switching

Multi-level switching uses two or more circuits controlled by different switches arranged in three ways as follows:

- Switching alternate lamps in each light fixture (see textbook figures 15-30 and 15-31)
- Switching alternate light fixtures
- Switching alternate rows of light fixtures

Multi-level switching saves energy by allowing some lamps to remain off when daylight levels are sufficient and allow for lower levels of lighting for occupant preference. In an individual office setting, a three-way multi-level switch can provide indirect lighting to a fixture with one lamp pointed upward, and direct light with the other two lamps pointed downward. This is shown diagrammatically in figure 15-14.

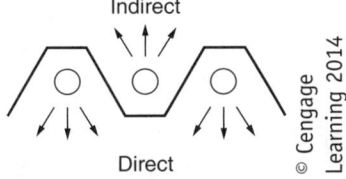

FIGURE 15-14 *Direct and indirect lighting.*

A switch used to control a three-lamp fixture in an office is shown in figure 15-15. This switch has a manual switch to turn the power on or off, a multi-level switch to control the three lamps, and a motion sensor to shut off the light if the room becomes unoccupied for a period of time.

Lighting 347

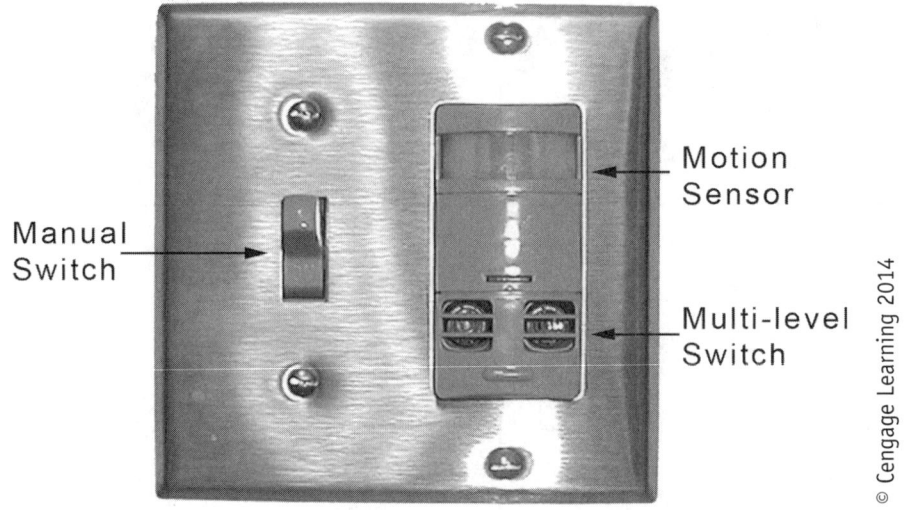

FIGURE 15-15 *Multi-function switch.*

Zoning

Zoning of lighting fixtures can divide a room into separate spaces. This can be desirable when daylighting varies across a room, as shown in figure 15-32 of the textbook. Zoning can be used to control lights that serve different functions, as shown in figure 15-16.

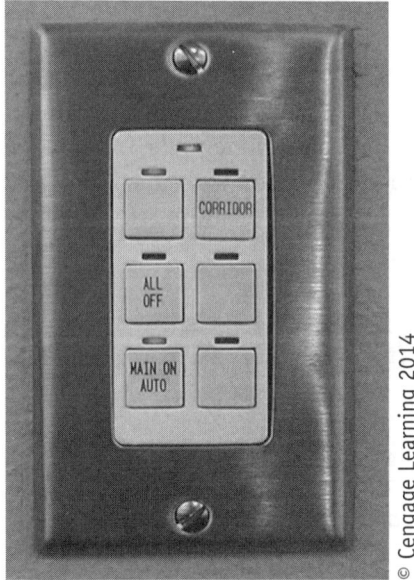

FIGURE 15-16 *Zoning switch.*

Zoning can also be used to divide spaces into separate areas and be combined with multi-level switching, as shown in figure 15-17.

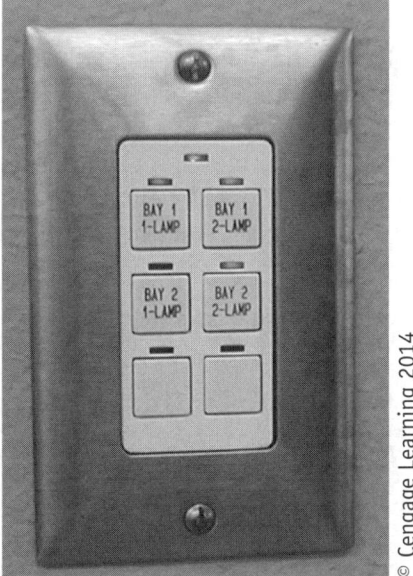

FIGURE 15-17 *Multi-level zoning switch.*

Dimming

Dimmers can be used to vary the amount of wattage delivered to a lighting fixture. There is better control in dimming incandescent fixtures than in fluorescent fixtures. A dimmable switch that controls multiple zones and has an LED display is shown in figure 15-18.

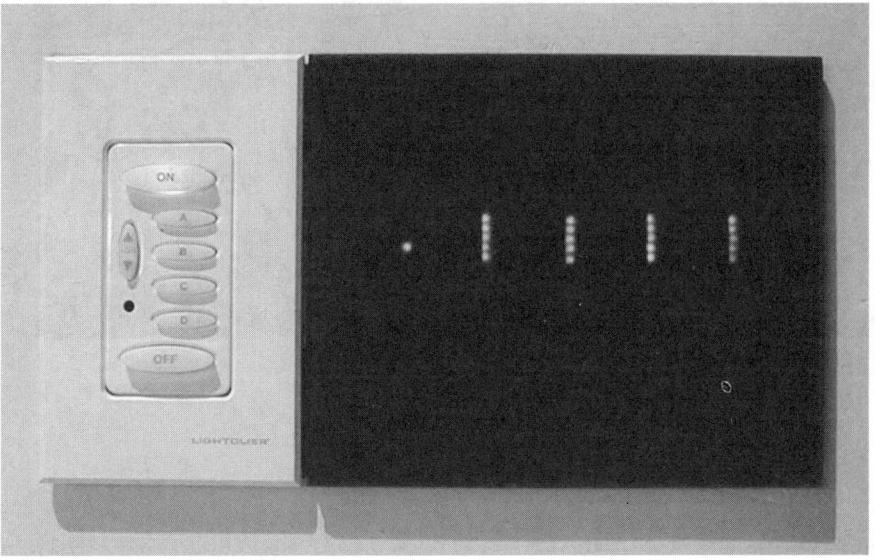

FIGURE 15-18 *Dimmer switch.*

Occupancy Sensors

Various types of occupancy sensors can be used to save energy by turning off lights in unoccupied spaces. They can use infrared or ultrasonic sensors to detect motion within a space. Infrared sensors require line of sight to detect motion, whereas ultrasonic sensors do not. The switch shown in figure 15-15 has an infrared sensor.

Lighting 349

Photosensors

Photosensors can be used to detect the level of daylighting and adjust the electric lighting appropriately. This can be one of the most effective ways to provide energy efficiency. The switches in figure 15-17 that are used to control zones are in combination with photosensors to automatically turn off one, two, or three lamps based on the amount of daylighting from the skylights and windows. As lighting-control systems become more elaborate, computers and high-performance lighting panels should be employed. Museums and art galleries are areas that require precise control of lighting levels. A high-performance lighting control panel with emergency circuits and computer controls is shown in figure 15-19.

FIGURE 15-19 *High performance lighting panel.*

Determining the Proper Quantity of Light

The Illuminating Engineers Society of North America (IESNA) publishes the *IESNA Lighting Handbook,* which establishes seven illuminance categories. These categories are summarized in table 15-7 with their recommended lighting levels.

Category	Visual task and spaces	Illuminance level	
		Lux	Footcandles
Orientation and simple visual task. Visual performance is largely unimportant. These tasks are found in public spaces where reading and visual inspection are only occasionally performed.			
A	Public spaces	30	3
B	Simple orientation for short visits	50	5
C	Working spaces where simple visual task are performed	100	10
Common visual task. Visual performance is important. These tasks are found in commercial, industrial, and residential applications. Recommended illuminance levels differ because of the characteristics of the visual task being illuminated.			
D	Performance of visual task of high contrast and large size	300	30
E	Performance of visual task of high contrast and small size, or visual task of low contrast and large size	500	50
F	Performance of visual task of low contrast and small size	1,000	100
Special visual task. Visual performance is critically important. These tasks are very specialized, including those very small or low-contrast critical elements. Recommended illuminance levels should be achieved with supplemental task lighting.			
G	Performance of visual task near threshold	3000–10,000	300–1,000

TABLE 15-7 *IESNA illuminance categories.*

The IESNA handbook contains a detailed matrix that prioritizes design issues for the various spaces. There are six general types of areas, including; interior, industrial, outdoor, sports/recreation, transportation, and safety/security. The design guide contains the recommended illuminance level for horizontal and vertical surfaces or the task plane, depending on the type of space. For interior spaces, the guide rates 16 design issues based on importance. The portion of the design guide for some interior spaces is summarized in table 15-8 with the recommended illuminance categories for horizontal and vertical surfaces.

Interior Location and Task	Illuminance Category	
	Horizontal	Vertical
Auditorium	B	A
Conference Rooms	D	B
Drafting/Graphic Arts	D	A
Educational Classrooms	E	D
Educational Gymnasium (Basketball)	F	D
Exhibition Halls	C	A
Food Service Kitchens	E	A
Health Care Emergency/Outpatient	E	C
Health Care Patient Rooms	B	A
Health Care Physical Therapy	D	B
Health Care Operating Room	G	E
Hotels	D	B
Houses of Worship	C	A
Libraries Stacks	D	D
Merchandising Display	E	C
Supermarkets	E	C
Museums	D	B
Office	E	B
Residential Kitchen	E	C
Toilets and Washrooms	B	A

TABLE 15-8 *IESNA lighting design guide Illuminance.*

Example 15.2 What are the recommended illuminance levels, in lux, for the desk and the marker board in a classroom?

The desk is a horizontal surface, and the marker board is a vertical surface. From table 15-8 for educational classrooms, the illuminance categories are E and D. From table 15-8, the illuminance levels are 500 and 300 lux.

The desk should have 500 lux and the marker board 300 lux.

Task Lighting

Task lighting should be used to supplement the general illumination where specific activities require higher lighting levels. The IESNA design guide subcategorizes specific tasks within general space, such as teller stations in a bank lobby or various types of reading or writing tasks. The values in table 15-8 are for the general illumination level in the space. It is more energy-efficient to provide a separate fixture for task lighting or to use multi-level switching. The maximum recommended ratio of task-to-ambient lighting levels is 10.

Problem Set 15.4 Lighting

Problem 15-29	What system consumes the largest portion of electricity in commercial buildings?
Problem 15-30	What percentage of total energy consumed in the United States is due to lighting in buildings?
Problem 15-31	What are the three basic steps in the lighting design process?
Problem 15-32	Why is the surface reflectance on a plaster wall painted white?
Problem 15-33	What are the three primary needs in lighting design?
Problem 15-34	What are the four basic types of lighting?
Problem 15-35	What is the basic measure of light energy flow?
Problem 15-36	What is the measure of light produced based on energy consumed?
Problem 15-37	What is the definition of the *average rated life* of a luminaire?
Problem 15-38	What is the least energy-efficient lighting type?
Problem 15-39	What is the primary difference in T5, T8, and T12 standard fluorescent lights?
Problem 15-40	What type of switch should be used to control one or more lighting sources from two locations?
Problem 15-41	What are three ways to use multi-level light switching?
Problem 15-42	Describe the switch in figure 15-17 in terms of zoning and multi-level switching.
Problem 15-43	What are the recommended illuminance levels, in lux, for a task of high contrast and large size?
Problem 15-44	What are the recommended illuminance levels, in lux, for a museum?
Problem 15-45	What are the recommended illuminance levels, in lux, for a supermarket?

CHAPTER 16
Planning for Plumbing

Skills List

After completing the problems in this chapter, you should be able to:

- Recognize the factors that affect quality and quantity of water supply
- Research plumbing codes to find materials and pluming components
- Understand plumbing processes and materials used in plumbing
- Identify symbols and abbreviations used to communicate plumbing
- Understand the supply and waste systems for plumbing
- Be aware of measures to conserve water in pluming design

BACKGROUND

Architects and engineers should have an understanding and be able to design plumbing systems in order to select components in an efficient manner. The supply and waste systems should be clearly communicated in the drawings and specifications.

Quality and Quantity of Water Supply

The basic plumbing systems are supply, waste, and venting. A simple diagram of these systems is shown in figure 16-2 of the textbook. The venting system is needed to allow the wastewater to drain freely and to vent waste gases, such as methane. The water supply from the municipal water system, that is suitable for drinking, is known as *potable water*. Normally, a backflow-prevention device similar to the one in figure 16-3 of the textbook is required. Some residences may use potable water from a well. In the design of green buildings, non-potable water can be used to water landscaping, wash cars, flush toilets, and in other places where potable water is not required. There are many proprietary rainwater-collection systems. Rainwater is normally collected from the roof, filtered, and stored in rigid containers similar to those in figure 16-1. This container system can be used in a vertical or horizontal orientation.

FIGURE 16-1   *Modular rainwater storage tanks.*

Another rainwater storage system that was developed in Australia uses flexible rain bladders to store water as shown in figure 16-2.

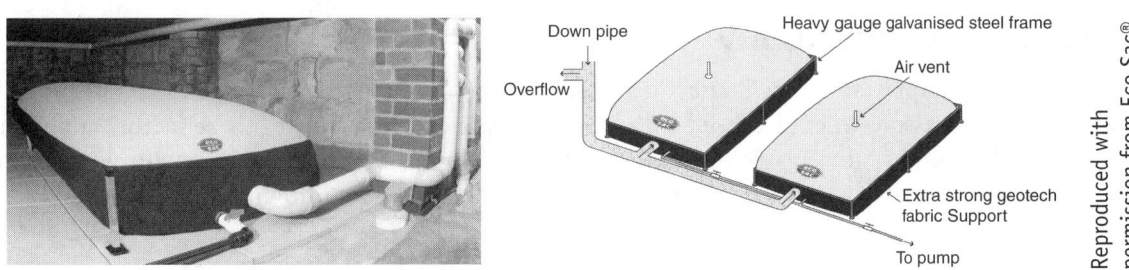

FIGURE 16-2   *Rainwater storage bladder.*

For any water supply system, the pressure and flow rate are the most important factors. The calculation of water supply was covered in chapter 7. The three main causes of reduction in water pressure, once it enters a building, are explained in the textbook and summarized as follows:

- Loss of static head to reach higher elevations in the building
- Friction loss due to interior pipe roughness, fittings, fixtures, and other devices
- Operation of the fixtures adding resistance

An isometric drawing of a residential plumbing system is shown in figure 16-4 of the textbook. The supply system is shown in color, the venting shown in white, and the waste system in black. Many of the minimum plumbing code requirements are shown on this drawing and are summarized as follows:

- Supply line 1/2 inch type-I rigid copper minimum
- Supply lines greater than 25 feet long or serving multiple fixtures must be 3/4 inch minimum
- Main supply line from service must be 1 inch type-L soft copper
- Whole-house shutoff valve and stop valves at each fixture required
- Sink drain 1-1/2 inch and toilet drain 3-inch minimum
- Branch drains 2 inch minimum with 1/8 inch per foot of slope
- Main drain stack 3 inch minimum and 4 inch outside to sewer system
- P-traps required at all fixtures
- Branch vent 2 inch and stack vent 3 inch minimum
- Drain and vent pipe are schedule 40 polyvinyl chloride (PVC) or acrylonitrile butadiene styrene (ABS)

Problem Set 16.1   Quality and Quantity of Water Supply

Problem 16-1   What are the two basic types of plumbing systems?
Problem 16-2   What are the two reasons to provide venting for plumbing?
Problem 16-3   What is the name of water that is suitable for drinking?
Problem 16-4   What are three ways that non-potable water can be used?
Problem 16-5   What are the three causes of water supply pressure loss inside a building?
Problem 16-6   What is the minimum size of a 30-foot-long water supply line?
Problem 16-7   What is the minimum size of a toilet drain?
Problem 16-8   What is used to prevent sewer gases from entering a fixture through the drain?

BACKGROUND

Architects and engineers should be able to research the local water supply system and building codes to determine requirements for water supply and other parts of the plumbing systems.

Location of the Water Supply

The location of the water supply source will affect the design of the water supply systems. In chapter 7, utilities, water supply, and water pressure were discussed with respect to site design. Many of these factors apply directly to the design of the plumbing system inside the building. The basic water pressure is the specific weight of water times the difference in height between the water source and final use location, known as the *static head*.

$$p = \gamma h$$

p = fluid pressure (psi or psf)
γ = specific gravity (lb/in^3 or lb/ft^3)
h = static head (in or ft)

Recall, that the basic water pressure of a one foot water column is 0.433 psi as follows:

$$1\ foot\ of\ water = 62.4\ \frac{lb}{ft^2} \left(\frac{1\ ft^2}{144\ in^2} \right) = 0.433\ \frac{lb}{in^2} \frac{1}{ft} = 0.433\ psi$$

Example 16.1 Determine the basic water pressure due to the height differential between a 48 foot water tower in sitting at an elevation of 1,000 feet and a house at an elevation of 996 feet.

The static head is the maximum static height of the water drop, which is the height of the water tower plus the difference in the elevations of the tower and the building.

$$h = 48 \text{ ft} + (1000 - 996) \text{ ft} = 52 \text{ ft}$$

The height can then be multiplied by the pressure for one foot of water to determine the pressure.

$$p = (52 \text{ ft}) 0.433 \frac{\text{psi}}{\text{ft}} = 22.5 \text{ psi}$$

The basic water pressure will drop as it travels from the source to the destination due to friction loss in the pipe and fittings. In chapter 7, the Hazen Williams formula used to calculate static head loss was introduced as follows:

$$h_f = \frac{10.44 \cdot L \cdot Q^{1.85}}{C^{1.85} \cdot d^{4.8655}}$$

h_f = head loss in feet of water
L = length of pipe in feet
Q = flow of water in gallons per minute (gpm)
C = Hazen-Williams coefficient (see textbook figure 7-41 or 16-5)
d = diameter of the pipe in inches

Refer to example 7-4 for use of the Hazen Williams formula to calculate head loss. Also, the fittings provide addition friction loss and can be converted to an equivalent length of pipe using a table similar to figure 7-42 of the textbook. Refer to example 7-5 for equivalent length calculations.

The final pressure at a location is known as the *dynamic pressure* and is equal to the total dynamic head times the specific fluid pressure. The total dynamic head is the static head minus the head loss. These are shown in the following two formulas from chapter 7:

$$TDH = \text{static head} - \text{head loss}$$

$$\text{dynamic pressure} = TDH \frac{0.433 \text{ psi}}{\text{ft}} = TDH \frac{\text{psi}}{2.31 \text{ ft}}$$

The water supply should be provided with shutoff valves on each side of the water meter and other fixture locations. A backflow protection device should be installed on the main supply line entering the building. A single line detail of a domestic water service entry is shown in figure 16-3 for a commercial project. A similar residential detail is shown in figure 16-6 of the textbook.

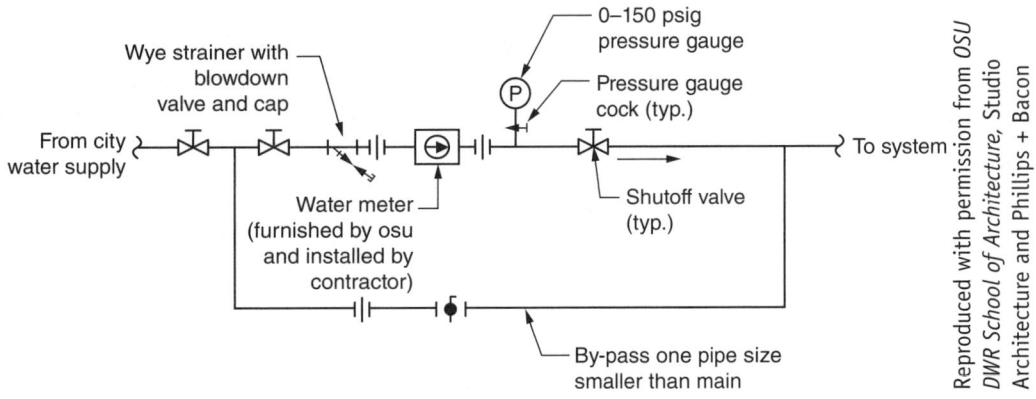

FIGURE 16-3 *Commercial water supply diagram.*

Codes

The International Building Code (ICC) and the International Plumbing Code (IPC) both contain requirements for plumbing in most buildings. The International Residential Code (IRC) covers the requirements for single-family residences, but mostly refers back to the IPC for fixture requirements. Chapter 29 of the ICC covers the required number of fixtures based on building occupancy category, while the entire IPC covers more detailed requirements of the plumbing system, such as pipe sizes, water pressures, and plumbing materials. To determine the number of total plumbing fixtures for a building use, table 16-1, which is based on the IBC table 2901.1. Only some of the occupancies are listed.

Classification	Occupancy	Description	Water Closets (Urinals) M	Water Closets (Urinals) F	Lavatories M	Lavatories F	Bathtubs or Showers	Drinking Fountains
Assembly	A1	Theaters	1/125	1/65	1 per 200		—	1 per 500
Assembly	A2	Restaurants	1 per 150		1 per 200		—	1 per 500
Assembly	A3	Religious	1/150	1/75	1 per 200		—	1 per 1,000
Business	B	Business, Banks, etc	1 per 25 for 1st 50, 1 per 50 beyond 50		1 per 40 for 1st 80, 1 per 80 beyond 80		—	1 per 100
Educational	E	Educational Facilities	1 per 50		1 per 50		—	1 per 100
Institutional	I-2	Hospitals	1 per room		1 per room		1 per 15	1 per 100
Institutional	I-4	Daycare	1 per 15		1 per 15		1	1 per 100
Mercantile	M	Stores	1 per 500		1 per 750		—	1 per 1000
Residential	R-1	Hotels, Motels	1 per unit		1 per unit		1 per unit	—

TABLE 16-1 *Minimum number of required plumbing fixtures.*

Urinals may be substituted for water closets but shall not exceed 67 percent of the count in occupancies A or E and 50 percent in all other occupancies. Water coolers or bottled water dispensers may be substituted for up to 50 percent of the drinking fountains. Specific details for each plumbing fixture can be found in table 16-2, which is based on IPC tables 604.3, 604.4, and 604.5.

Fixture	Maximum Flow Rate or Quantity	Required Water Supply		Minimum Pipe Size (inch)
		Flow Rate (gpm)	Pressure (psi)	
Bathtub	—	4	20	1/2
Dishwasher	—	2.75	8	1/2
Drinking Fountain	—	0.75	8	3/8
Hose Bibb	—	5	8	1/2
Lavatory	2.2 gpm at 60 psi	2	8	3/8
Shower	2.5 gpm at 80 psi	3	8	1/2
Sink	2.2 gpm at 60 psi	2.5	8	1/2
Urinal, flush value	1.0 gallon per flushing cycle	12	25	3/4
Water closet, tank	1.6 gallon per flushing cycle	3–6	20	3/8
Water closet, valve		25	35–45	1

TABLE 16-2 *Plumbing fixture and water supply criteria.*

Problem Set 16.2 Location of the Water Supply

Problem 16-9 What is the water pressure at the bottom of a 20-foot-deep lake?

Problem 16-10 What is the head loss for a property 3 miles from a water supply source? The water flows through a 12 inch steel pipe at 150 gpm.

Problem 16-11 What is the equivalent length of a 12 inch flanged pipe for the fittings shown in table 16-3?

Type of Fitting	Number of Fittings
Regular 90" elbow	4
Line flow tee	16
Branch flow tee	4
Gate valve	8
Globe valve	1

TABLE 16-3 *Pipe fittings.*

Problem 16-12 How many water closets are required to serve 200 occupants on an office building floor?

Problem 16-13 How many lavatories are required in the men's restroom serving a 2,000-seat theater?

Problem 16-14   How many urinals could be used in a men's restroom at an office building with 3 required water closets?

Problem 16-15   What is the maximum flow rate allowed for a tank-type water closet?

Problem 16-16   What is the required water pressure to supply a bathtub?

Problem 16-17   What size supply pipe is required to serve a shower?

Problem 16-18   Why does the required supply pipe size differ between a tank type and flush valve water closet?

BACKGROUND

Architects and engineers must be able to select the proper type of pipe for various plumbing systems.

Plumbing Materials and Process

The International Plumbing Code specifies the type of piping material that can be used for water service, water distribution, building drains, and vent pipes, both aboveground and underground, and sewer pipe. The water service is the pipe from the water main or other source of potable water supply, or from the meter when the meter is at the public right of way, to the water distribution system of the building served. The water distribution is the pipes within the structure or on the premises that conveys water from the water service pipe or from the meter when the meter is at the structure, to the points of utilization. The building drain is that part of the lowest piping of a drainage system that receives the discharge from soil, waste, and other drainage pipes inside and that extends 30 inches beyond the exterior walls of the building and conveys the drainage to the building sewer. The sewer is that part of the drainage system that extends from the end of the building drain and conveys the discharge to a public sewer, private sewer, individual sewage disposal system, or other point of disposal. Table 16-4 can be used for pipe selection and is based on the IPC tables 605.3, 605.4, 702.1, 702.2, and 702.3.

Material		Water Supply		Drainage and Venting		
		Service	Distribution	Aboveground	Underground	Sewer
Acrylonitrile butadiene styrene	ABS	x		x	x	x
Asbestos-cement		x			x	x
Brass		x	x	x		
Cast-iron				x	x	x
Chlorinated polyvinyl chloride	CPVC	x	x			
Concrete						x
Copper or copper-alloy pipe			x	x	x	
Copper or copper-alloy tubing		*	x	x	x	x	x

TABLE 16-4 Approved pipe material for plumbing systems.

Material		Water Supply		Drainage and Venting		Sewer
		Service	Distribution	Aboveground	Underground	
Cross-linked polyethylene	PEX	x	x			
PEX-AL-PEX		x	x			
PEX-AL-HDPE		x	x			
Ductile iron		x	x			
Galvanized steel		x	x	x		
Glass				x		
Polyethylene pipe	PE	x				x
Polyethylene tubing	PE	x				
PE-AL-PE		x	x			
Polypropylene	PP	x	x			
Polyolefin				x	x	
Polyvinyl chloride	PVC	x		x	x	x
Polyvinylidene fluoride	PVDF			x	x	x
Stainless steel (type 304)		x	x	x		x
Stainless steel (type 304L)		x	x			
Stainless steel (type 316)		x	x			
Stainless steel (type/316L)		x	x	x	x	x
Vitrified clay						x

TABLE 16-4 *Continued.*

*Refer to the code for the type of copper tubing allowed. PEX-AL-PEX is cross-linked polyethylene/aluminum/cross-linked polyethylene, PEX-AL-HDPE is cross-linked polyethylene/aluminum/high-density polyethylene, and PE-AL-PE is polyethylene/aluminum/polyethylene.

In the plumbing trade, the size of copper tubing is measured by its nominal diameter (average inside diameter). Some trades, such as, heating and cooling technicians for instance, use the outside diameter (OD) to designate copper tube sizes. Common wall-thicknesses of copper tubing are "Type K," "Type L," and "Type M," and sizes are given in table 16-5.

Nominal size	Outside Diameter (OD) (inches)	Inside Diameter (ID) (inches)		
		Type K	Type L	Type M
3/8	1/2	0.402	0.430	0.450
1/2	5/8	0.528	0.545	0.569
5/8	3/4	0.652	0.668	0.690
3/4	7/8	0.745	0.785	0.811
1	1 1/8	0.995	1.025	1.055
1 1/4	1 3/8	1.245	1.265	1.291
1 1/2	1 5/8	1.481	1.505	1.527
2	2 1/8	1.959	1.985	2.009
2 1/2	2 5/8	2.435	2.465	2.495
3	3 1/8	2.907	2.945	2.981

TABLE 16-5 *Copper tubing sizes for plumbing.*

Problem Set 16.3   Plumbing Materials and Process

Problem 16-19   *What are the approved uses of ABS pipe?*
Problem 16-20   *What are the approved uses of copper tubing?*
Problem 16-21   *What are the approved uses of PVC pipe?*
Problem 16-22   *What are the approved uses of PEX pipe?*
Problem 16-23   *What are the outside and inside diameters of 3/4 inch Type L copper tubing?*
Problem 16-24   *What is the thickest type of copper tubing?*

BACKGROUND

Architects and engineers should be able to clearly communicate the water supply system to the client and building trades.

Communication of the Water Supply System

The primary drawing used to communicate the plumbing design is known as the *plumbing plan*. The plumbing plan shows the fixture locations and the piping needed to service the system. Part of a plumbing plan is shown in figure 16-4 for a commercial building. Refer to figure 16-9 in the textbook for a residential plumbing plan.

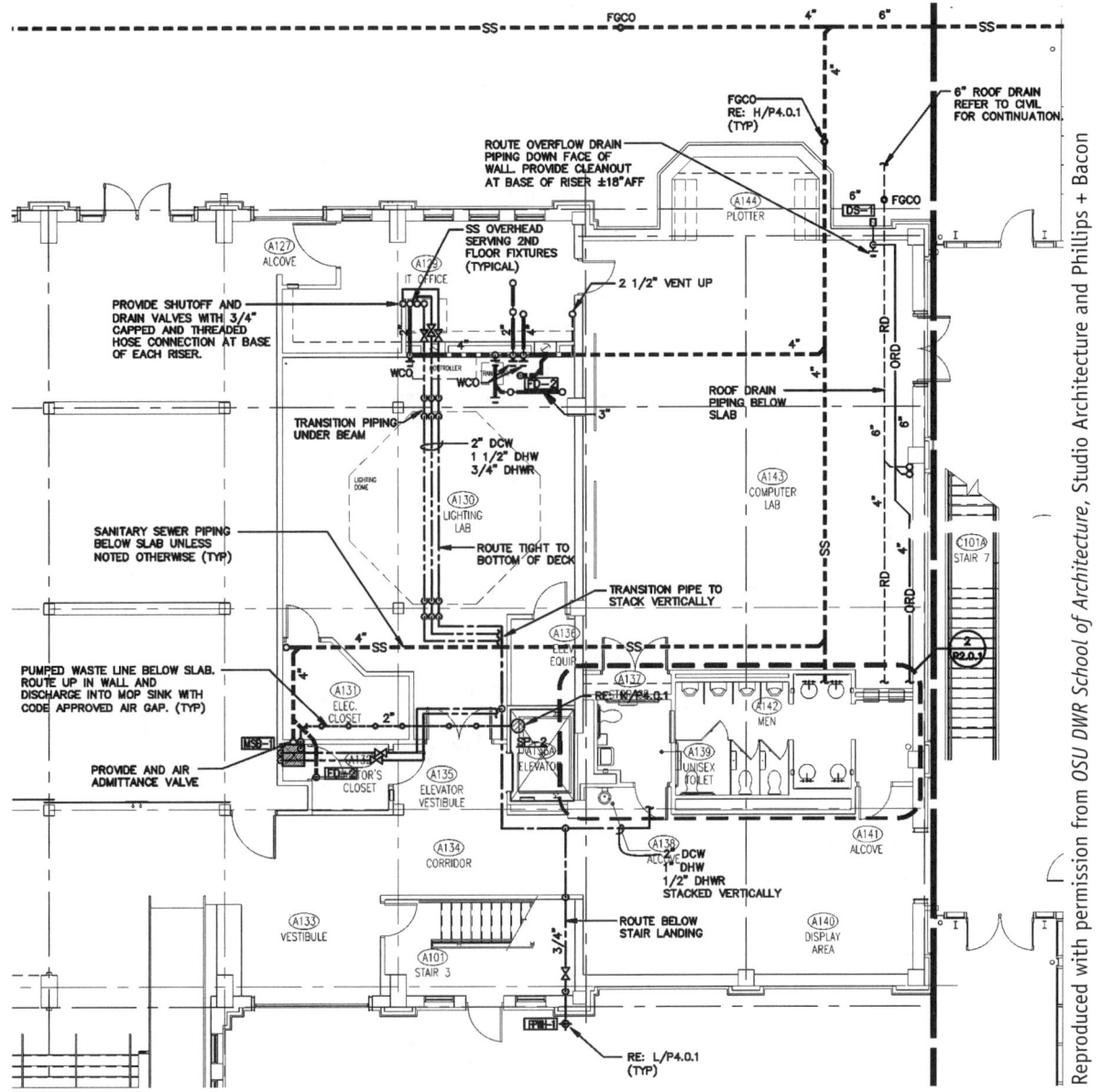

FIGURE 16-4 *Plumbing plan.*

All plumbing drawings should use standard symbols for both fixtures and piping. Typical residential plumbing fixture symbols and abbreviations are shown in figure 16-5. Typical commercial pipe and fittings are shown in figure 16-6.

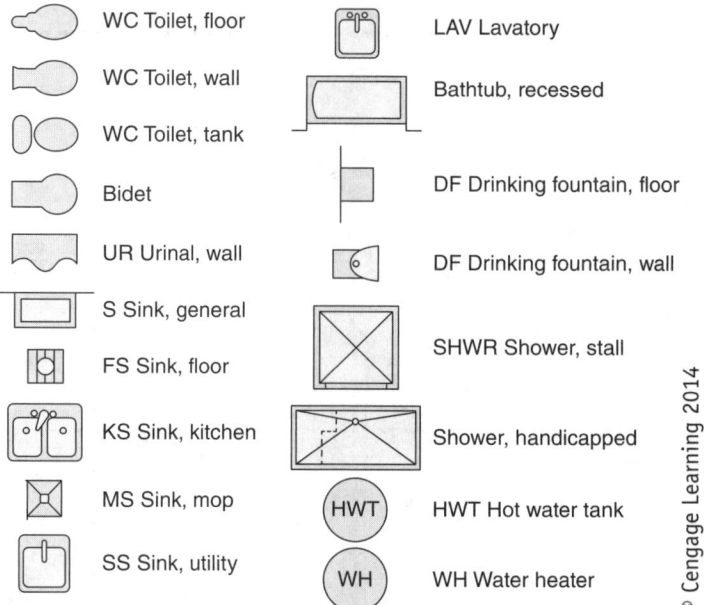

FIGURE 16-5 *Plumbing fixture symbols.*

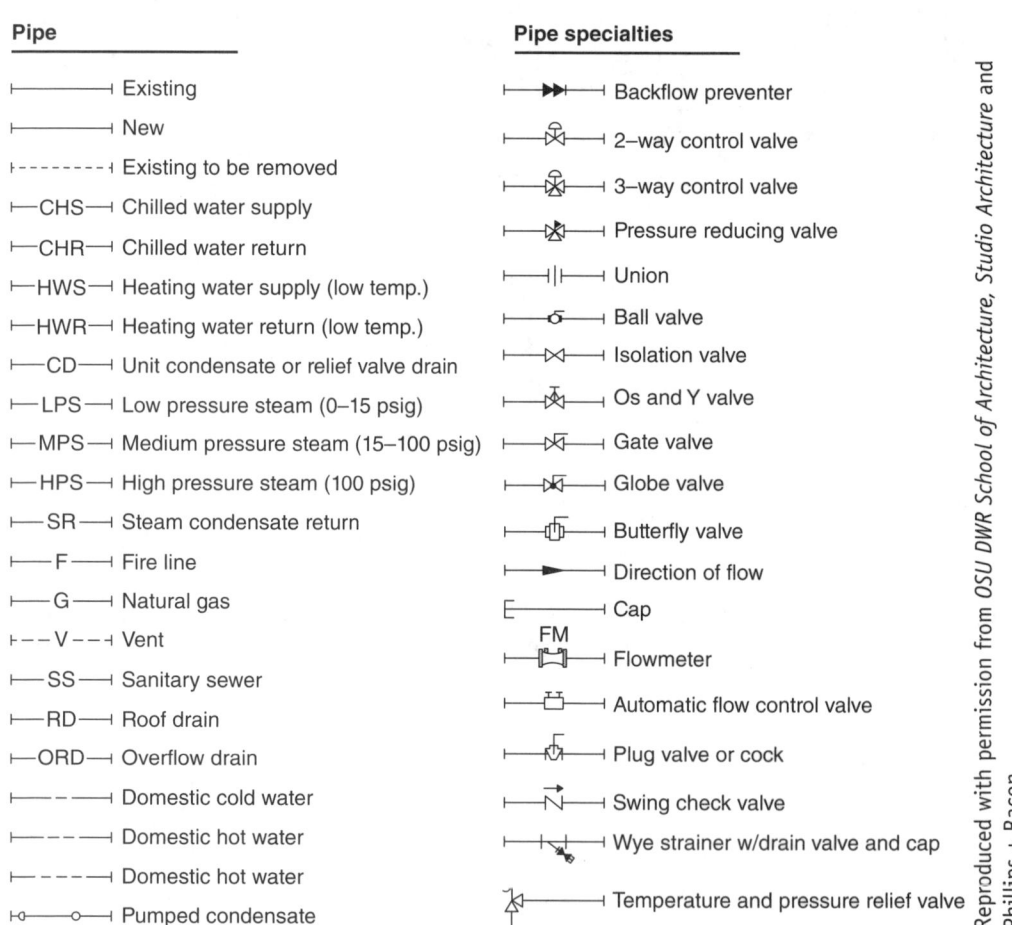

FIGURE 16-6 *Plumbing pipe and fitting symbols.*

Diagramming the Water System

If the scale of the plumbing plan is not large enough to show all the required information, a detailed or enlarged plumbing plan may be drawn, as shown in figure 16-7.

Chapter 16: Planning for Plumbing

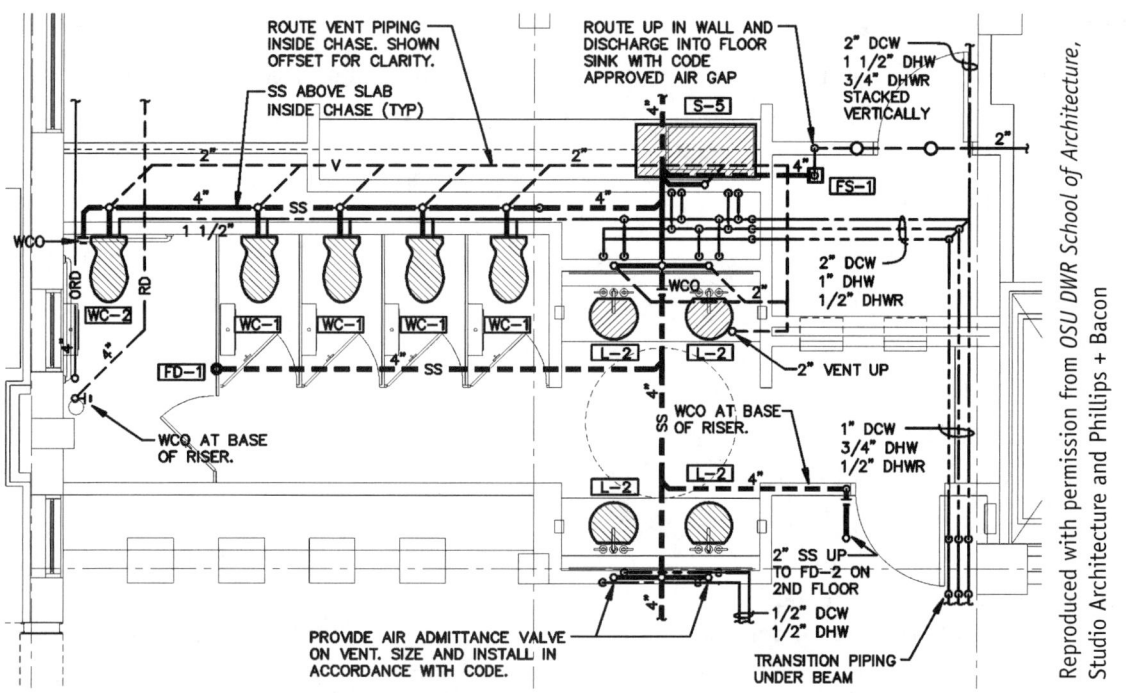

FIGURE 16-7 *Detailed plumbing plan.*

Other systems, such as the mechanical heating and air-conditioning, may require separate plumbing or piping plans, as shown in figure 16-8.

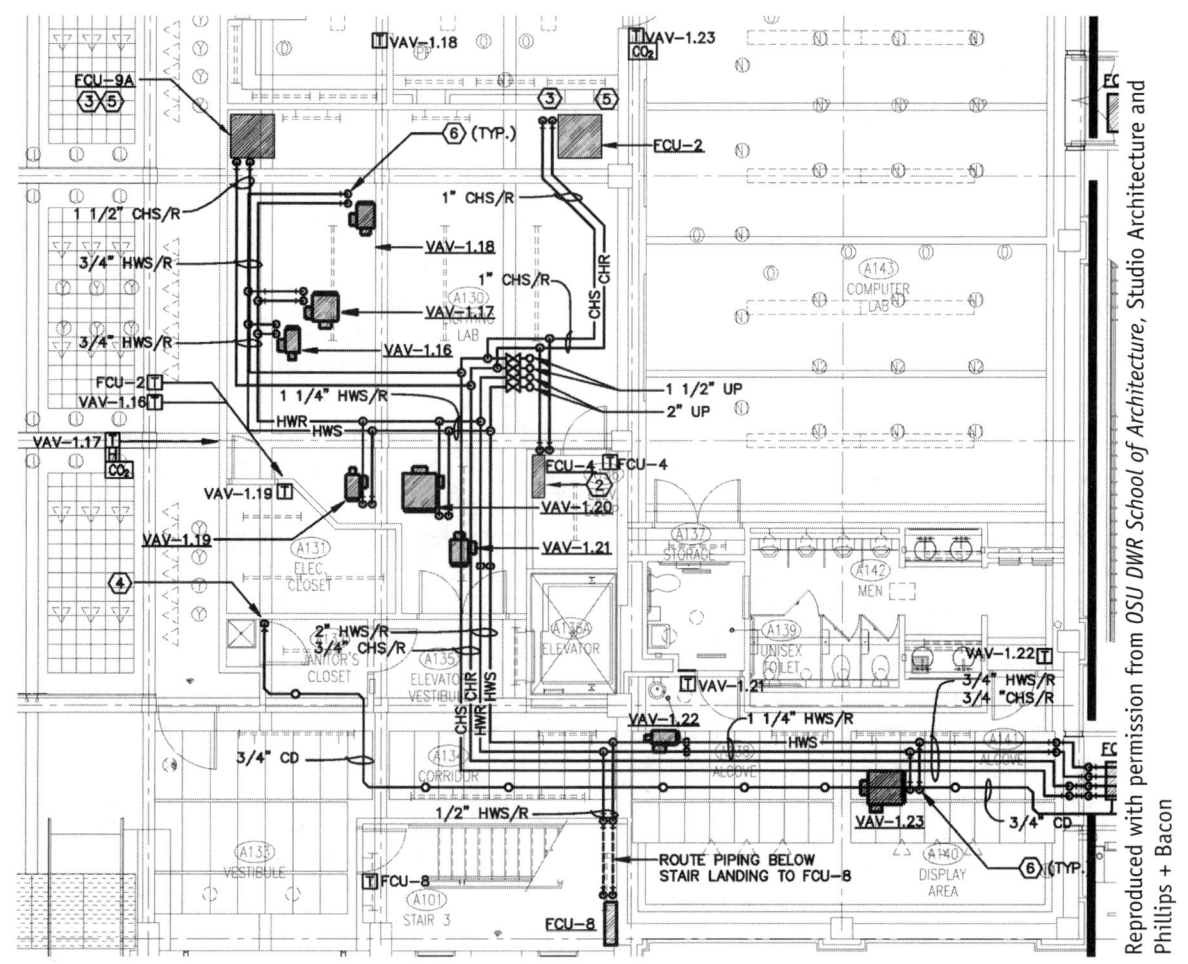

FIGURE 16-8 *Mechanical piping plan.*

The textbook covers the process of drawing a plumbing plan in figures 16-13 and 16-15.

Communication of the Water Supply System 365

Strategies for Water Conservation

Water has many uses, and consumption varies widely across the United States. The eight categories used by the United States Geological Survey (USGS) and the percentage of total use are summarized as follows:

- Public supply – 11 percent
- Domestic – 1 percent
- Irrigation – 31 percent
- Livestock – Less than 1 percent
- Aquaculture – 2 percent
- Industrial – 4 percent
- Mining – 1 percent
- Thermoelectric power – 49 percent

California, Texas, Idaho, and Florida account for more than one-fourth of all fresh and saline water withdrawn in the United States. If we look at only residential domestic use, we can start to find ways to conserve water in buildings. The percentage of water used for various uses in the home is shown in figure 16-9. Indoor use accounts for only about 40 percent of total home consumption, with the other 60 percent due to outdoor consumption. The Environmental Protection Agency (EPA) has a program titled Water Sense that gives recommended ways to conserve water in the home. Figure 16-10 is a list of indoor and outdoor water-saving techniques.

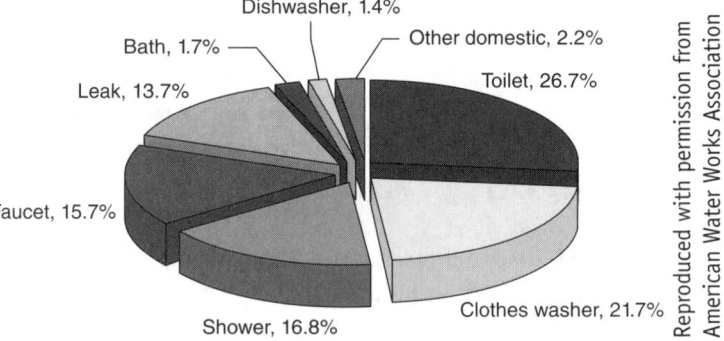

FIGURE 16-9 *Residential indoor water usage.*

Indoor Water Conservation

- Older toilets use between 3.5 and 7 gallons of water per flush. However, WaterSense® labeled toilets use at least 60 percent less water.
- A leaky toilet can waste about 200 gallons of water every day.
- A bathroom faucet generally runs at 2 gallons of water per minute. By turning off the tap while brushing your teeth or shaving, a person can save more than 200 gallons of water per month.
- High-efficiency washing machines can conserve large amounts of water. Traditional models can use 50 gallons or more of water per load, but newer, energy- and water-conserving models (front-loading or top-loading, non-agitator ones) use less than 27 gallons per load.
- Washing the dishes with an open tap can use up to 20 gallons of water, but filling the sink or a bowl and closing the tap saves 10 of those gallons.
- Keeping a pitcher of water in the refrigerator saves time and water instead of running the tap until it gets cold.
- Not rinsing dishes prior to loading the dishwasher could save up to 10 gallons per load.

Outdoor Water Conservation

- Many people water their lawns too often and for too long, oversaturating plants. It's usually not necessary to water grass every day. Instead, test your lawn by stepping on a patch of grass; if it springs back, it doesn't need water.
- Converting to a water-efficient landscape through proper choice of plants and careful design can reduce outdoor water use by 20 to 50 percent.
- If homeowners with irrigation systems hired WaterSense irrigation partners to perform regular maintenance, each household could reduce irrigation water by 15 percent or about 9,000 gallons annually—or the amount of water that would flow from a garden hose nonstop for nearly a whole day.
- Soil moisture sensors determine the amount of water in the ground available to plants. These sensors, when professionally installed and properly maintained, can potentially save a household more than 11,000 gallons of water used for irrigation annually.

FIGURE 16-10 *Indoor and outdoor water conservation – EPA.*

The cost of water varies, depending on the amount and type of usage. A 2009 survey of the 50 largest cities by Black & Veatch is summarized in table 16-6 based on amount and usage. From the last column, you can see that the more water is used, the cheaper it becomes.

Type	Usage (gallons)	Water	Sewer	Total	(per 1000 gallons)
Residential	3,750	15.35	20.03	35.38	9.43
	7,500	25.66	33.80	59.46	7.93
	15,000	50.09	63.28	113.36	7.56
Commercial	100,000	307.87	434.52	742.39	7.42
Industrial	10,000,000	27,765.61	41,011.83	68,777.44	6.88

TABLE 16-6 *Average monthly cost of water and wastewater.*

Other methods of water efficiency include harvesting rain and the use of gray water. Examples of rainwater storage systems were shown in figures 16-1 and 16-2. Gray water, or water from sinks, showers, and bathtubs, can be reused for irrigation or reclaimed. Both can lead to Leadership in Energy and Environmental Design (LEED) credits. The following is a summary of LEED criteria (1 point each) that could be achieved from a rain harvesting system:

- Limit disruption and pollution of natural water flows by managing storm water runoff.
- Limit disruption of natural water flows by eliminating stormwater runoff, increasing on-site infiltration, and eliminating contaminant.
- Use captured rain or recycled site water to reduce potable water consumption for irrigation by 50 percent over conventional means.
- Use only captured rain or recycled site water to eliminate all potable water use for site irrigation.
- Reduce the generation of wastewater and potable water demand, while increasing the local aquifer recharge.
- Maximize water efficiency within buildings to reduce the burden on municipal water supply and wastewater systems. Use 20 percent less water than the water use baseline calculated for the building (not including irrigation).
- Maximize water efficiency within buildings to reduce the burden on municipal water supply and wastewater systems. Use 30 percent less water than the water use baseline calculated for the building (not including irrigation).

Problem Set 16.4   Communication of the Water Supply System

Problem 16-25   Draw a restroom approximately 8 feet by 6 feet with a recessed bathtub, tank-type toilet, and lavatory on a 4-foot base cabinet.

Problem 16-26   Draw the water supply system pipes and fittings for the bathroom in problem 16-25.

Problem 16-27   How are existing and new pipes normally distinguished in a plumbing plan?

Problem 16-28   What size supply pipe is used for the water closets in figure 16-7?

Problem 16-29   What industry category uses the largest percentage of water?

Problem 16-30   What fixture inside a residence uses the largest percentage of water?

Problem 16-31   How much water use in a residence is used outdoors?

Problem 16-32   List three ways to reduce indoor water use in a residence.

Problem 16-33 List three ways to reduce outdoor water use in a residence.
Problem 16-34 What is the average cost of water and sewer in residences consuming approximately 7,500 gallons per month?
Problem 16-35 What is gray water, and how can it be used to reduce domestic water consumption?
Problem 16-36 List three ways that rain harvesting can achieve LEED credits in building construction?

BACKGROUND

Architects and engineers should be able to clearly communicate the waste removal system to the client and building trades.

Communication of the Waste Removal System

The waste removal system is separate from the water supply system but is normally drawn on the same plumbing plan. The key components are the drainage and venting pipes. In textbook figure 16-19, these two components are shown. In a plumbing plan, such as figure 16-7, the pipes are labeled SS for the drainage and V for the vent.

Venting of the Waste System

The venting system is used to allow water to drain freely through the drainage pipe and also vent sewer gases out of the system. Traps that hold a small amount of water at the drain prevent gases from flowing back into the occupied space or fixture. Fittings used in the drainage pipes should always allow for water to flow in the proper direction with wye and tee fittings properly installed (textbook figure 16-22). Special venting is required for an island sink or other fixture, as shown in figure 16-27 of the textbook.

Diagramming the Waste Removal System

The drawing of the drainage waste and venting (DWV) system is similar to the water supply and differs mainly in the flow direction. The textbook covers the process of drawing a plumbing plan in figures 16-23, 16-25, 16-26, and 16-28.

Problem Set 16.5   Communication of the Waste Removal System

Problem 16-37 What size drainage waste pipe is used for the water closets in figure 16-7?
Problem 16-38 What size vent pipe is used for the water closets in figure 16-7?
Problem 16-39 Make a drawing of a drainage waste junction with a wye fitting oriented in the proper direction.
Problem 16-40 What are the two reasons to provide venting for plumbing?

BACKGROUND

Architects and engineers should develop a site in such a way as to not adversely impact the flow of stormwater around and off the site. The topography should be carefully designed as to provide minimum impact on development and the surrounding properties.

Storm Water Drainage

Stormwater drainage was extensively covered in chapter 7: Site Design. The grading plan and stormwater runoff calculations were detailed in that chapter. A few additional problems are provided in this section.

Problem Set 16.6   Storm Water Drainage

Problem 16-41   What is the maximum slope for a lawn?

Problem 16-42   What is the difference in the runoff coefficients for sandy and clay soil lawns with a slope of 10 percent?

Problem 16-43   What would be the time of concentration for a site with a hydraulic length of 1,000 feet, an average slope of 5 percent, and a runoff coefficient of 0.50?

Problem 16-44   What is the precipitation intensity for a time of concentration of 60 minutes and a 10-year storm using figure 7-11?

Problem 16-45   Determine the stormwater runoff for a pre-development 10-year storm with a composite runoff coefficient of 0.2, a precipitation intensity of 4.5 in/hr, and a 20-acre site.

CHAPTER 17
Indoor Environmental Quality and Security

Skills List

After completing the problems in this chapter, you should be able to:

- Identify the various forms of energy

- Understand energy consumption in buildings

- Determine ventilation rates for indoor environmental quality

- Understand how heating and cooling loads are calculated

- Explain the methods of heat transfer

- Understand advantages of intelligent building systems and security

BACKGROUND

Architects and engineers should have an understanding of the various forms of energy used in buildings and how to minimize environmental impact from energy use. The supply of heating and cooling systems for buildings should provide comfort for the occupants and not adversely impact the environment.

Energy

The primary sources of energy and their uses according to the U.S. Energy Information Administration (EIA) were covered in the second section of chapter 15 and are summarized as follow:

- Primary energy sources for the major sectors of the economy for 2008 are shown in figure 15-1.
- Energy consumption in 2009 for the major sectors of the economy is shown in figure 15-2.
- Increasing energy consumption for the last five decades is shown in figure 15-3.
- Energy sources used in commercial buildings for 2003 are shown in figure 15-4.
- Energy usage for the systems within commercial buildings in 2003 is shown in figure 15-5.

Electrical makes up a majority of consumption, more than all other sources combined, even though the energy used for heating, ventilation, and air conditioning (HVAC) is 51 percent of the total used in commercial buildings. In residential buildings, the energy used for HVAC systems is 46 percent, as shown in figure 15-7. The use of renewable energy sources must be increased to reduce the use of dwindling, non-renewable fossil fuels. The amount of renewable energy sources currently used is only 7 percent of the total, as shown in figure 15-1. Renewable energy sources include: biomass, geothermal, hydroelectric, solar, and wind. These are summarized in figure 17-3 of the textbook. Non-renewable energy sources include: coal, natural gas, nuclear, and petroleum. The Environmental Protection Agency (EPA) recognizes the renewable energy sources as green power sources that are more environmentally friendly than non-renewable energy sources.

The American Society of Heating, Refrigerating, and Air Conditioning Engineers (ASHRAE) and other organizations like the EPA provide standards for indoor environmental quality (IEQ). Many problems can damage the IEQ, such as mold, PCBs, lead, asbestos, radon, mercury, formaldehyde, and other chemicals. Environmentally preferable products should be used to increase the IEQ. The EPA outlines the Environmentally Preferable Purchasing (EPP) program to help the federal government buy green products. The energy efficiency of a building is impacted by many factors, such as exterior environment, climate, building envelope, construction materials, window choices, number of occupants, occupant activities, interior features, and other considerations. Some of these will be included later in the selection of HVAC systems.

Problem Set 17.1   Energy

Problem 17-1   What is the smallest primary energy source in the United States?

Problem 17-2   What percentage of energy is consumed by the commercial sector in the United States?

Problem 17-3   How much has energy consumption increased over the last 50 years in the commercial sector of the United States?

Problem 17-4   What is the largest source of energy for commercial buildings in the United States?

Problem 17-5   What percentage of energy consumed by commercial buildings in the United States is for space heating?

Problem 17-6   What percentage of energy consumed by residential buildings in the United States is for space heating?

Problem 17-7   What are three energy sources that are classified as green power sources?

Problem 17-8   What are three energy sources that are not classified as green power sources?

Problem 17-9   What are three materials that can damage the indoor environmental quality of a building?

Problem 17-10   What are three factors that impact the energy efficiency of a building?

BACKGROUND

Architects and engineers must be aware of the many factors that influence the selection of the heating and cooling systems in buildings. The consumption of energy during the lifecycle of the building is the primary concern in this selection.

Factors Influencing Energy Consumption

The choice of heating and cooling systems to use is affected by many considerations. The size of the building is probably the first thing to consider. Small buildings usually have a single-occupancy use, and the selection is primarily based on building considerations. The selection of heating and cooling systems is summarized in table 17-1, which is modified from *Architect's Studio Companion: Rules of Thumb for Preliminary Design*, by Edward Allen and Joseph Iano. The selection of heating and cooling systems for large buildings is primarily based on the occupancy and will be covered later in this chapter. The abbreviated system list is given below and used in table 17-1.

- FA – Forced air
- HP – Heat pump furnace
- HH – Hydronic heating (heating only)
- AS – Active solar heating (heating only)
- EC – Evaporative cooler (cooling only)
- PU – Packaged terminal units or Through-the-wall units
- EB – Electric baseboard convectors (heating only)
- EF – Electric fan-forced unit heaters (heating only)
- RH – Radiant heating (heating only)
- WF - Wall furnace and direct-vent space heaters (heating only)
- HS – Heating stoves (heating only)
- PS – Passive solar heating (heating only)

When several considerations are included in the selection process, it is a good idea to prioritize the considerations. This can help to make the optimum selection.

Consideration	FA	HP	HH	AS	EC	PU	EB	EF	RH	WF	HS	PS
Combine heating and cooling	X	X				X						
Minimize first cost					X	X	X	X		X	X	
Minimize operating cost in very cold climates	X		X	X							X	X
Minimize operating cost in moderate climates		X		X	X					X	X	X
Maximize control of air velocity and air quality	X	X										
Maximize individual control over temperature			X			X	X	X				
Minimize system noise	X	X	X	X			X		X			X
Minimize visual obtrusiveness	X	X							X			
Maximize enjoyment of seasons											X	X
Minimize floor space used for the system					X	X	X	X	X	X		X
Minimize system maintenance	X						X	X	X	X		X
Avoid having a chimney		X				X	X	X	X			
Maximize the speed of construction						X	X	X		X		

TABLE 17-1 *Heating and cooling systems for small buildings.*

Example 17.1   Select a combined heating and cooling system that minimizes operating cost in very cold climates and a combined system that minimizes operating cost in moderate climates.

From table 17-1, the three combined systems are FA, HP, and PU. To minimize operating costs in very cold climates, select FA, forced air. To minimize operating cost in moderate climates select HP, heat pump furnace.

Other selection strategies can be used, such as the consideration recommended by ASHRAE. The considerations include: occupant comfort, energy efficiency, ease of use, service life, initial cost, lifecycle cost, and value added features. To reduce loads for any system used, the building should employ other strategies, such as maximizing daylight, ventilation and moisture control, and avoiding materials with volatile organic compounds (VOCs). Examples of VOCs are mold, PCBs, lead, asbestos, radon, mercury, and formaldehyde. A list of VOCs can be found online at the EPA's site www.epa.gov/iaq/voc.html. Figure 17-1 shows a skylight used to increase daylighting and to reduce heating loads due to lighting.

FIGURE 17-1   *Skylight photo.*

Energy Codes

Many building and energy codes should be used in the proper selection and design of mechanical systems. The following is a list of sources:

- ASHRAE – American Society of Heating, Refrigerating, and Air Conditioning Engineers
- ACCA – Air Conditioning Contractors of America
- ICC – International Code Council
- IECC – International Energy Conservation Code
- LEED – Leadership in Energy and Environmental Design
- EPA – Environmental Protection Agency
- IRC – International Residential Code

One method of the system design process is to use the ACCA manuals. Figure 17-2 is a flow chart given by the ACCA for the both residential and commercial design.

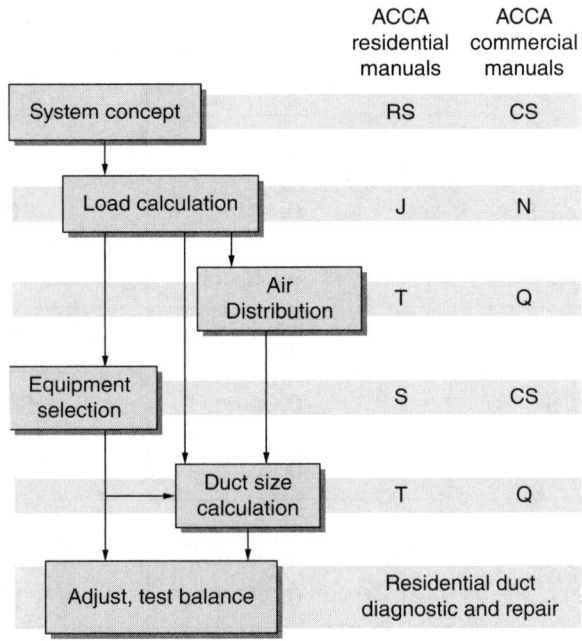

FIGURE 17-2 *ACCA system design process.*
Source: www.acca.org/design/

ASHRAE Standard 62.1 gives the ventilation rate procedure for acceptable indoor air quality, which is approved by the American National Standards Institute (ANSI). The basic formula for outdoor airflow is given below followed by table 17-2, which lists some of the values for various building occupancies.

$$V_{bz} = R_p \times P_z \times R_a \times A_z$$

A_z = zone floor area, the net occupiable floor area of the zone (ft^2)
P_z = zone population, the largest number of people expected to occupy the zone during typical usage

Note: If P_z cannot be accurately predicted during design, it shall be an estimated value based on the zone floor area and the default occupant density listed in the table

R_p = outdoor airflow rate required per person
R_a = outdoor airflow rate required per unit area

An approximate value for the ventilation rate can be found using the combined outdoor air rate times the number of occupants.

Occupancy Category	People Outdoor Air Rate, R_p cfm/person	Area Outdoor Air Rate, R_a cfm/ft²	Occupancy Density #/1,000 ft²	Combined Outdoor Air Rate cfm/person
Classrooms	10	0.12	25	17
Daycare	10	0.18	25	15
Restaurant Dining	7.5	0.18	70	10
Hotel Rooms	5	0.06	10	11
Office Space	5	0.06	5	17
Auditorium	5	0.06	150	5
Religious Worship	5	0.06	120	6
Residential Unit	5	0.06	1+1/bedroom	—
Retail Sales	7.5	0.12	15	16

TABLE 17-2 *Minimum ventilation rates in breathing zones.*

Example 17.2 Calculate the minimum ventilation rate for 10,000 ft² office space. Also determine the rate using the combined air rate.

From table 17-2 the R_p = 5 cfm/person, R_a = 0.06 cfm/ft², Occupant Density = 5/1,000 ft², and Combined Outdoor Air Rate = 17 cfm/person. The number of occupants is equal to the density times the floor area in 1,000 ft². In this case, that would be five occupants/1,000 ft² × 10,000 ft²/1,000 ft² = 50 occupants.

$$V_{bz} = R_p \times P_z + R_a \times A_z = 5 \text{ cfm/person} \times 50 \text{ person} + 0.06 \text{ cfm/ft}^2 \times 10,000 \text{ ft}^2 = 850 \text{ cfm}$$

The approximate value using the combined outdoor air rate would be 17 cfm/person × 50 person = 850 cfm.

The amount of ventilation required for heating and cooling systems also has an impact on the aesthetic design of the building. In commercial buildings, both fresh air and exhaust air have to enter and exit the building. The fresh air and exhaust must be separated. In figure 17-3 A, they cannot be separated, so the exhaust air is piped out in the lower right louver to the fan in figure 17-4. The fan must have a fairly high air velocity and is on base isolators to prevent vibrations in the roof. In figure 17-3 B & C, the fresh air and exhaust air are physically separated by the corner of the building.

The basic climate in which a building is located and constructed has a great impact on the type of mechanical system selected and energy efficiency. The map shown in figure 17-6 of the textbook identifies basic climatic zones in North America. Building construction materials also have a great impact on energy efficiency. Windows are one area of a building where energy efficiency is reduced. Figure 12-9 shows window label from the National Fenestration Rating Council (NFRC). The textbook shows a similar label in figure 17-7a. In figure 17-7b of the textbook, some of the values are shown on a window. The values from figure 12-9 are summarized as follows, and some of these will be used to calculate heat loss:

- U – Factor (U) measures how well a product prevents heat from escaping. The values generally fall between 0.20 and 1.20, where the lower the value, the better insulating provided.
- Solar Heat Gain Coefficient (SHGC) measures how well a product blocks heat caused by sunlight. The values fall between 0 and 1, where the lower the value, the less heat transmitted.

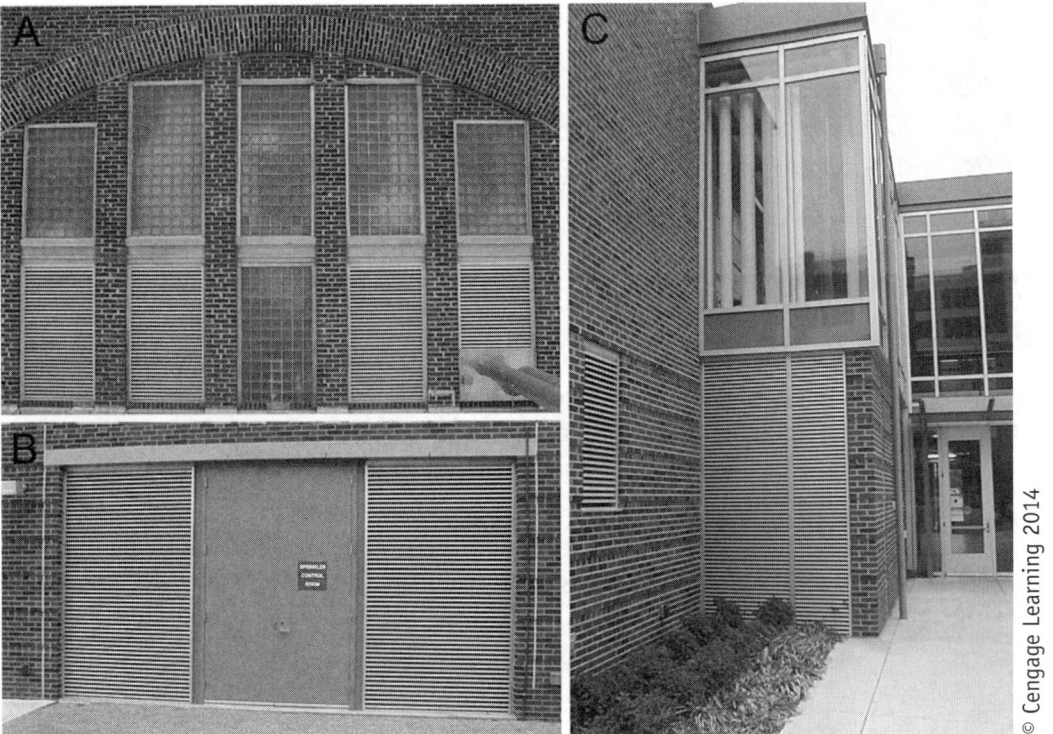

FIGURE 17-3 Louver photos.

- Visible Transmittance (VT) measures how much light comes through a product. The values fall between 0 and 1 where the higher the value, the more light is transmitted.
- Air Leakage (AL) is indicated by an air leakage rating in ft^3 of air per ft^2 of window. The lower the value, the less air is transmitted.
- Condensation Resistance (CR) measures the ability of a product to resist the formation of condensation on the interior surface of the product. The values fall between 0 and 100, where the higher the value, the better.

Problem Set 17.2   Energy Codes

Problem 17-11   Select a combined heating and cooling system that minimizes system maintenance.

Problem 17-12   Select a heating-only system that minimizes operating cost in very cold climates and minimizes system noise.

Problem 17-13   Which combined heating and cooling system requires a chimney?

Problem 17-14   Select a heating only system that minimizes visual obtrusiveness.

Problem 17-15   What strategies, other than mechanical system selection, can be used to reduce the system loads on a building?

Problem 17-16   What is a VOC? Give a few examples.

FIGURE 17-4 Exhaust fan photo.

Factors Influencing Energy Consumption

Problem 17-17 Calculate the minimum ventilation rate for a 10,000 ft² classroom. Also determine the rate using the combined air rate.

Problem 17-18 Calculate the minimum ventilation rate for a 40,000 ft² auditorium.

Problem 17-19 Describe the solar heat gain coefficient.

Problem 17-20 Describe how air leakage is determined in windows.

BACKGROUND

Architects and engineers should understand how heating and cooling loads are calculated. Since these loads are affected by the selection of the materials and system, alternatives should be explored for energy efficiency.

Calculating Heating and Cooling Loads

The heating and cooling loads on a building are derived from the loss or transmission of heat through the building envelope and gains in heat from internal and external sources. The overall heat balance equation is as follows:

$$\pm M = I + S \pm T \pm O$$

M = mechanical heating (−) or cooling (+)
I = internal gains, people, lights, appliances, etc.
S = solar heat gains
T = transmission of heat through the envelope loss (−) or gain (+) also (Q')
O = outside air load; heat loss (−) or gain (+) due to infiltration or ventilation

Some environmental and building factors can be controlled throughout the building's service life, and monitoring can help increase energy efficiency. Figure 17-5 shows a data logger that monitors environmental levels in a building.

The primary factor that a designer can control is the basic heat loss. The materials and systems may be changed to reduce this loss. The equation used to calculate the total heat transmission, Q', in British Thermal Units per hour (BTU/h) is as follows:

$$Q' = UA\Delta T$$

U = heat transfer coefficient (BTU/h°Fft²)
A = area of component through which heat is lost
ΔT = difference in temperature between interior and exterior surface of the component

The U-value is the reciprocal of the more commonly used R-value, which is the thermal resistance of a material ($U = 1/R$). When determining the resistance of an element made up of several layers of material, like a wall, the R-values may be added. Figure 17-32 in the textbook lists R-values for various construction materials.

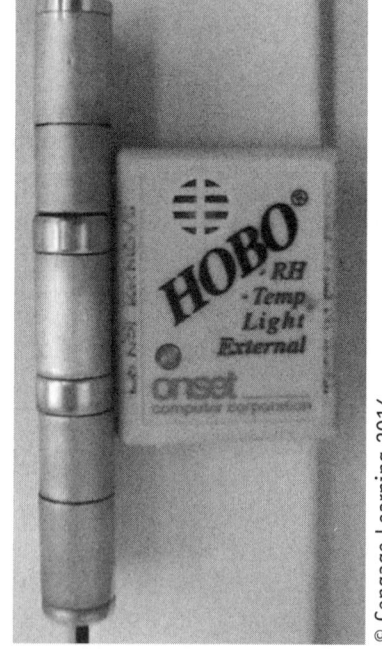

FIGURE 17-5 *Data logger photo.*

Example 17.3 Calculate the *R*-value for the wall section shown in figure 17-6 through the insulated portion of the wall. The wall has 4-inch common modular brick and 3/4 inch plywood exterior sheathing on the exterior side of the air gap.

From the R-value table 17-35 of the textbook, table 17-3 can be constructed:

Component	R-value
Outside air film	0.17
4 in. Common Modular brick	0.80
Air gap	1.00
3/4 in. exterior plywood sheathing	0.94
R19 Batt insulation	19.0
5/8 in gypsum wallboard	0.56
Inside air film	0.68
Total	23.15

TABLE 17-3 *R-value calculation.*

Example 17.4 Calculate the total heat loss for a single-story residence built on grade with a temperature gradient of 50°F. The plan size is 20 feet by 40 feet, and the wall height is 8 feet. The wall construction is *R*23, and the roof construction is *R*32. There are 200 sq. ft. of double insulated windows with 1/2 inch air space and 100 sq. ft. of 2 1/4 inch solid core flush doors.

The area of the roof is taken as the plan area of 20 feet × 40 feet = 800 ft². The total wall area is taken as the perimeter times the height or (20+20+40+40) feet × 8 feet = 960 ft². Table 17-4 below shows the collation of the areas calculated and the R-value from the table in Appendix C of the textbook:

Component	Area	R-value	U-Value	ΔT	$Q' = UA\Delta T$
Roof	800	23	0.0434	50	1,736
Wall excluding windows and door	660	32	0.0313	50	1,033
Windows	200	2.04	0.4902	50	4,902
Doors	100	3.70	0.2703	50	1,351
Total	—	—	—	—	9,022

TABLE 17-4 *Heat loss calculation.*

The heat loss total is 9022 BTU/h.

Calculating Heating and Cooling Loads

Problem Set 17.3 Calculating Heating and Cooling Loads

Problem 17-21   What are the four sources in determining the overall heat balance of a building?

Problem 17-22   What is the R-value for 3 1/2 inches of fiberglass batt insulation?

Problem 17-23   What is the R-value for a metal door with 2 inches of urethane insulation?

Problem 17-24   What is the R-value for a double insulated window with a 1/2 inch air gap and Low-E3 glass?

Problem 17-25   What would be the total R-value of the wall section in figure 17-6 if only 3 1/2 inches of batt insulation were used?

Problem 17-26   Determine the total R-value for the roof shown in figure 17-7.

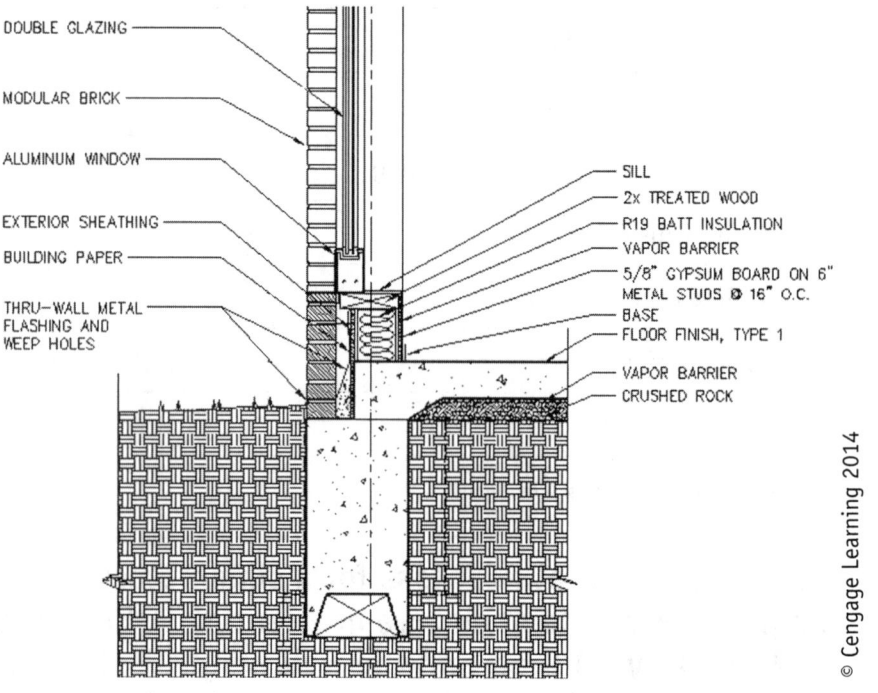

FIGURE 17-6 Wall section.

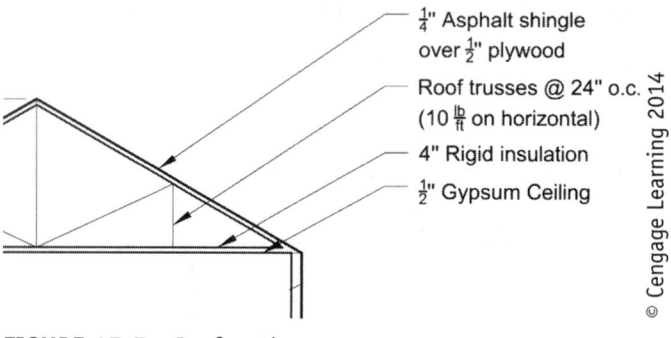

FIGURE 17-7 Roof section.

Problem 17-27   Determine the total R-value for the wall shown in figure 17-8.

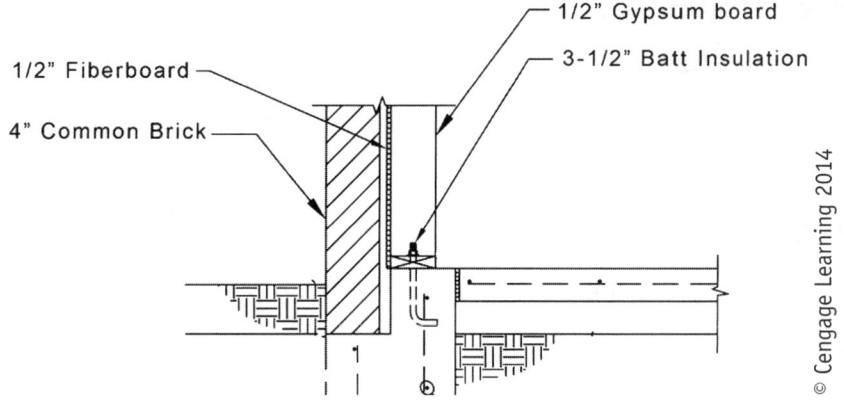

FIGURE 17-8 *Wall section.*

Problem 17-28   Determine the total R-value for the wall shown in figure 17-9.

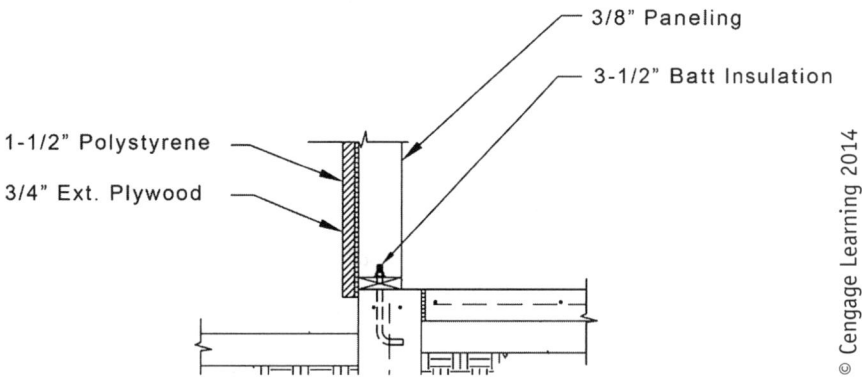

FIGURE 17-9 *Wall section.*

Problem 17-29   Calculate the total heat loss for a single-story residence built on grade with a temperature gradient of 45°F. The plan size is 20 feet by 30 feet, and the wall height is 8 feet. The wall construction is R24 and the roof construction is R36. There are 100 sq. ft. of double insulated windows with ¼ inch air space and 50 sq. ft. of 1 3/4 inch solid core flush doors.

Problem 17-30   Calculate the total heat loss for a single-story residence with a crawl space with a temperature gradient of 40°F. The plan size is 30 feet by 40 feet and the wall height is 10 feet. The wall construction is R24, floor construction is R16 and the roof construction is R32. There are 240 sq. ft. of double insulated windows with 3/4 inch air space and 120 sq. ft. of 1 3/4 inch solid core flush doors.

BACKGROUND

Architects and engineers should be aware of how heat is transferred. Heat transfer should provide a positive impact on building occupants.

Methods of Heat Transfer

Heating is normally achieved by heating of the air or heating of objects. An example of heating of the air is when heated air is distributed by a furnace through ductwork. An example of heating of an object is using heated water circulated in embedded tubing in concrete provided for radiant heating. The three basic mechanisms for heat transfer are as follows:

- Conduction – energy transferred from one solid material to another by contact
- Convection – thermal energy carried by a moving liquid or gas
- Radiation – thermal energy exchanged by electro-magnetic waves, like the sun

Figure 17-14 in the textbook shows how all three methods can impact interior space. These methods are warm air circulating in a room is convection, heat loss through a wall by conduction to the colder side, and solar heat entering a room by radiation.

Problem Set 17.4   Methods of Heat Transfer

Problem 17-31   Give an example of heat transfer by conduction.
Problem 17-32   Give an example of heat transfer by convection.
Problem 17-33   Give an example of heat transfer by radiation.
Problem 17-34   In what direction does conductive heat move?
Problem 17-35   In what direction does convective heat move?

BACKGROUND

Architects and engineers must be aware of the many types of heating and cooling systems in buildings. The selection of the systems should include considerations for size and occupancy of the building, in addition to environmental factors.

Heating Systems

For smaller buildings, table 17-1 may be used to select heating and cooling systems based on many considerations. For larger buildings, the primary consideration for selection is the occupancy. Table 17-4, which is modified from *Architect's Studio Companion: Rules of Thumb for Preliminary Design*, by Edward Allen and Joseph Iano, may be used to narrow mechanical system selection. For mixed-use occupancy, multiple systems could be considered if they are compatible.

Occupancy	Variable Air Volume	Single Duct Constant Air Volume	Multizone	Fan-Coil Terminals	PU Table 17-1
Apartments				X	X
Arenas, Exhibition Hall	X	X		X	
Auditoriums, Theaters	X	X	X		
Factories	X	X	X	X	
Hospital	X		X	X	
Hotels, Motels, Dorms	X			X	X
Laboratories	X	X			
Libraries	X		X		
Nursing Homes	X			X	X
Offices	X		X		
Places of Worship	X	X	X		
Schools	X			X	
Shopping Centers	X		X		
Stores	X		X		

TABLE 17-5 *Heating and cooling systems for large buildings.*

In the first three systems, air is conditioned with outside air at a central source and distributed by supply and return fans and ducts to occupied spaces. A central unit for a VAV system is shown in figure 17-10. The separate doors give access to the heating, cooling, and filtering units.

FIGURE 17-10 *Central mechanical unit photo.*

The differences in the three central source systems are as follows:

- Variable Air Volume (VAV) – At each zone, a thermostat controls room temperature by regulating the air volume discharged through the diffusers and air temperature is controlled at the central source (most widely used).
- Constant Air Volume (CAV) – A master thermostat controls the central heating and cooling coils to regulate building temperature, and the air volume is constant (large single-occupancy zone).
- Multizone – Several ducts from a central fan serve several zones. In one variation, dampers mix cold and hot air at the space to control temperature. In another, reheat coils at the unit regulate the temperature (not very common).

Some of the central systems require two separate ducts to distribute both hot and cold air to the space. The single duct VAV system can distribute either hot or cold air to the spaces. In figure 17-11, we see two main duct lines from a VAV system and the smaller distribution ducts.

FIGURE 17-11 *Distribution ductwork photo.*

The conditioned air provided to a space is dispersed through many types of diffusers. In figure 17-12, both linear and spot diffusers are shown. The spot diffuser in this case can be adjusted to control direction.

FIGURE 17-12 *Air diffusers photo.*

In fan-coil terminals, hot and/or chilled water is piped to a terminal unit. In packaged terminal units or through-the-wall units, heating and cooling is supplied at each unit with no additional piping or ducting. For all of these systems, fresh air and exhaust have to be supplied at the unit or ducted to the unit. For large buildings where heated and chilled water are required for the system, water is normally piped from a central boiler and chiller. In figure 17-13, supply and return of heating and chilled water are shown. A mechanical piping plan was shown with plumbing systems in figure 16-8.

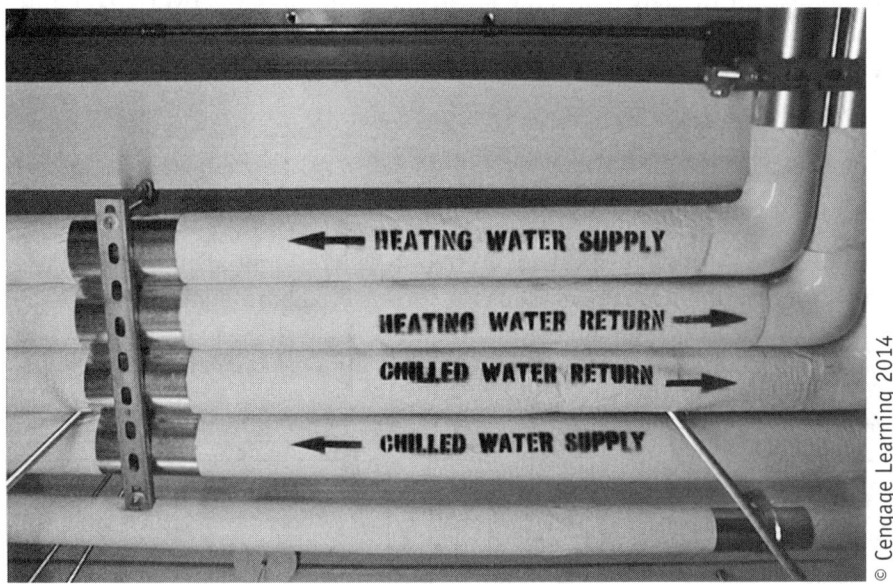

FIGURE 17-13   *Mechanical piping photo.*

Example 17.5   Select appropriate heating and cooling systems for a church. If only one system were to serve the sanctuary, what would be the best choice?

From table 17-4 there are three systems that could be selected: Variable air volume, constant air volume, and multizone. The best choice would be constant air volume for a large single-occupancy zone.

Heat from solar radiation

Solar heating of a building can be achieved in many ways. Passive solar systems, like a trombe wall shown in figure 17-16 of the textbook, store energy from the sun during the day and release it back into the room at night. Active systems, like photovoltaic devices (solar cells), change solar energy into electric that is used to power equipment or stored in batteries for later use. Solar panels, like the ones shown in figure 17-17 of the textbook, should have proper sun orientation.

Heat Pumps

Heat pumps typically use a water source instead of other liquids or gases in the refrigeration process. They still use an evaporator, compressor, and condenser to provide heat exchange. A ground source heat pump (GSHP), like the one shown in figure 17-18 of the textbook, uses a closed loop pipe system buried in the ground to provide a constant source. Robert C. Webber built the first GSHP system in the late 1940s, and Dr. James Bose at Oklahoma State University used this idea and developed the International Ground Source Heat Pump Association (IGSHPA). Dr. Bose has been instrumental in the advancement of the geothermal industry.

Boilers

Boilers are used to heat water for use in mechanical systems or for radiant heating through closed piping or radiators units. Temperature control of the water in radiant floor systems, like the one in figure 17-19 of the textbook, is done through mixing with cooler water. Another radiant system, shown in figure 17-20 of the textbook, is hot water circulated through a radiator, commonly using copper fins to transfer heat.

Furnace

Furnaces can be used to heat air directly that is then distributed through ductwork to an individual space. The furnace should be properly sized in accordance with the ACCA system shown in figure 17-2. The process requires the use of the following manuals for design:

- Manual J – Load calculation
- Manual S – Equipment selection
- Manual D – Duct size calculation

Figure 17-21 of the textbook shows the mechanical layouts in a residential building. Figure 17-22 of the textbook and figure 17-14 show partial mechanical plans for commercial buildings.

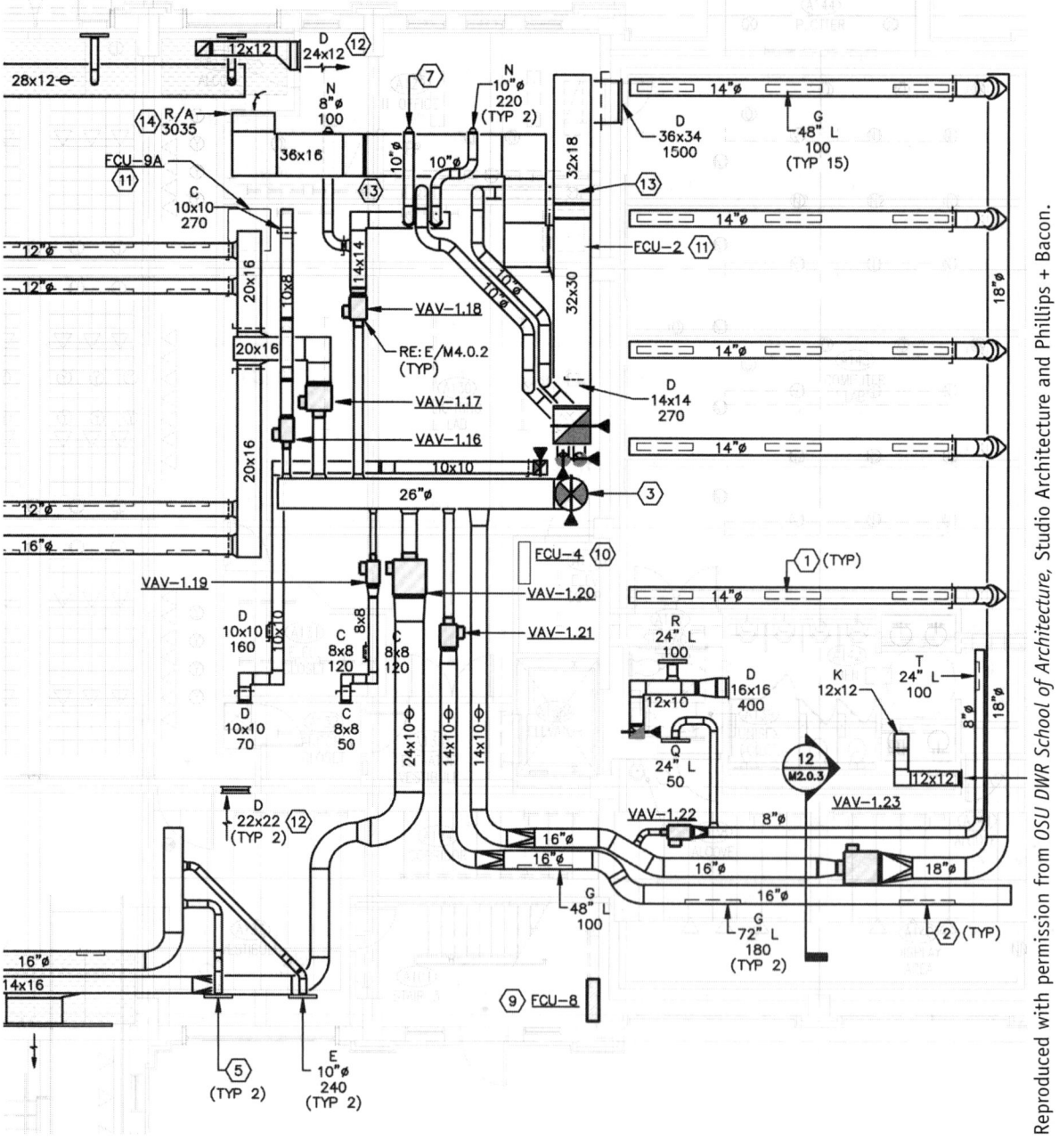

FIGURE 17-14 *Mechanical plan.*

Air Handling Units and Roof Top Units

Air Handling Units (AHU), like the ones shown in figure 17-23 of the textbook or the photo in figure 17-10, contain many components to heat and cool buildings. Typically, they have heating and cooling units, in addition to air-mixing and -filtering units. They can be contained in mechanical rooms or in outdoor areas, like a rooftop. When they are inside, they will require fresh air and exhaust grills, like those in figure 17-3.

Air Conditioning

Air-conditioning can be achieved through several different systems. All of the systems shown in table 17-5 can provide cooling. Normally, chillers are used to condition air by blowing air over coils that contain compressed refrigerant. Commercial chillers are rated in tons from about 15 to 1,000 tons and supply thousands to millions of BTU/h. Residential units are normally 3 to 5 tons and should provide for a minimum amount of efficiency. A ton of refrigeration is equal to 12,000 BTU/h. The EPA uses the seasonal energy efficiency ratio (SEER) to indicate mechanical unit efficiency. A SEER rating of about 20 indicates high energy efficiency, and minimum SEER rating is 13.

Ventilation

Ventilation for acceptable air quality was covered in an earlier section of this chapter. In addition to the fresh air requirements and air-quality factors, proper humidity should be maintained. Intake and exhaust air concepts are shown in figures 17-24 through 17-27 of the textbook.

Problem Set 17.5   Heating Systems

Problem 17-36   Select a heating and cooling system for apartment units without a hot or chilled water supply.

Problem 17-37   Select a heating and cooling system for individual hotel units with a hot or chilled water supply.

Problem 17-38   Select appropriate heating and cooling systems for a library. Which would be the best selection?

Problem 17-39   Select appropriate heating and cooling systems for a school. Which would be the best selection?

Problem 17-40   What is the most widely used heating and cooling system used in larger buildings?

Problem 17-41   What heating and cooling system is most efficient for one large single-occupancy space?

Problem 17-42   What is the primary difference between fan-coil and through-the-wall units?

Problem 17-43   Describe a single-duct VAV system in terms of how the air is conditioned and distributed.

Problem 17-44   What are the three parts of a ground source heat pump?

Problem 17-45   What is the primary purpose of a boiler used in buildings?

Problem 17-46   What are the three manuals used for the design of a furnace driven hot air system?

Problem 17-47   What is the primary purpose of a furnace used in buildings?

Problem 17-48   What is the minimum energy-efficiency rating for an air conditioning unit?

BACKGROUND

Architects and engineers should understand the various control systems that provide for occupant security. Computer-controlled systems provide for many systems to be controlled from one location.

Security and Protection

Many devices are used to provide security in buildings, including equipment for: sensing illumination, monitoring, occupancy warning, and control of access. With digital and wireless technology, they can be integrated in the whole design process along with the environmental systems.

Fire and Smoke Protection

Fire protection is one of the most important issues in building design. Fire codes and egress are covered in other sections of the book. The equipment used for fire and smoke protection includes fire and smoke alarms (figure 17-15), exit lighting, sprinklers, hose bibs, and extinguishers. Sprinkler systems have been included in many figures, such as 17-1, 17-11, and 17-12.

The criteria for sprinkler system design are specified by the International Building Code (IBC), International Fire Code (IFP), and the National Pire Protection Association (NFPA) codes. The following is a summary of the IBC requirements for when a sprinkler system is required in assembly occupancies:

- A-1 – The fire area exceeds 12,000 sq. ft., or has an occupant load of 300 or more, or is located on a floor other than a level of exit discharge, or contains a multi-theater complex.
- A-2 – The fire area exceeds 5,000 sq. ft., or has an occupant load of 100 or more, or is located on a floor other than a level of exit discharge.
- A-3 – The fire area exceeds 12,000 square feet, or has an occupant load of 300 or more, or is located on a floor other than a level of exit discharge.
- A-4 – The fire area exceeds 12,000 square feet, or has an occupant load of 300 or more, or is located on a floor other than a level of exit discharge.
- A-5 – The occupancies in the following areas: concession stands, retail areas, press boxes, and other accessory use areas in excess of 1,000 sq. ft.

FIGURE 17-15 *Fire alarm photo.*

Example 17.6   Is a sprinkler system required for 4,000 sq. ft. nightclub (A-2 occupancy) that serves 120 occupants?

Although the area is less than 5,000 sq. ft., the occupancy load exceeds 100, therefore a sprinkler system is required.

Command Center, Monitor and Control

Many devices and equipment are used to provide for controlled access and monitoring of a building. These include card scanners (figure 17-16), cameras (figure 17-17), special lighting, metal detectors, motion detectors, and communication systems.

FIGURE 17-16 *Card access photo.*

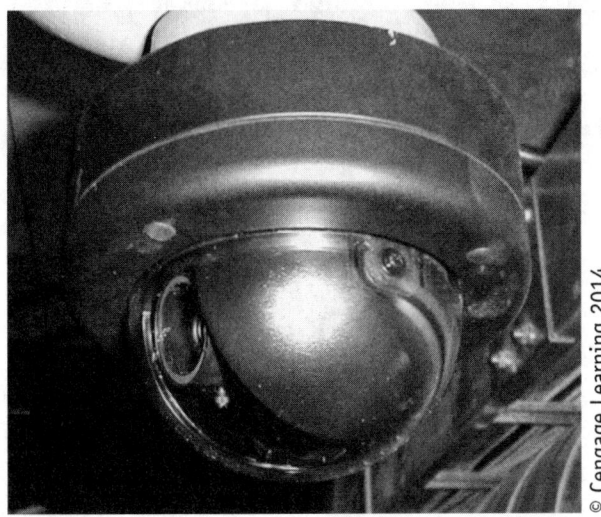

FIGURE 17-17 *Security camera photo.*

The type and amount of security varies with occupancy, location, codes, and desired protection. The U.S. General Services Administration (GSA) covers fire protection and security online at www.gsa.gov.

Intelligent Buildings

Building systems can be designed to change controls with variations in environmental and physical conditions. This can be done to improve energy efficiency, security, communication, and comfort. There are many systems that automate building systems controls including the following:

- Building automation systems (BAS)
- Energy management systems (EMS)
- Energy management and control systems (EMCS)
- Central control and monitoring systems (CCMS)
- Facilities management systems (FMS)

An example of intelligent systems would be changing thermal controls with changes in solar gain entering the space, or changes in lighting levels with changes in daylighting entering a space. The textbook covers additional examples.

Problem Set 17.6   Intelligent Buildings

Problem 17-49	Is a sprinkler system required for a 10,000 sq. ft. exhibition hall (A-3 occupancy) that serves 250 occupants?
Problem 17-50	Research and determine when a sprinkler system is required in a business occupancy.
Problem 17-51	Research and determine when a sprinkler system is required in an educational occupancy.
Problem 17-52	What are some devices used to provide for fire and smoke protection in a building?
Problem 17-53	What are some devices used to provide for control and monitoring in a building?
Problem 17-54	Give an example of an intelligent building system.

CHAPTER 18
Landscaping

Skills List

After completing the problems in this chapter, you should be able to:

- Perform a needs analysis for landscape design

- Perform and document a site analysis for conditions affecting landscaping

- Research local city ordinances and regulations pertaining to landscape design

- Determine the function of a landscape

- Understand the principles of landscape design

- Develop a conceptual landscape plan

- List sustainable landscaping techniques

- Understand the difference between conceptual and final landscape plans

BACKGROUND

Landscaping, when done properly, may be used for one or many reasons. Some uses for landscaping are increasing curb appeal, privacy, energy efficiency, security, shading and cooling, and traffic flow. The task in any landscaping project is to determine the design problem by performing needs and site analyses, researching any rules and regulations, and identifying the function of the landscaping.

An example of landscaping used as a tool for direction of traffic and pedestrian flow is shown in figure 18-1.

FIGURE 18-1 *Tree-lined street.*

Problem Identification

The process for landscape design is similar to building design, which is outlined below.

- Collect data regarding a client's needs, limitations of the site, and code restrictions or limitations
- Develop and revise a conceptual plan
- Develop and revise a landscape plan
- Select landscape materials
- Create a final plan

Needs Analysis

A needs analysis should be performed before any design begins, to determine what is needed in the project, and what constraints are present. Project needs will also vary, depending on the type of project (residential or commercial) and users of the space. All users should be consulted to ensure that all needs are considered and that all site areas are noted by use: public, private, or service. A list of intended activities for each area should be documented. Lastly, the client's preferences on colors, textures, plant species, level of maintenance, and special features should be reflected in the design.

Site Analysis

Site analysis was discussed extensively in chapters 5, 6, and 7, and site plans should include the following information:

- Property lines
- Easements and setbacks
- Topography of the site
- Location and size of structures
- Man-made surfaces
- Utility line locations
- Circulation patterns

Other site factors exist that may affect the landscape design. Research of the site will lead to a successful design, and the following should be considered:

- Climate
- Prevailing wind speed and direction
- Solar orientation
- Terrain
- Soil analysis (obtained from a soils report)

- Existing vegetation
- Noise, smells, and lights
- Surrounding property landscaping

An example of a site plan is shown in figure 7-4.

Rules and Regulations

Like most areas of a building project, landscaping may be controlled by rules and regulations and must be in compliance. Some regulations that may affect landscape design are listed below:

- Community landscape code
- State regulations
- Local and municipal ordinances and zoning
- Restrictive covenants
- Environmental regulations

A city landscaping requirement for Stillwater, Oklahoma, is shown in figure 7-3. Some common code requirements are:

- Stormwater drainage
- Tree ordinances/controls
- Street trees
- Street yard
- Buffer yard
- Street wall
- Parking screen and islands
- Trash screening

Figure 18-2 shows a parking lot with landscaping islands.

FIGURE 18-2 *Parking islands.*

Figures 18-3 through 18-10 show a tree-preservation application utilized by the City of Geneva, Illinois. Figures 18-3 and 18-4 outline the tree preservation application guidelines and general information.

City of Geneva
Tree Preservation Application Guidelines and General Information

- Submittal of a Tree Preservation Review Application is required for construction projects (see table below for most common types) on property that contain one or more trees over 10 inches in diameter (measured 4.5 feet above the ground) located within the Construction Activity Zone or Tree Preservation Zone (delineated by the applicant). A good rule of thumb is a 10 inch or larger tree on your property or any sized tree on adjacent public or private property within approximately 50 feet of the construction project. Please refer to Title 8-5C of the Geneva City Code for further information (available upon request or at www.geneva.il.us).

- The types of construction projects that require a Tree Preservation Review Application include the following:

New building	Lot re-grading
Demolition only	Detached garage
Building addition	Driveway, patio, deck, porch, tennis court, or similar structure
Swimming pool	New water/sewer/utility lines

  Completed Tree Preservation Review Applications shall be submitted to the Building Division located at 109 James Street. Contact the Building Division if you are unsure whether or not your project requires a Tree Preservation Review Application at 630-262-0280.

- A completed application packet is required to be submitted with the building permit application including the following information:

  Submit one original and 1 copy of the following:
  1. Application form with both statements of compliance signed
  3. Tree Location Sketch
  4. Tree Protection Plan (may be combined on one sheet with the Tree Location Sketch)
  5. Tree Protection Plan Description

- Once a completed application packet has been accepted by the Building Division, the City's Consulting Forester will review the plans and conduct an initial inspection of the site. The Consulting Forester will then either: Approve the Tree Protection Plan, approve the Tree Protection Plan with conditions, or deny the Tree Protection Plan. The Forester will provide you with a written review of your submitted Tree Protection Plan.

FIGURE 18-3 *Tree Preservation Application Guidelines.*
Source: City of Geneva, Illinois Tree Preservation Application.

- If the plan is approved, you can implement your protection measures and notify the Building Division when the protection measures are in place. The Consulting Forester will then return to the site and inspect the protection measures to ensure they conform to the approved plan.

- If the plan is approved with conditions, you can implement your protection measures and notify the Building Division when the protection measures are in place. The Consulting Forester will return to the site and inspect the protection measures to ensure they conform to the approved plan.

- If the plan is denied, you must follow the direction given by the Consulting Forester and resubmit the plans for a second review. Your project will not be permitted until the Tree Protection Plan has been approved.

- Typically, tree protection measures consist of 48 inch high plastic poly-type fencing (snow fencing) of a highly visible type placed around the critical root zone with signage warning workers to stay away. Additional protection measures may include providing a 3 to 6 inch layer of mulch over the critical root zone, root pruning, crown pruning, fertilizing, or other actions as required by the Consulting Forester.

- The Consulting Forester is generally available to review plans and inspect construction sites once a week. In most cases, the Consulting Forester will need to visit your property at least twice, once for an initial inspection of the trees to confirm the accuracy of the location sketch, and at least one follow up visit to inspect the protection measures. If a fully compliant plan is submitted and the required protection measures are put in place promptly, the Consulting Forester can approve the site for work within 10 days for most small projects such as decks, patios, small additions, sheds, etc. For larger projects such as large additions or new homes, the review time can be longer.

- No construction activity of any kind (except work directly related to tree protection) is allowed until the Tree Protection Plan and installed protection measures are approved by the Consulting Forester.

- A copy of the Tree Protection Plan and Tree Protection Plan Description shall be kept on site in a visible area at all times during construction. Failure to keep copies of these documents available on site will result in a stop work order being issued.

- All required tree protection measures shall be properly maintained and remain in the approved locations until either a final inspection has been passed, a final certificate of occupancy has been granted, or as otherwise required by the Consulting Forester. If it is found at anytime during construction, that the approved Tree Protection Plan is not being followed, the Building Division will issue a stop work order and work will be halted until the required protection measures have been replaced.

FIGURE 18-4 *Tree Preservation Application Guidelines.*
Source: City of Geneva, Illinois Tree Preservation Application.

Figures 18-5 and 18-6 give the tree protection plan directions.

City of Geneva
Tree Protection Plan Directions

The Tree Protection Plan may be shown as a combined tree location sketch/ tree protection plan or on a separate sheet. In either case follow these steps:

1. Draw the "footprint" of construction project you are applying for. Include dimensions to the nearest property lines and dimensions of each side of the structure.

2. Draw in the Construction Activity Zone (CAZ). Refer to the definitions sheet included in the packet for a definition of the CAZ.

    Be sure to include an area for materials storage, access from the street for workers and/or vehicles, an area for a dumpster, if needed, and any lot re-grading. The CAZ is an area that the applicant delineates. Once the CAZ is delineated, no construction activity of any kind is permitted outside this zone so plan this area carefully. Try to keep as many trees out of the CAZ as possible. Discuss with your contractor how much space they will need to perform the work and decide where the CAZ boundary can be shifted to avoid trees. Any trees located within the CAZ can be removed if warranted by construction. For those trees to be protected within the CAZ, best protection efforts are required.

    Use the following symbol to define the CAZ border: ────── CAZ ──────

3. Will any construction equipment or materials be stored on site? Attempt to store any construction equipment or materials away from the trees. Use the following symbol for any equipment or materials storage areas:

    | S |

4. Draw in the Tree Protection Zone (TPZ). Refer to the definitions sheet included in the packet for a definition of the TPZ.

    This is the area of your property where trees are required to be protected. No construction activity is permitted in this area of any kind. The City's Consulting Forester will help you determine what protection measures are needed to protect these trees.

    Use the following symbol for the TPZ: ────── TPZ ──────

FIGURE 18-5 *Tree Protection Plan Directions.*
Source: City of Geneva, Illinois Tree Preservation Application.

5. Are there any trees on adjacent private property (border trees) or in the public street right-of-way (municipal trees)? Border trees are defined as: Any off-site tree at the outside edge of and within fifteen (15) feet of a delineated, on-site construction activity zone that could be susceptible to damage during the construction process. Municipal trees are defined as: Any tree regardless of size or condition growing on any City property including all parkway trees, trees on medians and at other City owned locations. The TPZ should be shown on the plan to include all of these off-site trees. You are required to protect all border trees from damage caused by construction activity. The Consulting Forester is available to assist you in determining what protection measures are needed to protect these trees.

6. Using the following symbols, determine which trees are to be protected, removed, or relocated:

   Protected: (P)   Removed: ⊗   Relocated: ○ - - - → (○)

7. For protected trees, protective fencing will be required at a minimum. The protective fencing should be shown at the outside edge of the critical root zone. Refer to the definitions sheet included in this packet for a definition of the critical root zone. Use the following symbol to show protective fencing/critical root zone:

   — — — — — — — — — — — —

8. Add a number next to every 10 inch or larger tree within the CAZ, TPZ and any border or municipal tree within 50 feet of the project. This number will correspond to the number on the Tree Preservation Plan Description that is included in this packet.

9. Fill out the Tree Protection Plan Description page. Indicate which trees are to be protected, removed or relocated in the space provided and what protection measures are proposed for each tree.

This completes the location sketch/tree protection plan preparation. The Consulting Forester will review this material and recommend what protection measures are needed for each tree to be protected.

FIGURE 18-6 *Tree Protection Plan Directions.*
Source: City of Geneva, Illinois Tree Preservation Application.

Figure 18-7 gives the grid palette for the tree protection sketch.

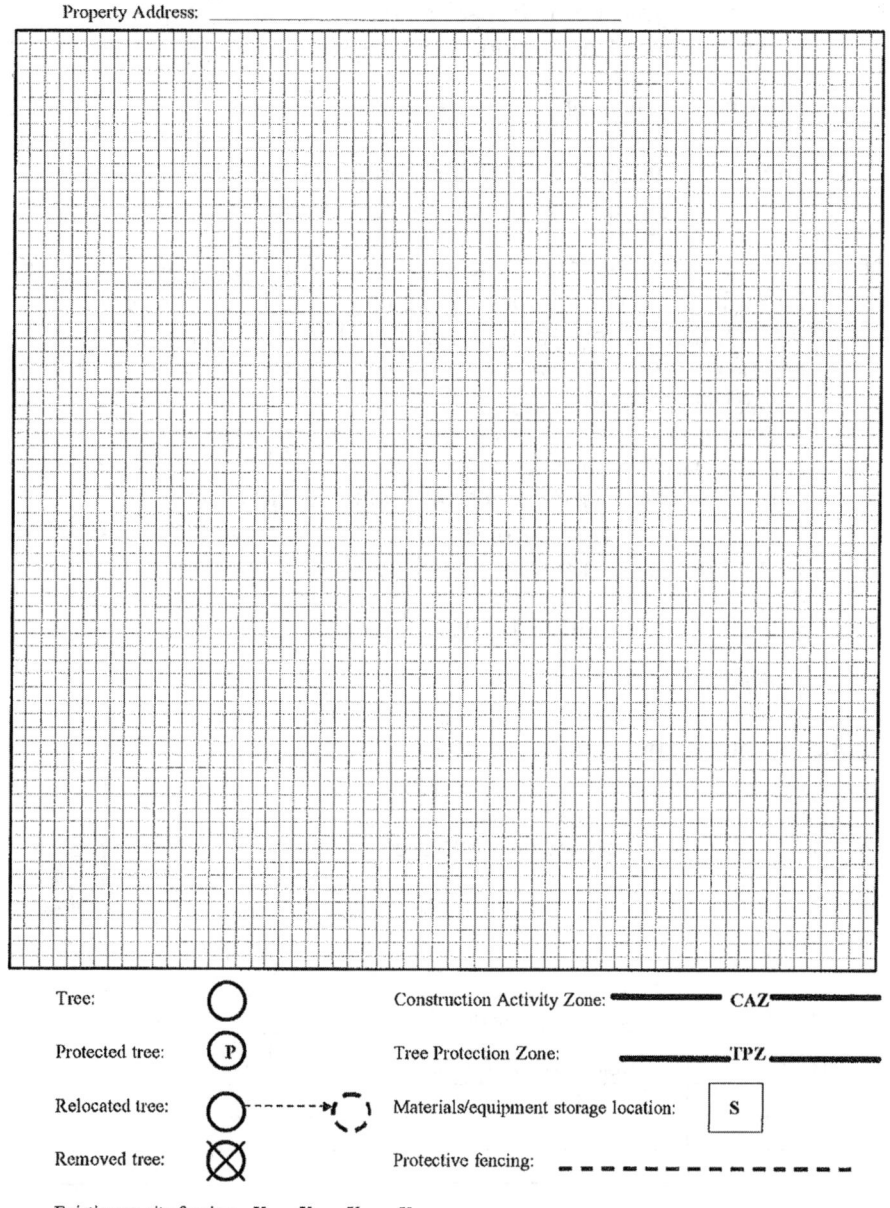

FIGURE 18-7 *Tree Location Sketch.*
Source: City of Geneva, Illinois Tree Preservation Application.

The City of Geneva requests that the condition of the trees shown on the tree survey/sketch be determined and documented per the following criteria.

Rating	Description	General Criteria
1	Excellent	The tree is typical of the species, has less than 10 percent deadwood in the crown that is attributable to normal causes, has no other observed problems, and requires no remedial action.
2	Good to Fair	The tree is typical of the species and/or has less than 20 percent deadwood in the crown and only 1 or 2 minor problems that are easily corrected with normal care.
3	Fair	The tree is typical of the species and/or has less than 30 percent deadwood in the crown, 1 or 2 minor problems that are not eminently lethal to the tree, and no significant decay or structural problems, but the tree must have remedial care above normal care in order to ensure continued health.
4	Fair to Poor	The tree is not typical of the species and/or has significant problems such as 30 to 50 percent deadwood in the crown, serious decay or structural defect, insects, disease, or other problems that can be eminently lethal to the tree or create a hazardous tree if not corrected in a short period of time or if the tree is subjected to additional stress.
5	Poor	The tree is not typical of the species and/or has over 50 percent deadwood in the crown, major decay or structural problems, is hazardous or is severely involved with insects, disease, or other problems that, even if aggressively corrected, would not result in the long-term survival of the tree.
6	Dead	Less than 10 percent of the tree shows signs of life.

TABLE 18-1 *Tree condition rating criteria.*

Figure 18-8 depicts the tree protection plan description.

City of Geneva

Tree Protection Plan Description

NOTE: The Consulting Forester may require that you post an approved Tree Protection Plan Description at the building site; if so it will be noted on the Letter of Transmittal.

Project Address:______________________________

Tree Description				Classification		Action				
Tree No.	DBH*	Species	Condition	Protected	Unprotected	Remove	Root Prune	Crown Prune	Fertilize	No Action

* Diameter Breast Height: The diameter of the trunk of the Tree measured in inches at a point 4.5 feet above ground line.

Prepared by:______________________________

FIGURE 18-8 *Tree Protection Plan Description.*

Source: City of Geneva, Illinois Tree Preservation Application.

Figures 18-9 and 18-10 give definitions pertaining to the tree-protection plan.

City of Geneva
Tree Preservation Definitions

A. <u>Arboricultural Specifications Manual</u>. A manual prepared and updated by the Community Development Department from time to time, which contains regulations and standards for the preservation, protection, planting, maintenance, relocation and removal of trees on private property.

B. <u>Border Tree.</u> Any off-site tree at the outside edge of and within fifteen (15) feet of a delineated, on-site construction activity zone that could be susceptible to damage during the construction process.

C. <u>Class A Tree (Key Specimen Tree)</u>. Any tree identified by the designated City Forester as having the following characteristics:
1. be in fair to excellent condition;
2. have a diameter at approx. four and one half feet above ground ("DBH") of 20" or larger;
3. have a species rating percentage of 80% or greater as listed in the most recent issue of <u>Species Ratings and Appraisal Factors for Illinois</u>, published by the International Society of Arboriculture.

The following trees shall be classified as Class A Trees:
Ash- green ash improved cultivars, white ash*
Bald cypress
Douglas fir
Filbert - Turkish filbert
Fir- white fir
Gingko- male cultivars
Hackberry- common hackberry
Hemlock- Canadian
Hickory- native shagbark*
Honey locust- thorn less cultivars
Kentucky coffeetree- males
Linden- Redmond linden
Maples- black maple, sugar maple species & improved cultivars
Oak- bur, chinquapin, red*, swamp white, white
Walnut- black*
*Recommended due to prominence in the native flora.

D. <u>Class B Tree (Significant Tree)</u>. Any tree identified by the designated City's Forester as having the following characteristics:
1. be in good to excellent condition;
2. have a DBH measurement of 10 inches or larger;

FIGURE 18-9 *Tree Preservation Definitions.*
Source: City of Geneva, Illinois Tree Preservation Application.

3. have a species rating percentage of 60% or greater as listed in the most recent edition of <u>Species Ratings & Appraisal Factors for Illinois</u> published by the International Society of Arboriculture.

The following trees shall be classified as Class B trees:
Ash- green and white improved cultivars, green species, blue
Bald cypress
Birch- River
Buckeye- Ohio and yellow
Douglas fir
Elm- Lacebark, hybrid resistant
Fir- White
Hemlock- Canadian
Hickory- Bitternut, shagbark
Magnolia- Cucumbertree, saucer
Maples- Black, sugar species & improved cultivars, hedge or field, Freeman, Amur,
Oak- White, bur, swamp white, chinquapin, Hill's, shingle, pin, chestnut, red
Pine- Eastern white
Spruce- Norway, white, Serbian, Colorado

E. <u>Construction Activity Zone</u>. All areas of a zoning lot, as delineated by a building permit applicant, where building construction, grading and associated activity is expected to occur. The construction activity zone shall be the minimum area of the lot necessary to perform the permitted construction, maneuver equipment and provide for reasonable storage of materials. No building construction activity, including movement and placement of equipment, site access and material storage shall be permitted outside the delineated construction activity zone.

F. <u>Critical Root Zone.</u> The area around a tree that must be left undisturbed in order to preserve a sufficient root mass to give a tree a reasonable chance of survival. This area is defined by a circle on the ground beneath a tree having its center point at the center of the trunk of the tree and having a radius equal to 1 foot for every inch of trunk diameter measured 4.5 feet above ground.

G. <u>Municipal Tree.</u> Any tree regardless of size or condition growing on any City property including all parkway trees, trees on medians and at other City owned locations.

H. <u>Protected Tree</u>. A tree shown on a Tree Protection Plan that is located within a Tree Preservation Zone and designated to be retained after construction is completed.

I. <u>Tree Location Sketch.</u> A graphic depiction (in plan view) showing the location, size, condition rating and species (if known) of all existing trees 10 inches in diameter (measured 4.5 feet above the ground), including Municipal Trees and Border Trees, that are located in a Tree Preservation Zone and designated by the owner to be removed, relocated or protected, and trees located in a Construction Activity Zone that area designated by the property owner to receive initial protection measures.

J. <u>Tree Preservation Zone.</u> All areas of a zoning lot located adjacent to and within 30 feet of the outside edge of a Construction Activity Zone, as delineated by a building permit applicant, where protected trees are present.

K. <u>Tree Protection Plan</u>. A plan having text and/or graphic illustrations showing the trees designated by the property owner to be removed, relocated or protected in Tree Preservation Zones, Critical Root Zones of Protected Trees, trees designated by the owner to receive initial protection measures in Construction Activity Zones, tree protection measures to be used, parties responsible for the maintenance of protection measures, temporary material storage areas and building/site improvements including new utility services.

FIGURE 18-10 *Tree Preservation Definitions.*
Source: City of Geneva, Illinois Tree Preservation Application.

Example 18.1 Provide a tree protection plan description for the proposed addition in figure 18-11 based on the City of Geneva's tree ordinance and the following notes:

- Tree 1: 30" dia. Oak, excellent condition, requires fertilizer
- Tree 2: 17" dia. Elm, good condition, requires pruning and fertilizer
- Tree 3: 25" dia. Linden, fair condition, no action required
- Tree 4: 10" dia. Ash, good condition
- Tree 5: 14" dia. Oak, condition unknown
- Tree 6: 22" dia. Hickory, good condition, requires fertilizer
- Tree 7: 8" dia. Hackberry, good condition, no action required
- Tree 8: 11" dia. Douglas Fir, excellent condition, no action required
- Tree 9: 19" dia. Birch, good condition, no action required

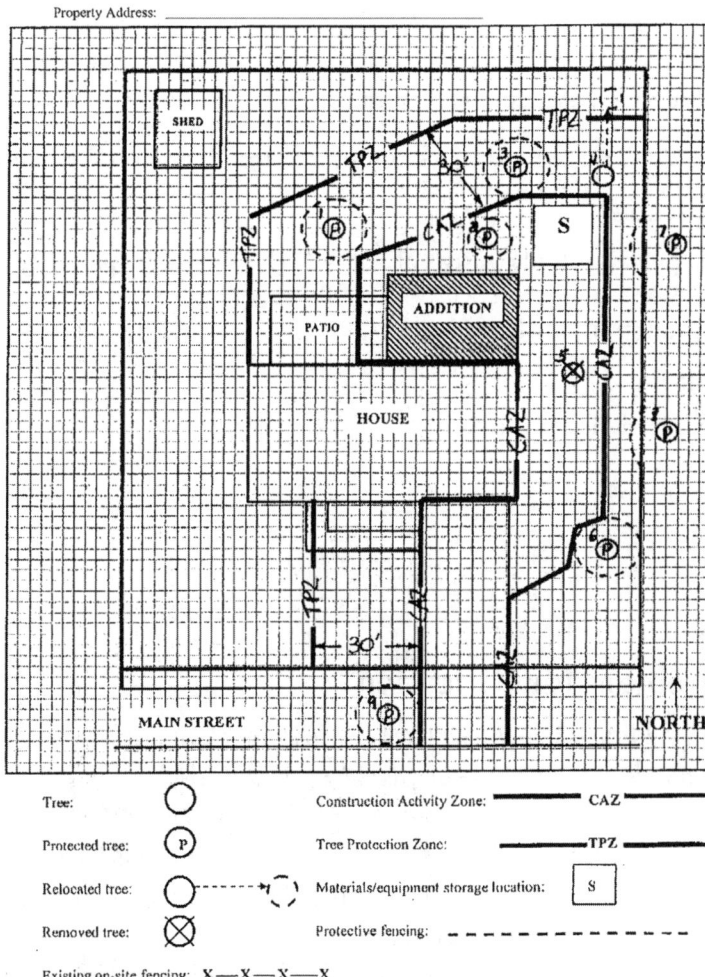

FIGURE 18-11 *Tree Location Sketch for Example 18.1.*
Source: City of Geneva, Illinois Tree Preservation Application.

Trees are documented and recorded in figure 18-12.

City of Geneva
Tree Protection Plan Description

NOTE: The Consulting Forester may require that you post an approved Tree Protection Plan Description at the building site; if so it will be noted on the Letter of Transmittal.

Project Address: __SAMPLE__

Tree No.	DBH*	Species	Condition	Protected	Unprotected	Remove	Root Prune	Crown Prune	Fertilize	No Action
1	30"	Oak	excellent	✓					✓	
2	17"	Elm	good	✓			✓	✓	✓	
3	25"	Linden	fair	✓						✓
4	10"	Ash	good		✓	relocate				
5	14"	Oak	?		✓	✓				
6	22"	Hickory	good	✓					✓	
7	8"	border tree: Hackberry	good	✓						✓
8	11"	border tree: Douglas Fir	excellent	✓						✓
9	19"	municipal tree: Birch	good	✓						✓

*Diameter Breast Height: The diameter of the trunk of the Tree measured in inches at a point 4.5 feet above ground line.

Prepared by: _____

FIGURE 18-12 *Tree Protection Plan Description.*
Source: City of Geneva, Illinois Tree Preservation Application.

Note: Trees 7 and 8 are border trees, and tree 9 is a municipal tree, based on their locations.

The Function of the Landscape

Landscaping may serve many purposes and consists of not only the softscape (plantings) but also the hardscape (e.g., fences, decks, walkways, walls, and terraces).

Some common landscape functions are listed below:

- Aesthetic value and enhanced livability
- Conservation and environmental protection
- Solar heat control
- Wind control
- Sound control
- Slope stability

Problem Set 18.1   Problem Identification

Problem 18-1   How is landscape design similar to site design and building design?

Problem 18-2   Whose needs and opinions should be considered in landscape design?

Problem 18-3   What are the three use categories in landscape design?

Problem 18-4   List three questions you might ask a client, to determine their landscaping preferences.

Problem 18-5   In what ways would maintenance affect landscaping decisions?

Problem 18-6   List four things that should be documented when visiting the site before design begins. Where are these recorded?

Problem 18-7   What information on the site plan will be useful in planning the landscape design?

Problem 18-8   Using textbook figure 18-4 as an example, document the site plan shown in figure 7-4 so that it may be used for landscape planning. Use site information on wind, noise, and existing conditions provided in figures 6-25, 6-26, 6-27, and 6-28.

Problem 18-9   List common code requirements that may affect landscape design.

Problem 18-10 Research your local city code. Do any specific landscape requirements exist?

Problem 18-11 According to the City of Geneva's Tree Preservation Application Guidelines, what types of construction projects require a review application?

Problem 18-12 According to the City of Geneva's Tree Preservation Application Guidelines, what does a typical tree protection measure consist of?

Problem 18-13 List five trees that are classified as Class A Trees by the City of Geneva.

Problem 18-14 What is a critical root zone?

For problems 18-15 and 18-16, refer to figure 18-13 and the following notes:

- Tree 1: 10" dia. Ash, good condition, no action required
- Tree 2: 11" dia. Ash, excellent condition, no action required
- Tree 3: 14" dia. Oak, good condition, no action required
- Tree 4: 16" dia. Magnolia, good condition, requires fertilizer and pruning
- Tree 5: 12" dia. Pine, condition unknown
- Tree 6: 18" dia. Hickory, good condition, no action required
- Tree 7: 10" dia. Maple, good condition, no action required
- Tree 8: 22" dia. Oak, excellent condition, fertilizer required

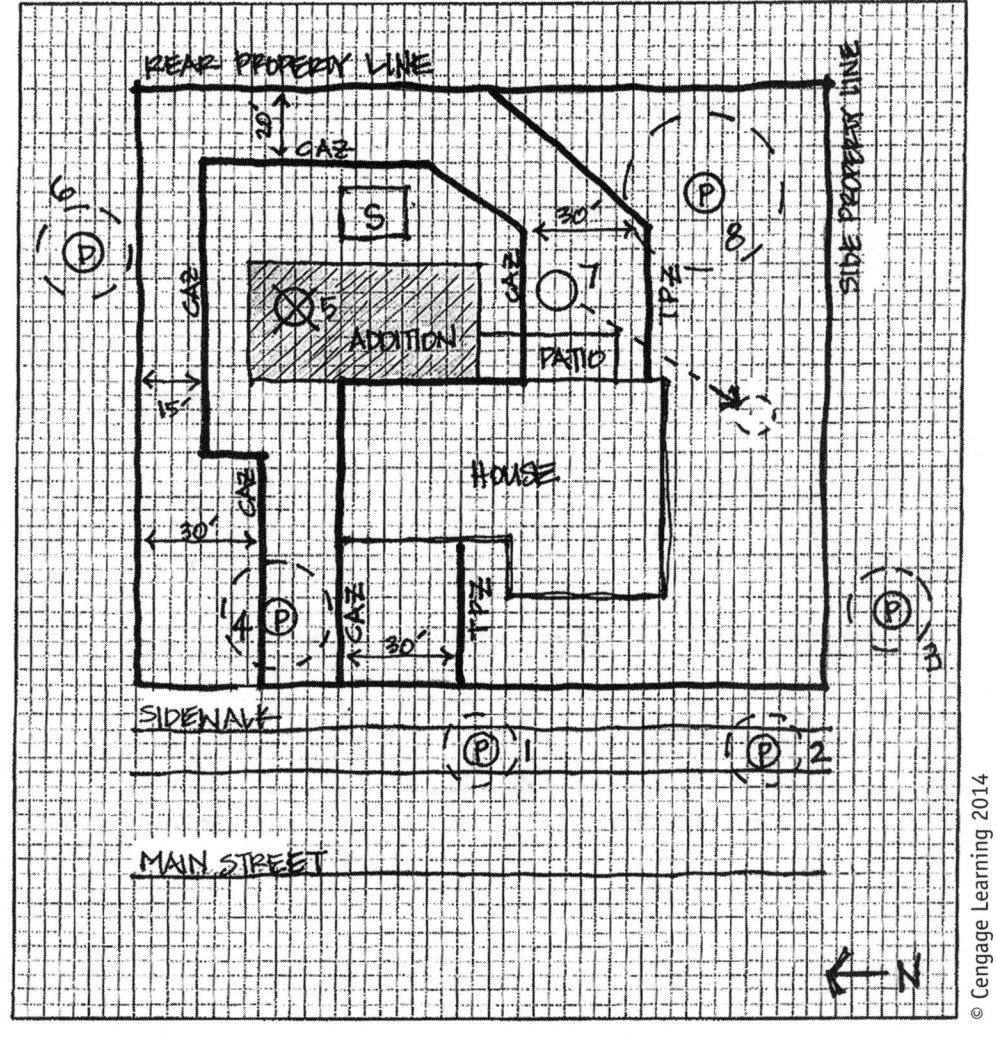

FIGURE 18-13 *Tree location sketch/protection plan for problems 18-15 and 18-16.*

Problem 18-15   Provide a tree protection plan and description based on the City of Geneva's requirements.

Problem 18-16   Provide a tree inventory based on the City of Pasadena's requirements (refer to textbook figure 18-5). Re-draw figure 18-13 with the appropriate tree protection zones.

For problems 18-17 through 18-20, refer to figure 18-14.

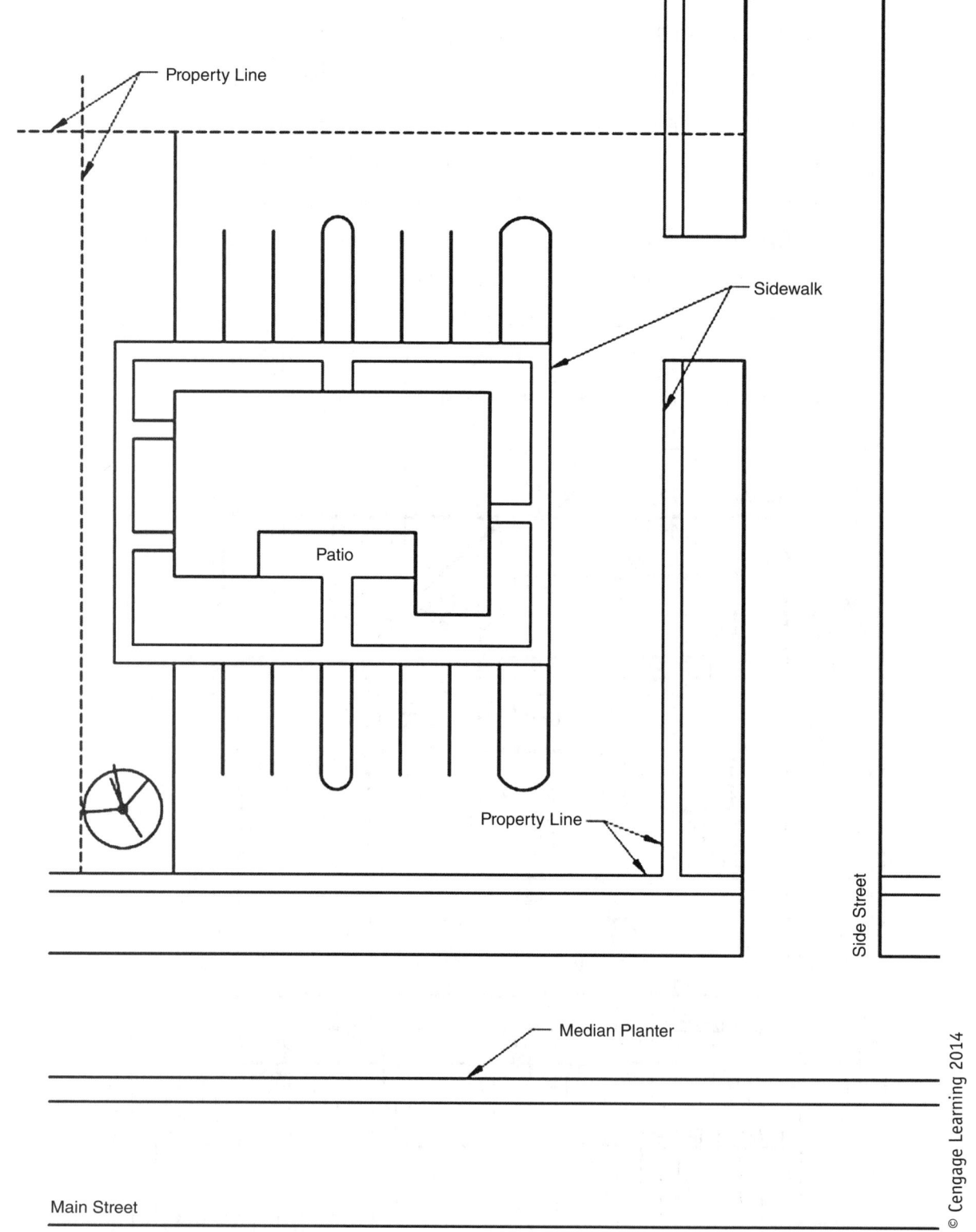

FIGURE 18-14   *Plan for street tree, street yard, buffer yard, and parking islands.*

404   Chapter 18: Landscaping

Problem 18-17   Sketch a street tree diagram.
Problem 18-18   Sketch a street yard diagram.
Problem 18-19   Sketch a buffer yard diagram.
Problem 18-20   Sketch a parking islands diagram.
Problem 18-21   How might landscaping provide trash screening?

For problems 18-22 through 18-24, use the following website provided in the textbook: www.arborday.org/trees/whattree/

Problem 18-22   An 80-foot tree with approximately 8-inch needle-like leaves, arranged in clusters of two to three, with thick-scaled cones, is spotted in South Carolina. Identify this tree based on its description.
Problem 18-23   A tree with few scale-shaped leaves and thick, green branches covered in spines, is found in Arizona. It also bears a large, red berry. Identify this tree based on its description.
Problem 18-24   Photograph a tree in your neighborhood, and write a description. Using your description, identify this tree species based on the Arbor Day website.

For problems 18-25 and 18-26, use the following website: www.arborday.org/trees/index.cfm

Problem 18-25   What is your hardiness zone? What are some of the most popular trees for this zone?
Problem 18-26   Referring to the tree planting and sizing guide, how far should a small tree be placed from the wall of a one-story building?
Problem 18-27   What is the definition of *softscape* in landscaping?
Problem 18-28   What is the definition of *hardscape* in landscaping?
Problem 18-29   Identify the softscape and hardscape elements in figure 18-15.
Problem 18-30   Identify the softscape and hardscape elements in figure 18-16.
Problem 18-31   Take photographs of the landscaping around your home. Document any examples of softscape and hardscape elements.
Problem 18-32   List four common functions of landscaping.
Problem 18-33   Find a photo example of each landscaping function listed in the textbook.
Problem 18-34   What is solar heat gain?
Problem 18-35   Describe coniferous, deciduous, and broadleaf trees.
Problem 18-36   What is the best type of landscaping barrier against a variable wind?
Problem 18-37   How far should a windbreak be located from the tallest row within the windbreak?
Problem 18-38   How can landscaping be effective at controlling noise?
Problem 18-39   In what ways can landscaping reduce or prevent erosion?
Problem 18-40   What function does the landscape serve in figure 18-17?

FIGURE 18-15   *Hardscape/softscape.*

FIGURE 18-16   *Hardscape/softscape.*

FIGURE 18-17 *Figure for problem 18-40.*

FIGURE 18-18 *Figure for problem 18-41.*

Problem 18-41 What function does the landscaping serve in figure 18-18?

Problem 18-42 What function does the landscaping serve in figure 18-19?

FIGURE 18-19 *Figure for problem 18-42.*

BACKGROUND

After the site has been studied and analyzed, conceptual planning begins to determine the best placement and design of the landscaping.

Conceptual Planning

Conceptual design is critical in determining which principles of design to follow, what works with the site, and what meets the needs of the clients.

Principles of Design

The textbook lists five design principles that should be considered when developing a concept for the landscaping design.

- Unity
- Repetition
- Balance
- Emphasis
- Spirit

When considering the unity of the design, the style should be consistent, and the color scheme well thought out. According to www.landscape-design-advisor.com, many design styles exist and should be researched when developing a concept. A few examples are listed below:

- Tuscan – Stone, old brick, wrought iron, and heavy wooden beams accented with live oak branches, Italian cypress, olive, and citrus trees as well as Tuscan shrubs
- English – Colorful blooms, green lawns, and winding pathways with creeping vines and large shade trees
- Tropical – Large leaves and intense, colorful flowers are staples as well as palm, bamboo, and banana trees

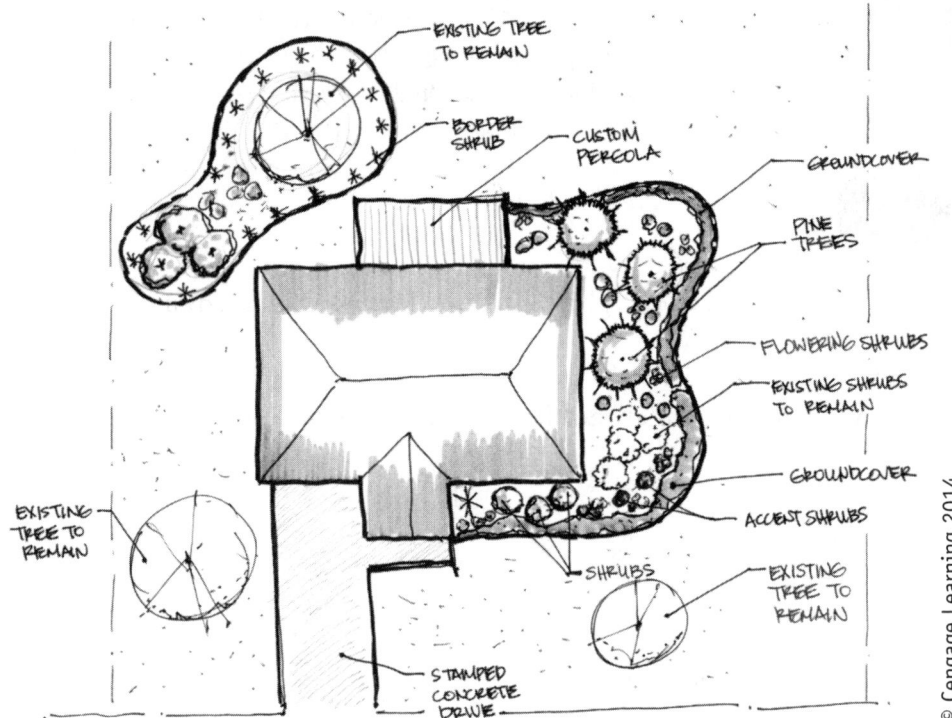

FIGURE 18-20 *Conceptual landscaping plan.*

- Asian – Rocks, stones, and boulders are generally accompanied by water features and Japanese maples, black pines, bamboo stalks, low-maintenance flowers, and bonsai trees.
- Contemporary – New materials, bold patterns, and clean lines
- Desert – Shaded patios, pools, fire pits and native plants that require very little water

In addition to a coherent style, any successful design should adhere to balance, repetition and emphasis. Balance may be achieved through a formal style that focuses on symmetry, or an informal balance that is asymmetrical. The repetition of the same elements throughout a design creates balance by using the same lines and color palette and similar textures and plants. Lastly, emphasis emerges when a focal point is planned.

Bubble Diagram

Bubble diagrams are often utilized in the planning of use areas in residential and commercial projects. Separate bubbles are used for areas determined during the needs analysis. Textbook figures 18-32 and 18-33 are examples of bubble diagrams.

Concept Plan

After the use areas are defined in a bubble diagram, the concept plan is developed. This plan shows a sketched layout of elements but generally does not include dimensions or specific plant species. An example of a concept plan is shown in figure 18-20.

Problem Set 18.2   Conceptual Planning

Problem 18-43   Consider the landscaping principles of design. Does the landscaping of your home meet any of these principles? Why or why not?

Problem 18-44   What are some examples of design styles?

Problem 18-45   Photograph and document an example of formal balance.

Problem 18-46   Photograph and document an example of informal balance.

Problem 18-47   Using text figure 18-32 as an example, construct a proposed bubble diagram for your home showing planned use areas.

Problem 18-48   Using text figure 18-33 as an example, construct a proposed bubble diagram for figure 7-4 indicating public and service entrances, play areas, and screening. The drive to the northwest is public, while the drive to the south is for service. Consider the top of the page to be north.

Problem 18-49   Construct a conceptual landscaping plan for your home.

Problem 18-50 Construct a conceptual landscaping plan for the home in figure 9-3. Consider the top of the page to be north. What principles of design were utilized?

Problem 18-51 Construct a conceptual landscaping plan for the home in figure 9-4. Consider the top of the page to be north. What principles of design were utilized?

BACKGROUND

Landscaping elements may be used for curb appeal and beauty but have several environmental advantages as well.

Sustainable Landscaping

Landscaping techniques may also provide environmental benefits to a project. The following techniques are listed in the textbook:

- Protect natural areas
- Reduce the use of turf
- Mulch planted areas
- Use native plant species
- Use landscaping elements to reduce energy consumption
- Use xeriscape

Natural areas are generally protected by local or federal governments and are generally identified during research of the site. Protecting these areas is critical to conservation of the natural ecosystem. Another technique, and possibly the easiest to accomplish, is to use less turf. Maintaining large areas of grass or turf requires watering, pesticides, and fertilizers that are detrimental to the environment. Once the landscaping is installed, mulch may be utilized to help retain water and increase curb appeal. Naturally growing plants and trees have naturally falling leaves and organic materials that enhance root growth and provide essential nutrients. Plants and trees growing in urban environments generally have less organic matter and poorer soils. A layer of mulch (consisting of 2 to 4 inches) helps provide organic material and simulates a more natural environment for transplanted trees and plants.

Some benefits to proper mulching are listed below:

- Maintains soil moisture by reducing evaporation
- Minimizes watering
- Helps controls weeds
- Provides insulation and helps keep soil warmer in the winter and cooler in the summer
- Improves soil aeration and drainage
- Improves curb appeal

Mulch is shown around the base of a tree in figure 18-21.

Selecting plants that are native to a region helps sustainability by reducing maintenance and by conserving water. The website www.plantnative.org hosts an extensive list of plant nurseries and native plants by state and region. Non-native plants or invasive plants may be harmful to an environment and may be researched at www.invasive.org.

Landscaping elements are used to reduce energy areas. The federal government offers some landscaping tips for energy conservation on its website www.energysavers.gov. Based on climate and region (refer to figure 18-22), the following tips are listed for each region:

- Temperate Region

    Maximize warming effects of the sun in the winter

    Maximize shade during the summer

FIGURE 18-21 *Mulch.*

408 Chapter 18: Landscaping

Deflect winter winds away from buildings

Funnel summer breezes toward the home

- Hot-Arid Region

    Provide shade to cool roofs, walls, and windows

    Allow summer winds to access naturally cooled homes

    Block or deflect winds away from air-conditioned homes

- Hot-Humid Region

    Channel summer breezes toward the home

    Maximize summer shade with trees that allow penetration of low-angle winter sun

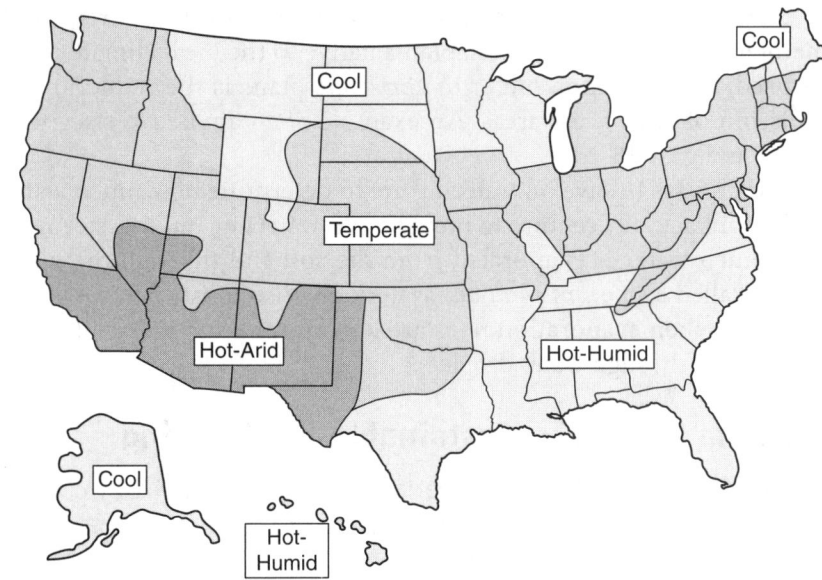

FIGURE 18-22 *Climate zones.*
Source: U.S. Department of Energy.

Avoid locating beds close to the home if they require frequent watering

- Cool Region

    Use dense windbreaks to protect the home from cold winter winds

    Allow the winter sun to reach south-facing windows

    Shade south and west windows and walls from the direct sun if summer overheating is a problem

Climate Zones

Shading might not only be an attractive way to landscape a home or business, but could provide a noticeable cost savings if done correctly. As discussed previously, solar heat control is one of the primary functions of landscaping. According to www.energysavers.gov, shading may reduce the surrounding air temperatures by as much as 9°F. Cool air settles near the ground, and air temperatures directly under trees can be as much as 25°F cooler than air temperatures above nearby blacktop pavement.

Knowing the sizes, shapes, and densities of trees and shrubs will assist in effective shading plans based on climate and microclimate. Deciduous trees block solar heat in the summer but allow for winter warming. When planted on the south side of the home, a deciduous tree with high branches and leaves provides maximum summertime shading. A deciduous tree with lower leaves and branches works well on the west side to combat low, afternoon sun angles. Evergreen trees and shrubs provide continuous shade and windbreaks.

In addition to roof and wall shading, trees, shrubs, and vines are used to shade ground and pavement around the home. Figure 18-23 shows a walkway shaded by vines.

Green roofs and living walls are becoming more and more prevalent as architectural elements in design. These architectural landscaping features are credited with cooling benefits as well. Several companies have begun marketing such landscaping elements. One, in particular, has been credited with the design of the longest living wall built in Kennett Square, Pennsylvania, in late 2010. GSky Plant Systems is based in Vancouver, British Columbia, and has worked on numerous projects all over North America. Some examples of projects can be found at www.GSky.com.

FIGURE 18-23 *Vine covered walkway.*

Xeriscaping limits water use and, in turn, minimizes required maintenance. To be successful, plants native to the local climate must be used. One common factor in xeriscape plans is the reduction or omission of lawn grass areas. An example of an applied xeriscape is shown in figure 18-24.

Other ways to save on watering are to determine how much water plants will actually require to prevent overwatering and to determine how much water is evaporated from the soil and through the plant leaves, called *evapotranspiration*. Watering is also most effective in the morning when evaporation rates are low, and the sun is less likely to burn plants through water droplets.

FIGURE 18-24 *Xeriscape.*

Problem Set 18.3   Sustainable Landscaping

Problem 18-52 Using the website www.plantnative.org, compile a list of at least two native trees, shrubs, flowering perennials, ferns, vines, and grasses that would be appropriate for your region or state.

Problem 18-53 Referring to figure 18-22, what tips does the website www.energysavers.gov give for your region?

Problem 18-54 What does the website www.energysavers.gov say about using landscaping as a windbreak?

Problem 18-55 What type of tree is used for shading against the summer sun?

Problem 18-56 What type of tree provides continuous, year-round shading?

Problem 18-57 What type energy-saving landscape technique is shown in figure 18-25?

Problem 18-58 What type energy-saving landscape technique is shown in figure 18-26?

Problem 18-59 What type energy-saving landscape technique is shown in figure 18-27?

FIGURE 18-25 *Figure for problem 18-57.*

FIGURE 18-26 *Figure for problem 18-58.*

FIGURE 18-27 *Figure for problem 18-59.*

BACKGROUND

The Details: Choosing Landscape Elements

Once the general concept has been decided, materials and watering systems are selected based on the design and required maintenance.

Hardscape

As previously discussed, the hardscape consists of paving, fencing, and benches in the landscape design. The textbook lists several materials and options for hardscape elements.

- Paving
    - Poured concrete
    - Asphalt
    - Pervious concrete
    - Paver blocks
    - Gravel
    - Urbanite
- Decks
    - Wood
    - Plastic lumber made from recycled plastic and wood
- Retaining walls
    - Cast-in-place concrete
    - Concrete masonry units
    - Brick
    - Stone
    - Heavy timber
    - Sheet piles

Softscape

The softscape consists of the plantings in the landscape design. The concept plan should show each of these plants, which have a distinct purpose, and a respect for the environment and natural conditions.

Irrigation

Irrigation is a necessity for any landscape design, even xeriscape plans. Some options for irrigation are rainwater harvesting and greywater systems. Rainwater harvesting consists of collecting excess rainwater in rain barrels, cisterns, or ponds. This was also discussed in Chapter 16, and rainwater collection systems may be seen in figures 16-1 and 16-2. Greywater is household wastewater that might be collected from bathroom sinks, showers, or bathtubs, and that might be used again to water outdoor plants.

Problem Set 18.3   Sustainable Landscaping

Problem 18-60   What types of materials might be used in paving designs?
Problem 18-61   What types of materials might be used for retaining walls?
Problem 18-62   List and describe the two types of irrigation systems outlined in the text.

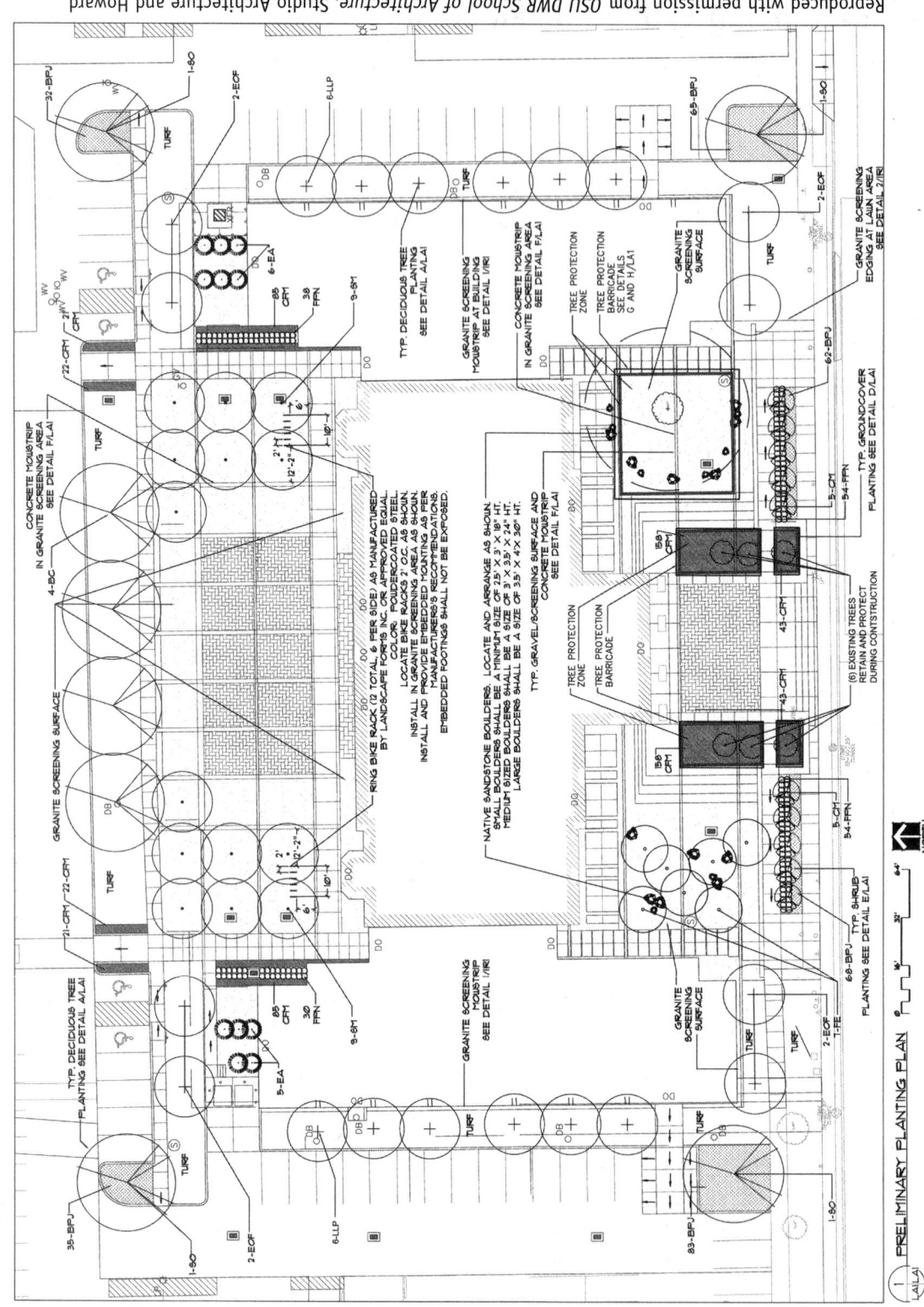

FIGURE 18-28 *Landscape plan.*

BACKGROUND

After the concept and materials have been finalized, a final plan is developed. This plan is generally included in the construction documents and shows dimensions, plant types, and spacings.

Final Landscape Plan

The final landscape plan for the Donald W. Reynolds School of Architecture at Oklahoma State University is shown in figure 18-28, while figure 7-15 shows paving requirements around the building. The final landscaping plan depicts new plantings as well as existing trees within a tree-protection zone.

Figures 18-29 through 18-33 are the planting details that accompany figure 18-28.

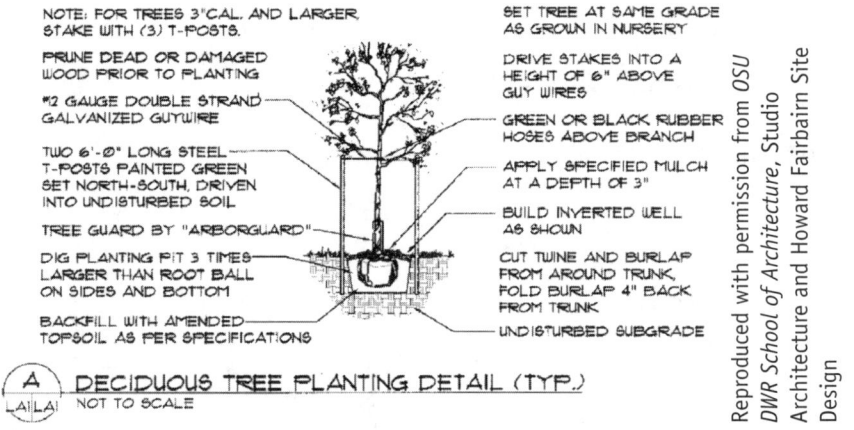

FIGURE 18-29 *Deciduous tree planting detail.*

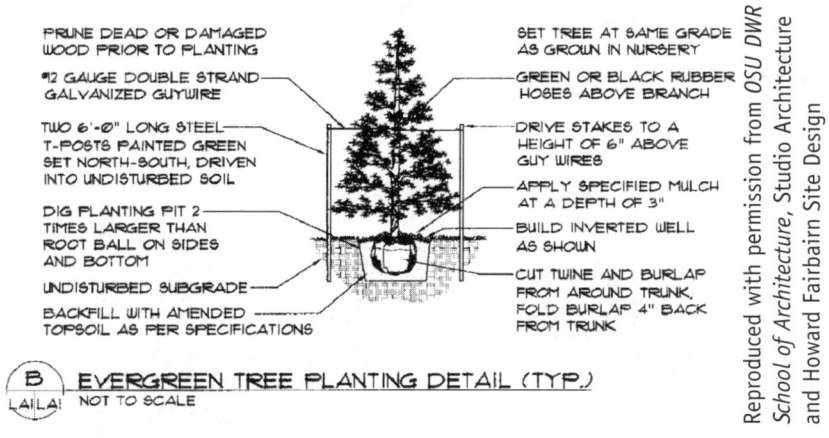

FIGURE 18-30 *Evergreen tree planting detail.*

As one might see in figure 18-28, many abbreviations are used on the landscape plan to conserve space and simplify the drawing. These abbreviations are often defined within a plant schedule. An example is shown in figure 18-34.

In addition to the plans and details, specifications or notes may be required. A sample set of general landscaping notes are shown in figure 18-35.

Problem Set 18.4 Final Landscape Plan

Problem 18-63   Referring to figure 18-29, describe how the tree is secured after planting.
Problem 18-64   Referring to figure 18-30, how big should the planting pit be?
Problem 18-65   What do figures 18-30, 18-31, and 18-33 say about pruning?

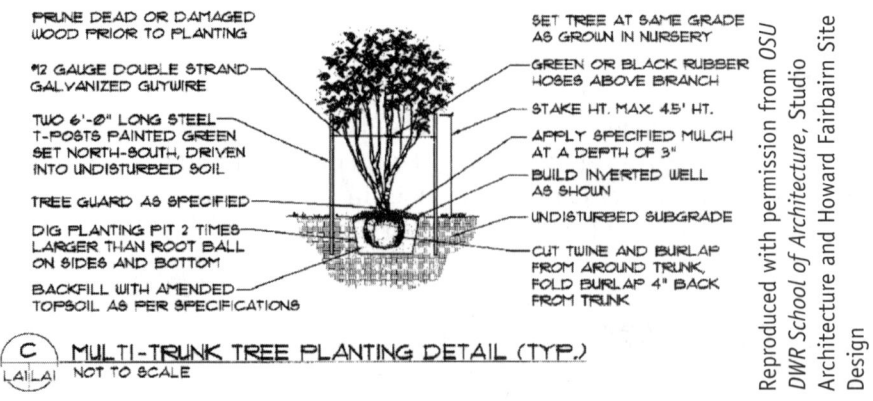

FIGURE 18-31   *Multi-trunk tree planting detail.*

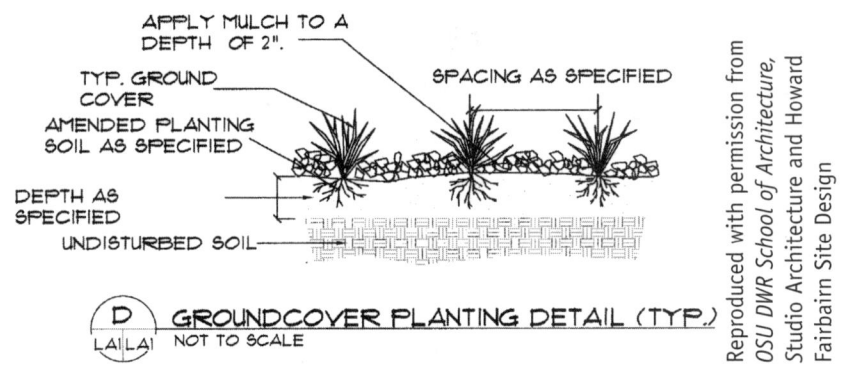

FIGURE 18-32   *Groundcover planting detail.*

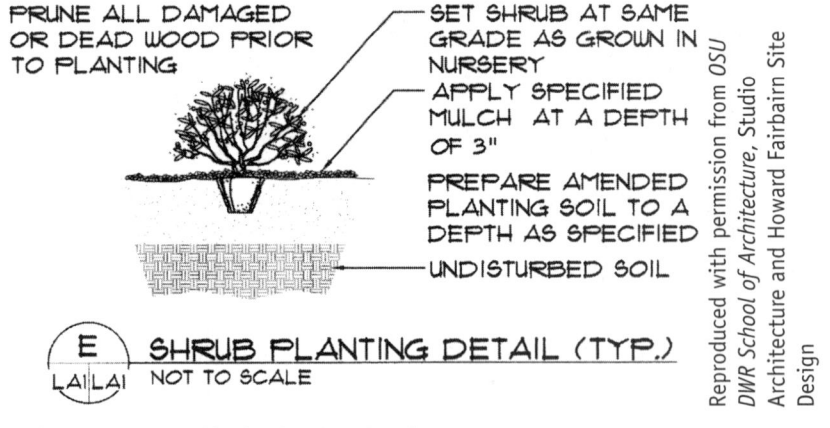

FIGURE 18-33   *Shrub planting detail.*

For problems 18-66 through 18-69, refer to figure 18-34.
Problem 18-66   How many shumard oaks are specified? What is the symbol? What is the tree pit size?
Problem 18-67   How many frontier elms are specified? What is the symbol? What is the tree pit size?
Problem 18-68   What does the abbreviation MT stand for?
Problem 18-69   What is the spacing requirement of the blue pacific juniper?

PLANT SCHEDULE

SYM.	QTY.	BOTANICAL NAME	COMMON NAME	SIZE	COND.	SPCG.	ROOT BALL SIZE	TREE PIT SIZE
CM	10	LAGERSTROEMIA INDICA 'RED VELOUR'	RED VELOUR CRAPEMYRTLE	7-8' HT.	BB/C	AS	24"-28"	48"-52"
EA	11	THUJA OCCIDENTALIS 'SMARAGD'	EMERALD ARBORVITAE	7-8' HT.	BB/C	AS	24"-26"	48"-52"
FE	7	ULMUS PARVIFOLIA X CARPINIFOLIA	FRONTIER ELM	2.5" CAL.	BB/C	AS	24"-26"	48"-52"
SM	14	ACER TRUNCATUM	SHANTUNG MAPLE	2.5" CAL.	BB/C	AS	24"-26"	48"-52"
LLP	12	PINUS TAEDA	LOBLOLLY PINE	7-8' HT.	BB/C	AS	24"-28"	56"-64"
BC	5	TAXODIUM DISTICHUM	BALD CYPRESS	3" CAL.	BB/C	AS	28"-32"	56"-64"
EOF	8	QUERCUS ROBUR 'FASTIGIATA'	ENGLISH COLUMNAR OAK	3" CAL.	BB/C	AS	28"-32"	56"-64"
SO	4	QUERCUS SHUMARDII	SHUMARD OAK	3" CAL.	BB/C	AS	28"-32"	56"-64"
FPN	176	NANDINA DOMESTICA NANA 'FIREPOWER'	'FIREPOWER' NANDINA	2 GAL.	C	AS	NA	NA
CFM	658	OPHIOPOGON JABURAN 'HOCF'	CRYSTAL FALLS MONDO GRASS	1 GAL.	C	18" o.c.	NA	NA
BPJ	345	JUNIPERUS CONFERTA 'BLUE PACIFIC'	'BLUE PACIFIC' JUNIPER	1 GAL.	C	30" o.c.	NA	NA

AS = AS SHOWN, BB = BALLED AND BURLAPPED, C = CONTAINER, CAL. = CALIPER, GAL. = GALLON, HT = HEIGHT, MIN. = MINIMUM, MT = MULTI-TRUNK, N.A. = NOT APPLICABLE, O.C. = ON CENTER

FIGURE 18-34 *Plant schedule.*

For problems 18-70 through 18-73, refer to figure 18-35.

Problem 18-70 What type of topsoil blend is requested as a replacement for the existing topsoil?

Problem 18-71 How much of a soil recess is specified? Why?

Problem 18-72 How long is the plant material guaranteed? How many times may it be replaced?

Problem 18-73 What should disturbed turf areas be replaced with?

PLANT NOTES

1. CONTRACTOR IS RESPONSIBLE FOR LOCATING AND AVOIDING ALL SITE UTILITIES.
2. FIELD ADJUST PLANT LOCATIONS AS NECESSARY, COORDINATE WITH THE LANDSCAPE ARCHITECT AND ARCHITECT.
3. REMOVE EXISTING SOIL IN BED AREAS TO A DEPTH OF 12". REPLACE WITH A TOPSOIL BLEND OF 5-PARTS TOPSOIL, 1-PART WASHED SAND AND 1-PART COMPOSTED ORGANIC MATERIAL.
4. 8" TOPSOIL CROWN IN ALL BED AREAS TO INSURE PROPER DRAINAGE. WHERE SHRUB BED IS ADJACENT TO THE BUILDING, SLOPE THE BED AWAY FROM THE BUILDING.
5. RECESS SOIL LEVELS OF ALL PLANTING BEDS 2" BELOW PAVING EDGE TO ALLOW FOR 3" DEPTH OF MULCH.
6. NO SUBSTITUTIONS SHALL BE MADE WITHOUT PRIOR APPROVAL FROM LANDSCAPE ARCHITECT.
7. APPLY PRE-EMERGENT HERBICIDE (PREEN OR EQUAL) TO BED AREAS AFTER PLANTING AND APPLY MULCH.
8. PLANT MATERIAL TO BE GUARANTEED FOR A PERIOD OF (1) YEAR, BEGINNING ON THE DATE OF ACCEPTANCE BY LANDSCAPE ARCHITECT THAT INSTALLATION IS COMPLETE.
9. PLANTS WILL BE REPLACED (1) TIME ONLY, AFTER PROJECT HAS BEEN ACCEPTED.
10. CONTRACTOR TO USE ALL EXCESS SOIL ON SITE IN BERMING AREAS.
11. ALL SHRUB BEDS SHALL BE BORDERED WITH CONCRETE MOW STRIP UNLESS ADJACENT TO PAVING.
12. MAINTAIN ALL PLANT MATERIAL, SPRIGGING, AND SODDING UNTIL FINAL ACCEPTANCE BY THE OWNER.
13. COORDINATE THE PLACEMENT OF LANDSCAPE WITH THE IRRIGATION SYSTEM, SLEEVES, ROTARY HEADS, SPRAY HEADS, PIPING, DRIPPER LINE, ETC.
14. COORDINATE THE PLACEMENT OF TREES, SHRUBS AND GROUND COVER WITH THE LANDSCAPE ARCHITECT PRIOR TO INSTALLATION.
15. REPAIR ALL DISTURBED TURF AREAS WITH 'U-3' BERMUDA GRASS SOD.

FIGURE 18-35 *Plant notes.*

CHAPTER 19
Visual Communication of Design Intent

Skills List

After completing the problems in this chapter, you should be able to:

- Understand the difference between orthographic, paraline, and perspective drawings

- Draw paraline and perspective drawings

- Add enhancements to drawings to show realism, depth, and texture

- List various types of green building software

- Understand the different types of built scale models and their contribution to presentation

BACKGROUND

Visual communication of a design might be the most important step of the design process, since in many instances the progression of the project depends on an effective presentation. Designers must be able to clearly demonstrate their ideas to clients, committees, and the general public. Fund raising or permitting may also depend on the ability to portray the design intent.

Presentation Graphics Support Design Visualization

The selection and execution of presentation drawings are critical when putting together a graphical representation of a design. Drawings may include paraline or perspective drawings, and each has its own purpose in a presentation.

Pictorial Drawings and Design Visualization

Orthographic drawings are two-dimensional drawings, such as plans, elevations, and sections. Some examples are shown in figure 19-1. These drawings are important and are utilized heavily in construction drawings as many floor plans are presented in a 2-D drawing. However, multiple views are required to understand the full concept of the building.

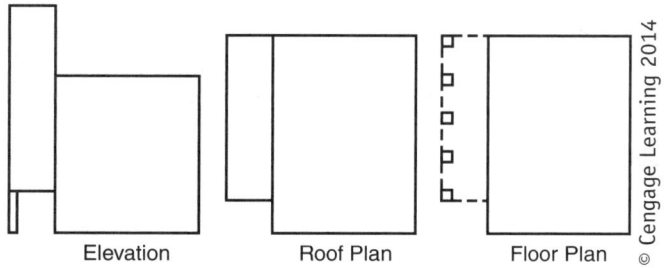

FIGURE 19-1 *Line drawings.*

In presentations, single-view drawings are utilized to illustrate the form in three dimensions. These drawings depict a more realistic representation and are sometimes called *pictorial drawings*. Two types of pictoral drawings are paraline and perspective. Paraline drawings consist of isometrics, plan obliques, and elevation obliques. In paraline drawings, the lines remain parallel. The first, isometric drawings, give all three visible sides the same emphasis. The same form detailed in figure 19-1 is shown in three dimensions in figure 19-2.

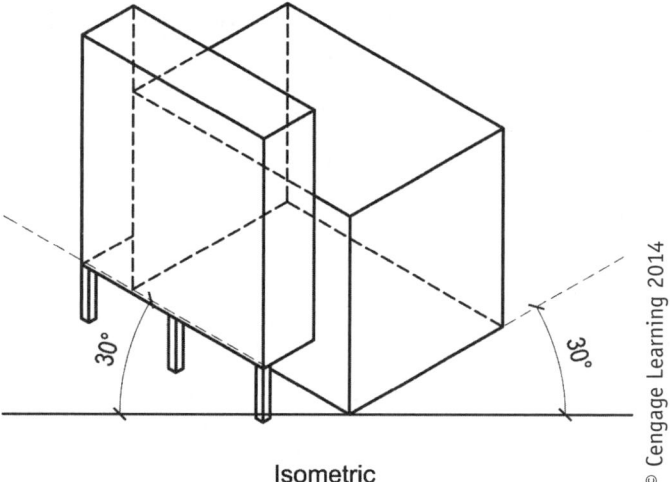

FIGURE 19-2 *Isometric.*

Another type of single-view drawing is the oblique. Obliques are drawn as elevation or plan obliques. Plan obliques are drawn with a 45°–45° or a 30°–60° rotation and give more emphasis to the horizontal planes than isometric drawings, due to the higher angle of view. In the 45°–45° rotation, the vertical planes are given equal emphasis, while a 30°–60° rotation favors one vertical plane. When drawing an elevation oblique, a vertical plane remains parallel to the drawing surface and is shown to scale. The form is projected back at a variable angle. A 30°–60° plan oblique and a 60° elevation oblique are shown in figure 19-3.

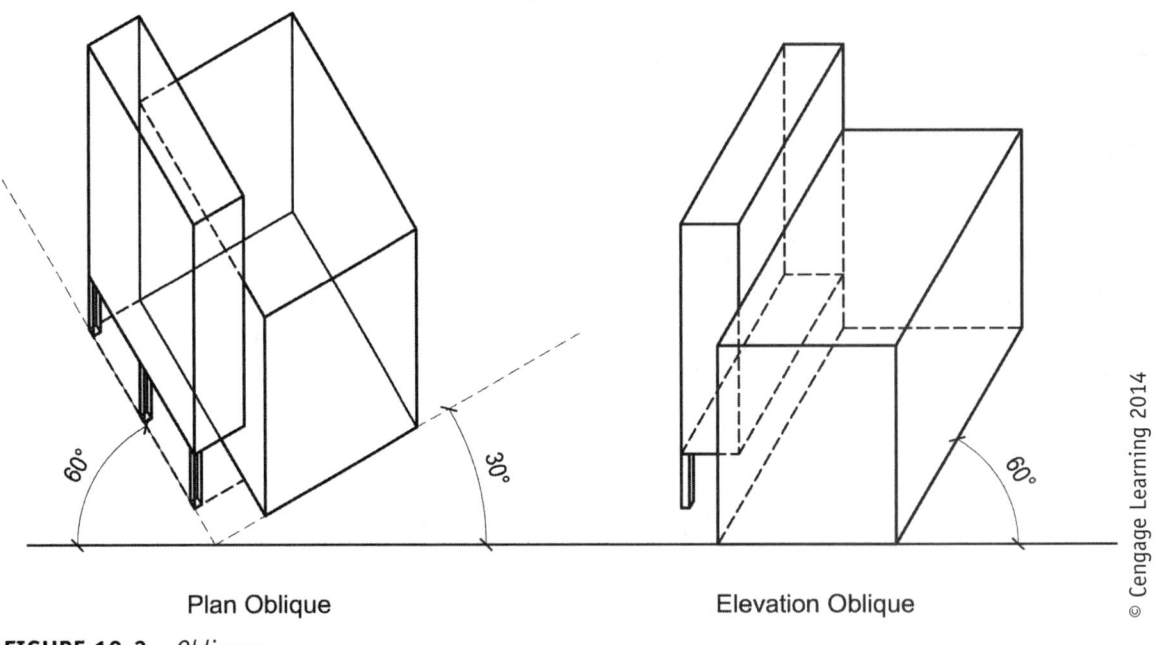

FIGURE 19-3 *Obliques.*

Perspective drawings are utilized to gain a more realistic idea of what the building or form will look like, since they are drawn from a natural point of view. Perspectives are able to create a sense of space and depth by converging parallel lines to vanishing points. Perspectives are drawn with respect to the observer's point of view, and they change with the angle of view and the height and distance from the form to the picture plane (P.P.). Two main categories exist when dealing with perspective drawings, one-point and two-point. Both types of perspectives are drawn with the observer's sight line, or center of view, coinciding with the station point (S.P.). Where the cone of vision intercepts the horizon line (H.L.), a vanishing point (V.P.) is formed. Figures 19-4 and 19-5 illustrate one-point perspectives. Figure 19-4 is drawn with the station point centered on the main structure. Figure 19-5 shifts the station point down and to the right of the original location. Note how this change affects the angle of the columns.

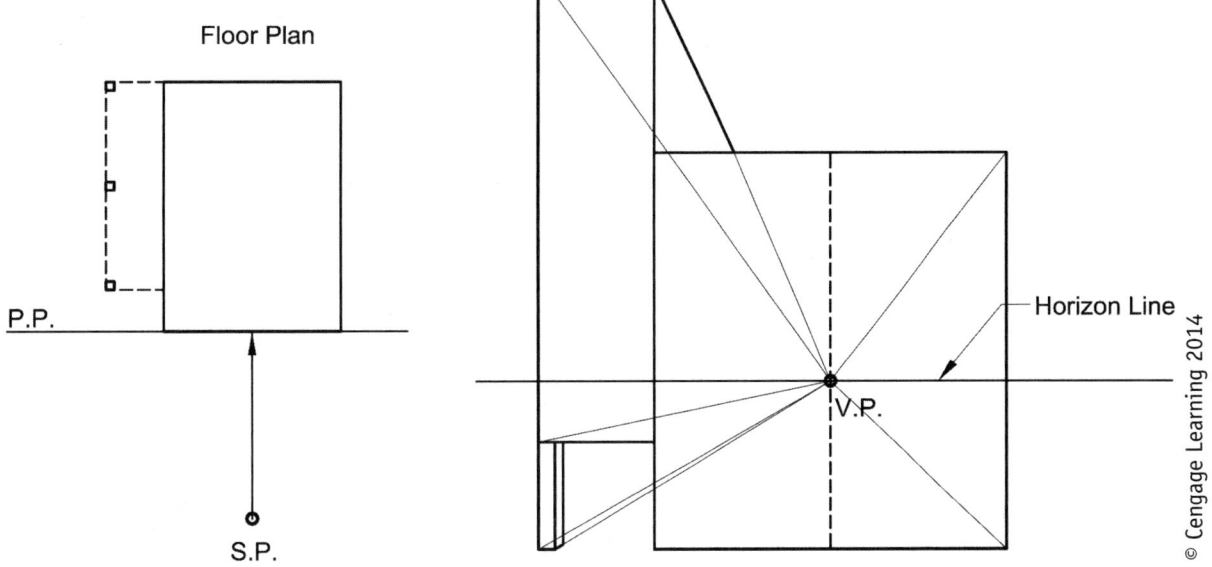

FIGURE 19-4 *Single-point perspective.*

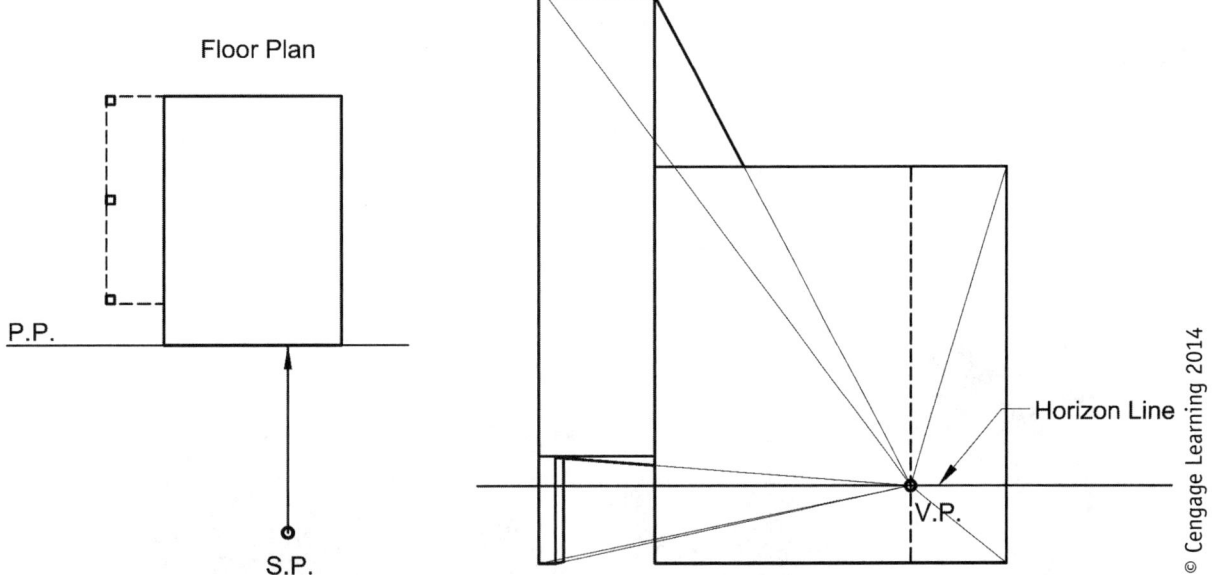

FIGURE 19-5 *Single-point perspective.*

Two-point perspectives are common in presentation drawings and generally illustrate how the building will fit into its surroundings. These drawings are constructed as shown in figure 19-6. The building plan is rotated at a variable degree from the picture plane. Lines parallel to the plan are drawn off of the station point. Where these lines intersect the picture plane, a vanishing point is formed.

Presentation Graphics Support Design Visualization 419

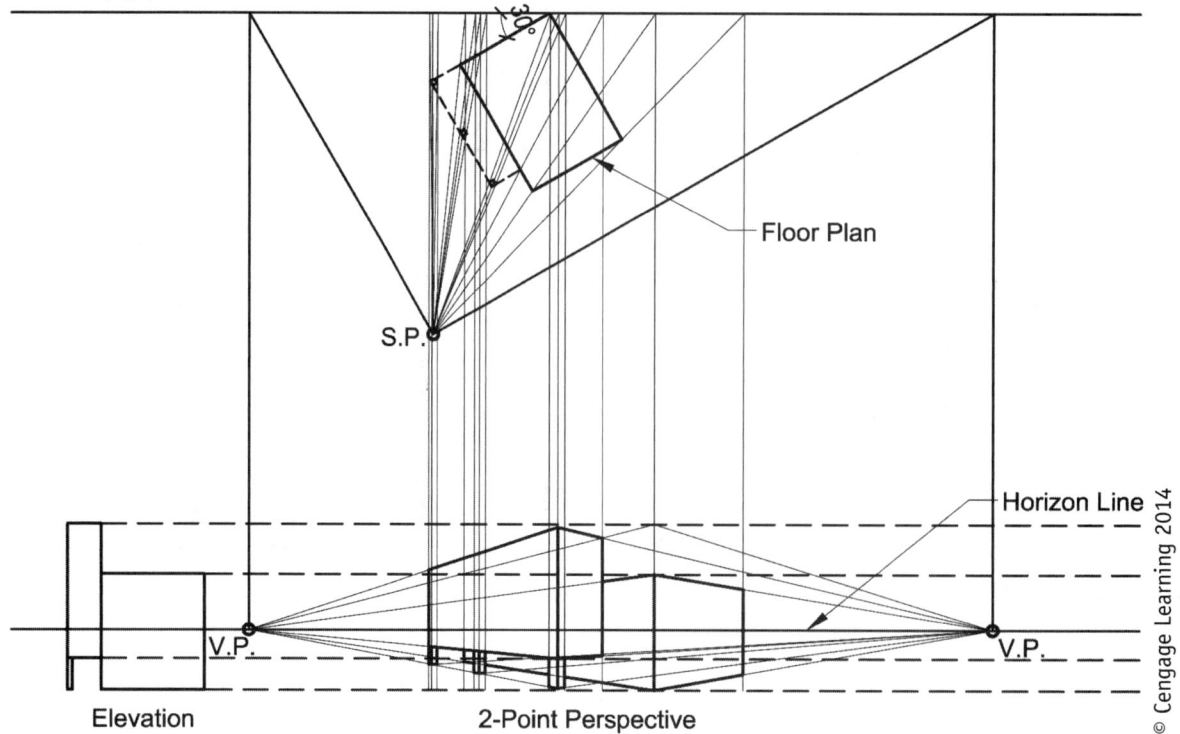

FIGURE 19-6 *Two-point perspective.*

Pictorial drawings may be drawn by hand or by computer. An example of a hand drawn presentation is shown in figure 19-7.

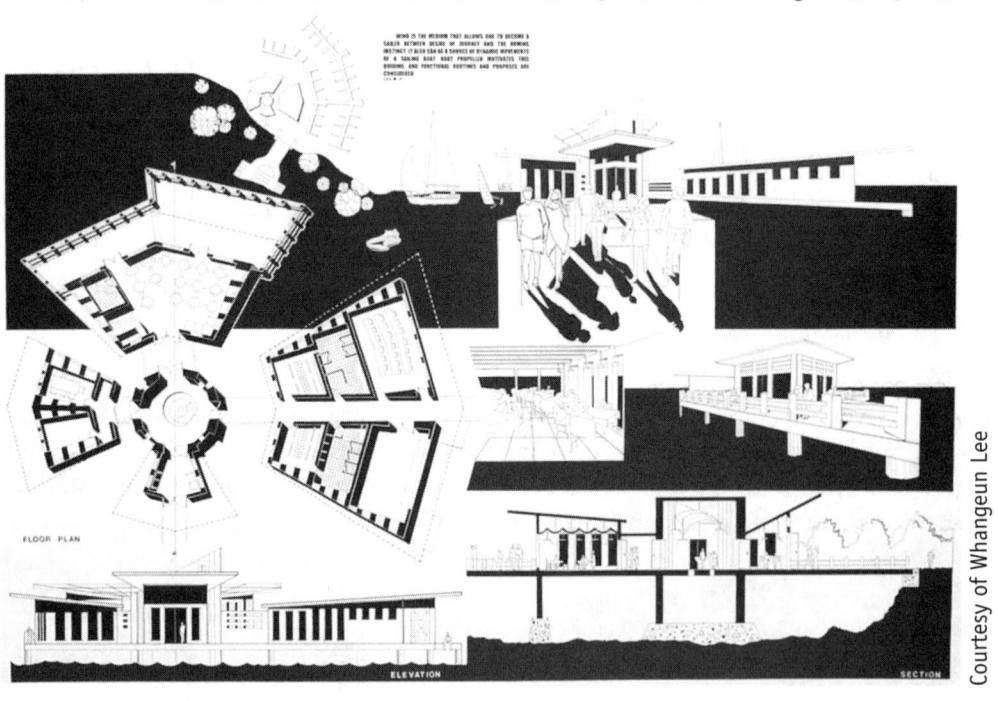

FIGURE 19-7 *Hand drawn presentation.*

With the advancement of computers, it has become much easier to produce presentation graphics and drawings. These drawings are commonly referred to as *computer renderings*. They also carry a heavy advantage over hand drawings in that they can be easily modified. Figure 19-8 shows an example of a computer presentation.

420   Chapter 19: Visual Communication of Design Intent

FIGURE 19-8 *Computer generated presentation.*

Problem Set 19.1 Presentation Graphics Support Visualization

Problem 19-1 What are orthographic drawings, and how do they differ from perspective drawings?

Note: Refer to figure 19-9 for problems 19-2 through 19-6.

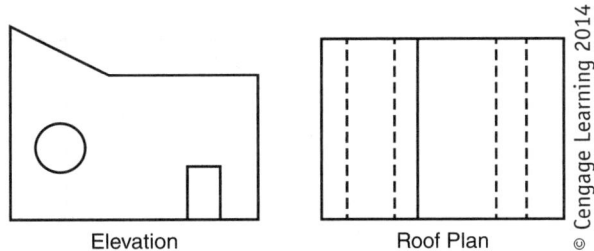

FIGURE 19-9 *Line drawing.*

Presentation Graphics Support Design Visualization

Problem 19-2　　Draw the corresponding 30° isometric for figure 19-9.
Problem 19-3　　Draw the corresponding 30°–60° plan oblique for figure 19-9.
Problem 19-4　　Draw the corresponding 60° elevation oblique for figure 19-9.
Problem 19-5　　Draw a corresponding one-point perspective for figure 19-9.
Problem 19-6　　Draw a corresponding two-point perspective for figure 19-9.
Problem 19-7　　Construct a two-point perspective for the shape shown in figures 19-1 through 19-6 with the given plan rotation and station point in figure 19-10.

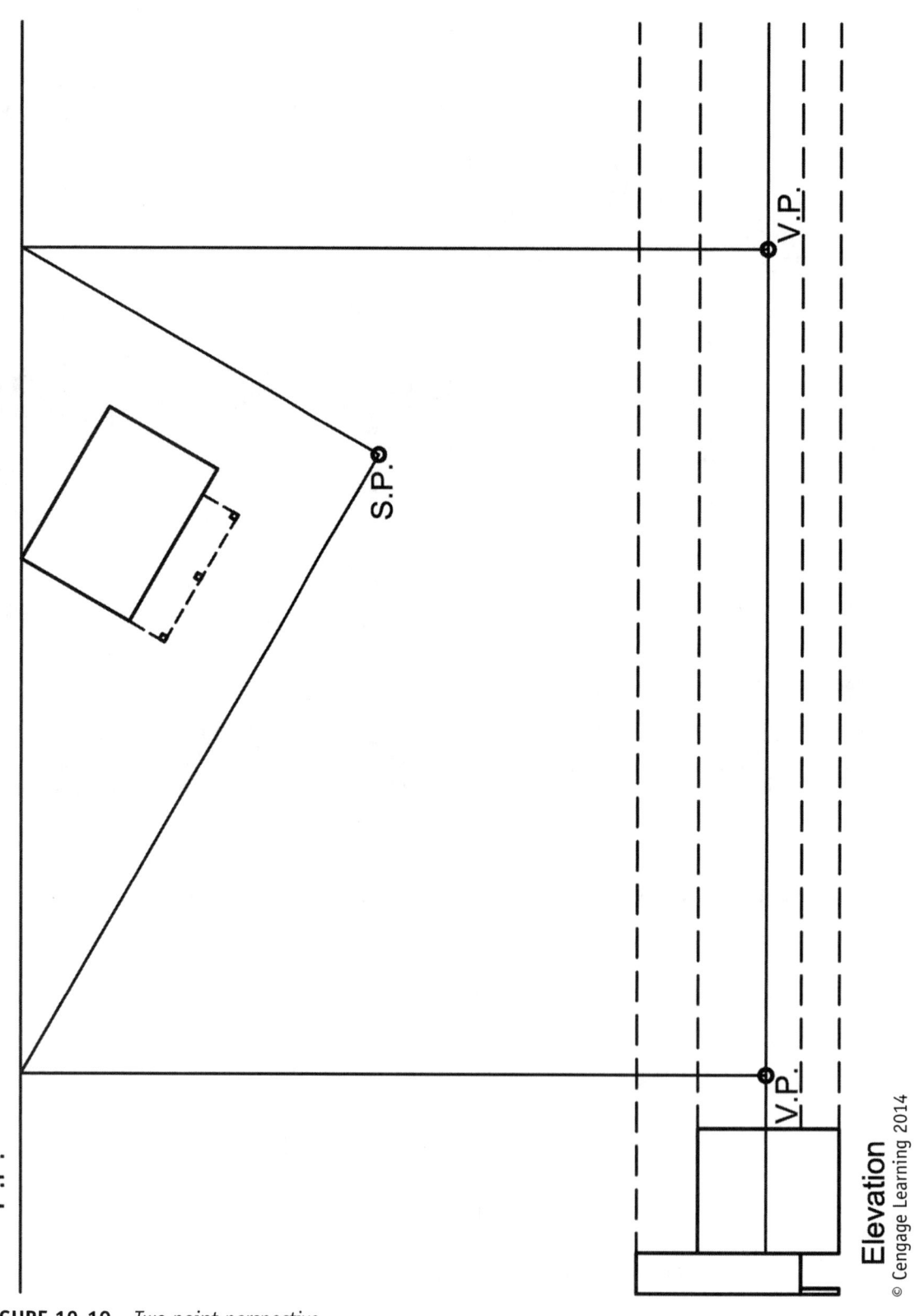

FIGURE 19-10　　*Two-point perspective.*

422　Chapter 19: Visual Communication of Design Intent

BACKGROUND

Before assembling presentation drawings, one must consider the audience and the objective of the presentation.

Audience and Purpose Influence Presentation Graphics

The intended audience of a presentation will greatly influence the type of drawings and the level of detail. For example, an audience lacking a design background may need more detailed and enhanced drawings to obtain a clear idea of the design intent. The enhancements include color, shades, shadows, trees, and people.

Colorful Elevation Views

An elevation is a type of orthographic drawing that depicts one side of a building at a time. Color may be added not only to show realism but also to show material and texture. A color elevation is shown in figure 19-10 of the textbook.

Enhancing Realism of Presentation Graphics

A few small touches may drastically change the appearance of presentation drawings. Shades, shadows, trees, people, and furniture add to the realism of any drawing and help convey scale and context to the viewer.

Shades and shadows are very effective at showing depth and are utilized in orthographic and perspective drawings. A shade occurs on the backside of an object lit by the sun and is free from any direct light, while a shadow is a dark projection of an object onto another surface. Figure 19-11 depicts shade and shadow of an object lit by the sun from the left.

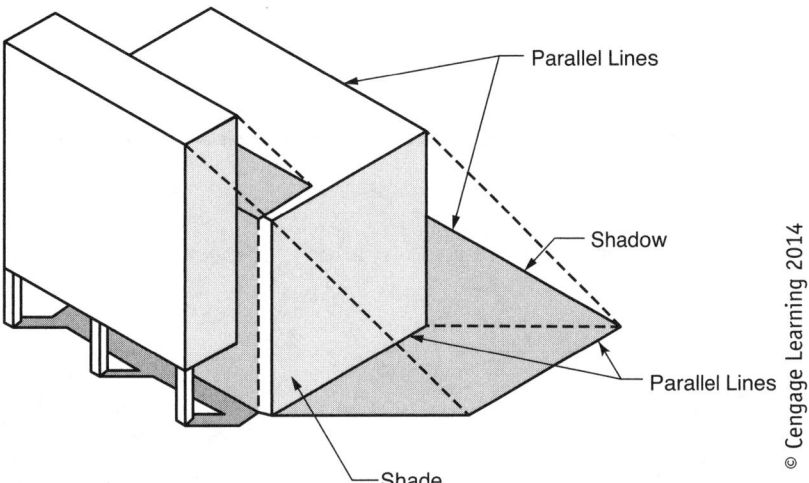

FIGURE 19-11 *Shades and shadows.*

Trees, people, cars, and furniture are useful in adding scale and use to a drawing. These enhancements vary in detail and should match the drawing detail level of the building. A careful balance should be achieved by adding enough landscaping and life for realism, without creating a distraction from the building. Figures 19-12 and 19-13 illustrate people and trees with varying ranges of detail.

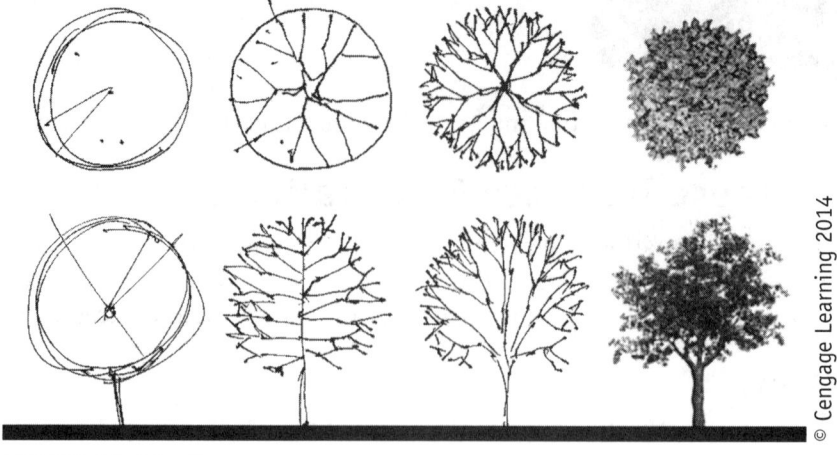

FIGURE 19-12 *Trees.*

FIGURE 19-13 *People.*

Figure 19-14 illustrates a "before" perspective and an "after" perspective with shades, shadows, trees, and people.

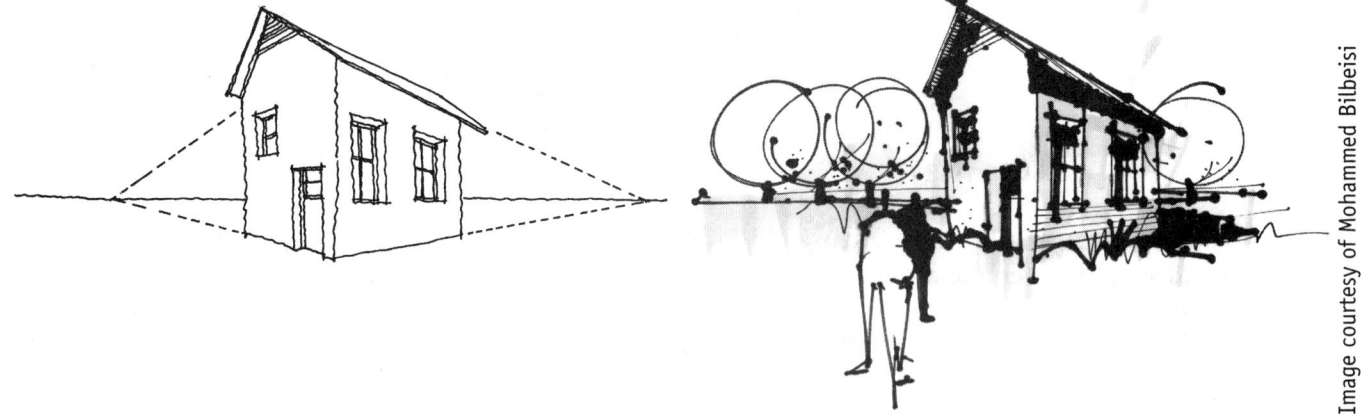

FIGURE 19-14 *Perspectives.*

Building Design Software Illustrates Climate Conditions

With the focus on climate and efficiency becoming more prevalent, building design software is showing an increased interest as well. Building information modeling (BIM) software, such as Revit® Architecture, allows for buildings to be modeled with wall and window types and finishes. This software easily calculates mechanical loads when the inside and outside air temperatures are known.

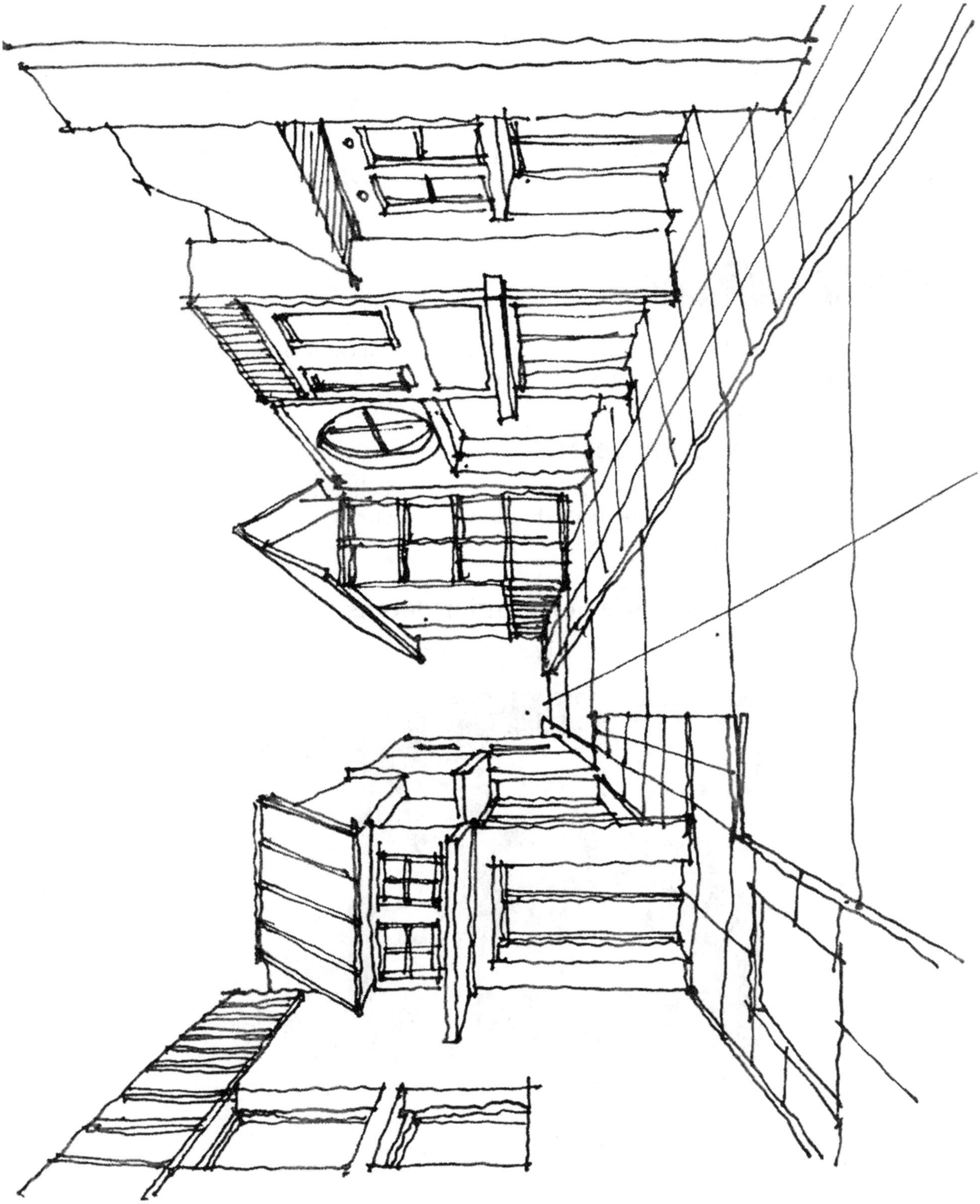

FIGURE 19-15 *City street perspective.*

In 1999, a non-profit organization, The Green Building XML schema (gbXML), was funded and developed to transfer information to and from engineering models. More information may be found at www.gbxml.org.

Problem Set 19.2 Audience and Purpose Influence Presentation Graphics

Note: Refer to figure 19-15 for problems 19-7 and 19-8.

Problem 19-7   What type of perspective is figure 19-15?
Problem 19-8   Enhance the sketch in figure 19-15 by adding color, shades, shadows, people, and trees.
Problem 19-9   From the gbXML website, what are three examples of compatible BIM software?
Problem 19-10   From the gbXML website, what are three examples of analysis software?

BACKGROUND

As with exterior details and renderings, interior design may be communicated through renderings and perspectives and enhanced with color, furniture, and people.

Communication of Interior Details

Interior details are expressed with drawings, such as, the perspective shown in figure 19-16 and at times with material boards, similar to the one in figure 19-17, that provide samples of floor and wall coverings that represent the color and texture of the intended design.

Problem Set 19.3 Communication of Interior Details

Problem 19-11   What type of perspective is figure 19-16?

FIGURE 19-16 *Interior computer perspective.*

Problem 19-12   When would an interior perspective be beneficial?

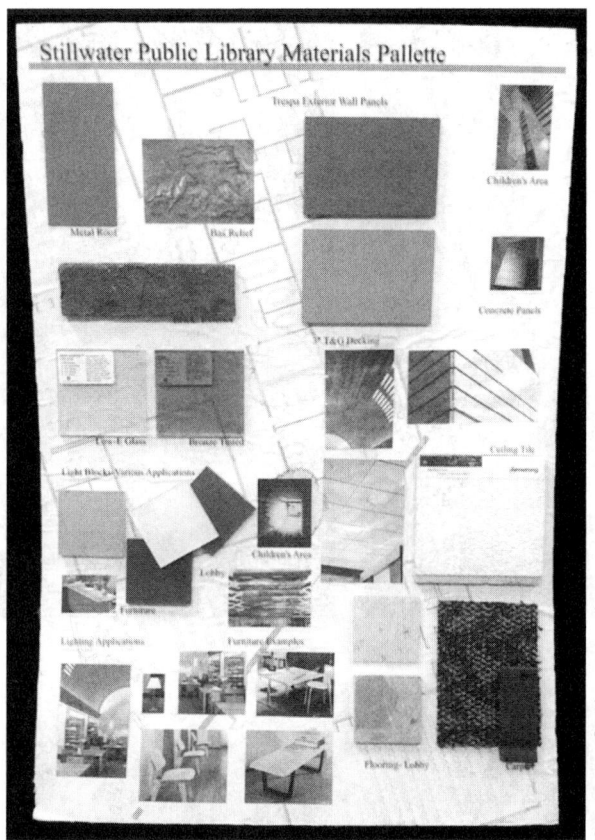

FIGURE 19-17 *Materials board.*

BACKGROUND

In addition to presentation drawings, models are also used to communicate ideas to the client or the public. Models can easily portray scale, use, texture, and context just as efficiently, if not more so, than drawings.

Models Provide Perspective and Analysis

Two types of models exist that are helpful in portraying ideas in the presentation phase; computer and built scale models. Computer models were already discussed, and the product is shown in figure 19-8. Built scale models consist of study models and detailed, professional models. These can be exterior, interior, landscaping, or urban. Study models are generally rough models that show massing and are built with cardboard or foam. Professional models are built with more expensive materials that exhibit texture, landscaping, and people. These models may be built by hand or by use of machines and are generally built to a desired scale, e.g., 1/4" = 1'-0", 1/8" = 1'-0". Figure 19-18 shows a laser cutter utilized in model building.

FIGURE 19-18 *Laser cutter.*

Models Provide Perspective and Analysis   427

Figure 19-19 shows examples of a study model and a professional scale model representing the same design. Figure 19-20 shows these models placed within an urban site model to show context.

FIGURE 19-19 *Models.*

FIGURE 19-20 *Models.*

Problem Set 19.4 Models Provide Perspective and Analysis

Problem 19-13 What are the two types of built scale models?

Problem 19-14 Why might built scale models be more effective than drawings?

Problem 19-15 Build a scale model of figure 19-1.

CHAPTER 20
Formal Communication and Analysis

Skills List

After completing the problems in this chapter, you should be able to:

- List the relevant steps associated with the development of a formal building proposal

- Understand audience influence and focus

- Organize documents and support materials

- Be familiar with the process included in a presentation

BACKGROUND

Making a building or project become a reality may depend heavily on a proper presentation of the proposal.

Formal Presentation of a Building Proposal

There are several steps that may be required when compiling a formal building proposal. These steps include:

- Researching
- Brainstorming for optimum content and format
- Knowing the topic well to decrease nervousness and build confidence
- Preparing an outline to provide structure and clarity
- Identifying supporting documentation and visuals (Chapter 19: Visual Communication of Design Intent)
- Practicing the presentation
- Providing evidence to support desired outcome
- Performing post presentation analysis

Problem Set 20.1   Formal Presentation of a Building Proposal

Problem 20-1   Referring to the steps included in a formal building proposal, are there any additional steps that may be required?

Problem 20-2   If the desired outcome is not achieved, what may be done to remedy the situation?

BACKGROUND

Understanding the presentation purpose, the intended audience, and the logistics are the first steps in building a presentation.

Presentation Purpose, Audience, and Logistics

Often, small details can make or break a presentation. One should begin by considering the goal of the presentation. Setting goals early will help drive the bulk of the decision making regarding the presentation.

Another key element to the presentation is the audience. Size, expertise, and needs are all elements that are to be considered prior to the meeting. Use appropriate vocabulary for the audience, and define any terms that may not be common.

Once the audience is defined, the logistics of the meeting should be clearly understood. The size of the room, available technology, lighting, location, and time allotted will all need to be considered.

In order for a successful presentation, potential mishaps should be thought out in advance. If relying on technology, become acquainted with the equipment beforehand. Determine whether the appropriate software is available to support your presentation and that it is current and compatible. Have a plan for using the computer, projector, and network, and perform a trial run if possible. Have a hard copy of the presentation on hand as a safety measure.

Problem Set 20.2   Presentation Purpose, Audience, and Logistics

Problem 20-3   If the presentation is computer-based, what logistics might you consider?

Problem 20-4   Would it be appropriate to provide refreshments during a presentation? Why or why not?

Problem 20-5   What is an acronym, and is it acceptable to use in a presentation?

BACKGROUND

After adequate research on the building proposal is complete, brainstorming over content and format of the presentation begins.

Brainstorming Content and Format

Many hours are often dedicated to brainstorming the presentation's content and format to ensure that the correct approach is achieved.

Problem Set 20.3   Brainstorming Content and Format

Problem 20-6   According to the text, what is a good source for relevant project information?
Problem 20-7   What team members might be included in the brainstorming sessions?

BACKGROUND

Once research and brainstorming are complete, organization of the support materials is critical to guiding the presentation.

Gather and Prepare Presentation and Support Materials

The text provides the following list of information that may be included in a building proposal:

- Project concept
- Goals and objectives
- Project needs
- Permit requirements
- Business strategy
- Building and site design
- Environmental and community impacts
- Timeline and schedule
- Budget
- Sustainability
- Resiliency
- Building lifecycle
- Future expansion options

Covering all of these topics in a presentation might not be an effective use of time or audience attention. Preparing an agenda with critical topics will help build a successful and relevant presentation. Presentation graphics, visual media, and models will also aid in portraying the intent and are discussed extensively in Chapter 19: Visual Communication of Design Intent. Narrowing down information and graphics will depend primarily on the audience. The text provides the following list for potential audiences:

- City, town, village, county, or local planning boards
- Zoning and code officials
- Community members
- Prospective builders
- Future building occupants

- Building owners
- Financial lenders
- Environmental rights organization
- Historical association
- Chamber of commerce
- Federal funding commission
- Community development agency

Organization of Documents

Being well organized with your presentation, as well as your documentation, is extremely important to the success of the presentation. Clarity and readability should be obtained from font, spacing, and size. This is true for handouts as well as visual aids. If providing handouts, a table of contents, page numbers, and an index will make following the presentation easier and will prevent audience frustration. Slideshows are often the basis for visual aids. The slideshow should be viewed before the presentation for clarity and readability from all possible audience viewpoints.

The purpose of utilizing visual aids is to:

- Gain interest
- Increase presentation effectiveness
- Highlight key points
- Clarify a concept or idea
- Support design visualization
- Show data and study results
- Enhance audience understanding and retention

FIGURE 20-1 *Project presentation.*

Problem Set 20.4 Gather and Prepare Presentation and Support Materials

Problem 20-8   What is a value-added approach? List some examples.

Note: refer to the list of concerns on text page 716 for problems 20-9 through 20-12.

Problem 20-9   Consider a presentation given to members of the community. What concerns might arise? What aspects of the building proposal are critical to the presentation? What visual aids might be most effective?

Problem 20-10   Consider a presentation given to the building owner. What concerns might arise? What aspects of the building proposal are critical to the presentation? What visual aids might be most effective?

Problem 20-11   Consider a presentation given to the historical association. What concerns might arise? What aspects of the building proposal are critical to the presentation? What visual aids might be most effective?

Problem 20-12   Consider a presentation given to prospective builders. What concerns might arise? What aspects of the building proposal are critical to the presentation? What visual aids might be most effective?

Problem 20-13   What strategies are used in organizing documentation?

Problem 20-14   How do visual aids help a presentation?

BACKGROUND

Much time and effort is contributed to the preparation of a presentation, but can be all for loss if the ideas and intents are not communicated effectively.

Communication/Implementation of the Presentation

The Rehearsal

Once the presentation is finalized, several rehearsals may be necessary to obtain a level of comfort for the presenter and fluidity for the presentation. When the presenter is relaxed and comfortable, they appear more knowledgeable and put the audience at ease. It is also critical to not read the presentation, but to make eye contact with the audience and use notes only as a reference. A rehearsal will also give a realistic time consideration so the presentation may be extended or shortened accordingly.

The Presentation

Countless hours and efforts generally go into preparing for the actual presentation, so allowing plenty of time to set up and test equipment is always recommended to ensure a glitch-free presentation. After everything is in place, the remaining time may be used to greet the audience. This will give the presenter and audience time to become better acquainted. The formal presentation generally begins with a welcome and introductions. Providing a small background on each of the presenters helps the audience to understand the role of the team and their involvement in the project. From rehearsal, each member of the team knows their role in the presentation, when visual aids are utilized, and who controls equipment.

Communicating Features and Benefits

The emphasis of a presentation should be clear to the audience. The presenter generally states the purpose, benefit, value, and goals at the beginning and again at the end of the presentation.

Presentation Components

A presentation should have a clear introduction, body, and conclusion. This flow is planned and rehearsed in advance to limit confusion and disorganization. The goals and benefits should be clearly defined. Presenters may also address anticipated questions within the presentation and allow for an allotted time for questions from the audience.

Problem Set 20.5   Communication/Implementation of the Presentation

Problem 20-14   Why is rehearsal important to the success of a presentation?
Problem 20-15   What would be some clear indicators that a presentation lacked rehearsing?
Problem 20-16   What is the general order of a presentation?
Problem 20-17   Why is it important to arrive early?
Problem 20-18   At what point in the presentation should the main goal be emphasized?

BACKGROUND

The end of a presentation may be used to emphasize key points, generate enthusiasm, and answer questions. This is also the time to gain any feedback desired from the audience.

Presentation Closure

The end of a presentation should restate the presentation's goals and purpose. This is also the time to address any audience questions. Time for questions should be worked into the presentation schedule. The team member with the most experience to answer the question accurately should be appointed to answer questions. If a question may arise that does not have a clear response, more time should be requested to research an accurate answer rather than providing an inaccurate response or guess. After all questions are answered, the presenter should indicate any action from the audience. At times, audience feedback is desired and may be obtained from a questionnaire.

Problem Set 20.6   Presentation Closure

Problem 20-19   How are audience questions addressed?
Problem 20-20   The text indicates some phrases used to indicate the closure of a presentation. List some examples.
Problem 20-21   What are some appropriate ways to collect audience feedback?

BACKGROUND

Collecting and acting on audience feedback will help the success of the project.

Feedback Supports Presentation Analysis

It is beneficial to have a team meeting after a presentation to capture the feedback on the presentation from the audience. Notes on what questions were asked, what comments were made, and audience reaction are very important when analyzing the presentation. If audio or video recordings were taken, it may also be helpful to review these, and note strong points and areas needing improvement.

Problem Set 20.7   Feedback Supports Presentation Analysis

Problem 20-22   List ways to collect and document audience feedback of a presentation.

Modification or Supplementation to Achieve Desired Outcome

After the presentation and feedback are analyzed, brainstorming occurs once again to make sure all avenues are considered before a final outcome is achieved. At times, ideas are developed or deficiencies are realized during the presentation, and these are addressed before the final design is submitted.